Plant Development: Role of Gene Regulatory Networks

Plant Development: Role of Gene Regulatory Networks

Editor: Isabel Nelson

STATES
ACADEMIC PRESS
www.statesacademicpress.com

States Academic Press,
109 South 5th Street,
Brooklyn, NY 11249, USA

Visit us on the World Wide Web at:
www.statesacademicpress.com

ISBN: 978-1-63989-420-8 (Hardback)

Cataloging-in-Publication Data

Plant development : role of gene regulatory networks / edited by Isabel Nelson.
p. cm.
Includes bibliographical references and index.
ISBN 978-1-63989-420-8
1. Plant genetic regulation. 2. Gene regulatory networks. 3. Plant cellular control mechanisms.
4. Plant molecular genetics. I. Nelson, Isabel.
QK981.4 .P53 2022
572.865 2--dc23

Table of Contents

Preface

Every book is a source of knowledge and this one is no exception. The idea that led to the conceptualization of this book was the fact that the world is advancing rapidly; which makes it crucial to document the progress in every field. I am aware that a lot of data is already available, yet, there is a lot more to learn. Hence, I accepted the responsibility of editing this book and contributing my knowledge to the community.

Gene regulatory networks are composed of molecular regulators such as transcription factors which bind short, non-coding DNA sequences. These sequences are also called cis-regulatory elements and are located in the promoter region of a gene. Transcription regulators form an interconnected regulatory network which integrates endogenous and environmental signals into changes in gene expression. Gene regulatory networks provide an insight about organization of plant cells and their functions. They consist of thousands of connections between the target genes and transcription factors. Together, they regulate many cellular functions. They also help in understanding the regulation of genes under various conditions. The book studies, analyses and upholds the pillars of gene regulatory networks and its utmost significance in modern times. From theories to research to practical applications, case studies related to all contemporary topics of relevance to this field have been included herein. As this field is emerging at a rapid pace, the contents of this book will help the readers understand the modern concepts and applications of the subject.

While editing this book, I had multiple visions for it. Then I finally narrowed down to make every chapter a sole standing text explaining a particular topic, so that they can be used independently. However, the umbrella subject sinews them into a common theme. This makes the book a unique platform of knowledge.

I would like to give the major credit of this book to the experts from every corner of the world, who took the time to share their expertise with us. Also, I owe the completion of this book to the never-ending support of my family, who supported me throughout the project.

Editor

1

Overview of OVATE FAMILY PROTEINS, A Novel Class of Plant-Specific Growth Regulators

Shucai Wang[1], Ying Chang[2] and Brian Ellis[3]*

[1] Key Laboratory of Molecular Epigenetics of MOE and Institute of Genetics and Cytology, Northeast Normal University, Changchun, China, [2] College of Life Science, Northeast Agricultural University, Harbin, China, [3] Michael Smith Laboratories, The University of British Columbia, Vancouver, BC, Canada

OVATE FAMILY PROTEINS (OFPs) are a class of proteins with a conserved OVATE domain. OVATE protein was first identified in tomato as a key regulator of fruit shape. OFPs are plant-specific proteins that are widely distributed in the plant kingdom including mosses and lycophytes. Transcriptional activity analysis of *Arabidopsis* OFPs (AtOFPs) in protoplasts suggests that they act as transcription repressors. Functional characterization of OFPs from different plant species including *Arabidopsis*, rice, tomato, pepper, and banana suggests that OFPs regulate multiple aspects of plant growth and development, which is likely achieved by interacting with different types of transcription factors including the KNOX and BELL classes, and/or directly regulating the expression of target genes such as *Gibberellin 20 oxidase* (*GA20ox*). Here, we examine how OVATE was originally identified, summarize recent progress in elucidation of the roles of OFPs in regulating plant growth and development, and describe possible mechanisms underpinning this regulation. Finally, we review potential new research directions that could shed additional light on the functional biology of OFPs in plants.

Edited by:
José M. Romero,
Instituto de Bioquímica Vegetal y Fotosíntesis, Spain

Reviewed by:
Gorou Horiguchi,
Rikkyo University, Japan
Gonzalo Villarino,
North Carolina State University, USA

***Correspondence:**
Shucai Wang
wangsc550@nenu.edu.cn

Keywords: OVATE, OVATE FAMILY PROTEINS, fruit shape, transcription factor, plant growth and development, *Arabidopsis*, rice, pepper

INTRODUCTION

More than 100 years ago, it was proposed that the pear-shaped fruit phenotype in tomato (*Solanum lycopersicum*) might be genetically controlled by a single recessive locus, namely *ovate* (Hedrick and Booth, 1907; Price and Drinkard, 1908). However, it was only at the end of the last century that Ku et al. (1999) established, by construction and analysis of near-isogenic lines (NILs), that the *ovate* locus could account for both the pear shape and the elongated fruit shape in tomato. The *ovate* locus was later mapped to chromosome 2 (Ku et al., 2001), and the *ovate* gene was finally cloned in 2002 (Liu et al., 2002). Amino acid sequence analysis showed that OVATE is different from all the previously characterized plant genetic regulators, indicating that OVATE FAMILY PROTEINS (OFPs) represent a novel class of plant regulators (Liu et al., 2002).

Subsequent studies revealed that OFPs are widely distributed in the plant kingdom, and that they regulate multiple aspects of plant growth and development (Gui and Wang, 2007; Rodríguez et al., 2011; Tsaballa et al., 2011; Wang et al., 2011; Huang et al., 2013).

THE 100-YEAR PATHWAY TO THE IDENTIFICATION OF THE *OVATE* GENE

About a century ago, a series of genetic studies in tomato suggested that fruit size and shape in this economically important species are quantitatively inherited (Hedrick and Booth, 1907; Price and Drinkard, 1908; Grane, 1915; Lindstrom, 1926, 1927, 1929, 1932; MacArthur, 1926; Tanksley, 2004). More specifically, pear-shaped fruit form in tomato was proposed to be controlled by a single recessive quantitative trait locus (QTL), which was named *pyriform* (*pr*) (Hedrick and Booth, 1907; Price and Drinkard, 1908). Later, *pr* was found to co-segregate with the locus causing oblate- to oval-shaped fruit, and was therefore renamed *ovate* (*o*). In the late 1920s, based on its linkage to the *dwarf* locus, *ovate* was placed on chromosome 2 (Lindstrom, 1926, 1927, 1929; MacArthur, 1926).

Nearly 70 years later, Ku et al. (1999) conducted a more detailed molecular marker-based analysis of fruit shape in *Lycopersicon*. By crossing *Lycopersicon esculentum* cv. Yellow Pear (TA503), a variety of tomato bearing small, pear-shaped fruit, with *Lycopersicon esculentum pimpinellifolium* (LA1589), a wild tomato species bearing round-shaped fruit, and examining the F2 population, they found that the pear-shaped fruit phenotype is largely controlled by a major QTL on chromosome 2. This observation was confirmed by analyzing F2 populations from crosses between TA503 and a round-fruited introgression tomato line IL2-5, which carried the distal portion of the chromosome 2 from the *L. pennellii* genome. Based on these results, they proposed that the QTL detected on chromosome 2 corresponds to the *ovate* locus (Ku et al., 1999).

High-resolution mapping of the *ovate* region on chromosome 2 using a total of 3000 near-isogenic lines (NILs) derived from TA503 and *L. pennellii*, allowed Ku et al. (2001) to place *ovate* adjacent to a known marker. By screening tomato BAC (bacterial artificial chromosome) and binary BAC libraries with the known marker-derived probe, and mapping the ends of the selected BAC clones, they were able to identify two *ovate*-containing BAC clones (Ku et al., 2001). A combination of sequencing and fine mapping analysis of one of these BAC clones obtained further narrowed the *ovate* locus to a 55 kb fragment that contained eight open reading frames (ORFs) (Liu et al., 2002). To identify the *OVATE* gene, Liu et al. (2002) amplified and compared the corresponding 55 kb fragments from TA496, a round-fruited wild type *L. esculentum* genotype, and TA493 (*L. esculentum* cv. Heinz 1706), an *ovate* genotype. They identified a G_{TA496}-to-T_{TA493} nucleotide polymorphism (SNP) in one of ORFs. This SNP created an early stop codon in the *ovate* genotype, leading to a 75-aa truncation in the C-terminus of the predicted protein. Sequence comparison of the corresponding ORF in several pear-shaped tomato varieties including TA503, LA791 (*L. esculentum* cv. Longjohn), and LA0025, as well as complementation of the pear-shaped fruit phenotype in TA503 by over-expression of the genomic DNA fragment containing the ORF and its 5′ and 3′ untranslated regions from TA496, confirmed the identity of the tomato *OVATE* gene.

Amino acid sequence analysis showed that the OVATE protein contains an ~70-aa carboxyl-terminal domain, referred to as the OVATE domain, that is conserved in both *Arabidopsis* and rice. The premature termination occurring in the *ovate* genotype eliminates most of this conserved OVATE domain (Liu et al., 2002). The OVATE protein also contains a putative bipartite nuclear localization signal (NLS), and two putative Von Willebrand factor type C (VWFC) domains required for protein–protein interaction, features that distinguish OVATE from any of the previously identified plant genetic regulators (Liu et al., 2002).

Thanks to fast development of new sequencing technologies over the recent years, the genomes of many more plant species have now been fully sequenced, which has greatly facilitated the identification of protein homologs and phylogenetic studies in plants. Based on amino acid sequence similarity analysis, OFPs are found exclusively in plants (Hackbusch et al., 2005; Gui and Wang, 2007; Wang et al., 2007, 2011; Rodríguez et al., 2011; Tsaballa et al., 2011; Huang et al., 2013). By using the amino acid sequences of OFPs in *Arabidopsis* and the OVATE protein in tomato to search genomes of 13 land plants including *Solanum lycopersicum*, *Solanum tuberosum*, *Mimulus guttatus*, *Arabidopsis*, *Vitis vinifera*, *Populus trichocarpa*, *Prunus persica*, *Carica papaya*, *Aquilegia coerulea*, rice (*Oryza sativa*), *Zea mays*, *Selaginella moellendorffii*, and *Physcomitrella patens*, Liu et al. (2014) found that OFPs are distributed in all the plants examined, including the seedless vascular plant *Selaginella moellendorffii*, and the non-vascular plant *Physcomitrella patens*. In contrast, no OFPs were identified in chlorophytes, a division of the green algae (Liu et al., 2014). Interestingly, monocot species appeared to have more OFP family members than eudicots; for example, *Zea mays* had 45 OFPs, and rice had 33 OFPs, whereas tomato had 26 OFPs and *Vitis vinifera* had only 9 (Liu et al., 2014). It was subsequently reported that 31 genes in rice encode full-length OFPs (Yu et al., 2015). It should also be noted that Liu et al. (2014) found that the *Arabidopsis* genome contains 19 OFP-encoding genes, rather than 18 as reported previously (Hackbusch et al., 2005; Wang et al., 2011).

ROLES OF OFPs IN PLANT GROWTH AND DEVELOPMENT

Although OFPs are widely distributed in the plant kingdoms, their biological functions in plants remain largely unknown. However, limited studies in several different plant species including tomato, *Arabidopsis*, pepper (*Capsicum annuum*) and banana (*Musa acuminata*) have revealed that OFPs proteins are involved in regulation of several aspects of plant growth and development.

Ovule Development

Characterization of a T-DNA insertion mutant for AtOFP5 suggests that this member of the *Arabidopsis* OFP family may be involved in the regulation of ovule development. When Pagnussat et al. (2007) failed to isolate homozygous plants from a T-DNA insertion line of *AtOFP5* (SALK_ 010386), they examined female gametophyte development in pistils of the *ofp5*−/+ heterozygotes. This revealed that, out of the 256 embryo sacs studied, ∼38% of them seemed to collapse at the FG2 female gametophyte development stage, when two-nucleate cells are normally produced, and ∼8% of the embryo sacs showed abnormal micropylar cells with two egg cells, indicating that *AtOFP5* is required for proper female gametophyte development. Because some embryo sacs in the *ofp5*- genetic background had two egg cells, AtOFP5 may be specifically involved in the regulation of a cell fate switch from synergid to egg cell development (Pagnussat et al., 2007).

Vascular Development

OsOFP2, a rice OFP gene, was found to be expressed mainly within the vasculature in all growth stages examined in rice, as shown by GUS staining in *OsOFP2pro:GUS* transgenic rice plants. Rice plants over-expressing *OsOFP2* under the control of the *35S* promoter showed reduced height and exhibited altered leaf morphology, seed shape, and positioning of vascular bundles in the stems. Transcriptome analysis of the *OsOFP2* over-expressing rice plants showed that the expression of genes associated with vascular development, lignin biosynthesis, and hormone homeostasis was altered, consistent with a role for *OsOFP2* in the regulation of vascular development (Schmitz et al., 2015).

Fruit Shape

Consistent with OVATE's role as a key regulator of fruit shape in tomato (Liu et al., 2002; Azzi et al., 2015; Wang et al., 2015; Wu et al., 2015), the *OVATE* gene is expressed mainly in reproductive organs. Its transcripts can be detected in tomato flowers 10 days before anthesis and begin to decrease in developing fruit 8 days after anthesis (Liu et al., 2002). *CaOvate*, an *OVATE* family member in pepper that shares 63% identity with the tomato OVATE protein, was also found to be involved in the regulation of fruit shape (Tsaballa et al., 2011). However, unlike in tomato, differences in fruit shape in two pepper cultivars, cv. "Mytilini Round" and cv. "Piperaki Long," are associated with different patterns of *CaOvate* expression, rather than with an ORF mutation; i.e., the expression of *CaOvate* is higher in cv. "Mytilini Round" than in cv. "Piperaki Long." Down-regulation of *CaOvate* in cv. " Mytilini Round" produced through virus-induced gene silencing (VIGS) thus changed its fruit to a more oblong form (Tsaballa et al., 2011). Beyond the Solanaceae, several QTLs controlling fruit shape in melon (*Cucumis melo*) have also been identified, and several melon OFP homologs (*CmOFP*) were found to co-map with seven fruit shape QTLs, indicating that OFPs in melon may also control fruit shape (Monforte et al., 2014).

It should be noted that the G-to-T mutation in *OVATE* is not associated with a single fruit shape phenotype in tomato. In some genetic backgrounds, the mutation leads to pear-shape fruits, but in other backgrounds, fruit shape remains largely unchanged, indicating that the *OVATE* locus may interact with other regulators to control a specific fruit shape (Tanksley, 2004; Gonzalo and van der Knaap, 2008). Indeed, two suppressor loci for *OVATE*, *suppressor of OVATE1* (*sov1*) and *sov2*, have been identified in tomato (Rodríguez et al., 2013).

Grafting of *Capsicum* cv. "Mytilini Round" scion on the long-shaped cultivar, cv. "Piperaki Long" rootstock, also resulted in heritable fruit shape changes in the scion, and these fruit phenotypic changes were retained through two generations of seed-derived progeny (Tsaballa et al., 2013). However, only a slight difference in *CaOvate* gene expression accompanied this change in the fruit shape, and simple sequence repeat (ISSR) analysis of the progenies of the scion fruits showed that their genetic profile closely resembled the parental scion genetic profile. This suggests either that only slight changes in the expression of *CaOvate* are enough to induce fruit shape changes, or that *CaOvate* interacts with other genes in an epistatic manner to regulate fruit shape in pepper (Tsaballa et al., 2013). It is also possible that epigenetic modification plays a role in the regulation of fruit shape in pepper.

DNA Repair

When Hackbusch et al. (2005) tried to characterize *Arabidopsis ofp1* mutants, they failed to recover any homozygous plants from three independent lines with T-DNA inserted either in the single exon or in the 5′ upstream region of *AtOFP1*. They concluded that *AtOFP1* must be required for essential processes in gametophyte or sporophyte development. However, they did not observe any apparent morphological aberrations in either the pollen or ovules in plants heterozygous for the T-DNA insertion. When they further investigated male and female transmission of the T-DNA insertion by reciprocally crossing the mutants with wild type plants, they found that when the heterozygous mutant lines were used as pollen donors, no T-DNA insertion in the *AtOFP1* gene was detected in the F1 plants. The authors therefore concluded that AtOFP1 is essential for male gamete and pollen function (Hackbusch et al., 2005).

In contrast to this report, Wang et al. (2007) successfully obtained homozygous mutant plants from *ofp1-2* and *ofp1-3*, two mutant lines derived from the same T-DNA insertion lines used by Hackbusch et al. (2005). These authors reported that, in the *ofp1-2* mutant, the T-DNA was actually inserted in the 5′-UTR of *AtOFP1* rather than in the exon as indicated by the T-DNA Express database (http://signal.salk.edu/cgi-bin/tdnaexpress). The expression of *AtOFP1* in the *ofp1-2* mutant was largely unaffected. Wang et al. (2007) also identified an *ofp1-1* mutant, a true loss-of-function mutant line for *AtOFP1* derived from a transposon insertion event, and found that this mutant is morphologically similar to the wild type plant. These results are inconsistent with the earlier proposal that *OFP1* plays a critical role in male gamete transmission or pollen function.

Although the *ofp1-1* mutant is largely similar to wild type plant (Wang et al., 2007), *ofp1-1* mutant seedlings were reported to be more sensitive to treatment with methyl methanesulfonate (MMS), a DNA-damaging reagent. The *ofp1-1* mutants also showed relative lower non-homologous end-joining (NHEJ) activity *in vivo*. Because the NHEJ pathway is thought to be involved in the repair of DNA double-strand breaks (DSBs), these results suggest that OFP1 may play a role in this DNA repair pathway (Wang et al., 2010).

Secondary Cell Wall Formation

Although all the single and double *AtOFP* gene mutants identified in *Arabidopsis* are morphologically similar to wild type plants (Wang et al., 2011), careful examination of the anatomy of stem inflorescence cross-sections showed that *ofp4* mutants exhibited an *irregular xylem* (*irx*) phenotype, marked by increased thickness of interfascicular fiber cell walls, and decreased wall thickness of vessel and xylary fiber cells (Li et al., 2011). This phenotype is similar to that of *knat7*, a loss-of-function mutant of the *KNOTTED ARABIDOPSIS THALIANA7* transcription factor gene (Brown et al., 2005). Further analysis showed that the phenotype of the *ofp4 knat7* double mutant was similar to that of the *ofp4* or *knat7* single mutants, and that the *AtOFP4* over-expression phenotype (kidney-shaped cotyledons, and round and curled leaves) (Wang et al., 2011), was suppressed in a *knat7* mutant background. Taken together, these results suggest that AtOFP4 is involved in the regulation of secondary cell wall formation, and that its function is at least partially dependent on *KNAT7*.

Unlike *ofp4*, xylem and interfascicular fiber morphology in the *ofp1-1* mutant is indistinguishable from that in wild type plants (Li et al., 2011). However, the *AtOFP1* over-expression phenotype is similar to that of *AtOFP4* over-expression plants (Wang et al., 2011), and is also suppressed in a *knat7* mutant background, suggesting that AtOFP1 may also be involved in the regulation of secondary cell wall formation (Li et al., 2011). However, the possibility cannot be ruled out that this behavior may simply reflect the close homologous relationship between AtOFP1 and AtOFP4.

Fruit Ripening

A role for OFPs in regulating fruit ripening has so far only been reported for banana. The banana OFP1 (MaOFP1) was identified in a yeast two-hybrid (Y2H) study when using the banana MADS-box protein MuMADS1 as bait to screen a 2-day-postharvest (DPH) banana fruit cDNA library (Liu et al., 2015). Quantitative RT-PCR (qRT-PCR) analysis showed that both *MuMADS1* and *MaOFP1* were highly expressed in 0 DPH fruit, but had relative low levels of expression in the stem. However, different expression patterns of *MuMADS1* and *MaOFP1* were observed in different tissues and developing fruits. qRT-PCR analysis also showed that expression of *MuMADS1* and *MaOFP1*was differentially regulated by ethylene, i.e., expression of *MuMADS1* was induced by exogenously applied ethylene, and suppressed by the ethylene competitor 1-methylcyclopropene (1-MCP), whereas expression of *MaOFP1* was suppressed by ethylene, and induced by 1MCP. These results indicate that MuMADS1 and MaOFP1 may play antagonistic roles in ethylene-induced postharvest fruit ripening in banana (Liu et al., 2015).

Pleiotropic Effects

Although some evidence suggests that OFPs may have specific effects on different aspects of plant growth and development, as described above, other evidence suggests that OFPs can also have complex pleiotropic effects on plant growth and development. For example, in addition to resulting in round fruit, over-expression of *OVATE* in tomato also produced a number of abnormal phenotypes, including reduced size of floral organs, dwarf plants with smaller compound-leaf size and rounder leaflets (Liu et al., 2002). Similarly, over-expression of some *Arabidopsis OFP* genes also resulted in inhibited plant growth and development (Hackbusch et al., 2005; Wang et al., 2011).

Analysis of transgenic plants over-expressing *AtOFPs* in *Arabidopsis* also reveals that they regulate multiple aspects of plant growth and development in this species. Based on the similarity of the phenotypes observed, AtOFPs could be grouped into different groups. Over-expression of *AtOFP1*, *AtOFP2*, *AtOFP4*, *AtOFP5* and *AtOFP7* resulted in similar phenotypes including kidney-shaped cotyledons, as well as round and curled leaves, and these AtOFPs were designated as Class I AtOFPs. Over-expression of *AtOFP6* and *AtOFP8*, on the other hand, resulted in a different phenotype including flat, thick and cyan leaves, and these were therefore designated as Class II AtOFPs. Over-expression of *AtOFP13*, *AtOFP15*, *AtOFP16* and *AtOFP18* led to another distinct phenotype including blunt-end siliques, and these AtOFPs were designated as Class III AtOFPs. Plants over-expressing all other AtOFPs examined were largely indistinguishable from wild type plants (Wang et al., 2011). Interestingly, this functional-based classification is largely consistent with the subgroups of the AtOFPs identified in phylogenetic analysis (**Figure 1**). Because *AtOFP19* and *AtOFP20* are newly identified OFPs in *Arabidopsis* (Liu et al., 2014), their functions have not yet been examined. Consistent with their functions in regulating multiple aspects of plant growth and development, expression of most of the *OFPs* in tomato, rice and *Arabidopsis* was detectable in all the tissues and organs examined (Wang et al., 2011; Liu et al., 2014; Schmitz et al., 2015). Nevertheless, differential expression patterns are also observed; for example, nearly half of the *OsOFPs* were found to be more highly expressed during the stages of rice seed development (Yu et al., 2015).

It is interesting that all the AtOFPs shown through genetic manipulations, as described above, to have a specific role in regulating plant growth and development are from the Class I subfamily. All knockout mutants identified so far for other *AtOFP* genes, including *AtOFP8*, *AtOFP10*, *AtOFP15*, and *AtOFP16*, are morphologically similar to wild type plants (Wang et al., 2011). A double mutant between two class III OFP genes, *ofp15 ofp16* is also indistinguishable from wild type plants (Wang et al., 2011).

MECHANISMS UNDERLYING THE REGULATION OF PLANT GROWTH AND DEVELOPMENT BY OFPs

Regulation of Target Gene Expression

The evidence available so far supports a model in which OFPs regulate plant growth and development by directly influencing expression of their target genes, and/or through interaction with other transcription factors. Both AtOFP1 and AtOFP5 were found to associate with the cytoskeleton in transient expression

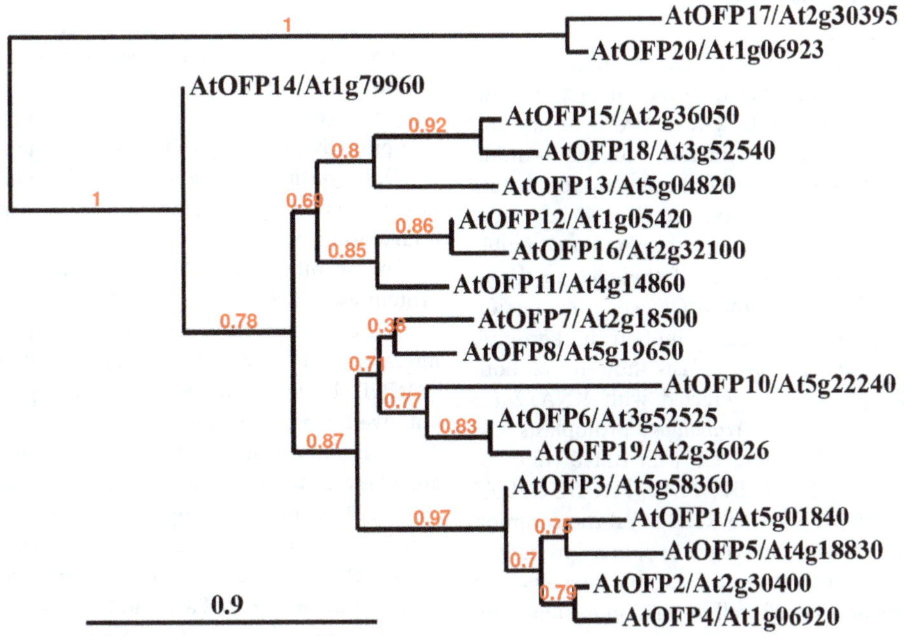

FIGURE 1 | Phylogenetic analysis of *Arabidopsis* OVATE FAMILY PROTEINS (AtOFPs). The entire amino acid sequences of AtOFPs were obtained from Phytozome (https://phytozome.jgi.doe.gov/pz/portal.html) and used to generate the phylogenetic tree by using "One Click" mode with default settings on Phylogeny (www.phylogeny.fr). Branch support values are indicated above the branches, and Bar indicates branch length.

assays in tobacco leaves (Hackbusch et al., 2005). However, similar to the tomato OVATE protein, AtOFP1 also possesses a putative NLS (Liu et al., 2002; Wang et al., 2007), and in both transient transfection assays in *Arabidopsis* protoplasts, and stable transformed *Arabidopsis* plants, AtOFP1 was found to localize in the nucleus (Wang et al., 2007). AtOFP1 also contains an LxLxL motif (Wang et al., 2007), which is also found in Aux/IAA proteins and ERF transcription factors and is required for their transcriptional repression functions (Ohta et al., 2001; Hiratsu et al., 2003; Tiwari et al., 2004). Transfection assays in *Arabidopsis* protoplasts showed that AtOFP1 repressed reporter gene expression when recruited to the promoter region of the *Gal4:GUS* reporter gene by a fused Gal4 DNA binding domain (GD), suggesting that AtOFP1 may function *in vivo* as a transcriptional repressor (Wang et al., 2007). Although some of the AtOFPs do not contain a LxLxL motif, all the AtOFPs examined repressed *Gal4:GUS* reporter gene expression in transfected protoplasts, indicating that AtOFPs in *Arabidopsis* are a novel transcription repressor family (Wang et al., 2011).

Targets of OFP Proteins

A gain-of-function mutant of AtOFP1, *ofp1-1D*, as well as *AtOFP1* over-expressing plants, all showed reduced lengths in their aerial organs, including hypocotyl, rosette leaves, cauline leaves, inflorescence stem, floral organs and siliques (Hackbusch et al., 2005; Wang et al., 2007). Detailed analysis showed that this growth phenotype was the result of a reduction in cell elongation, rather than in cell division (Wang et al., 2007), suggesting that AtOFP1 represses cell elongation. Hackbusch et al. (2005) had earlier shown that expression of *AtGA20ox1*, a gene encoding a key enzyme in gibberellic acid (GA) biosynthesis, was reduced in plants over-expressing AtOFP1, and Wang et al. (2007) showed, by use of chromatin immunoprecipitation assays, that the *AtGA20ox1* promoter is a direct target of AtOFP1.

GA20ox is likely also a target of OFPs in pepper and rice. In pepper, two lines of evidence suggest that *GA20ox1* is likely a target gene of the OFP homolog, CaOvate. First, RT-PCR analysis showed that although *GA20ox1* has a similar expression level at most growth stages in both cv. "Mytilini Round" and cv. "Piperaki Long," its expression was significantly higher in cv. "Piperaki Long" fruits 10 days after anthesis. In addition, expression of *GA20ox1* was elevated after VIGS suppression of *CaOvate* expression (Tsaballa et al., 2011, 2012). Similarly, rice *GA20ox7* is likely a target gene of OsOFP2, since expression of *GA20ox7* was reduced in transgenic rice plants over-expressing *OsOFP2* (Schmitz et al., 2015).

Although *GA20ox1* is probably a direct target of AtOFP1 (Wang et al., 2007), and reduction in the expression of rice *GA20ox7* is correlated with lower gibberellin content in transgenic rice plants (Schmitz et al., 2015), exogenous GA can only partially restore the defects in cell elongation in *Arabidopsis* transgenic plants over-expressing AtOFP1 (Wang et al., 2007). This indicates that AtOFP1 may have other target genes, and indeed, microarray-based gene expression assays showed that a total of 129 genes were down-regulated at least twofold when *AtOFP1-GR* (glucocorticoid receptor) transgenic plants were treated with DEX (dexamethasone), which will allow the AtOFP1-GR protein to relocate into the nucleus (Wang et al., 2011). However, it remains unknown if any of these genes is directly targeted by AtOFP1.

Interaction of OFPs with Other Transcription Factors

So far nearly all OFPs with known functions were found to regulate plant growth and development via interaction with homeodomain proteins (**Table 1**). In a yeast two-hybridization screen, nine AtOFPs were found to interact with the 3-amino acid loop extension homeodomain (TALE) transcription factors KNOX and BELL (BEL1-like homeodomain) (Hackbusch et al., 2005). Consistent with the observation that the *ofp4 knat7* double mutant phenotype was similar to that of the *ofp4* or *knat7* single mutants, and that the *AtOFP1* and *AtOFP4* over-expression phenotype was suppressed in a *knat7* mutant background, transfection assays in *Arabidopsis* protoplasts showed that both AtOFP1 and AtOFP4 physically interacted with KNAT7 (Li et al., 2011). Transfection assays in *Arabidopsis* protoplasts also showed both OFP1 and OFP4 interacted with BLH6 (Liu and Douglas, 2015). Considering that MYB75/PAP1 (PRODUCTION OF ANTHOCYANIN PIGMENT 1), a R2R3 MYB transcription factor that has been shown to regulate phenylpropanoid biosynthesis in *Arabidopsis* (Borevitz et al., 2000), interacts with KNAT7 to modulate secondary cell wall deposition in stems and seed coat in *Arabidopsis* (Bhargava et al., 2013), it is likely that AtOFPs, MYB transcription factors, and TALE homeodomain proteins form one or more multi-protein complexes to regulate secondary cell wall formation. Recently, AtOFP1 was also found to interact with the BELL transcription factor BLH3 to regulate the vegetative to reproductive phase transition in *Arabidopsis* (Zhang et al., 2016).

Homologs of *Arabidopsis* KNOX and BELL proteins have also been fund to interact with OFPs and to regulate secondary cell wall formation in other plant species. For example, GhKNL1, a homeodomain protein in cotton (*Gossypium hirsutum*) is found to be preferentially expressed in developing fibers at the stage of secondary cell wall biosynthesis, and ectopic expression of *GhKNL1* can partially rescue the cell wall defective phenotype of the *Arabidopsis knat7* mutant. Yeast two-hybrid assays showed that GhKNL1 interacts with GhOFP4, as well as with AtOFP1, AtOFP4, and AtMYB75 (Gong et al., 2014). In rice, OsOFP2 was found to interact with putative vascular development KNOX and BELL proteins, so it is also likely that OsOFP2 can modulate KNOX-BELL function in this species to regulate diverse aspects of development, including vascular development (Schmitz et al., 2015).

AtOFP5 has been shown to be involved in the regulation of female gametophyte development. The "two egg cells" phenotype in *eostre-1*, a female gametophyte mutant, was caused by elevated expression of the BELL transcription factor gene *BLH1* in the embryo sac, and this phenotype is dependent upon the function of the class II knox gene, *KNAT3*, (Pagnussat et al., 2007). Disruption of AtOFP5, a known interactor of KNAT3 and BLH1, also partially phenocopies the *eostre* mutation (Pagnussat et al., 2007), suggesting that the roles of *AtOFP5* in female gametophyte development may also involve interactions of AtOFP5 with TALE homeodomain proteins.

On the other hand, Y2H screening using tomato OVATE protein as bait did not lead to the identification of any KNOX or BELL transcription factors. Instead, the screen revealed interactions of OVATE with 11 out of 26 members of the TONNEAU1 Recruiting Motif (TRM) superfamily, including the putative tomato ortholog of AtTRM17/20 (van der Knaap et al., 2014). Because some of the TRMs can bind microtubules and are likely centrosomal components in plant cells (Drevensek et al., 2012), this result suggests that, in addition to interacting with KNOX or BELL transcription factors, and acting as transcriptional repressors to repress target gene expression, OFPs may also interact with TRMs and microtubules to regulate plant growth and development (van der Knaap et al., 2014).

CONCLUDING REMARKS AND FUTURE PERSPECTIVES

Studies in recent years reveal that OFPs are widely distributed in the plant kingdom, and that they help regulate multiple aspects of plant growth and development in various species (Liu et al., 2002; Hackbusch et al., 2005; Gui and Wang, 2007; Wang et al., 2007, 2011; Rodríguez et al., 2011; Tsaballa et al., 2011; Huang et al., 2013). Beyond this general observation, however, it will be important to establish the extent to which OFP family members are functionally conserved in different taxa. As described above, it would appear that some OFP homologs may have conserved functions; for example, OVATE protein was identified as a key regulator of fruit shape in tomato (Liu et al., 2002), while fruit shape in pepper is also at least partially controlled by an OFP homologue, CaOvate (Tsaballa et al., 2011). However, tomato and *Capsicum* are closely related genera, so it remains to be established whether fruit shape in other plants is also controlled by homologous OFPs. In addition, a major QTL, *fs10.1*, has been shown to control fruit elongation in *Capsicum*, but it is unclear if it represents an *OFP* gene (Borovsky and Paran, 2011). Similarly,

TABLE 1 | OVATE FAMILY PROTEINS (OFPs) interact with homeodomain proteins to regulate plant growth and development.

OFPs	Interactors	Functions in plants	Reference
AtOFP1	KNAT7, BLH6	Secondary cell wall formation	Li et al., 2011; Liu and Douglas, 2015
AtOFP1	BLH3	Phase transition	Zhang et al., 2016
AtOFP4	KNAT7, BLH6	Secondary cell wall formation	Li et al., 2011; Liu and Douglas, 2015
AtOFP5	KANT3, BLH1	Female gametophyte development	Pagnussat et al., 2007
GhOFP4	GhKNL1	Secondary cell wall formation	Gong et al., 2014
OsOFP2	OsKNAT7, BLH6-like1, bHLH6-like2	Vascular development	Schmitz et al., 2015
MaOFP1	MuMADS1	Fruit ripening	Liu et al., 2015

several QTLs controlling fruit shape in apple have been identified, but so far it is unclear whether any of them is associated with an *OFP* gene (Cao et al., 2015). Once these genomes have been fully sequenced, the candidate regions can be explored for the possible presence of OFP homologs.

Further evidence is also required to establish the extent to which the function of OFPs in controlling secondary cell formation might be evolutionarily conserved. In *Arabidopsis*, interaction of specific OFPs with homeodomain proteins strongly influences secondary cell wall formation (Li et al., 2011; Liu and Douglas, 2015). In cotton (*Gossypium hirsutum*), GhKNL1 has been reported to interact with GhOFP4, but it remains to be demonstrated that GhOFP4 is also involved in the regulation of secondary cell wall formation in cotton fibers.

So far only *AtGA20ox1* has been shown to be a direct target of AtOFP1. However, the observation that expression of >100 genes was down-regulated in DEX-treated *AtOFP1-GR* transgenic plants (Wang et al., 2011), raises the possibility that some of these genes may also serve as direct targets of AtOFP1.

The possibility cannot be excluded that OFPs may regulate plant growth and development in ways other than directly targeting expression of particular genes or interaction with other transcription factors. Phenotype similarity-based characterization may make it possible to detect the operation of a previously unsuspected mechanism. For example, transgenic *Arabidopsis* over-expressing Class III AtOFP genes display blunt siliques, a phenotype also observed in the mutants *erecta*, a loss-of-function mutant of *ERECTA* (*ER*), a putative receptor protein kinase, and *agb1-1*, a loss-of-function mutant of the heterotrimeric G-protein β subunit gene *AGB1* (Torii et al., 1996; Lease et al., 2001). The *agb1-1* mutant was originally identified as an *erecta-like* mutant, *elk4*, during the characterization of an ER signaling pathway, and was renamed as *agb1-1* after *ELK4* was found to encode AGB1. Their results indicate that AGB1 can influence silique morphology via an ER-type Leucine-rich repeat (LRR) receptor-like kinase signaling pathway. Considering that an ER LRR receptor-like kinase signaling pathway has been shown to regulate multiple aspects of plant growth and development, including organ shape (Shpak et al., 2003), and that the transcription factor SPEECHLESS has been shown to be phosphorylated by a MAP kinase signaling pathway (Lampard et al., 2008), it will be worthwhile to investigate whether Class III AtOFPs might be phosphorylated by kinases operating downstream of an ER LRR receptor-like kinase signaling pathway, and whether such post-translational modification of OFPs plays a role in regulating silique morphology.

AUTHOR CONTRIBUTIONS

SW conceived the review topic, BE and YC participated in the discussion of topic, SW drafted the manuscript, BE reversed the manuscript, and all the authors approved the final version of manuscript.

ACKNOWLEDGMENTS

This work was supported by the National Natural Science Foundation of China (31470297, 31370221), and a Discovery Grant from the Natural Science and Engineering Research Council of Canada (BEE). The funders had no role in study design, data collection and analysis, decision to publish, or preparation of the manuscript.

REFERENCES

Azzi, L., Deluche, C., Gévaudant, F., Frangne, N., Delmas, F., Hernould, M., et al. (2015). Fruit growth-related genes in tomato. *J. Exp. Bot.* 66, 1075–1086. doi: 10.1093/jxb/eru527

Bhargava, A., Ahad, A., Wang, S., Mansfield, S. D., Haughn, G. W., Douglas, C. J., et al. (2013). The interacting MYB75 and KNAT7 transcription factions modulate secondary cell wall deposition both in Stems and seed coat in *Arabidopsis*. *Planta* 237, 1199–1211. doi: 10.1007/s00425-012-1821-9

Borevitz, J. O., Xia, Y., Blount, J., Dixon, R. A., and Lamb, C. (2000). Activation tagging identifies a conserved MYB regulator of phenylpropanoid biosynthesis. *Plant Cell* 12, 2383–2393. doi: 10.2307/3871236

Borovsky, Y., and Paran, I. (2011). Characterization of fs10.1, a major QTL controlling fruit elongation in Capsicum. *Theor. Appl. Genet.* 123, 657–665. doi: 10.1007/s00122-011-1615-7

Brown, D. M., Zeef, L. A., Ellis, J., Goodacre, R., and Turner, S. R. (2005). Identification of novel genes in *Arabidopsis* involved in secondary cell wall formation using expression profiling and reverse genetics. *Plant Cell* 17, 2281–2295. doi: 10.1105/tpc.105.031542

Cao, K., Chang, Y., Sun, R., Shen, F., Wu, T., Wang, Y., et al. (2015). Candidate gene prediction via quantitative trait locus analysis of fruit shape index traits in apple. *Euphytica* 206, 381–391. doi: 10.1007/s10681-015-1488-y

Drevensek, S., Goussot, M., Duroc, Y., Christodoulidou, A., Steyaert, S., Schaefer, E., et al. (2012). The *Arabidopsis* TRM1-TON1 interaction reveals a recruitment network common to plant cortical microtubule arrays and eukaryotic centrosomes. *Plant Cell* 24, 178–191. doi: 10.1105/tpc.111.089748

Gong, S. Y., Huang, G. Q., Sun, X., Qin, L. X., Li, Y., Zhou, L., et al. (2014). Cotton KNL1, encoding a class II KNOX transcription factor, is involved in regulation of fibre development. *J. Exp. Bot.* 65, 4133–4147. doi: 10.1093/jxb/eru182

Gonzalo, M. J., and van der Knaap, E. (2008). A comparative analysis into the genetic bases of morphology in tomato varieties exhibiting elongated fruit shape. *Theor. Appl. Genet.* 116, 647–656. doi: 10.1007/s00122-007-0698-7

Grane, M. B. (1915). Heredity of types of inflorescence and fruits in tomato. *J. Gen.* 5, 1–11. doi: 10.1007/BF02982149

Gui, B., and Wang, Y. (2007). Cloning and sequence analysis of ovate orthologous gene in tobacco (*Nicotiana tabacum* L.). *Plant Physiol. Commun.* 43, 1050–1056.

Hackbusch, J., Richter, K., Müller, J., Salamini, F., and Uhrig, J. F. (2005). A central role of *Arabidopsis thaliana* ovate family proteins in networking and subcellular localization of 3-aa loop extension homeodomain proteins. *Proc. Natl. Acad. Sci. U.S.A.* 102, 4908–4912. doi: 10.1073/pnas.0501181102

Hedrick, U. P., and Booth, N. O. (1907). Mendelian characters in tomato. *Proc. Am. Soc. Hort. Sci.* 5, 19–24.

Hiratsu, K., Matsui, K., Koyama, T., and Ohme-Takagi, M. (2003). Dominant repression of target genes by chimeric repressors that include the EAR motif, a repression domain, in *Arabidopsis*. *Plant J.* 34, 733–739. doi: 10.1046/j.1365-313X.2003.01759.x

Huang, Z. J., Van Houten, J., Gonzalez, G., Xiao, H., and van der Knaap, E. (2013). Genome-wide identification, phylogeny and expression analysis of SUN, OFP and YABBY gene family in tomato. *Mol. Genet. Genomics* 288, 111–129. doi: 10.1007/s00438-013-0733-0

Ku, H. M., Doganlar, S., Chen, K. Y., and Tanksley, S. D. (1999). The genetic basis of pear-shaped tomato fruit. *Theor. Appl. Genet.* 9, 844–850. doi: 10.1007/s001220051304

Ku, H. M., Liu, J., Doganlar, S., and Tanksley, S. D. (2001). Exploitation of *Arabidopsis*-tomato synteny to construct a high-resolution map of the ovate-containing region in tomato chromosome 2. *Genome* 44, 470–475. doi: 10.1139/g01-024

Lampard, G. R., Macalister, C. A., and Bergmann, D. C. (2008). *Arabidopsis* stomatal initiation is controlled by MAPK-mediated regulation of the bHLH SPEECHLESS. *Science* 322, 1113–1116. doi: 10.1126/science.1162263

Lease, K. A., Wen, J., Li, J., Doke, J. T., Liscum, E., and Walker, J. C. (2001). A mutant *Arabidopsis* heterotrimeric G-protein beta subunit affects leaf, flower, and fruit development. *Plant Cell* 13, 2631–2641. doi: 10.1105/tpc.13.12.2631

Li, E., Wang, S., Liu, Y., Chen, J. G., and Douglas, C. J. (2011). OVATE FAMILY PROTEIN4 (OFP4) interaction with KNAT7 regulates secondary cell wall formation in *Arabidopsis thaliana*. *Plant J.* 67, 328–341. doi: 10.1111/j.1365-313X.2011.04595.x

Lindstrom, E. W. (1926). Linkage inheritance in tomatoes. *Iowa State Col. J. Sci.* 1, 3–13.

Lindstrom, E. W. (1927). The inheritance of ovate and related shapes of tomato fruit. *J. Agri. Res.* 34, 961–985.

Lindstrom, E. W. (1929). Fruit size and shape genes on the first chromosome of the tomato. *Proc. Iowa Acad. Sci.* 36, 189–190.

Lindstrom, E. W. (1932). First-chromosome genes in the tomato. *Genetics* 17, 351–357.

Liu, D., Sun, W., Yuan, Y., Zhang, N., Hayward, A., Liu, Y., et al. (2014). Phylogenetic analyses provide the first insights into the evolution of OVATE family proteins in land plants. *Ann. Bot.* 113, 1219–1233. doi: 10.1093/aob/mcu061

Liu, J., van Eck, J., Cong, B., and Tanksley, S. D. (2002). A new class of regulatory genes underlying the cause of pear-shaped tomato fruit. *Proc. Natl. Acad. Sci. U.S.A.* 99, 13302–13306. doi: 10.1073/pnas.162485999

Liu, J., Zhang, J., Hu, W., Miao, H., Zhang, J., Jia, C., et al. (2015). Banana Ovate family protein MaOFP1 and MADS-box protein MuMADS1 antagonistically regulated banana fruit ripening. *PLoS ONE* 10:e0123870. doi: 10.1371/journal.pone.0123870

Liu, Y., and Douglas, C. J. (2015). A role for OVATE FAMILY PROTEIN1 (OFP1) and OFP4 in a BLH6-KNAT7 multi-protein complex regulating secondary cell wall formation in *Arabidopsis thaliana*. *Plant Signal. Behav.* 10:e1033126. doi: 10.1080/15592324.2015.1033126

MacArthur, J. W. (1926). Linkage studies with the tomato. *Genetics* 11, 387–405.

Monforte, A. J., Diaz, A., Caño-Delgado, A., and van der Knaap, E. (2014). The genetic basis of fruit morphology in horticultural crops: lessons from tomato and melon. *J. Exp. Bot.* 65, 4625–4637. doi: 10.1093/jxb/eru017

Ohta, M., Matsui, K., Hiratsu, K., Shinshi, H., and Ohme-Takagi, M. (2001). Repression domains of class II ERF transcriptional repressors share an essential motif for active repression. *Plant Cell* 13, 1959–1968. doi: 10.1105/tpc.13.8.1959

Pagnussat, G. C., Yu, H. J., and Sundaresan, V. (2007). Cell-fate switch of synergid to egg cell in *Arabidopsis eostre* mutant embryo sacs arises from misexpression of the BEL1-like homeodomain gene BLH1. *Plant Cell* 19, 3578–3592. doi: 10.1105/tpc.107.054890

Price, H. C., and Drinkard, A. W. (1908). Inheritance in tomato hybrids. *Va. Agric. Exp. Stn. Bull.* 177, 17–53.

Rodríguez, G. R., Kim, H. J., and van der Knaap, E. (2013). Mapping of two suppressors of OVATE (sov) loci in tomato. *Heredity* 3, 256–264. doi: 10.1038/hdy.2013.45

Rodríguez, G. R., Muños, S., Anderson, C., Sim, S. C., Michel, A., Causse, M., et al. (2011). Distribution of SUN, OVATE, LC, and FAS in the tomato germplasm and the relationship to fruit shape diversity. *Plant Physiol.* 156, 275–285. doi: 10.1104/pp.110.167577

Schmitz, A. J., Begcy, K., SaratG, H., and Walia, H. (2015). Rice Ovate Family Protein 2 (OFP2) alters hormonal homeostasis and vasculature development. *Plant Sci.* 241, 177–188. doi: 10.1016/j.plantsci.2015.10.011

Shpak, E. D., Lakeman, M. B., and Torii, K. U. (2003). Dominant negative receptor uncovers redundancy in the *Arabidopsis* ERECTA Leucine-rich repeat receptor-like kinase signaling pathway that regulates organ shape. *Plant Cell* 15, 1095–1110. doi: 10.1105/tpc.010413

Tanksley, S. D. (2004). The genetic, developmental, and molecular bases of fruit size and shape variation in tomato. *Plant Cell* 16(Suppl. 1), S181–S189. doi: 10.1105/tpc.018119

Tiwari, S. B., Hagen, G., and Guilfoyle, T. J. (2004). Auxin/IAA proteins contain a potent transcriptional repression domain. *Plant Cell* 16, 533–543. doi: 10.1105/tpc.017384

Torii, K. U., Mitsukawa, N., Oosumi, T., Matsuura, Y., Yokoyama, R., Whittier, R. F., et al. (1996). The *Arabidopsis* ERECTA gene encodes a putative receptor protein kinase with extracellular leucine-rich repeats. *Plant Cell* 8, 735–746. doi: 10.1105/tpc.8.4.735

Tsaballa, A., Athanasiadis, C., Pasentsis, K., Ganopoulos, I., Nianiou-Obeidat, I., and Tasftaris, A. (2013). Molecular studies of inheritable grafting induced changes in pepper (*Capsicum annuum*) fruit shape. *Sci. Hortic.* 149, 2–8. doi: 10.1016/j.scienta.2012.06.018

Tsaballa, A., Pasentsis, K., Darzentas, N., and Tsaftaris, A. S. (2011). Multiple evidence for the role of an Ovate-like gene in determining fruit shape in pepper. *BMC Plant Biol.* 11:46. doi: 10.1186/1471-2229-11-46

Tsaballa, A., Pasentsis, K., and Tsaftaris, A. S. (2012). The role of a Gibberellin 20-oxidase gene in fruit development in pepper (*Capsicum annuum*) *Plant Mol. Biol. Rep.* 30, 556–565.

van der Knaap, E., Chakrabarti, M., Chu, Y. H., Clevenger, J. P., Illa-Berenguer, E., Huang, Z., et al. (2014). What lies beyond the eye: the molecular mechanisms regulating tomato fruit weight and shape. *Front. Plant Sci.* 5:227. doi: 10.3389/fpls.2014.00227

Wang, L., Li, J., Zhao, J., and He, C. (2015). Evolutionary developmental genetics of fruit morphological variation within the Solanaceae. *Front. Plant Sci.* 6:248. doi: 10.3389/fpls.2015.00248

Wang, S., Chang, Y., Guo, J., and Chen, J. G. (2007). *Arabidopsis* Ovate Family Protein 1 is a transcriptional repressor that suppresses cell elongation. *Plant J.* 50, 858–872. doi: 10.1111/j.1365-313X.2007.03096.x

Wang, S., Chang, Y., Guo, J., Zeng, Q., Ellis, B. E., and Chen, J. G. (2011). *Arabidopsis* Ovate Family Proteins, a novel transcriptional repressor family, control multiple aspects of plant growth and development. *PLoS ONE* 6:e23896. doi: 10.1371/journal.pone.0023896

Wang, Y. K., Chang, W. C., Liu, P. F., Hsiao, M. K., Lin, C. T., Lin, S. M., et al. (2010). Ovate family protein 1 as a plant Ku70 interacting protein involving in DNA double-strand break repair. *Plant Mol. Biol.* 74, 453–466. doi: 10.1007/s11103-010-9685-5

Wu, S., Clevenger, J. P., Sun, L., Visa, S., Kamiya, Y., Jikumaru, Y., et al. (2015). The control of tomato fruit elongation orchestrated by sun, ovate and fs8.1 in a wild relative of tomato. *Pant Sci.* 238, 95–104. doi: 10.1016/j.plantsci.2015.05.019

Yu, H., Jiang, W., Liu, Q., Zhang, H., Piao, M., Chen, Z., et al. (2015). Expression pattern and subcellular localization of the Ovate protein family in rice. *PLoS ONE* 10:e0118966. doi: 10.1371/journal.pone.0118966

Zhang, L., Zhang, X., Ju, H., Chen, J., Wang, S., Wang, H., et al. (2016). Ovate family protein1 interaction with BLH3 regulates transition timing from vegetative to reproductive phase in *Arabidopsis*. *Biochem. Biophys. Res. Commun.* 470, 492–497. doi: 10.1016/j.bbrc.2016.01.135

Conflict of Interest Statement: The authors declare that the research was conducted in the absence of any commercial or financial relationships that could be construed as a potential conflict of interest.

2

The Role of *SHI/STY/SRS* Genes in Organ Growth and Carpel Development is Conserved in the Distant Eudicot Species *Arabidopsis thaliana* and *Nicotiana benthamiana*

*Africa Gomariz-Fernández, Verónica Sánchez-Gerschon, Chloé Fourquin and Cristina Ferrándiz**

Instituto de Biología Molecular y Celular de Plantas, Consejo Superior de Investigaciones Científicas–Universidad Politécnica de Valencia, Valencia, Spain

Edited by:
Federico Valverde,
Consejo Superior de Investigaciones Científicas, Spain

Reviewed by:
Natalia Pabón-Mora,
University of Antioquia, Colombia
Charlie Scutt,
Centre National de la Recherche Scientifique, France

***Correspondence:**
Cristina Ferrándiz
cferrandiz@ibmcp.upv.es

Carpels are a distinctive feature of angiosperms, the ovule-bearing female reproductive organs that endow them with multiple selective advantages likely linked to the evolutionary success of flowering plants. Gene regulatory networks directing the development of carpel specialized tissues and patterning have been proposed based on genetic and molecular studies carried out in *Arabidopsis thaliana*. However, studies on the conservation/diversification of the elements and the topology of this network are still scarce. In this work, we have studied the functional conservation of transcription factors belonging to the SHI/STY/SRS family in two distant species within the eudicots, *Eschscholzia californica* and *Nicotiana benthamiana*. We have found that the expression patterns of *EcSRS-L* and *NbSRS-L* genes during flower development are similar to each other and to those reported for Arabidopsis *SHI/STY/SRS* genes. We have also characterized the phenotypic effects of *NbSRS-L* gene inactivation and overexpression in Nicotiana. Our results support the widely conserved role of *SHI/STY/SRS* genes at the top of the regulatory network directing style and stigma development, specialized tissues specific to the angiosperm carpels, at least within core eudicots, providing new insights on the possible evolutionary origin of the carpels.

Keywords: carpel evolution, *Eschscholzia californica*, gynoecium, *Nicotiana benthamiana*, style and stigma, virus-induced gene silencing (VIGS), SHI STY SRS factors

INTRODUCTION

Organ development is directed by gene regulatory networks (GRNs) that control the temporal and spatial expression of downstream effectors responsible for creating morphogenetic outputs. GRNs are composed mainly of transcription factors and other regulators of gene expression such as microRNAs or chromatin-modifying elements, and signaling molecules such as mobile peptides or hormones, that interact extensively at different levels to generate morphogenetic patterns (Davidson and Levine, 2008). Over the last few decades, a wealth of genetic and molecular studies has uncovered major players that participate in organ formation and patterning in plants, and more recently, system biology approaches have complemented these studies to generate global and more detailed pictures of the regulatory networks that underlie organ development. Most of these studies have been carried out in the model species *Arabidopsis thaliana*, where, for example, meristem formation, root development, or floral organ specification are increasingly understood from this global perspective (Alvarez-Buylla et al., 2007; De Lucas and Brady, 2013; O'Maoileidigh et al., 2014; Tian and Jiao, 2015; Davila-Velderrain et al., 2016). In addition, the rapidly increasing toolkit of genetic, molecular and functional resources across the plant kingdom is also expanding this knowledge to many other species and boosting evo-devo approaches to understand the evolution of plant form (Vialette-Guiraud et al., 2016).

Angiosperms are the largest and most diverse group of land plants. Carpels are the ovule-bearing structures of angiosperms, representing a major evolutionary innovation for this group that was probably key for their success. The carpels provide a confined casing to protect the ovules in their development. In the flower they may occur as single carpels, multiple unfused carpels or a syncarpic structure resulting from multiple fused carpels, where each individual structure is termed pistil. While morphological diversity of pistils is huge across angiosperms, they mostly share a basic organization plan. Apically, specialized cells form the stigma, which receives, discriminates and helps to germinate the pollen grains. The stigma is connected to the ovary through the style, generally a tube-like structure containing transmitting tissues that serve to grow and direct the pollen tubes toward the ovules. The ovary occupies a basal position, housing the ovules and, upon fertilization, it becomes a fruit, which may or may not incorporate additional parts of the flower, and serves to protect the developing seeds and later to facilitate seed dispersal (Ferrandiz et al., 2010). GRNs directing pistil patterning have been studied in Arabidopsis. Several genetic and hormonal factors required for the specification of carpel identity or the development of the specialized pistil tissues have been identified in the last few years, as well as some of their interactions and regulatory hierarchies. From these studies, GRNs directing the different functional modules in the Arabidopsis carpels have been proposed, and, although we are still far from completing an integrated network that provides a comprehensive view of spatial–temporal pistil morphogenesis, we increasingly understand how the basic blocks that compose a functional pistil are formed (Ferrandiz et al., 2010; Reyes-Olalde et al., 2013; Chávez Montes et al., 2015; Ballester and Ferrandiz, 2016; Marsch-Martinez and de Folter, 2016).

The evolutionary importance of carpels has also prompted many different labs to explore questions mainly focused on the evolutionary origin of carpels and the conservation of the genetic functions that specify carpel identity (Bowman et al., 1989; Bradley et al., 1993; Davies et al., 1999; Pan et al., 2010; Yellina et al., 2010; Dreni et al., 2011; Fourquin and Ferrandiz, 2012). However, only more recently these studies are being extended to other components of the emerging GRN proposed to direct pistil patterning. The increasing availability of plant genome sequences has allowed to reconstruct phylogenies for many of the gene families involved in carpel development with good taxonomic sampling, and thus to propose hypotheses on the possible evolution of the pistil GRNs (Pabon-Mora et al., 2014; Pfannebecker et al., 2017a,b). However, it is necessary to complement these works by carrying out functional studies in different taxa, which are still scarce. In this context, it is especially interesting to explore the functional conservation of the elements that drive style and stigma specification, since these tissues are only found across angiosperms and intimately linked to the evolutionary origin of carpels.

In Arabidopsis, it has been shown that correct auxin signaling is essential to establish apical-basal polarity in the Arabidopsis pistil, for correct development of the style and stigma and to ensure apical closure (Sundberg and Ostergaard, 2009; Larsson et al., 2013; Larsson et al., 2014; Moubayidin and Ostergaard, 2014). Two families of transcription factors are essential for style and stigma formation, and, at least in part, they exert these functions by regulating auxin synthesis, transport and response. The four *NGATHA (NGA)* genes belong to the RAV clade of the B3-domain transcription factor family and act redundantly to specify style and stigma identity. The Arabidopsis *nga* quadruple mutants completely fail to form these apical tissues and are female sterile, but develop fairly normal ovaries (Alvarez et al., 2009; Trigueros et al., 2009). The SHI/STY/SRS family of zinc-finger transcription factors, named after the members of this family in Arabidopsis SHORT INTERNODES (SHI), STYLISH (STY), and SHI RELATED SEQUENCE (SRS), play similar roles in pistil development, and multiple combinations of mutants in four or more members of the family cause almost identical phenotypes to those of quadruple *nga* mutants (Kuusk et al., 2002, 2006). NGA and SHI/STY/SRS factors share similar patterns of expression and common targets, and appear to work as master regulators of the GRN directing style and stigma development. In fact, simultaneous overexpression of *NGA3* and *STY1* in Arabidopsis is sufficient to direct ectopic style tissue formation to the whole surface of the ovary (Kuusk et al., 2002, 2006; Sohlberg et al., 2006; Trigueros et al., 2009; Staldal et al., 2012; Martinez-Fernandez et al., 2014). The role of NGA orthologs in pistil development is strongly conserved in distant species such as the basal eudicot *Eschscholzia californica* or the asterid core eudicot *Nicotiana benthamiana*, where *NGA* inactivation leads to the absence of style and stigma differentiation (Fourquin and Ferrandiz, 2014). Several reports show that the connection of *SHI/STY/SRS* genes to auxin signaling is also conserved across land plants, as well as a general role in controlling plant architecture and possibly other hormone pathways (Eklund et al., 2010a,b; Zawaski et al., 2011; Islam et al., 2013; Youssef et al., 2017). However, the role of *SHI/STY/SRS* genes in pistil development has not been explored in detail outside Brassicaceae, although it has been described that barley mutants in the *Lks2* gene, a member of the SHI/STY/SRS family, have short awns and defective stigmas (Yuo et al., 2012).

In this work, we have studied whether members of the SHI/STY/SRS family have conserved roles in style and stigma development in *E. californica* and *N. benthamiana*. This study supports that, as shown for *NGA* genes, SHI/STY/SRS factors are essential for the development of the apical tissues of pistils at least in the core eudicots and that their downstream effectors leading to apical-basal patterning in the carpels are also likely conserved.

RESULTS

Identification of *SHI/STY/SRS* Genes in *E. californica* and *N. benthamiana*

The *SHI/STY/SRS* gene family in Arabidopsis comprises 10 members. Published phylogenies show that they belong to a plant-specific family, where homologs can be found already in the moss *Physcomitrella patens* or the lycophyte *Selaginella moellendorffii* (Eklund et al., 2010b; Pfannebecker et al., 2017a). SHI/STY/SRS factors share two highly conserved domains: a C3HC3H RING zinc finger domain and a IGGH motif that appears to be specific to this family (Fridborg et al., 2001), but outside these conserved regions the protein sequences are highly divergent (**Supplementary Table S1**).

To search for *SHI/STY/SRS* homologs in *E. californica*, we designed degenerate primers based on the sequence the RING-like and IGGH conserved domains of *SHI/STY/SRS* genes from other species. Only one putative *SHI/STY/SRS* gene, named *EcSRS-L*, was amplified from cDNA of *E. californica* flowers. The complete coding sequence of *EcSRS-L* was subsequently amplified by TAIL PCR and by the use of an adapted oligo-dT primer. The predicted EcSRS-L protein sequence possessed the typical RING domain and IGGH motif, but was different from the EscaSTY-L protein sequence recently published (Pfannebecker et al., 2017a), that we were not able to amplify with this strategy, probably because it contains a variant of the IGGH domain that does not align with our degenerate primers (**Supplementary Figure S1**). Eleven SRS-related sequences were identified by searching the most recent draft of *N. benthamiana* genome (**Figure 1**) (Bombarely et al., 2012), all of them coding for predicted proteins that contained the RING and IGGH domains.

Using the identified *E. californica* and *N. benthamiana* predicted protein homologs and the Arabidopsis protein sequences of the SHI/STY/SRS family members we performed comparative sequence analyses, using the Neighbor joining algorithm built into the Mega7 software (**Figure 1**). The resulting tree had an overall topology similar to that of other previously published, although some differences were found mainly affecting less-supported clades and probably due to the use of different tree reconstruction methods and datasets. The clade comprising AtSTY1, AtSHI, and AtSRS8 was related to one formed by five Nicotiana predicted proteins. Four additional Nicotiana proteins grouped in a well-supported clade related to AtSRS5 and AtSRS7, while both EcSRS-L and EscaSTY-L grouped in the same clade as AtSRS3. Finally, two Nicotiana predicted proteins clustered as an outgroup with AtLRP. Low support of some branches did not allow to unequivocally establish direct relationships of some of the *N. benthamiana* and *E. californica* factors with the Arabidopsis SHI/STY/SRS proteins, but the tree strongly suggested that the duplication events that resulted in the high number of homologs found in Arabidopsis and Nicotiana were independent, and also that in Nicotiana, a recent allotetraploid, two copies of each gene were usually present.

Functional studies in Arabidopsis have shown that members of the family have redundant functions and that the degree of this redundancy does not depend strongly on how similar are their sequences. Thus, *sty1-1* is the only single mutant that shows an abnormal phenotype in gynoecium development, but combinations with mutant alleles in either *LRP*, *SRS5*, *STY2*, or *SHI* genes, which belong to different subclades in the family and have, respectively, a similarity score of 30, 34, 42, and 54% with STY1, similarly enhance the *sty1-1* defects in carpel development (Kuusk et al., 2002, 2006). For this reason, and since none of the *NbSTY/SHI/SRS* genes showed a clear orthology relationship with *AtSTY1*, we decided to focus for this study on three *N. benthamiana* genes: Niben101Scf06228g01001.1, a member of a sister clade to the one formed by the Arabidopsis *AtSRS5* and *AtSRS7* genes, and a pair of closely related genes that grouped together with the AtSTY1/SHI/SRS8 clade (Niben101Scf00158g06008.1 and Niben101Scf08273g05002.1, **Figure 1**). For simplicity, we renamed these as *NbSRS-L1*, *NbSRS-L2*, and *NbSRS-L3* (**Figure 1**). In addition, we also chose for functional characterization the *EcSRS-L* gene that we were able to amplify from *E. californica* flowers. An alignment of all the sequences of the predicted proteins used in this study, also including the Arabidopsis AtSTY1, AtSRS3, AtSRS5, and EscaSTY-l factors is shown in **Supplementary Figure S1**.

SHI/STY/SRS Gene Expression Patterns Are Similar in Arabidopsis, *E. californica*, and *N. benthamiana*

The expression pattern of the *SRS-L* genes identified in *E. californica* and *N. benthamiana* was characterized by RNA *in situ* hybridization on young flower buds.

EcSRS-L transcripts were weakly detected in young flowers from early stages of development. In very young *E. californica* buds, *EcSRS-L* was detected in the emerging stamen and carpel primordia (**Figure 2A**). At later stages, *EcSRS-L* mRNA accumulated weakly in anthers and in the placenta and ovules (**Figures 2B,C**). *EcSRS-L* expression could also be detected in the most distal cells of the developing gynoecium, in the presumptive domain that would further develop into the style and the stigma (**Figures 2B–D**). This expression pattern generally resembled those described for Arabidopsis *SHI/STY/SRS* genes, which showed distal accumulation in the developing gynoecium and, at least in the case of *AtSTY1*, expression in ovules (Kuusk et al., 2002).

In *N. benthamiana*, the three *SHI/STY/SRS* genes included in this study showed similar expression patterns, although not identical. *NbSRS-L1* was detected in very young floral buds, before floral organs in inner whorls were initiated (**Figure 2E**). At later stages, *NbSRS-L1* expression concentrated in the distal end of growing petals and of the gynoecium, as well as in the placentae and the anthers (**Figure 2F**). After style fusion, *NbSRS-L1* was mainly detected in ovules and pollen grains, while only weakly present in the transmitting tissues at the central domain of the style and in the stigma (**Figures 2G,H**). *NbSRS-L2* showed a more expanded expression in floral organs. In young buds, *NbSRS-L2*

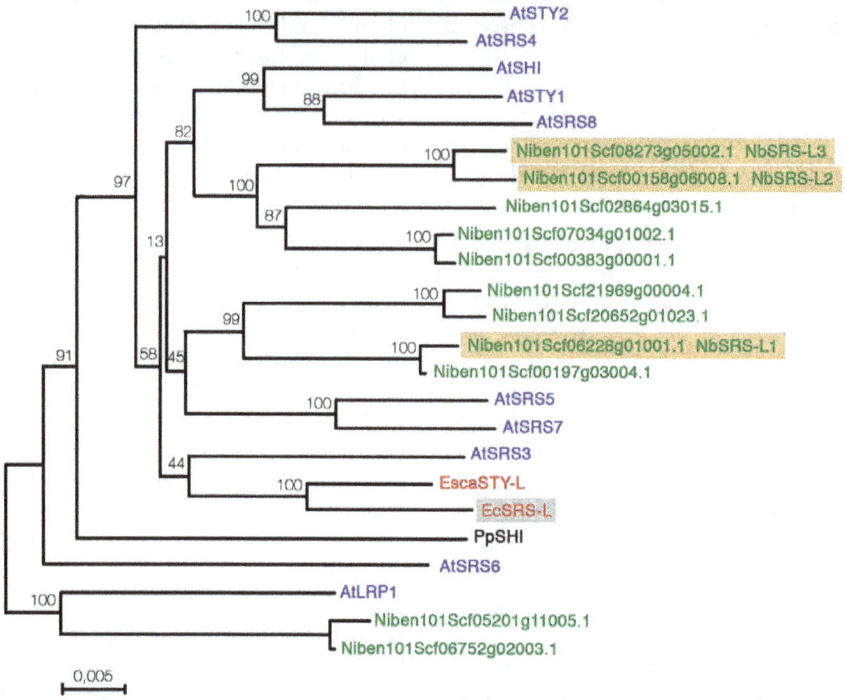

FIGURE 1 | Distance analysis between *Nicotiana benthamiana*, *Eschscholzia californica*, and *Arabidopsis thaliana* SHI/STY/SRS factors. Sequences were aligned with Muscle and the phylogenetic tree was generated using MEGA7 software using the Neighbor-Joining method. The optimal tree is shown. The percentage of replicate trees in which the associated protein sequences clustered together in the bootstrap test (2000 replicates) are shown next to the branches.

expression was associated to all incipient floral organ primordia (**Figure 2I**). At later stages, expression was detected in growing petals, stamens and gynoecia, mainly at the distal end (**Figure 2J**). After style fusion, *NbSRS-L2* was expressed mainly in ovules and it could be also found in the central domain of the style, the ovary wall and the distal end of growing petals (**Figures 2K,L**). *NbSRS-L3* expression pattern was most similar to that of *NbSRS-L1* (**Figures 2M–O**), although at later stages expression in the inner style was stronger than for *NbSRS-L1* (**Figure 2P**).

In summary, *NbSRS-L1, NbSRS-L2,* and *NbSRS-L3* genes showed remarkably similar expression patterns during flower development, being detected in all floral organ primordia from early stages of development and then mainly associated to the distal end of petals and carpels, anthers, placentae, and ovule primordia, as well as the inner style in more advanced stages. Again, these expression patterns resembled those described for several Arabidopsis *SHI/STY/SRS* genes (Kuusk et al., 2002, 2006).

Overexpression of *NbSHI/STY/SRS* Genes in *Arabidopsis thaliana* Mimics the Effect of the Ectopic Expression of the Endogenous Arabidopsis *STY* Genes

Eschscholzia californica, N. benthamiana, and Arabidopsis *SHI/STY/SRS* genes showed similar expression patterns, suggesting that they might have conserved functions. However, outside the RING and IGGH domains, low similarity could be found among *E. californica, N. benthamiana,* and Arabidopsis SHI/STY/SRS factors (**Supplementary Figure S1** and **Table S1**). Nevertheless, extensive functional redundancy of Arabidopsis genes has been shown in spite of sequence divergence (Kuusk et al., 2006), suggesting that the RING and IGGH domains might be sufficient for their shared function. To explore whether the *NbSRS-L* genes under study had similar functional properties among them and also to their Arabidopsis homologs, we transformed Arabidopsis wild type and *sty1 sty2* mutants with constructs for the overexpression of the three *NbSRS-L* genes and observed the corresponding phenotypes of at least 30 individual T1 plants for each construct and background. Overexpression of the three different *NbSRS-L* genes caused similar phenotypic alterations in carpel and fruit development when transformed into wild type Arabidopsis plants, resembling those observed for previously reported lines constitutively expressing *AtSTY1* or *AtSTY2* (Kuusk et al., 2002) (**Figure 3A**).

35S::NbSRS-L1 and 35S::NbSRS-L3 caused milder defects, mainly affecting the shape and length of the style, which was reduced and did not elongate as in wild type fruits, and the overall shape of the fruit, which appeared wider and blunt at the distal end in the lines with weaker phenotypic defects (**Figures 3A,B**, category 1), or shorter than wild type, wider specially at the apical end, and with an irregular ovary surface (**Figures 3A,B,G,H**, category 2) in the lines with stronger defects. *NbSRS-L2* overexpression caused related but more dramatic phenotypic alterations in fruit development. About one third of the lines showed fruit phenotypes resembling those of the most

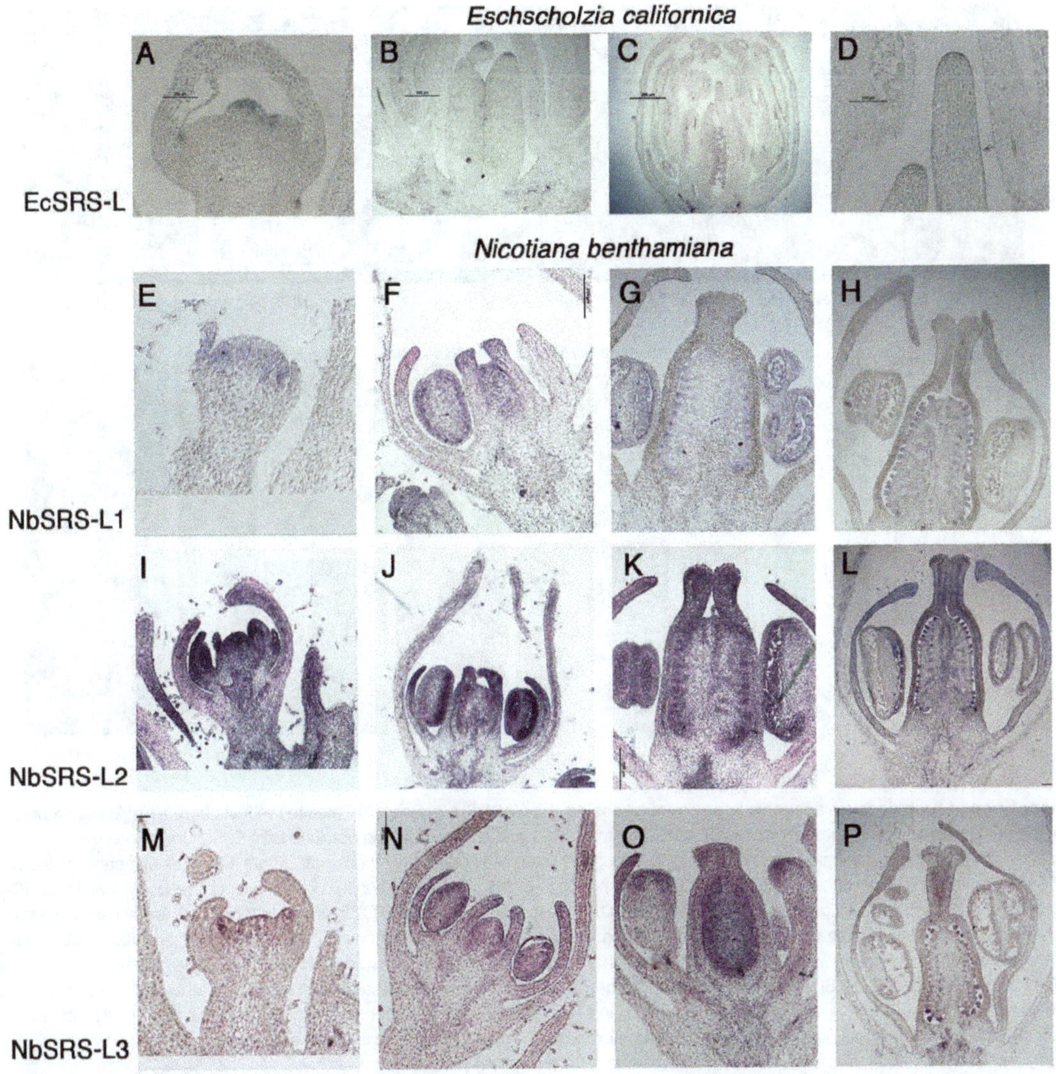

FIGURE 2 | *In situ* expression analyses of *SRS-L* genes in *E. californica*, and *N. benthamiana flowers*. Control hybridizations with sense probes are shown in **Supplementary Figure S2**. **(A–D)** Longitudinal sections of *E. californica* flowers probed with *EcSRS-L*. **(A)** At stage 3, *EcSRS-L* mRNA is detected in a few cells at the tip of the incipient pistil primordium and of sepal and petal primordia. **(B)** At stage 6, expression in the gynoecium is restricted to the most apical cells of the incipient style. **(C)** Later in development, expression of *EcSRS-L* is detected in ovule primordia, the anthers and in the most apical cells of the developing style. **(D)** Close up of the apical style shown in **(C)**. **(E–P)** Longitudinal sections of *N. benthamiana* flowers probed with *NbSRS-L* genes **(E–H)**, *NbSRS-L2* **(I–L)** and *NbSRS-L3* **(M–P)**. The three genes show remarkably similar expression patterns, although *NbSRS-L2* appears to be expressed at higher levels. **(E–H)** *NbSRS-L1* mRNA is detected in the floral primordium at very early stages **(E)**, and later it accumulates in the anthers, the developing placenta and the distal end of petals and carpels **(F)**. **(G,H)** After all floral organs differentiate, *NbSRS-L1* is mainly detected in ovule primordia, the style, the anthers and the distal end of growing petals. **(I,J)** *NbSRS-L2* is strongly expressed throughout flower development, mainly accumulating in distal positions in floral organs, placenta, ovules, anthers and the style. **(M,N)** The expression pattern of *NbSRS-L3* is very similar to that of *NbSRS-L1*, although the expression in the style appears stronger at later stages **(O,P)**.

affected *NbSRS-L1* or *NbSRS-L3* overexpressors (**Figures 3A,B**), but most of the transgenic lines showed further enhanced defects that were subdivided in two additional categories (**Figures 3A,B**, categories 3 and 4). Fruits in category 3 had short and inflated ovaries (**Figure 3A**). The cell in the valves had similar shapes to wild type but were not oriented in parallel to the apical-basal axis of fruit growth, but formed an angle with the replum (**Figures 3C,I**). In addition, the valve margins at the apical end were not clearly defined, the style was short and expanded laterally, and the demarcation between the style and the valves was not discernible (**Figures 3C,D,J**). Finally, about 40% of the 35S::NbSRS-L2 lines had short fruits of tapered shape, with short ectopic style cells developing extensively at the lateral domains of the apical half of the valves (**Figures 3E,F**), and cells closer to the replum that retained valve identity but elongated forming a wider angle with the apical-basal axis (**Figure 3E**). When the same constructs were introduced into Arabidopsis *sty1 sty2* mutants, all of them were able to partially or fully rescue the phenotypic defects of *sty1 sty2* stigma and style (**Figures 4A–F**), although at different degrees.

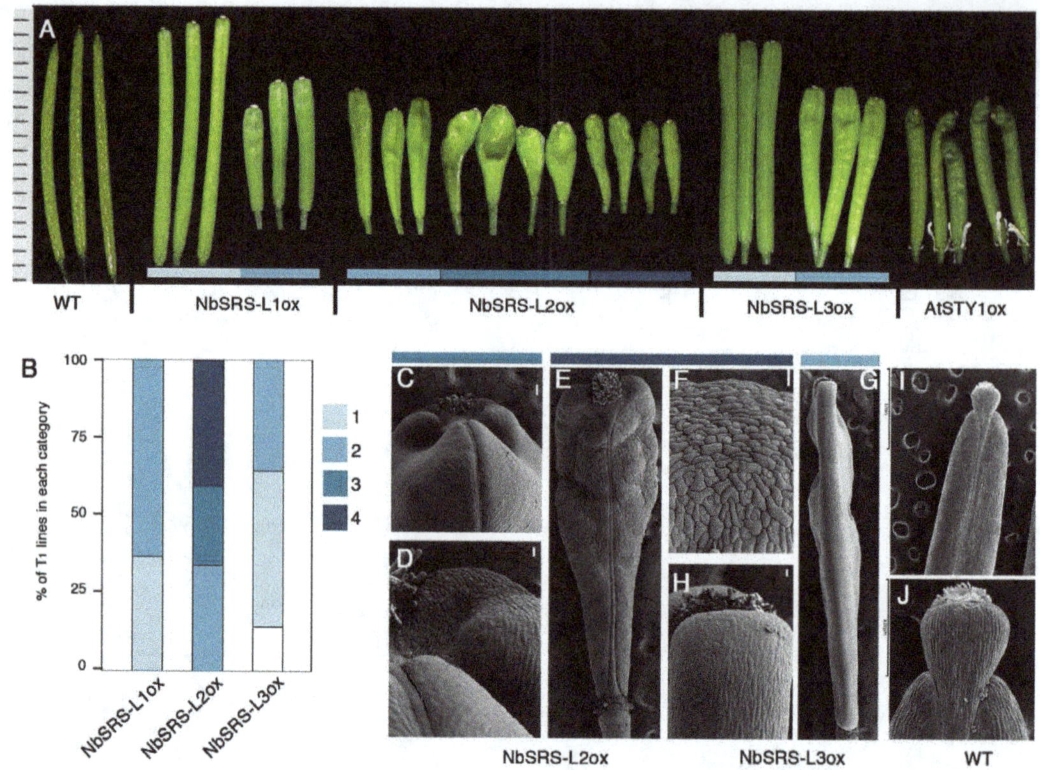

FIGURE 3 | Phenotypes of Arabidopsis transgenic lines overexpressing the *NbSRS-L* genes. (A) Three to five fruits from a single plant representative of the different phenotypic categories and genotypes are shown. The AtSTYox fruits belong to a transgenic line obtained from E. Sundberg's lab (Kuusk et al., 2002). **(B)** Proportions of the different phenotypic categories among the T1 transgenic lines obtained for each construct, where 1 (light blue) represents the weakest phenotype and 4 (dark blue) the strongest phenotype. White color corresponds to plants showing no conspicuous differences respect to wild type. **(C–J)** Scanning electron micrographs of fruits from selected genotypes and categories. **(C,D)** Fruits from 35S::NbSRS-L2 lines in category 3. **(C)** Apical part of the fruit. **(D)** Close up of the style/valve area. While cells retain their identity, there is no clear demarcation between the two zones. **(E)** Fruit from a 35S::NbSRS-L2 line in category 4. **(F)** Close up of the apical part of the fruit shown in **(E)**. All the cells have a crenelated surface with the typical morphology of style cells and no valve cells can be distinguished. **(G)** Lateral view of a 35S::NbSRS-L3 fruit in category 2. The valves grow laterally to the same level as the stigma and style and have a square morphology. **(H)** Close up of the apical valve cells of the fruit shown in panel **(G)**. All cells have valve identity and no ectopic formation of style cells can be observed. **(I)** Wild type fruit. **(J)** Close up of the apical domain, where the demarcation of stigma, style, and valves is clearly seen. Bars in **(C–G)**: 50 μm.

Altogether, these results indicated that the three *N. benthamiana* SHI/STY/SRS factors under study had similar molecular properties among them and also, at least, to the Arabidopsis *STY1* or *STY2* genes in spite of high sequence divergence outside the RING and IGGH domains.

A similar approach was undertaken with the *EsSRS-L* gene identified in this study and 30 independent 35S::EcSRS-L T1 lines were generated in wild type and *sty1 sty2* backgrounds. However, *EcSRS-L* overexpression was not able to produce neither any phenotypic effect in the wild type background, nor complementation of the *sty1 sty2* mutant phenotype (**Supplementary Figure S3**).

Silencing of *NbSHI/STY/SRS* Genes in *Nicotiana benthamiana* Using VIGS Alters Floral Organ Growth and Style and Stigma Development

The results of overexpressing the three *NbSRS-L* genes in Arabidopsis wild type or *sty1 sty2* plants suggested that the Nicotiana NbSRS-L factors were functionally equivalent to AtSTY1 or AtSTY2 to a large extent, but did not clarify whether these genes had similar roles in Nicotiana development to those described for the *SHI/STY/SRS* genes in Arabidopsis. To investigate the function of the three *NbSRS-L* genes in this study, we used Virus Induced Gene Silencing (VIGS) to reduce their transcript levels (Ratcliff et al., 2001; Wege et al., 2007; Fourquin and Ferrandiz, 2012). We generated three different TRV constructs designed to downregulate *NbSRS-L1*, *NbSRS-L2*, or *NbSRS-L3*. Twelve plants were inoculated with each construct independently and 5–20 flowers from each plant were chosen for further characterization (**Table 1**).

Nicotiana benthamiana has pentamerous flowers, where the perianth is composed of a calyx of five sepals, and five fused petals forming a tubular corolla. The five stamens are epipetalous and the gynoecium is bicarpellate and formed by a short bilocular ovary with central placentation and an elongated style, measuring more than 3 cm at anthesis, capped by a round flat stigma (**Figures 5A–F**). Plants inoculated with any of the three TRV2-NbSRS-L vectors showed similar phenotypic

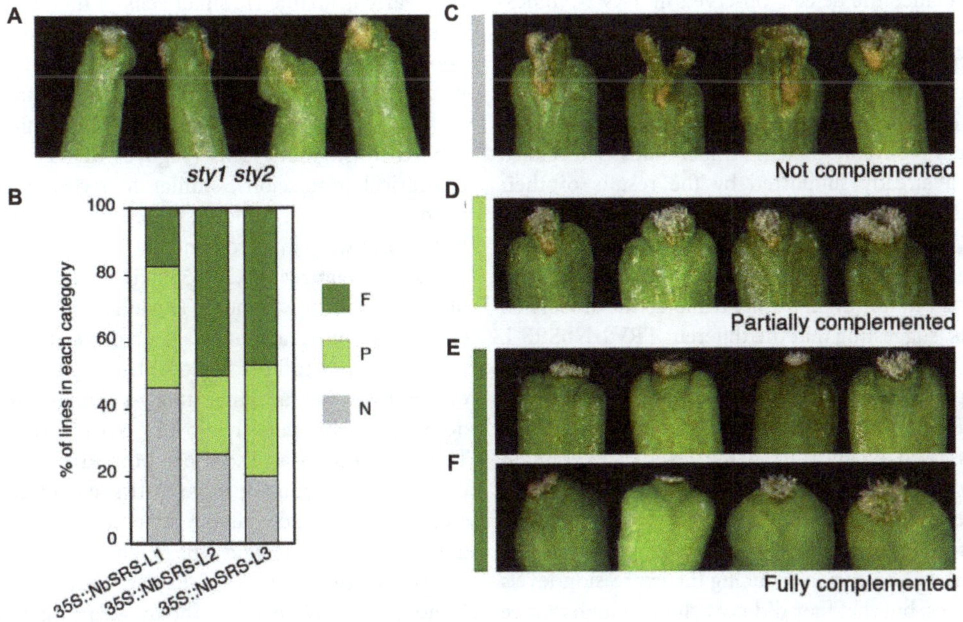

FIGURE 4 | Complementation phenotypes of Arabidopsis *sty1 sty2* mutants transformed with 35S::NbSRS-L constructs. (A) Style and stigma phenotype of *sty1 sty2* mutants. **(B)** Proportions of the different phenotypic categories among the T1 transgenic lines obtained for each construct, where N (gray) represents non-complemented lines, P (light green) partially complemented lines and F (dark green) fully complemented lines. **(C–F)** Examples of the different complementation categories among the transgenic lines.

TABLE 1 | Summary of the VIGS experiments on *Nicotiana benthamiana* and *Eschscholzia californica* plants.

VIGS construct	N° plants inoculated (dead)	N° plants with phenotype	N° flowers/plant observed	N° flowers/plant with phenotype	N° total flowers with phenotype	% Flowers with phenotype
N. benthamiana						
TRV2-NbSRS-L1	12 (5)	7	10–20	4–14	82/128	64%
TRV2-NbSRS-L2	12 (2)	8	7–20	1–6	22/143	15%
TRV2-NbSRS-L3	12 (2)	9	5–20	2–8	30/137	22%
TRV-NbSRS-L1/2	12 (3)	9	15–20	8–18	120/172	70%
TRV-NbSRS-L1/2/3	12 (3)	10	3–20	2–14	63/99	64%
Empty vector	6 (1)	0	20	0	0/100	0%
E. californica						
TRV2-EcSRS-L	120 (11)	0	1–5	0	0	0%
TRV2-EcPDS	70 (22)	48	n.d.	n.d.	48/70*	68%*
Empty vector	60 (16)	0	2	0	0/88	0%

**Bleaching caused by PDS inactivation.*

alterations, although TRV2-NbSRS-L1 had a greater efficiency (64% of the flowers showed abnormal development) when compared with TRV2-NbSRS-L2 (15%) or TRV2-NbSRS-L3 (22%) (**Table 1**). These defects affected upper corolla development, and most frequently consisted of alterations in the shape, the color or the symmetry of the petal lobes (**Figures 5G,L,P,M**). Style and stigma development were also strongly affected. Style length was reduced, and very frequently the style was unfused or even split at the apical end, where the stigma adopted irregular shapes (**Figures 5H–K,N,O,Q–S**). Aiming to enhance the phenotypic defects caused by *NbSRS-L* inactivation, 12 plants were also inoculated simultaneously with TRV2-NbSRS-L1/TRV2-NbSRS-L3 vectors or with TRV2-NbSRS-L1/TRV2-NbSRS-L2/TRV2-NbSRS-L3 vectors. These combined treatments produced phenotypic effects in 70 and 64% of the observed flowers, respectively, therefore at levels comparable to those resulting of TRV2-NbSRS-L1 inoculation alone. However, in addition to the phenotypes already described for the individual constructs, these combinations induced novel alterations mainly related to floral organ number. In plants treated with TRV2-NbSRS-L1/TRV2-NbSRS-L3, the calix was frequently formed by six sepals (**Figure 5T**). Gynoecia formed by four fused carpels were also observed, sometimes even capped by multiple styles derived from each carpel (**Figures 5U,W**), while in other flowers two bicarpellated fused pistils were formed (**Figure 5V**). Simultaneous inoculation with the three vectors did not further enhance the defects observed in TRV2-NbSRS-L1/TRV2-NbSRS-L3 treated plants (**Figures 5X–Z"**), except for occasional development of abnormal anthers (**Figure 5Z'**) and stronger defects in corolla development (**Figures 5X–Z**).

The similar effect of all the VIGS treatments tested suggested that the three *NbSRS-L* genes under study had related functions, an idea already supported by the results of their heterologous constitutive expression in Arabidopsis. However, since it was likely that these related functions were redundant, it was surprising that the inoculation with combinations of different constructs did not dramatically enhance the associated phenotypic defects, suggesting that the different TRV2-NbSRS-L vectors could be targeting more than one gene in the family and therefore inducing transitory inactivation of several *NbSRS-L* genes at once. The efficiency and specificity of each VIGS treatment was assessed by measuring the level of expression of *NbSRS-L1, NbSRS-L2,* and *NbSRS-L3* on flowers from two treated plants for each treatment that showed inactivation-related phenotypes. The results from these experiments showed that the VIGS treatments were effective in reducing the expression levels of the *NbSRS-L* genes, but that they did not achieve a high degree of specificity and therefore the observed phenotypes were likely caused by simultaneous inactivation of several *NbSRS-L* genes (**Supplementary Figure S4**).

The effect of transient inactivation of *EcSRS-L* in *E. californica* was also tested by inoculation of plants with a TRV2-EcSRS-L construct. One hundred and twenty plants were inoculated but none of them showed evident phenotypic defects (**Table 1**), in spite of the effective inactivation caused by the VIGS treatment as quantified by qRT-PCR (**Supplementary Figure S5**), indicating that either the *EcSRS-L* gene was fully redundant with the *EscaSTY* gene described in (Pfannebecker et al., 2017a), that the residual expression of *EcSRS-L* in VIGS treated plants was sufficient to provide its function, or that the phenotype associated with *EcSRS-L* inactivation was not affecting morphogenesis in a conspicuous way.

Overexpression of *NbSHI/STY/SRS* Genes in *Nicotiana benthamiana* Cause Ectopic Style and Stigma Development

To study the effect of ectopic expression of the *NbSRS-L* genes under study in *N. benthamiana* we generated transgenic plants in which we introduced the 35S::NbSRS-L1 and 35S::NbSRS-L2 constructs previously used for heterologous expression in Arabidopsis. Twenty-three independent transgenic lines were obtained for 35S::NbSRS-L1 and 9 for 35S::NbSRS-L2, and around half of them showed conspicuous phenotypic changes in gynoecium morphology (**Figures 6A,E,J**). Both 35S::NbSRS-L1 and 35S::NbSRS-L2 plants showed similar phenotypes, although, as already observed in Arabidopsis, the constitutive expression of *NbSRS-L2* had more dramatic effects.

The pistils of plants overexpressing *NbSRS-L1* had stigma and style of normal morphology, only shorter (**Figure 6E**). The demarcation between style and ovary was not defined as in wild type (**Figure 6B**), but showed a gradual transition between the elongated zone with columnar cells typical of the style and the nearly isodiametric small cells of the ovary (**Figures 6E–G**). The valve margins in the ovary, which in the wild type are poorly defined and adjacent (**Figure 6C**), in the 35S::NbSRS-L1 pistils appeared more pronounced and separated by several cell files, that in the upper part of the ovary bulged forming ridges capped by stigmatic cells (**Figures 6E–H**). Occasionally, small cylindrical protrusions, similar to styles and terminated by a rounded stigma, grew out of these ridges (**Figure 6F**). Finally, the gynophore was elongated and showed a clear demarcation from the ovary (**Figures 6D,I**). 35S::NbSRS-L2 pistils showed similar phenotypes, only enhanced. The junction between style and ovary showed stronger alterations and more frequently gave rise to what appeared to be ectopic styles (**Figures 6J–L**). The surface of the ovary was irregular and grew unevenly, with intermixed domains of different style/ovary cell identity (**Figures 6J,L,M**). The gynophore was also more pronounced and separated from the ovary by a bulging ridge of tissue (**Figure 6N**). Anatomical sections confirmed that the style and stigma of the transgenic lines were similar to wild type (**Figures 7A–F**). The ovaries of the lines overexpressing *NbSRS-L1* had thickened walls with between 2 and 4 more cell layers in the mesocarp than the wild type, and the ridges bulging at the valve margins clearly showed mixed identity of ovary and style/stigma tissues (**Figures 7G,H,J,K**). In the 35S::NbSRS-L2 lines, the upper part of the ovary showed thickened and irregular walls that did not resemble clearly the ovary wall morphology of the wild type (**Figure 7I**); the basal part of the ovary was less affected, but still showed irregularities in shape (**Figure 7L**).

In addition to changes in gynoecium morphology, overexpression of *NbSRS-L* genes caused strong alterations in leaf development, especially in 35S::NbSRS-L2 lines. The leaves of the transgenic plants were darker than wild type and growth was constricted in the margins, causing the blade to adopt a hood-like appearance (**Figure 8**).

DISCUSSION

In this work, we have studied the functional conservation of members of the *SHI/STY/SRS* gene family in the asterid core eudicot *N. benthamiana* and the basal eudicot *E. californica.* We have determined the expression patterns of the genes under study during flower development, characterized the floral phenotypes caused by their downregulation, and the effects of their overexpression both in the heterologous system *A. thaliana* and in *N. benthamiana.* Our work supports the conserved function of *STY/SHI/SRS* genes in apical gynoecium development and their position at the top of the GRN directing style and stigma development in core eudicots.

SHI/STY/SRS Function in Apical Gynoecium Development Is Conserved in Core Eudicots

The *SHI/STY/SRS* genes of Arabidopsis act redundantly to specify style and stigma development, as revealed by the phenotype of multiple combinations of mutations in members of the family (Kuusk et al., 2006). The expression of different

FIGURE 5 | Phenotypes of *N. benthamiana* plants inoculated with different pTRV2-NbSRS-L constructs. (A–F) Wild type *N. benthamiana* flower at anthesis. **(A)** Top view. Five expanded white petals and the central stigma surrounded by five stamens are clearly seen. **(B)** Top view of the dissected calix, formed by five green narrow sepals. **(C)** Lateral view of the flower, where some petals al sepals have been removed to show the ovary and the long style, that ends in a stigma leveled with the anthers. **(D)** Lateral view of the gynoecium. **(E,F)** Top and lateral view of the stigma. **(G)** NbSRS-L1-VIGS flower. The petals are reduced in size and the stigma is not visible at the center. **(H)** Lateral view of the basal part of the flowers, where some petals have been removed, to expose the short style. **(I)** NbSRS-L1-VIGS gynoecium. **(J,K)** Two examples of the defects found in the stigmas of NbSRS-L1-VIGS pistils. **(L)** NbSRS-L2-VIGS flower. The corolla is not symmetrical because one petal is severely underdeveloped. The stigma is not visible. **(M)** Top view of the dissected calix of a NBSRS-L2-VIGS flower. One of the sepals is reduced in size and deformed. **(N)** NbSRS-L2-VIGS gynoecium. **(O)** Close-up of the stigma shown in **(N)**, where the abnormal shape is visible. **(P)** Top view of a NbSRS-L3-VIGS flower. Note the unexpanded greenish corolla. **(Q)** Lateral view of a NbSRS-L3-VIGS flower where some petals have been removed to expose the short style. **(R)** Lateral view of a NbSRS-L3-VIGS gynoecium. **(S)** Example of defects observed in NbSRS-L3-VIGS styles. **(T–W)** Indeterminate phenotypes of flowers treated with TRV2-NbSRS-L1 and TRV2-NbSRS-L3 constructs simultaneously. **(T)** 6-sepal calix. **(U)** 4-carpel ovary. **(V)** Bi-pistillate flower. **(W)** Four style-like protrusions in a 4-carpel pistil. **(X–Z")** Flowers inoculated simultaneously with the three TRV2-NbSRS-L constructs. **(X–Z)** Examples of mature flowers with severely unexpanded corollas. **(Z')** Lateral view of a flower where a defective anther is visible. **(Z")** Split and supernumerary stigma.

SHI/STY/SRS genes in Arabidopsis has been analyzed by mRNA *in situ* hybridization or the use of promoter::GUS reporters. These experiments show largely overlapping expression patterns, generally associated with domains of auxin accumulation in all lateral organs and in the apical tissues of the developing gynoecium, consistent with the high level of redundancy found in the family (Fridborg et al., 2001; Kuusk et al., 2002, 2006; Eklund et al., 2011). In this work, we have characterized the expression of a *SHI/STY/SRS* gene from the basal eudicot *E. californica* and of three members of the family from the core eudicot *N. benthamiana*, observing a significant conservation in the pattern of mRNA accumulation in the flowers of these two species and of Arabidopsis. In all cases, we have detected expression in the distal domain of growing floral organs in

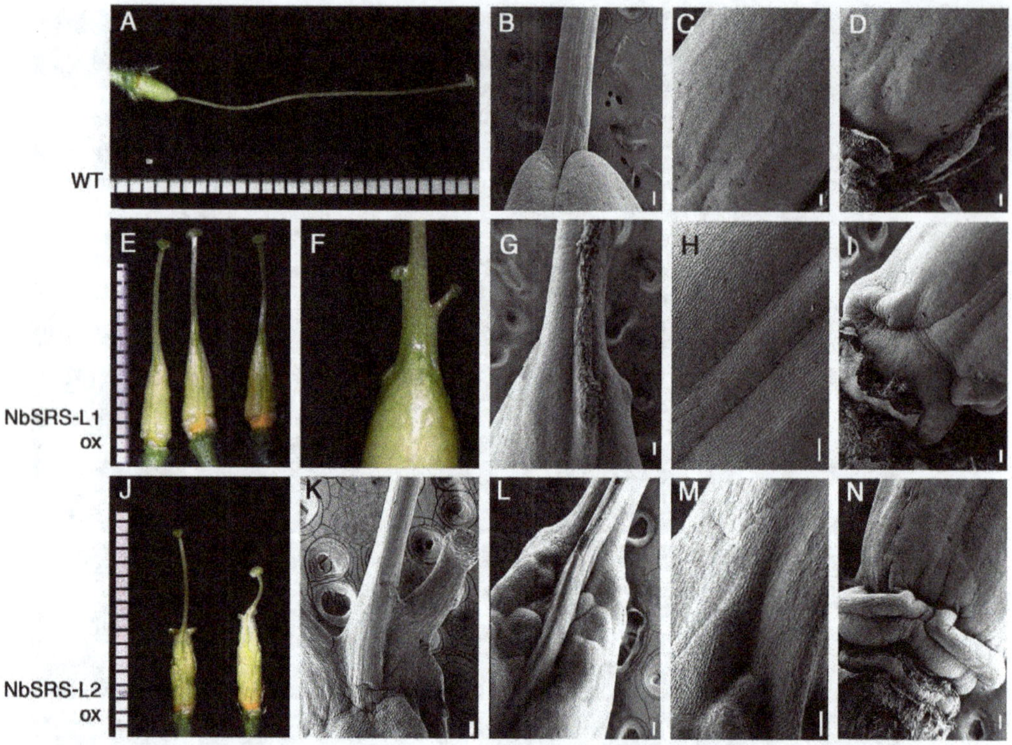

FIGURE 6 | Phenotypes of pistils of transgenic *N. benthamiana* lines overexpressing *NbSRS-L1* or *NbSRS-L2* genes. (A–D) Wild type pistils. **(A)** Lateral view. **(B–D)** Scanning Electron Micrographs of different parts of the wild type pistil. **(B)** Style-ovary junction. **(C)** Medial view of the ovary. The valve margins are adjacent and appear as a subtle crease at the center. **(D)** View of the gynophore at the basal end of the ovary. **(E–I)** 35S::NbSRS-L1 pistils. **(E)** Three pistils from different transgenic plants showing similar phenotypes. **(F)** Close-up of the ovary-style junction. Small protuberances capped with stigmatic tissue can be observed. **(G–I)** Scanning Electron Micrographs of 35S::NbSRS-L1 pistils. **(G)** Close-up of the ovary-style junction. Note the poor delimitation of both domains and the formation of ectopic stigmatic tissue. **(H)** Medial view of the ovary. The valve margins, which appear as constricted longitudinal ridges, leave a bulging area of several cell files between them resembling the Arabidopsis replum. **(I)** View of the elongated gynophore of a 35S::NbSRS-L1 pistil. **(J–N)** 35S::NbSRS-L2 pistils. **(J)** Two pistils from different transgenic plants showing similar phenotypes. **(K–N)** Scanning Electron Micrographs of 35S::NbSRS-L1 pistils. **(K)** Close-up of the ovary-style junction. Note the expansions of style tissue capped with stigmatic cells. **(L)** Close-up of the ovary-style junction of a different pistil. Note the bulging protrusion in medial position and the highly irregular ovary surface. The demarcation of style and ovary is not defined. **(M)** Close-up of another style-ovary junction where the mixed identity of style and ovary cells is observed. **(N)** View of the gynophore, where abnormal development of tissue is observed. Bars in micrographs: 100 μm.

young floral buds, the ovule primordia and the apical domain of the pistil, indicating that the regulatory regions of these genes contain conserved elements, despite the evolutionary distance of the corresponding species and the differences among paralogs within the same genome. Interestingly, it has been shown that a regulatory element present in the promoters of most Arabidopsis *SHI/STY/SRS* genes, a GCC-box bound by transcription factors of the AP2/ERF family, is required for the expression of *SHI/STY/SRS* genes in the distal domain of lateral organs including the apical tissues of the developing pistils (Eklund et al., 2011). A recent work which includes a comprehensive phylogeny of the *SHI/STY/SRS* family also reports the search for GCC-boxes in the promoters of selected genes from this study, finding that it is present in almost all of the angiosperm sequences but not in more basal taxa such as those including mosses, lycophytes or conifers (Pfannebecker et al., 2017a). Not surprisingly, we could also detect conserved GCC-boxes within the 1 kb upstream promoter sequences of the three Nicotiana genes, for which genomic sequences were available, supporting their importance in mediating apical gynoecium expression.

Heterologous constitutive expression of *N. benthamiana SHI/STY/SRS* genes in Arabidopsis caused phenotypic effects similar to those of overexpressing the endogenous genes (Kuusk et al., 2002; Kim et al., 2010), including ectopic style tissue formation, loss of definition between the style and the ovary, and an overgrowth of the valves at the apical end of the pistil, a morphology typically associated to auxin accumulation and that mimics exogenous treatment with auxin (Ståldal et al., 2008). Despite the sequence divergence between the sequences of Arabidopsis and Nicotiana proteins, which is high outside the conserved RING and IGGH domains, and the occurrence of independent duplication events in each species, the overexpression phenotypes associated to the three *NbSRS-L* genes were similar, as well as the ability of the different NbSRS-L factors to complement the *sty1 sty2* mutant phenotype. These experiments support that many of the SHI/STY/SRS proteins are basically equivalent in molecular function, even if outside the conserved RING and IGGH domains they show low sequence similarity, suggesting that most of their interactions with DNA targets and other proteins would be mediated by these conserved

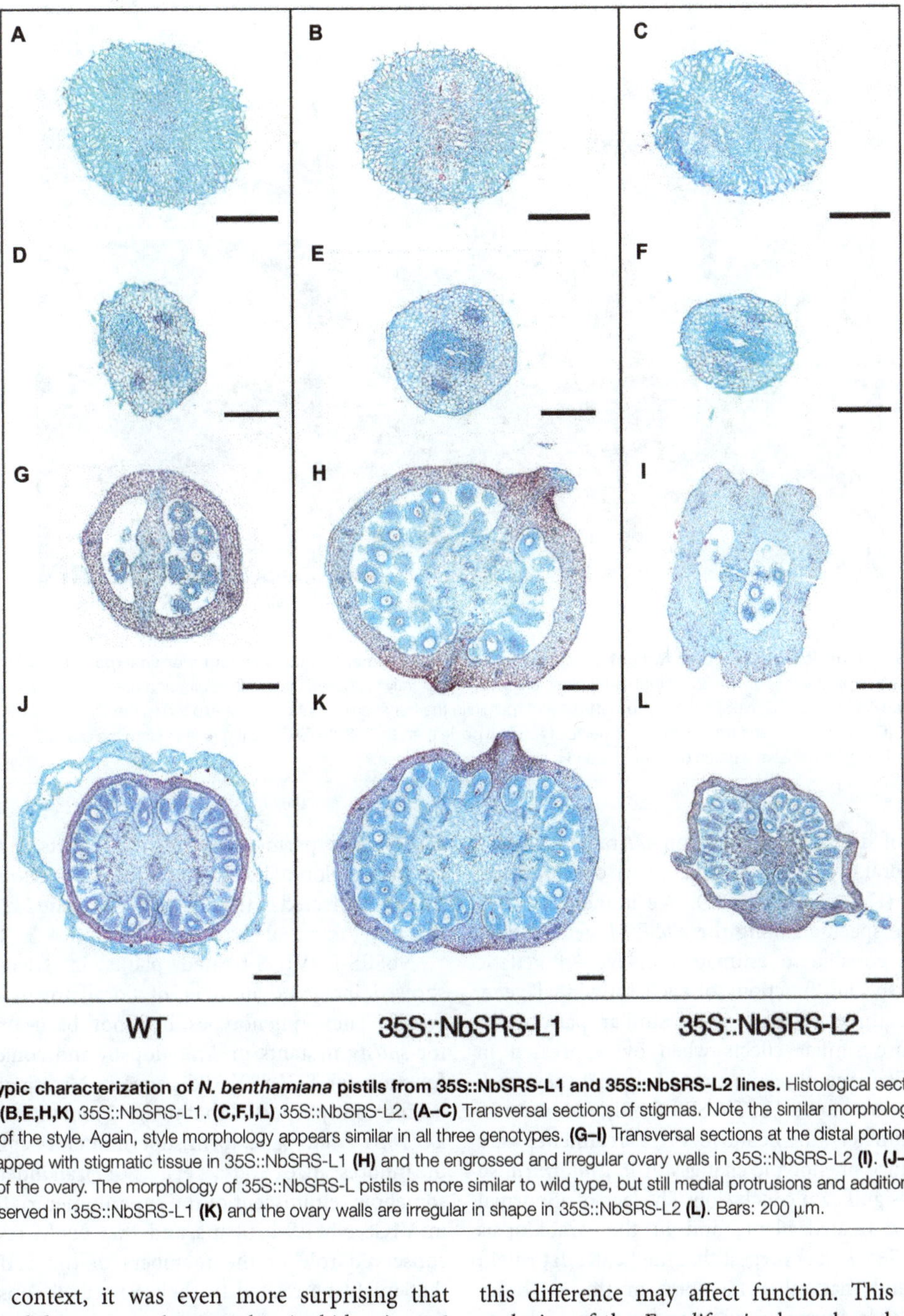

FIGURE 7 | Phenotypic characterization of *N. benthamiana* pistils from 35S::NbSRS-L1 and 35S::NbSRS-L2 lines. Histological sections of anthesis pistils. **(A,D,G,J)** Wild type. **(B,E,H,K)** 35S::NbSRS-L1. **(C,F,I,L)** 35S::NbSRS-L2. **(A–C)** Transversal sections of stigmas. Note the similar morphology in all cases. **(D–F)** Transversal sections of the style. Again, style morphology appears similar in all three genotypes. **(G–I)** Transversal sections at the distal portion of the ovary. Note the medial protrusions capped with stigmatic tissue in 35S::NbSRS-L1 **(H)** and the engrossed and irregular ovary walls in 35S::NbSRS-L2 **(I)**. **(J–L)** Transversal sections of the basal domain of the ovary. The morphology of 35S::NbSRS-L pistils is more similar to wild type, but still medial protrusions and additional cell layers in the ovary wall can be observed in 35S::NbSRS-L1 **(K)** and the ovary walls are irregular in shape in 35S::NbSRS-L2 **(L)**. Bars: 200 μm.

domains. In this context, it was even more surprising that the *EcSRS-L* gene did not complement the Arabidopsis *sty1 sty2* mutant phenotype, neither caused phenotypic defects when overexpressed in Arabidopsis wild type plants. A close inspection of the RING and IGGH domains showed high sequence similarity with that of other members of the family, although the last Cys residue in the zinc finger domain is preceded by a Pro residue in the *E. californica* proteins while in most members of the family this position is most frequently a charged or polar residue like Asp, Glu, His, or Gln. Given the structural properties of Pro residues, which have an exceptional conformational rigidity that usually impacts protein secondary structure and protein–protein interactions (Morris et al., 1992), it is possible that this difference may affect function. This Pro residue is not exclusive of the *E. californica* homologs, but it was also found in 5 of the 91 predicted proteins included in the phylogenetic study by Pfannebecker et al. (2017a). However, since none of these homologs has been functionally characterized, it would be necessary to perform equivalent analyses with some of them to test this hypothesis.

Silencing of the *NbSRS-L* homologs by VIGS also supported their conserved role in style and stigma development at least in *N. benthamiana*. The pistils of VIGS-NbSRS-L plants displayed a range of phenotypic defects that affected strongly these tissues, mostly shortening of the style and split and abnormal stigma formation. In addition, other phenotypic defects associated

FIGURE 8 | Overexpression of *NbSRS-L* genes in *N. benthamiana* alters leaf development by constricting margin expansion. (A) Wild type (left) and 35S::NbSRS-L2 (right) plants of similar age. The 35S::NbSRS-L2 plant is darker, has slightly increased internode length and aberrant leaf shape. **(B)** Adaxial view of a fully expanded wild type (left) and a 35S::NbSRS-L1 (right) leaf. The leaf margin in the transgenic plant does not expand to the same extension as the blade, and the leaf is no longer flat. **(C)** Adaxial (top) and abaxial (bottom) views of a wild type (left) and a 35S::NbSRS-L2 (right) leaf. Note the dramatically constricted leaf margin of the transgenic leaf, which shows a much darker color and a highly irregular surface.

previously with loss of SHI/STY/SRS function in Arabidopsis, like general defects in floral organ growth and anther development, were also observed (Kuusk et al., 2006). We found the VIGS treatments not to be specific among the *NbSRS-L* genes under study, so it is not possible to estimate the level of genetic redundancy or the specific functions of each individual gene. However, since the three of them show similar patterns of expression and cause similar effects when overexpressed in Arabidopsis or in Nicotiana, it seems most likely that they work redundantly as described for SHI/STY/SRS family members in Arabidopsis. Unfortunately, the EcSRS-L-VIGS treatments were not sufficient to induce phenotypic changes in *E. californica*, so we cannot infer any significant conclusion. The lack of abnormal phenotypes in VIGS-treated plants and in the Arabidopsis plants expressing *EcSRS-L* may suggest that the EcSRS-L protein is not functional and that other members of this or other families provide the style-stigma specification function. Only one additional *SHI/STY/SRS* gene has been described in *E. californica*, which presents high level of homology with the *EcSRS-L* gene in this study, including the distinctive Pro residue in the RING domain and thus unlikely to be functionally divergent (Pfannebecker et al., 2017a). However, the *E. californica* genome has not been sequenced yet, and it is possible that it encodes other members of the family that may have a role in style and stigma development. It would be interesting to extend this type of studies to other basal dicots and even more basal taxa within angiosperms to conclusively support the conservation of the role of SHI/STY/SRS factors in style and stigma differentiation across angiosperms. However, the reports of abnormal style and stigma development in the monocot *Hordeum vulgare short awn* mutants, affected in the *SHI/STY/SRS* gene *Lks2*, strongly support this idea (Yuo et al., 2012).

NbSRS-L VIGS-treated plants in Nicotiana occasionally showed increased number of floral organs or multicarpellate pistils. These phenotypes had not been previously reported for *shi/sty* mutants in Arabidopsis, and could reflect a specific function for *SHI/STY/SRS* genes in Nicotiana. However, it has been recently reported that the *vrs2* mutants in barley, affected in a *SHI/STY/SRS* gene, form supernumerary spikelet meristems in the inflorescence (Youssef et al., 2017), an indeterminate behavior somehow reminiscent of the supernumerary floral organs found in VIGS-NbSRS-L plants and that could reveal an additional conserved role of the members of the SHI/STY/SRS family, that might be related to their described conserved function in hormone homeostasis (Fridborg et al., 2001; Eklund et al., 2010b, 2011; Kim et al., 2010; Zawaski et al., 2011; Youssef et al., 2017).

A Conserved GRN for Carpel Development

The phenotypes of Nicotiana 35S::NbSRS-L lines parallel those described for Arabidopsis 35S::STY1 plants. In addition to support functional conservation among *SHI/STY/SRS* genes at the top of the GRN involved in style and stigma specification, these results also suggest that the set of targets for these factors in both species is also very conserved. Thus, the downstream pathways leading to style and stigma development are probably similar, despite the dramatic differences of style and stigma morphologies in both species. Moreover, in addition to directing ectopic development of these tissues, the overall changes in pistil morphology, such as the less conspicuous demarcation between ovary and style, or the irregular proliferation of cells in the ovary walls causing crinkled ovary surface, suggest that the process of patterning the pistil into different territories and functional domains responds to similar cues in species with highly diverse pistil morphologies.

Our study also reveals that, as in Arabidopsis, *NGA* and *SHI/STY/SRS* genes also share their functions in pistil development in Nicotiana. *NbNGA* and *NbSRS-L* genes show very similar expression patterns in flower development and lead to highly related phenotypes when downregulated by VIGS (Fourquin and Ferrandiz, 2014). However, while *NbNGA* inactivation caused a complete lack of stigma and style tissues in Nicotiana pistils, *NbSRS-L* downregulation only produced milder defects in these same tissues, which could suggest that the contribution of NbSRS-L factors to this function is less important. However, VIGS treatments were not as efficient for *NbSRS-L* inactivation as for the reported VIGS-NbNGA studies, and the number of *SHI/STY/SRS* genes in the *N. benthamiana* genome is also higher than the number of *NbNGA* homologs, and therefore it is likely that the observed phenotypes of the NbSRS-L-VIGS plants do not reflect the effect of a full loss of SHI/STY/SRS function (Fourquin and Ferrandiz, 2014). Interestingly, the phenotypes of *N. benthamiana* 35S::NbSRS-L lines also resemble some aspects of the effect of overexpressing the *AtNGA* genes in Arabidopsis, such as an enlarged and bulging replum (the region between the valve margins) or the elongation of the gynophore, also supporting the conserved position of both classes of factors at the top of the GRN directing style and stigma development where they would converge in common targets and functions (Trigueros et al., 2009).

NGA genes belong to the RAV subfamily of B3-domain transcription factors present in all land plants, but they possess three characteristic conserved domains only found in angiosperm NGA proteins (Fourquin and Ferrandiz, 2014; Pfannebecker et al., 2017b). *SHI/STYSRS* homologs are found already in bryophytes, although the GCC-box conferring carpel expression has only been found in the promoters of angiosperm *SHI/STY/SRS* genes (Pfannebecker et al., 2017a). This suggests that the *NGA* and the *SHI/STY/SRS* genes may have acquired angiosperm-specific functions linked to the evolutionary origin of the style and the stigma, angiosperm-specific tissues themselves.

In addition to *NGA* and *SHI/STY/SRS* genes, other factors involved in style and stigma development have been shown to have conserved functions in several angiosperm species. Members of the PLE-subclade of MADS-box genes have been related to style and stigma development in different dicot species (Colombo et al., 2010; Fourquin and Ferrandiz, 2012; Heijmans et al., 2012). Likewise, orthologs of *CRABS CLAW (CRC)* gene, a member of the YABBY family of transcription factors required for correct style development and apical gynoecium closure in Arabidopsis, also have a conserved role in these functions in a wide range of angiosperm species (Yamaguchi et al., 2004; Fourquin et al., 2005, 2007, 2014; Lee et al., 2005; Ishikawa et al., 2009; Orashakova et al., 2009; Yamada et al., 2011). Taking all this evidence together, we could propose an emerging evolutionary conserved GRN directing carpel patterning which would include *NGA*, *SHI/STY/SRS*, *CRC*, and *PLE*-like genes as a core of factors linked to style and stigma development, although we still need to investigate in more depth the conservation of their regulatory relationships. In addition, it would be interesting to expand these studies to other genes with a described role in apical pistil development in Arabidopsis, such as *HECATE*, *NO TRANSMITTING TRACT* or *HALF FILLED* (Crawford et al., 2007; Gremski et al., 2007; Crawford and Yanofsky, 2011), which have not been functionally characterized in other taxa, and to connect this apical gynoecium network with those directing valve margin, dehiscence or fruit patterning, which also have been addressed lately (Ferrandiz and Fourquin, 2014; Pabon-Mora et al., 2014).

MATERIALS AND METHODS

Plant Material and Growth Conditions

Eschscholzia californica Cham. and *Nicotiana benthamiana* L. plants were grown in the greenhouse, at 22°C: 18°C (day : night), with a 16 : 8 h, light : dark photoperiod, in soil irrigated with Hoagland no. 1 solution supplemented with oligoelements (Hewitt, 1966). *E. californica* germplasm used in this study (accession no. PI 599252) was obtained from the National Genetic Resources Program (United States).

For Arabidopsis transformation, *EcSRS-L* and *NbSRS-L* coding sequences were amplified with primers EcSTYF/EcSTYR (EcSRS-L), NbSTY1For/NbSTY1Rev (NbSRS-L1), NbSTY2For/NbSTY2Rev (NbSRS-L2), NbSTY3For/NbSTY3Rev (NbSRS-L3), cloned into PCR8/GW/TOPO (Invitrogen) and then transferred by Gateway reactions into the pMCD32 destination vector (Curtis and Grossniklaus, 2003). Each vector was introduced into *Agrobacterium tumefaciens* PMP90 for Arabidopsis transformation using the floral dip protocol (Clough and Bent, 1998) into the wild type Columbia background or the *sty1-1 sty2-1* double mutants (Kuusk et al., 2002). T1 plants were selected based on kanamycin selection. *N. benthamiana* transformation was done using the same vectors and following previously described standard procedures (Clemente, 2006). Three to five T1 individuals from each transformation were tested to confirm that the transgene was being expressed by qRT-PCR. See **Supplementary Table S2** for primer sequences.

Cloning and Sequence Analysis

The partial coding sequence of *EcSRS-L* gene was isolated by RT-PCR on cDNA of flowers of *E. californica* using the degenerate primers EcSTYdF2/EcSTYdR2 designed from the conserved motifs of SRS homologs from other species. The 3′ end of EcSRS-L was then isolated by reverse transcription polymerase chain reaction (RT-PCR) using the primers PoppyRT and RT (sequence added to the oligodT primer used for retrotranscription). To amplify the *NbSRS-L* genes, we designed primers from the annotated genomic sequences retrieved from solgenomics.net (same pairs as described for cloning the CDS in the previous subheading).

The deduced amino acid sequences alignments were analyzed using the MUSCLE tool built in the Macvector 15.1 software (MacVector Inc., Cary, NC, United States). The phylogenetic tree was inferred using the Neighbor-Joining method (Saitou and Nei, 1987). The optimal tree with the sum of branch length = 4.39003061 is shown. The percentage of replicate trees in which the associated taxa clustered together in the bootstrap test (2000 replicates) are shown next to the branches. The tree is drawn to scale, with branch lengths in the same units as those of the evolutionary distances used to infer the phylogenetic tree. The evolutionary distances were computed using the p-distance method and are in the units of the number of amino acid differences per site. The analysis involved 24 amino acid sequences. All ambiguous positions were removed for each sequence pair. There was a total of 723 positions in the final dataset. Evolutionary analyses were conducted in MEGA7 (Kumar et al., 2016).

In Situ Hybridization

RNA *in situ* hybridization with digoxigenin-labeled probes was performed on 8 μm paraffin sections of *E. californica* and *N. benthamiana* buds, as described by Ferrándiz et al. (2000). The RNA antisense and sense probes were generated from a 294 bp fragment of the EcSRS-L cDNA (positions 1-294), from a 438 bp fragment of the NbSRS-L1 cDNA (positions 607-1045), from a 407 bp fragment of the NbSRS-L2 cDNA (positions 1-407) and from a 536 bp fragment of the NbSRS-L3 cDNA (positions 1-536). Each fragment was cloned into the pGemT-Easy vector (Promega), and sense and antisense probes were synthesized using the corresponding SP6 or T7 polymerases.

Virus-Induced Gene Silencing (VIGS)

The same regions of EcSRS-L, NbSRS-L1, and NbSRS-L3 coding sequences used for *in situ* hybridization were used for the VIGS experiments, while for NbSRS-L2 a fragment of 537 bp (positions 1-537) was used. A Xbal restriction site was added to the 5′ end of the PCR fragment and a BamH1 restriction site was added to the 3′ end. The amplicon was digested by Xba 1 and Bam H1 and cloned into a similarly digested pTRV2 vector. The resulting plasmids were confirmed by digestion and sequencing, before being introduced into the *Agrobacterium tumefaciens* strain GV3101. The agroinoculation of *E. californica* seedlings was performed as described (Pabon-Mora et al., 2012). The agroinoculation of *N. benthamiana* leaves was performed as previously described (Dinesh-Kumar et al., 2003).

Quantitative RT-PCR

Total RNA was extracted from flowers in anthesis with the RNeasy Plant Mini kit (Qiagen). Four micrograms of total RNA were used for cDNA synthesis performed with the First-Strand cDNA Synthesis kit (Invitrogen) and the qPCR master mix was prepared using the iQTM SYBR Green Supermix (Bio-Rad). The primers used to amplify EcSRS-L1 (qEcSTYFor and qEcSTYRev), NbSRS-L1 (qNbSTY1For and qNbSTY1Rev), NbSRS-L2 (qNbSTY2For and qNbSTY2Rev), and NbSRS-L3 (qNbSTY3For and qNbSTY3Rev) did not show any cross-amplification. Results were normalized to the expression of the *ACTIN* gene of *E. californica*, amplified by EcACTFor and EcCTRev, and to the Elongation Factor 1 (*EF1*) gene of *N. benthamiana* (accession no. AY206004), amplified by qNbEF1For and qNbEF1Rev (Fourquin and Ferrandiz, 2014). The efficiencies in the amplification of the genes of interest and the corresponding reference gene were similar. Three technical and two biological replicates were performed for each sample. The PCR reactions were run and analyzed using the ABI PRISM 7700 Sequence detection system (Applied Biosystems Inc., Life Technologies Corp., Carlsbad, CA, United States). See **Supplementary Table S2** for primer sequences.

Scanning Electron Microscopy (SEM) and Histology

Plants treated with VIGS were analyzed by cryoSEM on fresh tissue under a JEOL JSM 5410 microscope equipped with a CRIOSEM instrument CT 15000-C (Oxford Instruments[1]). Samples were collected for histological analyses, fixed in FAA (3.7% formaldehyde, 5% acetic acid, 50% ethanol) under vacuum and embedded into paraffin. Sections 10 l m thick were stained in 0.2% toluidine blue solution, and observed under a Nikon Eclipse E-600 microscope[2].

AUTHOR CONTRIBUTIONS

CFe conceived the project and designed the experiments together with CFo. AG-F, VS-G, CFo, and CFe performed the experiments. CFe wrote the paper.

FUNDING

This work was supported by the Spanish MINECO/FEDER grants no BIO2012-32902 and BIO2015-64531-R to CFe. AG-F was supported by a predoctoral contract of the Generalitat Valenciana (ACIF/2013/044).

ACKNOWLEDGMENTS

We thank David Parejo and Victoria Palau (IBMCP) for greenhouse support, Marisol Gascón (IBMCP) for technical advice in microscopy and personnel at Servicio de Microscopía Electrónica (UPV) for help and technical assistance. *E. californica* germplasm used in this study was obtained from the National Genetic Resources Program (United States) and Arabidopsis *sty1-1 sty2-1* and 35S::STY1 seeds from Eva Sundberg. We acknowledge support of the publication fee by the CSIC Open Access Publication Support Initiative through its Unit of Information Resources for Research (URICI).

[1]http://www.oxford-instruments.com
[2]http://www.nikoninstruments.com

SUPPLEMENTARY MATERIAL

The Supplementary Material for this article can be found online at: http://journal.frontiersin.org/article/10.3389/fpls.2017.00814/full#supplementary-material

FIGURE S1 | Amino acid alignment of the SHI/STY/SRS homologs included in this study. The conserved C3HC3H RING Zinc finger and IGGH domains are highlighted in blue. In red, the atypical Pro residues in the *E. californica* proteins.

FIGURE S2 | Sense probes for *in situ* hybridization. (A) *EcSRS-L* sense probe on a transversal sections of an *E. californica* flower **(B–E)** *NbSRS-L* sense probes on transversal sections of *N. benthamiana* flowers **(B–C)** NbSRS-L1. **(D)** NbSRS-L2. **(E)** NbSRS-L3.

FIGURE S3 | Fruits of transgenic Arabidopsis plants transformed with the 35S::EsSRS-L construct in wildtype (left) or *sty1 sty2* **(right)** backgrounds.

FIGURE S4 | Expression level by real-time PCR analysis of *NbSRS-L1* (orange bars), *NbSRS-L2* (pink bars) and *NbSRS-L3* (blue bars) in TRV2-NbSRS-L1, TRV2-NbSRS-L2 or TRV2-SRS-L3 flowers. The error bars depict the s.e. based on two biological replicates. For technical problems, *NbSRS-L2* levels could not be analyzed in TRV2-SRS-L3 flowers.

FIGURE S5 | Pistils of *E. californica* wildtype or EsSRS-L-VIGS treated plants (top panels). Expression level by real-time PCR analysis of *EcSRS-L* in TRV2-EcSRS-L flowers. The error bars depict the s.e. based on two biological replicates.

TABLE S1 | Protein similarity and identity matrix between Arabidopsis, *E. californica* and *N. benthamiana* SHI/STY/SRS proteins generated by MUSCLE (PAM 200).

TABLE S2 | Primers used in this study.

REFERENCES

Alvarez, J. P., Goldshmidt, A., Efroni, I., Bowman, J. L., and Eshed, Y. (2009). The NGATHA distal organ development genes are essential for style specification in *Arabidopsis*. *Plant Cell* 21, 1373–1393. doi: 10.1105/tpc.109.065482

Alvarez-Buylla, E. R., Benitez, M., Davila, E. B., Chaos, A., Espinosa-Soto, C., and Padilla-Longoria, P. (2007). Gene regulatory network models for plant development. *Curr. Opin. Plant Biol.* 10, 83–91. doi: 10.1016/j.pbi.2006.11.008

Ballester, P., and Ferrandiz, C. (2016). Shattering fruits: variations on a dehiscent theme. *Curr. Opin. Plant Biol.* 35, 68–75. doi: 10.1016/j.pbi.2016.11.008

Bombarely, A., Rosli, H. G., Vrebalov, J., Moffett, P., Mueller, L. A., and Martin, G. B. (2012). A draft genome sequence of *Nicotiana benthamiana* to enhance molecular plant-microbe biology research. *Mol. Plant Microbe Interact.* 25, 1523–1530. doi: 10.1094/MPMI-06-12-0148-TA

Bowman, J. L., Smyth, D. R., and Meyerowitz, E. M. (1989). Genes directing flower development in *Arabidopsis*. *Plant Cell* 1, 37–52. doi: 10.1105/tpc.1.1.37

Bradley, D., Carpenter, R., Sommer, H., Hartley, N., and Coen, E. (1993). Complementary floral homeotic phenotypes result from opposite orientations of a transposon at the plena locus of *Antirrhinum*. *Cell* 72, 85–95. doi: 10.1016/0092-8674(93)90052-R

Chávez Montes, R. A., Herrera-Ubaldo, H., Serwatowska, J., and de Folter, S. (2015). Towards a comprehensive and dynamic gynoecium gene regulatory network. *Curr. Plant Biol.* 3-4, 3–12. doi: 10.1016/j.cpb.2015.08.002

Clemente, T. (2006). *Nicotiana (Nicotiana tobaccum, Nicotiana benthamiana)*. *Methods Mol. Biol.* 343, 143–154. doi: 10.1385/1-59745-130-4:143

Clough, S. J., and Bent, A. F. (1998). Floral dip: a simplified method for *Agrobacterium*-mediated transformation of *Arabidopsis thaliana*. *Plant J.* 16, 735–743. doi: 10.1046/j.1365-313x.1998.00343.x

Colombo, M., Brambilla, V., Marcheselli, R., Caporali, E., Kater, M., and Colombo, L. (2010). A new role for the SHATTERPROOF genes during *Arabidopsis* gynoecium development. *Dev. Biol.* 337, 294–302. doi: 10.1016/j.ydbio.2009.10.043

Crawford, B. C. W., Ditta, G., and Yanofsky, M. F. (2007). The NTT gene is required for transmitting-tract development in carpels of *Arabidopsis thaliana*. *Curr. Biol.* 17, 1101–1108. doi: 10.1016/j.cub.2007.05.079

Crawford, B. C. W., and Yanofsky, M. F. (2011). HALF FILLED promotes reproductive tract development and fertilization efficiency in *Arabidopsis thaliana*. *Development* 138, 2999–3009. doi: 10.1242/dev.067793

Curtis, M. D., and Grossniklaus, U. (2003). A gateway cloning vector set for high-throughput functional analysis of genes in planta. *Plant Physiol.* 133, 462–469. doi: 10.1104/pp.103.027979

Davidson, E. H., and Levine, M. S. (2008). Properties of developmental gene regulatory networks. *Proc. Natl. Acad. Sci. U.S.A.* 105, 20063–20066. doi: 10.1073/pnas.0806007105

Davies, B., Motte, P., Keck, E., Saedler, H., Sommer, H., and Schwarz-Sommer, Z. (1999). PLENA and FARINELLI: redundancy and regulatory interactions between two *Antirrhinum* MADS-box factors controlling flower development. *EMBO J.* 18, 4023–4034. doi: 10.1093/emboj/18.14.4023

Davila-Velderrain, J., Martinez-Garcia, J. C., and Alvarez-Buylla, E. R. (2016). Dynamic network modelling to understand flowering transition and floral patterning. *J. Exp. Bot.* 67, 2565–2572. doi: 10.1093/jxb/erw123

De Lucas, M., and Brady, S. M. (2013). Gene regulatory networks in the Arabidopsis root. *Curr. Opin. Plant Biol.* 16, 50–55. doi: 10.1016/j.pbi.2012.10.007

Dinesh-Kumar, S. P., Anandalakshmi, R., Marathe, R., Schiff, M., and Liu, Y. (2003). Virus-induced gene silencing. *Methods Mol. Biol.* 236, 287–294. doi: 10.1385/1-59259-

Dreni, L., Pilatone, A., Yun, D., Erreni, S., Pajoro, A., Caporali, E., et al. (2011). Functional analysis of all AGAMOUS subfamily members in rice reveals their roles in reproductive organ identity determination and meristem determinacy. *Plant Cell* 23, 2850–2863. doi: 10.1105/tpc.111.087007

Eklund, D. M., Cierlik, I., Staldal, V., Claes, A. R., Vestman, D., Chandler, J., et al. (2011). Expression of Arabidopsis SHORT INTERNODES/STYLISH family genes in auxin biosynthesis zones of aerial organs is dependent on a GCC box-like regulatory element. *Plant Physiol.* 157, 2069–2080. doi: 10.1104/pp.111.182253

Eklund, D. M., Staldal, V., Valsecchi, I., Cierlik, I., Eriksson, C., Hiratsu, K., et al. (2010a). The *Arabidopsis thaliana* STYLISH1 protein acts as a transcriptional activator regulating auxin biosynthesis. *Plant Cell* 22, 349–363. doi: 10.1105/tpc.108.064816

Eklund, D. M., Thelander, M., Landberg, K., Staldal, V., Nilsson, A., Johansson, M., et al. (2010b). Homologues of the *Arabidopsis thaliana* SHI/STY/LRP1 genes control auxin biosynthesis and affect growth and development in the moss *Physcomitrella* patens. *Development* 137, 1275–1284. doi: 10.1242/dev.039594

Ferrandiz, C., and Fourquin, C. (2014). Role of the FUL-SHP network in the evolution of fruit morphology and function. *J. Exp. Bot.* 65, 4505–4513. doi: 10.1093/jxb/ert479

Ferrandiz, C., Fourquin, C., Prunet, N., Scutt, C., Sundberg, E., Trehin, C., et al. (2010). Carpel development. *Adv. Bot. Res.* 55, 1–74. doi: 10.1016/S0065-2296(10)55001-8

Ferrándiz, C., Gu, Q., Martienssen, R., and Yanofsky, M. F. (2000). Redundant regulation of meristem identity and plant architecture by FRUITFULL, APETALA1 and CAULIFLOWER. *Development* 127, 725–734.

Fourquin, C., and Ferrandiz, C. (2012). Functional analyses of AGAMOUS family members in *Nicotiana benthamiana* clarify the evolution of early and late roles of C-function genes in eudicots. *Plant J.* 71, 990–1001. doi: 10.1111/j.1365-313X.2012.05046.x

Fourquin, C., and Ferrandiz, C. (2014). The essential role of NGATHA genes in style and stigma specification is widely conserved across eudicots. *New Phytol.* 202, 1001–1013. doi: 10.1111/nph.12703

Fourquin, C., Primo, A., Martinez-Fernandez, I., Huet-Trujillo, E., and Ferrandiz, C. (2014). The CRC orthologue from *Pisum sativum* shows conserved functions in carpel morphogenesis and vascular development. *Ann. Bot.* 114, 1535–1544. doi: 10.1093/aob/mcu129

Fourquin, C., Vinauger-Douard, M., Chambrier, P., Berne-Dedieu, A., and Scutt, C. P. (2007). Functional conservation between CRABS CLAW orthologues from widely diverged angiosperms. *Ann. Bot.* 100, 651–657. doi: 10.1093/aob/mcm136

Fourquin, C., Vinauger-Douard, M., Fogliani, B., Dumas, C., and Scutt, C. P. (2005). Evidence that CRABS CLAW and TOUSLED have conserved their roles in carpel development since the ancestor of the extant angiosperms. *Proc. Natl. Acad. Sci. U.S.A.* 102, 4649–4654. doi: 10.1073/pnas.0409577102

Fridborg, I., Kuusk, S., Robertson, M., and Sundberg, E. (2001). The Arabidopsis protein SHI represses gibberellin responses in Arabidopsis and barley. *Plant Physiol.* 127, 937–948. doi: 10.1104/pp.010388

Gremski, K., Ditta, G., and Yanofsky, M. F. (2007). The HECATE genes regulate female reproductive tract development in *Arabidopsis thaliana*. *Development* 134, 3593–3601. doi: 10.1242/dev.011510

Heijmans, K., Ament, K., Rijpkema, A. S., Zethof, J., Wolters-Arts, M., Gerats, T., et al. (2012). Redefining C and D in the petunia ABC. *Plant Cell* 24, 2305–2317. doi: 10.1105/tpc.112.097030

Hewitt, Y. (1966). *Sand and Water Culture Methods used in the Study of Plant Nutrition*. Farnham: Commonwealth Agricultural Bureau.

Ishikawa, M., Ohmori, Y., Tanaka, W., Hirabayashi, C., Murai, K., Ogihara, Y., et al. (2009). The spatial expression patterns of DROOPING LEAF orthologs suggest a conserved function in grasses. *Genes Genet. Syst.* 84, 137–146. doi: 10.1266/ggs.84.137

Islam, M. A., Lütken, H., Haugslien, S., Blystad, D.-R., Torre, S., Rolcik, J., et al. (2013). Overexpression of the AtSHI gene in poinsettia, *Euphorbia pulcherrima*, results in compact plants. *PLoS ONE* 8:e53377. doi: 10.1371/journal.pone.0053377

Kim, S. G., Lee, S., Kim, Y. S., Yun, D. J., Woo, J. C., and Park, C. M. (2010). Activation tagging of an *Arabidopsis* SHI-RELATED SEQUENCE gene produces abnormal anther dehiscence and floral development. *Plant Mol. Biol.* 74, 337–351. doi: 10.1007/s11103-010-9677-5

Kumar, S., Stecher, G., and Tamura, K. (2016). MEGA7: molecular evolutionary genetics analysis version 7.0 for bigger datasets. *Mol. Biol. Evol.* 33, 1870–1874. doi: 10.1093/molbev/msw054

Kuusk, S., Sohlberg, J. J., Long, J. A., Fridborg, I., and Sundberg, E. (2002). STY1 and STY2 promote the formation of apical tissues during *Arabidopsis* gynoecium development. *Development* 129, 4707–4717.

Kuusk, S., Sohlberg, J. J., Magnus Eklund, D., and Sundberg, E. (2006). Functionally redundant SHI family genes regulate *Arabidopsis* gynoecium development in a dose-dependent manner. *Plant J.* 47, 99–111. doi: 10.1111/j.1365-313X.2006.02774.x

Larsson, E., Franks, R. G., and Sundberg, E. (2013). Auxin and the *Arabidopsis thaliana* gynoecium. *J. Exp. Bot.* 64, 2619–2627. doi: 10.1093/jxb/ert099

Larsson, E., Roberts, C. J., Claes, A. R., Franks, R. G., and Sundberg, E. (2014). Polar auxin transport is essential for medial versus lateral tissue specification and vascular-mediated valve outgrowth in Arabidopsis gynoecia. *Plant Physiol.* 166, 1998–2012. doi: 10.1104/pp.114.245951

Lee, J.-Y., Baum, S. F., Oh, S.-H., Jiang, C.-Z., Chen, J.-C., and Bowman, J. L. (2005). Recruitment of CRABS CLAW to promote nectary development within the eudicot clade. *Development* 132, 5021–5032. doi: 10.1242/dev.02067

Marsch-Martinez, N., and de Folter, S. (2016). Hormonal control of the development of the gynoecium. *Curr. Opin. Plant Biol* 29, 104–114. doi: 10.1016/j.pbi.2015.12.006

Martinez-Fernandez, I., Sanchis, S., Marini, N., Balanza, V., Ballester, P., Navarrete-Gomez, M., et al. (2014). The effect of NGATHA altered activity on auxin signaling pathways within the Arabidopsis gynoecium. *Front. Plant Sci.* 5:210. doi: 10.3389/fpls.2014.00210

Morris, A. L., MacArthur, M. W., Hutchinson, E. G., and Thornton, J. M. (1992). Stereochemical quality of protein structure coordinates. *Proteins* 12, 345–364. doi: 10.1002/prot.340120407

Moubayidin, L., and Ostergaard, L. (2014). Dynamic control of auxin distribution imposes a bilateral-to-radial symmetry switch during gynoecium development. *Curr. Biol.* 24, 2743–2748. doi: 10.1016/j.cub.2014.09.080

O'Maoileidigh, D. S., Graciet, E., and Wellmer, F. (2014). Gene networks controlling *Arabidopsis thaliana* flower development. *New Phytol.* 201, 16–30. doi: 10.1111/nph.12444

Orashakova, S., Lange, M., Lange, S., Wege, S., and Becker, A. (2009). The CRABS CLAW ortholog from California poppy (*Eschscholzia californica*, Papaveraceae), EcCRC, is involved in floral meristem termination, gynoecium differentiation and ovule initiation. *Plant J.* 58, 682–693. doi: 10.1111/j.1365-313X.2009.03807.x

Pabon-Mora, N., Ambrose, B. A., and Litt, A. (2012). Poppy APETALA1/FRUITFULL orthologs control flowering time, branching, perianth identity, and fruit development. *Plant Physiol.* 158, 1685–1704. doi: 10.1104/pp.111.192104

Pabon-Mora, N., Wong, G. K., and Ambrose, B. A. (2014). Evolution of fruit development genes in flowering plants. *Front. Plant Sci.* 5:300. doi: 10.3389/fpls.2014.00300

Pan, I. L., McQuinn, R., Giovannoni, J. J., and Irish, V. F. (2010). Functional diversification of AGAMOUS lineage genes in regulating tomato flower and fruit development. *J. Exp. Bot.* 61, 1795–1806. doi: 10.1093/jxb/erq046

Pfannebecker, K. C., Lange, M., Rupp, O., and Becker, A. (2017a). An evolutionary framework for carpel developmental control genes. *Mol. Biol. Evol.* 34, 330–348. doi: 10.1093/molbev/msw229

Pfannebecker, K. C., Lange, M., Rupp, O., and Becker, A. (2017b). Seed plant specific gene lineages involved in carpel development. *Mol. Biol. Evol.* 34, 925–942. doi: 10.1093/molbev/msw297

Ratcliff, F., Martin-Hernandez, A. M., and Baulcombe, D. C. (2001). Technical advance. Tobacco rattle virus as a vector for analysis of gene function by silencing. *Plant J.* 25, 237–245. doi: 10.1046/j.0960-7412.2000.00942.x

Reyes-Olalde, J. I., Zuniga-Mayo, V. M., Chavez Montes, R. A., Marsch-Martinez, N., and de Folter, S. (2013). Inside the gynoecium: at the carpel margin. *Trends Plant Sci.* 18, 644–655. doi: 10.1016/j.tplants.2013.08.002

Saitou, N., and Nei, M. (1987). The neighbor-joining method: a new method for reconstructing phylogenetic trees. *Mol. Biol. Evol.* 4, 406–425.

Sohlberg, J. J., Myrenas, M., Kuusk, S., Lagercrantz, U., Kowalczyk, M., Sandberg, G., et al. (2006). STY1 regulates auxin homeostasis and affects apical-basal patterning of the Arabidopsis gynoecium. *Plant J.* 47, 112–123. doi: 10.1111/j.1365-313X.2006.02775.x

Staldal, V., Cierlik, I., Chen, S., Landberg, K., Baylis, T., Myrenas, M., et al. (2012). The *Arabidopsis thaliana* transcriptional activator STYLISH1 regulates genes affecting stamen development, cell expansion and timing of flowering. *Plant Mol. Biol.* 78, 545–559. doi: 10.1007/s11103-012-9888-z

Ståldal, V., Sohlberg, J., Eklund, D., Ljung, K., and Sundberg, E. (2008). Auxin can act independently of CRC, LUG, SEU, SPT and STY1 in style development but not apical-basal patterning of the *Arabidopsis* gynoecium. *New Phytol.* 180, 798–808. doi: 10.1111/j.1469-8137.2008.02625.x

Sundberg, E., and Ostergaard, L. (2009). Distinct and dynamic auxin activities during reproductive development. *Cold Spring Harb. Perspect. Biol.* 1:a001628. doi: 10.1101/cshperspect.a001628

Tian, C., and Jiao, Y. (2015). A systems approach to understand shoot branching. *Curr. Plant Biol.* 3–4, 13–19. doi: 10.1016/j.cpb.2015.08.001

Trigueros, M., Navarrete-Gómez, M., Sato, S., Christensen, S. K., Pelaz, S., Weigel, D., et al. (2009). The NGATHA genes direct style development in the *Arabidopsis* gynoecium. *Plant Cell* 21, 1394–1409. doi: 10.1105/tpc.109.065508

Vialette-Guiraud, A. C., Andres-Robin, A., Chambrier, P., Tavares, R., and Scutt, C. P. (2016). The analysis of gene regulatory networks in plant evo-devo. *J. Exp. Bot.* 67, 2549–2563. doi: 10.1093/jxb/erw119

Wege, S., Scholz, A., Gleissberg, S., and Becker, A. (2007). Highly efficient virus-induced gene silencing (VIGS) in California Poppy (*Eschscholzia californica*): an evaluation of VIGS as a strategy to obtain functional data from non-model plants. *Ann. Bot.* 100, 641–649. doi: 10.1093/aob/mcm118

Yamada, T., Yokota, S. Y., Hirayama, Y., Imaichi, R., Kato, M., and Gasser, C. S. (2011). Ancestral expression patterns and evolutionary diversification of YABBY genes in angiosperms. *Plant J.* 67, 26–36. doi: 10.1111/j.1365-313X.2011.04570.x

Yamaguchi, T., Nagasawa, N., Kawasaki, S., Matsuoka, M., Nagato, Y., and Hirano, H. Y. (2004). The YABBY gene DROOPING LEAF regulates carpel specification and midrib development in *Oryza sativa*. *Plant Cell* 16, 500–509. doi: 10.1105/tpc.018044

Yellina, A. L., Orashakova, S., Lange, S., Erdmann, R., Leebens-Mack, J., and Becker, A. (2010). Floral homeotic C function genes repress specific B function genes in the carpel whorl of the basal eudicot California poppy (*Eschscholzia californica*). *Evodevo* 1:13. doi: 10.1186/2041-9139-1-13

Youssef, H. M., Eggert, K., Koppolu, R., Alqudah, A. M., Poursarebani, N., Fazeli, A., et al. (2017). VRS2 regulates hormone-mediated inflorescence patterning in barley. *Nat. Genet.* 49, 157–161. doi: 10.1038/ng.3717

Yuo, T., Yamashita, Y., Kanamori, H., Matsumoto, T., Lundqvist, U., Sato, K., et al. (2012). A SHORT INTERNODES (SHI) family transcription factor gene regulates awn elongation and pistil morphology in barley. *J. Exp. Bot.* 63, 5223–5232. doi: 10.1093/jxb/ers182

Zawaski, C., Kadmiel, M., Ma, C., Gai, Y., Jiang, X., Strauss, S. H., et al. (2011). SHORT INTERNODES-like genes regulate shoot growth and xylem proliferation in *Populus*. *New Phytol.* 191, 678–691. doi: 10.1111/j.1469-8137.2011.03742.x

Conflict of Interest Statement: The authors declare that the research was conducted in the absence of any commercial or financial relationships that could be construed as a potential conflict of interest.

3

Evolution and Expression Patterns of *CYC/TB1* Genes in *Anacyclus*: Phylogenetic Insights for Floral Symmetry Genes in Asteraceae

*María A. Bello[1], Pilar Cubas[2], Inés Álvarez[1], Guillermo Sanjuanbenito[1] and Javier Fuertes-Aguilar[1]**

[1] *Plant Evolutionary Biology Group, Real Jardín Botánico (CSIC), Madrid, Spain,* [2] *Department of Plant Molecular Genetics, Centro Nacional de Biotecnología, CSIC-Universidad Autónoma de Madrid, Madrid, Spain*

Edited by:
*José M. Romero,
University of Seville, Spain*

Reviewed by:
*Elena M Kramer,
Harvard University, USA
Yin-Zheng Wang,
Chinese Academy of Sciences, China*

***Correspondence:**
*Javier Fuertes-Aguilar
jfuertes@rjb.csic.es*

Homologs of the *CYC/TB1* gene family have been independently recruited many times across the eudicots to control aspects of floral symmetry The family Asteraceae exhibits the largest known diversification in this gene paralog family accompanied by a parallel morphological floral richness in its specialized head-like inflorescence. In Asteraceae, whether or not *CYC/TB1* gene floral symmetry function is preserved along organismic and gene lineages is unknown. In this study, we used phylogenetic, structural and expression analyses focused on the highly derived genus *Anacyclus* (tribe Anthemidae) to address this question. Phylogenetic reconstruction recovered eight main gene lineages present in Asteraceae: two from *CYC1*, four from *CYC2* and two from *CYC3*-like genes. The species phylogeny was recovered in most of the gene lineages, allowing the delimitation of orthologous sets of *CYC/TB1* genes in Asteraceae. Quantitative real-time PCR analysis indicated that in *Anacyclus* three of the four isolated *CYC2* genes are more highly expressed in ray flowers. The expression of the four *AcCYC2* genes overlaps in several organs including the ligule of ray flowers, as well as in anthers and ovules throughout development.

Keywords: ***Anacyclus*, Asteraceae, *CYC2* diversification, *CYCLOIDEA*, *CYC/TB1*, floral symmetry**

INTRODUCTION

Asteraceae is the largest family of vascular plants with more than 23,600 species distributed among 13 different lineages (Panero et al., 2014). By contrast, its closest relatives, Calyceraceae and Goodeniaceae, have around 54 and 404 species, respectively (Carolin, 2007; Pozner et al., 2012). The evolutionary success of Asteraceae is strongly associated with its head-like inflorescence or capitulum (Broholm et al., 2014). Variations in perianth morphology, symmetry and sexuality of the flowers along the radial axis of this inflorescence have been used to discriminate different types of capitula (i.e., bilabiate, ligulate, radiate, discoid, and disciform; Jeffrey, 1977; Bremer, 1994). This diversity has important consequences, such as differential attractiveness to pollinators, increased rate of outcrossing and fitness by the presence of peripheral ray flowers (e.g., *Senecio*; Chapman and Abbott, 2009), or differential rates of seed germination across the capitulum (e.g., *Anacyclus*; Torices et al., 2013).

Floral symmetry is one of the most striking morphological variations within the capitulum. The evolution of floral symmetry in angiosperms is controlled by a restricted set of gene families (Luo et al., 1996; Costa et al., 2005; Broholm et al., 2008; Hileman, 2014). Among these, the TCP factor *CYCLOIDEA* (*CYC*) (Luo et al., 1996) has been recruited for independent transitions but then further modified or lost in conjuction with reversals to radial symmetry. The TCP gene family (Cubas et al., 1999) has expanded 600–800 million years ago, forming the class I and class II TCP subfamilies (Navaud et al., 2007). The TCP class II subfamily is formed by the *CYCLOIDEA/TEOSINTE BRANCHED1* (*CYC/TB1*) and *CINCINNATA* (*CIN*) clades (Martín-Trillo and Cubas, 2010). The *CYC/TB1* or ECE clade is specific to angiosperms and underwent duplication before the core eudicot diversification, producing three main groups: *CYC1*, *CYC2*, and *CYC3* (Howarth and Donoghue, 2006; Citerne et al., 2013). *CYC3*, and particularly *CYC1* genes, have a major role in the regulation of axillary bud outgrowth (Aguilar-Martinez et al., 2007; Finlayson, 2007), whereas *CYC2* genes control the growth patterns of flower meristems and regulate the establishment of bilateral symmetry of flowers in asterids and rosids (Busch and Zachgo, 2009; Preston et al., 2011; Nicolas and Cubas, 2015).

Investigation of Asteraceae *CYC/TB1* gene family evolution performed in *Gerbera* (Mutisieae), *Helianthus* (Heliantheae), and *Senecio* (Senecioneae) revealed that *CYC2* genes play a critical role in controlling ray flower identity (Broholm et al., 2008, 2014; Chapman et al., 2008; Kim et al., 2008; Juntheikki-Palovaara et al., 2014), and changes in temporal and spatial expression of *CYC2* genes are associated with modifications of the floral symmetry pattern (Busch and Zachgo, 2007; Gao et al., 2008; Zhou et al., 2008). The acquisition of bilateral flowers in Asteraceae involved parallel recruitment of the diversified *CYC2* genes, according to recent phylogenetic reconstruction of *CYC/TB1* genes (Chapman et al., 2008, 2012; Tähtiharju et al., 2012).

We characterized the *CYC/TB1* genes of the Mediterranean *Anacyclus* (Anthemideae), inferred their phylogenetic relationships and expression patterns, and compared them with the identified models in *Gerbera*, *Helianthus* and *Senecio* in order to test if the bilateral symmetry evolved early in the family due to cooption of CYC2 clade genes. *Anacyclus* L. with homogamous (i.e., all flowers bisexual, tubular and pentamerous flowers) and heterogamous capitula (with modified female, bilateral and trimerous ray flowers surrounding the bisexual, tubular and pentamerous disc flowers) (Bello et al., 2013) is a suitable model with which to carry out this comparative analysis because it is mostly annual, diploid (2n = 18, $x = 9$; Ehrendorfer et al., 1977; Humphries, 1981) and is suitable for *ex situ* cultivation. Also, the floral development of *Anacyclus*, *Gerbera* and *Helianthus* is similar, excepting the delayed development of the ray flowers relative to the disc flowers in *Anacyclus,* the exclusive presence of transitional flowers in *Gerbera,* the sterile condition of the ray flowers in *Helianthus* and the late zygomorphy of the disc flowers in *Anacyclus* (Laitinen et al., 2006; Tähtiharju et al., 2012; Bello et al., 2013). Moreover, unusual heterogamous capitula with peripheral "trumpet" flowers in natural populations of *A. clavatus* (Desf.) Pers. and *A. valentinus* L. documented from southern Spain (Bello et al., 2013) are very similar to *tub* (Berti et al., 2005) and *turf* (Chapman et al., 2012) mutant individuals of *Helianthus.* These "trumpet" flowers differ from the typical ray flowers in their tubular five-lobed perianth with radial symmetry and the labile presence of stamens (Bello et al., 2013).

As *Anacyclus* represents the derived and highly diversified tribe Anthemideae, where at least 20 genera have species presenting inflorescences with and without ray flowers, we extended the phylogenetic range of the *CYC/TB1* studies. Although there are previous partial Asteraceae *CYC/TB1* phylogenetic reconstructions (Chapman et al., 2008, 2012; Kim et al., 2008; Tähtiharju et al., 2012), a framework to visualize the entire diversification scenario in the family based on nucleotide variation is lacking. With the inclusion of several eudicot *CYC/TB1* sequences in our analysis, together with those available from Asteraceae and the isolated CYC-like genes from *Anacyclus*, we have reconstructed a wider lineage profile and propose it as a model system for the classification and identification of paralogous and orthologous groups of *CYC/TB1* genes in Asteraceae. Having this phylogenetic framework, we have explored if *CYC/TB1* diversification involved positive selection, differential rates of evolution or differential expression patterns of the paralogs in Anacyclus.

MATERIALS AND METHODS

Plant Materials

Seeds and entire plants of wild *Anacyclus clavatus* (IA 2006, Soto del Real, Madrid), *A. valentinus* (LM 4435, Altea, Valencia), *Matricaria aurea* (IA 1995, El Retiro, Madrid) and *Matricaria chamomilla* (IA 1996, living collection Real Jardín Botánico, Madrid) were collected in 2008–2009 and treated as indicated in Bello et al. (2013). *A. clavatus* with trumpet phenotypes were obtained from seeds collected in May 2012 (one population from Carchuna, Granada), sowed in November 2010 and harvested in May 2013.

Cycloidea Gene Analysis

CYC-like genes of *Anacyclus* and *Matricaria* species were amplified from genomic DNA and cDNA with previously published (Chapman et al., 2008) and own-designed primers (**Table S1**). The DNeasy® and RNeasy® Plant Mini kits from Qiagen® were used for DNA and RNA extraction, respectively. RNA extraction of individual plant tissues was performed after their dissection, fixation and disruption in liquid nitrogen. RNA concentration was measured by spectrophotometry (NanoDrop 1000 v3.7, Thermo Fisher Scientific Inc.) and adjusted among tissues. cDNA synthesis was performed with the Invitrogen™ ThermoScript™ RT-PCR system kit. Semi quantitative RT-PCR of CYC-like amplicons from young roots (10-cm-long), leaves (2-cm-long), peduncules (1-cm-long), capitula (ca. 1 cm diameter), inflorescence bracts (ca. 0.5-cm-long), young/full expanded ray flowers (1–2.5-cm-long) and young/mature disc flowers (0.5–1.5-cm-long) of *A. clavatus* was carried out three times using different CYC and actin primer sets (**Table S1**). PCR amplification was performed with Ready-To-Go PCR beads (Illustra™) using a general program as follows: 95°C/5 min followed by 35 cycles of 95°C/30 s, annealing temperature for 30 s and 72°C/45 s, and a final extension of 72°C/7 min (annealing temperatures are listed in **Table S1**). Amplified sequences were cloned using the Promega pGEM®-T Easy vector system (JM109 competent cells) and sequenced on a 3730 DNA Analyzer (Center for Research Support CAI, Universidad Complutense, Madrid) and an ABI 3700 instrument (STAB VIDA DNA sequencing service, Oeiras, Portugal). Inverse PCR and RACE techniques were used to amplify longer sequences from *Anacyclus*. For inverse PCR, DNA from *A. clavatus* and *A. valentinus* was extracted (Doyle and Doyle, 1987) and digested with 1.5 μL of restriction enzyme (*Bam*HI, *Pst*I, *Eco*RI, *Hind*III, *Xho*I, *Nco*I; New England Biolabs® Inc.) in a reaction containing 1 μL of DNA, 2 μL of buffer, and 15.5 μL water. After 3 h of incubation (37°C) and 10 min of enzyme deactivation (65°C), the reactions were diluted (280 μL water) and ligation was performed using 26 μL of DNA, 3 μL of ligase buffer and 1 μL of ligase. Ligation products were amplified with specific primer pairs (**Table S2**) and cloned for sequencing. For 3′ RACE amplification of *CYC* genes from young capitula of *A. clavatus,* the SMARTer™ RACE cDNA Amplification Kit and Advantage® 2 PCR Taq polymerase (Clontech Laboratories, Inc.) were used with the specific primer Ha2c_11 (**Table S1**).

Phylogenetic Analysis

Selected clones isolated from *Anacylus* (79) and *Matricaria* (16) were aligned with other *CYC*-like genes from other species of Asteraceae (44) and other eudicots (54) including Calyceraceae and Goodeniaceae (**Table S3**). Initial alignments were performed with Geneious Pro 5.5.5 (Biomatters, http://www.geneious.com/; Kearse et al., 2012) using the default options of Geneious, MUSCLE and ClustalW. Nucleotides were aligned considering the codon arrangement in the amino acid alignment. Nucleotide (4) and amino acid (1) matrices were assembled (**Table S4**). The models of evolution for CYC amino acid (JTT + I + G, −lnL = 11,683.44) and nucleotide matrices (GTT + I + G, −lnL = 22,128.4089) were estimated using ProtTest 3 (Darriba et al., 2011), jModelTest 2 (Darriba et al., 2012) and Modeltest V 3.8 (Posada and Crandall, 1998). For Bayesian analyses (Ronquist et al., 2012), the matrices were analyzed with Mr Bayes 3.2.2 on XSEDE as implemented in the CIPRES Science Gateway (http://www.phylo.org/sub_sections/portal; Miller et al., 2010). For the amino acid matrix, the model was set to fixed (Jones), the rates to gamma distribution, the number of generations to 5,000,000, the number of chains to 1, and the sample frequency to 2,000. The nucleotide analyses were performed with and without removal of the third codon. In two other data sets, ambiguously aligned positions were removed and *CYC2* genes from Asterales were analyzed independently (**Table S4**). For these analyses, the number of chains was set to four with 15,000,000 generations, a sample frequency of 2,000 and a diagnostic frequency of 5,000. The selected outgroup for all analyses was *AcTBLb* (*Acorus calamus*) except the *CYC2* genes from Asterales dataset that used *SlCYC1* (*Solanum lycopersicum*) as outgroup. In all cases, the post-burn in trees were selected after discarding 25% of the trees. Final toplologies were visualized with figTree v1.1.2 (http://tree.bio.ed.ac.uk/software/figtree). Mapping of the CYC amino acids along the trees was done with Mesquite 3.0 (Maddison and Maddison, 2011) using parsimony optimization. The amino acids were tracked on a post burn-in tree (tree 4,800,000) resulting from the amino acid Bayesian analysis. Maximum likelihood (ML) analysis was performed with GARLI 2.1 (Bazinet et al., 2014) and the bootstrap support was estimated with a 1,000 replicate search in Bootstrap RAxML (Stamatakis et al., 2008) in the CIPRES portal.

Diversifying/Purifying Selection of CYC Genes

Recombination Detection Program RDP v4.36 (Heath et al., 2006) was used to identify potential cases of recombination that could affect the estimate for selection, implementing the RDP (Martin and Rybicki, 2000) and MaxChi (Smith, 1992) methods. To detect individual sites subject to episodic diversifying selection, the CYC nucleotide matrix was analyzed under the mixed effects model of evolution (MEME) and the fixed effects likelihood approaches (FEL) in Datamonkey (Delport et al., 2010). In MEME, the distribution of the rate ω varies from site to site and also from branch to branch in a site, capturing the footprints of episodic and pervasive positive selection, whereas in FEL the synonymous and non-synonymous rates are fitted at each site with no variation along branches (Kosakovsky Pond et al., 2005; Murrel et al., 2012).

Diversification Rates Analysis

We used BAMM (Rabosky et al., 2014) to estimate rates of diversification across different gene lineages across the CYC phylogeny. The general model in this Bayesian method assumes that phylogenetic trees may have been shaped by a heterogeneous mixture of different evolutionary rates of gene diversification and extinction. Our working hypothesis was that, given a balanced sampling across paralogs of CYC2 in Asteraceae, which exhibit a number of paralogs larger than in other eudicot families, we could detect a significant heterogeneity across branches of arising paralogs in Anthemideae. We allowed each regime to be characterized by a distinct time-varying speciation process, where the diversification rate varies exponentially through time. The model of exponential change has been used in taxon diversification studies and is also expected as an approximation to diversity-dependent changes in gene diversification rates through time (Rabosky, 2014). We accounted for incomplete taxon sampling using the analytical correction implemented in BAMM, assuming that our sampling included 95% of extant *Anacyclus* CYC diversity. Visualization was performed using R scripts available through the R package BAMMTOOLS (Rabosky, 2014; Rabosky et al., 2014).

Expression Analysis by Quantitative PCR

Expression of the *CYC* genes was compared in wild rayed and "trumpet" inflorescences of *A. clavatus* using young plant tissues: roots (10-cm-long), leaves (2-cm-long), capitula stage 1 (≤0.5 cm diameter), capitula stage 2 (>1 cm diameter), mature ligules (>3-cm-long, full expanded) and closed disc flowers (0.5-mm-long). Three different individuals of wild and "trumpet" *A. clavatus* were included in the analysis, as well as three technical replicates of each tissue. RNA extraction was performed as described above. cDNA synthesis was carried out with the Transcriptor Universal cDNA Master kit® (Roche) using the following conditions: 25°C/5, 55°C/10, and 85°C/5 min. RNA concentration was adjusted among tissues, adding up to 15 ng/µL per reaction. The qPCR was run on a LightCycler 2.0 using 4 µL of Sensimix Capillary Kit, 0.2 µL of Sybr green and 0.75 µL of MyFi™ DNA Polymerase (Bioline) together with designed lineage-specific CYC primers (final concentration 0.2 µM, **Table S5**). Positive (genomic DNA) and negative (without nucleic acids) controls were included in each qPCR run to test the resultant crossing-point (Cp) values. Discarded Cp values included those higher than the Cp of the negative control, values above 35, and dissimilar melting temperatures compared against the positive control. Actin was used as the reference gene and was amplified in all tissues. The primer efficiency (E) was calculated from the amplification of three replicates of the "capitula stage 2" tissues, contrasting the logarithm of the fluorescence against the Cp and applying $E = 10^{(1/\text{slope})}$ (Pfaffl, 2001; **Table S6**). E was calculated with the wild and trumpet tissues (**Table S5**). For the relative quantification, all Cp values were normalized against the Cp from "capitula stage 1" tissue (Cp control) and the E target $^{\Delta C}$ /E control $^{\Delta C}$ ratio calculated (Pfaffl, 2001). For a graphical format of the *CYC* gene expression, the average of three tissue replicates of this ratio was calculated together with the standard deviation. The Kolmogorov-Smirnov test was used to evaluate the probability distribution of the expression average ration in trumpet and wild individuals ($\alpha = 0.05$, **Table S7**). A *t*-test paired two sample for means was conducted to test if there is significant difference between the mean expression of AcCYC2 genes in trumpet and wild individuals ($\alpha = 0.05$, **Tables S8, S9**).

RNA *In situ* Hybridization

A non-radioactive *in situ* protocol using RNA probes was followed using wild inflorescences of *A. clavatus* at different developmental stages. Capitula of different stages with a maximum of 1 cm of diameter were dissected under a Leica M165FC stereo microscope and fixed in 4% formaldehyde with 0.1% Tween-20 and 0.1% Triton X-100 (Jackson, 1991). Gene-specific probes for the *CYC2* clade genes *2A*, *2B*, *2C*, and *2D* were amplified using the following general program: 95°C/5 min, 35 cycles with 95°C/30 s, annealing temperatures for 1 min, 72°C/2 min, and a final extension of 72°C/12 min (**Table S10**). A sense probe was amplified with the 2b set of primers and used in further analyses. Antisense and sense probes were tested with M13 primers in combination with the *CYC2* primers. Probes were cloned into pBluescript II SK and linearized with BamHI before digoxigenin labeling (Coen et al., 1990), which was performed with anti-digoxigenin-AP, Fab fragments, T7 RNA polymerase and deoxynucleoside triphosphates (Roche). Tissue pretreatment, hybridization, washing and antibody staining steps followed Coen et al. (1990) and Fobert et al. (1994). The reaction to visualize the hybridized probes was incubated for 24–48 h at room temperature (~23°C). Sections were mounted with DePeX mounting medium (Serva) and observed and photographed using a Leica DMR microscope with an Olympus DP70 camera. Images were edited and organized in Adobe Photoshop CS4.

RESULTS

Phylogenetic Reconstruction Reveals Eight *CYC/TB1* Gene Lineages in Asteraceae

The *CYC/TB1* phylogeny reconstruction reveals new and previously identified gene lineages in Asteraceae (**Figure 1A**). We define a lineage as a group formed by genes from different Asteraceae species representing an orthologous gene set. Although the identified *CYC/TB1* lineages are unevenly sampled for all the Asteraceae species, each of them is congruent with the species phylogeny (**Figure 1**). For tribe Anthemideae, we identified cDNAs from 10 putative *CYC/TB1* genes (**Figure 1**) from *A. clavatus* (*AcCYC1a, AcCYC1b, AcCYC2a, AcCYC2a1, AcCYC2b, AcCYC2c, AcCYC2c1, AcCYC2d, AcCYC3a,* and *AcCYC3b*), three from *A. valentinus* (*AvCYC2b, AcCYC2c1,* and *AvCYC2d*), four from *Matricaria aurea* (*MaCYC2a1, MaCYC2c, MaCYC2c1,* and *MACYC2d*), and three from *M. chamomilla* (*McCYC2b, McCYC2c1,* and *McCYC2d*). We also identified several allelic variants for each gene (**Table S1, Figure 1**). The coding sequences (CDS) range between 783 and 900 bps. In the Bayesian and ML analyses reconstructed from the amino acid and nucleotide data, eight main lineages were recovered (**Figure 1A**): two formed by "CYC1-type" genes (*CYC1a, CYC1b*), four by *CYC2* genes (*CYC2a, CYC2b, CYC2c, CYC2d*) and two by *CYC3* genes (*CYC3a, CYC3b*). *CYC2a1* and *CYC2c1* are nested lineages in *CYC2a* and *CYC2c*, respectively. *CYC2a1* represents a gene diversification (**Figure 1**) congruent with the genera phylogeny. The *CYC2c1* genes display diversification of *Anacyclus* and *Matricaria* genes only (**Figure 1**). The *CYC2a* lineage involves *Gerbera* and *Anacyclus* genes, whereas in *CYC2c*, genes from different species display a diversification congruent with the species phylogeny (**Figure 1A**). *CYC2b* and *CYC2d* represent relatively well-sampled gene lineages congruent with the species phylogeny.

CYC2 and *CYC3* genes form monophyletic groups, but their relationship with other clades is unresolved or poorly supported (**Figure 1A, Table S11**). By contrast, genes formerly identified as "*CYC1*" do not form a single clade. Some "*CYC1*-like" genes, including Solanaceae and the Asteraceae 1A, 1B lineages, are more closely related to *CYC2* genes than to other "*CYC1*-like" eudicot loci (**Figure S1**, Tree V). In all topologies, the earliest divergent *CYC2* genes are those from Adoxaceae, Brassicaceae, Caprifoliaceae, Gesneriaceae, Goodeniaceae, Plantaginaceae, Solanaceae, and Vitaceae (**Figure 1**). The remaining *CYC2* members form a well-supported clade with genes from Asteraceae and Calyceraceae (**Table S11**). Hereafter, this clade will be called the Ast/Cal clade (**Figure 1A**). A close relationship between *CYC2b* and *CYC2c* is seen in all resultant trees excepting tree V from the GARLI analysis (**Figure S1D**). All *CYC2* genes from Asteraceae are placed in the CYC2a–CYC2d lineages except some sequences from *Dasyphyllum, Gerbera* and *Helianthus*, which display unstable locations when different topologies are compared (**Figure 1, Figure S1, Table S3**).

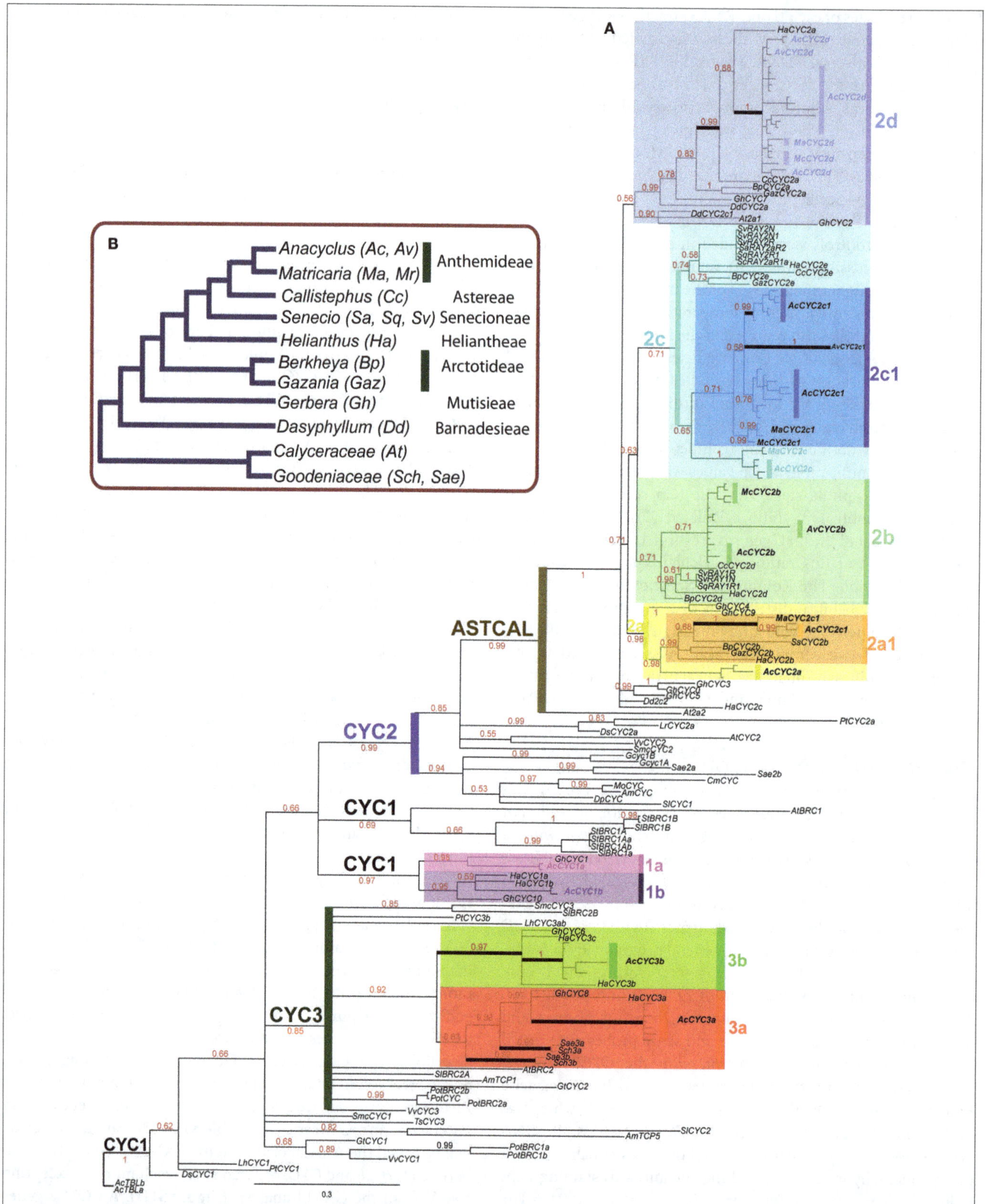

FIGURE 1 | (A) Majority rule consensus tree from Bayesian analysis of the nucleotide dataset with excluded third codon positions (tree II) with the main *CYC/TB1* gene lineages identified by colored boxes. Genes from Anthemideae are in bold. Branches affected by episodes of positive selection are indicated by black thick lines. Numbers in red represent posterior probabilities above 0.50. **(B)** Summary tree of the phylogeny of Asteraceae (modified from Panero et al., 2014) illustrating the relationships among the genera included in this study and their corresponding tribes.

CYC/TB1 Genes Display a Pervasive Purifying Selection with Bursts of Episodic Positive Selection

We tested whether the *CYC/TB1* genes included in the phylogeny have been under positive selection. After rejection of recombination in the *CYC/TB1* complete nucleotide matrix by RDP analysis ($P = 0.05$), the dataset was evaluated by MEME and FEL. The analysis with MEME, with higher log-likelihood values than FEL, suggested episodic positive selection for 42 codons (**Table S12**). The output from FEL indicated negative and neutral evolution for 100 and 146 codons, respectively. Neither MEME nor FEL suggested evidence of pervasive positive selection.

Although most amino acid positions with episodic positive selection within the TCP domain lie in the basic domain and in the loop, additional sites were detected within helix 1 and adjacent to helix 2. The maximum likelihood estimate (MLE) of the synonymous rate α is always higher than the non-synonymous rate $\beta-$ ($\beta \leq \alpha$) except in a few conservative sites ($\alpha = \beta-$) between TCP and ECE domains (**Table S12**). On the other hand, identified sites with high selective constrain ($\beta- = 0$) occur in the TCP domain and outside the ECE and R domains. The proportion of the branches evolving at unconstrained non-synonymous rate $\beta+$ is always small ($q+ < 38\%$) in comparison with branches where the synonymous substitutions prevail ($q- > 63\%$). Examination of the magnitude of the Empirical Bayes Factor (EBF) and the single nucleotide substitutions on different branches of the MEME output trees (not shown) revealed that several clades were affected by episodic positive selection (**Figure 1**). From the identified *CYC/TB1* lineages, the CYC3b clade is the only one affected by episodic positive selection just before its ortholog diversification. Although there is a pervasive purifying selection trend in the *CYC/TB1* genes here analyzed, there are episodes of positive selection not associated with the diversification of the main orthologous gene sets here identified, excepting *CYC3b*.

Inference of Diversification Rate Shifts

The phylogenetic data coupled with the BAMM model is designed to automatically detect changes of speciation rates. The BAMM analyses converged well as indicated by high ESS values (ESS log-likelihood = 1244.05, ESS number of shifts = 1333.27). BAMM failed to detect any significant rate-shift configuration associated with *CYC* lineage diversification, and the best-fit model to the phylogeny was one involving a homogeneous process of near constant per-lineage diversification rates except for lineage 2c1 restricted to *Anacyclus* (**Figure S2**). Running the analyses with the different prior settings did not change the overall pattern.

AcCYC2b, *AcCYC2c*, and *AcCYC2d* Genes Are Highly Expressed in Wild but Not in Trumpet Ray Flowers

There was little or no expression of CYC2 genes in vegetative tissues compared with flowers or inflorescences (**Figure 2**), except for *AcCYC2a* gene expression in roots and leaves of trumpet individuals. In ray and disc flowers, the expression of *AcCYC2a* was not different between wild and trumpet individuals (**Figures 2A–C**). *AcCYC2a* was expressed at slightly higher levels in young capitula (Cap1) and ray flowers than in mature capitula (Cap2) and disc flowers (**Figure 2A**). Expression of *AcCYC2a* in wild individuals (not shown) was high (97.45%) in young ray flowers with unexpanded ligules (ca. 0.5-cm-long). In addition, the relative expression of *AcCYC2b, AcCYC2c,* and *AcCYC2d* genes was much higher in ray flowers of the wild individuals than in other tissues (**Figures 2D–F**). In trumpet individuals, the expression was below 20% in young capitula (**Figures 2D–E**), mature capitula (**Figure 2E**), rays (**Figure 2D**), and disc flowers (**Figures 2E–F**). The average target EexpΔCt/ control EexpΔCt results for each gene follow a normal distribution according to the Kolmogorov-Smirnov test ($\alpha = 0.05$, **Table S11**). The *t*-test indicates that the expression of the *AcCYC2* genes in wild and trumpet is significantly different for all genes ($\alpha = 0.05$, **Table S12**).

Pattern of *AcCYC2* Gene Expression in Early Flower Development

To investigate the mRNA distribution of *AcCYC2* genes during capitulum and early flower development we carried out *in situ* hybridizations with digoxigenin labeled RNA probes complementary to these genes. *AcCYC2b* and *AcCYC2c* transcripts could not be detected in our experiments. In contrast, *AcCYC2d* was strongly expressed in young disc flower meristems and during floral organ initiation (**Figures 3A,B**). *AcCYC2d* mRNA also accumulated in young developing stamens and ovules (**Figures 3A,C**). Likewise, *AcCYC2a* transcripts were detectable in developing disc flowers, both in the developing stamens and the ovules (**Figures 3D,E**). At this floral stage *AcCYC2d* and *AcCYC2a* signals were clearly excluded from developing corolla lobes (**Figures 3A,D**). Sense probes of these genes gave no detectable signal in sections of similar tissues (**Figure S3**).

DISCUSSION

CYC/TB1 Diversification in Asteraceae Reflects Species Phylogeny

The phylogenetic reconstruction shown in this study (**Figure 1**) is congruent with previous partial analyses of *CYC/TB1*-like genes of Asteraceae (Chapman et al., 2008, 2012; Kim et al., 2008; Tähtiharju et al., 2012). The CYC2 orthologous sets proposed by Tähtiharju et al. (2012) correspond to our 2a–2d lineages, with the difference that we found two lineages within 2a (2a1 and 2a2) and a new diversification within Anthemideae (2c1 and 2c2). Within CYC3, we also identified the two orthologous groups 3a and 3b reported by Tähtiharju et al. (2012), and in CYC1 we inferred the lineages 1a and 1b. Although congruence between *CYC-like* genes and species trees is not easy to interpret (e.g., in Dipsacaceae; Carlson et al., 2011), our gene tree reconstruction for each of the *CYC* lineages is broadly consistent with the species tree phylogeny (Panero and Funk, 2008; **Figure 1B**). For example, in the lineage 2b, the topology of the genes from [(*Berkheya*

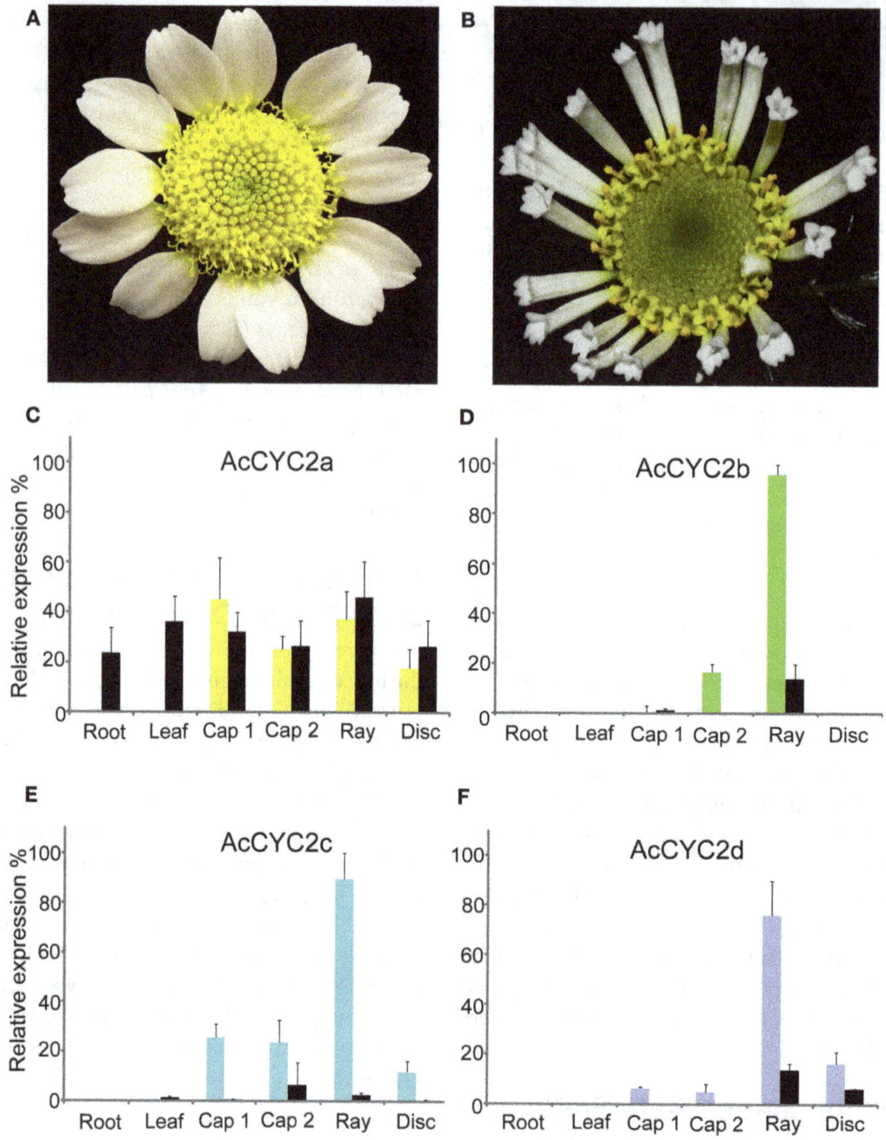

FIGURE 2 | Inflorescences of wild (A) and trumpet **(B)** individuals of *Anacyclus clavatus*. **(C–F)** Relative expression levels of *A. clavatus* genes *AcCYC2a–AcCYC2d* in vegetative and reproductive tissues of wild (colored bars) and trumpet (black bars) individuals. Bars represent relative differences in gene expression level in the tested tissues and are the average of three biological replicates. Error bars show the standard deviation. Cap 1, capitulum stage 1; Cap 2, capitulum stage 2; Ray, peripheric ray flower (wild type, bilateral) or tubular ("trumpet-type," actinomorphic); Disc, disc flowers. Photographs by M.A. Bello **(A)** and R. Riina **(B)**.

(*Helianthus* (*Senecio*+*Callistephus*))) (*Anacyclus*+*Matricaria*)] is congruent with the species phylogeny of Asteraceae (**Figure 1B**). The same consistency is maintained in lineages 1b, 2a1, 2a2, 2b, 2c, and most of 2d, where the only incongruence with the species tree is the closer relationship of *HaCYC2a* from *Helianthus* to *Anthemideae* (*AcCYC2d, AvCYC2d, MaCYC2d,* and *McCYC2d*) rather than to *Callisthephus* (*CcCYC2a*) genes (**Figure 1**).

Despite this general congruence, the distribution of the *CYC2* genes indicates specific gene gain/loss events in relatively well-sampled genera, such as *Gerbera, Helianthus* and *Senecio*. Tracking the *CYC2* paralogous genes of *Anacyclus* and *Matricaria*, we found that they are distributed in six gene clades, whereas *Gerbera* genes are absent from 2a1, 2b, and 2c clades (however, there is a phylogenetically unstable clade composed of *GhCYC0* and *GhCYC3* with an unresolved position with the CYC2 lineages, which could be potentially included in any of the CYC2 lineages; **Figure 1**, **Figure S1**). *Helianthus* genes are absent from the 2a1 clade and *HaCYC2c* relationships are unresolved, while *Senecio RAY1* and *RAY2* are present only in the 2b and 2c clades. Aside from the obvious lack of gene and species sampling that can alter the *CYC/TB1* diversification pattern, it would be important to test whether the differential gene diversification of the species is linked to morphological or

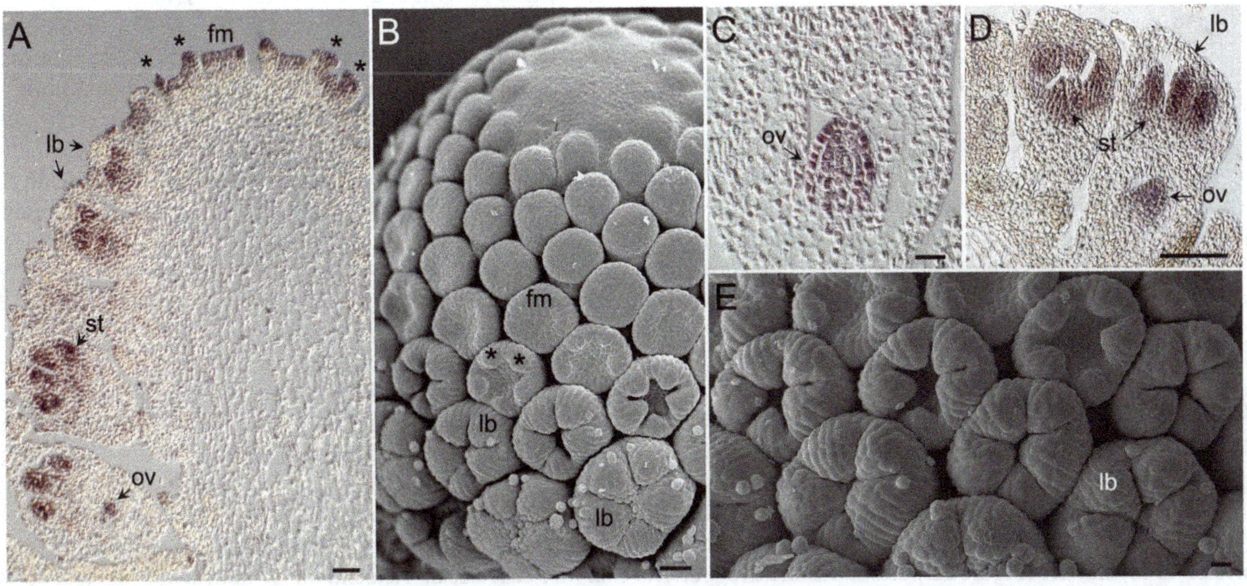

FIGURE 3 | Morphology of *Anacyclus* capitulum and florets, and tissue-specific expression of *AcCYC2* genes during capitulum development. (A) Section of a Anacyclus capitulum hybridized with a digoxigenin-labeled probe complementary to *AcCYC2d*. Youngest flower meristems (fm) can be seen at the top, older flowers to the left and right. *AcCYC2d* transcripts accumulate in young flower meristems and in meristems that are initiating corolla lobe primordia (*). In older flowers mRNA is detectable in developing stamens (st) and ovules (ov). No signal is detectable in developing corolla lobes (lb). **(B)** Scanning electron microscopy (SEM) image of a capitulum of an age comparable to that in **(A)**. **(C)** Close up of an ovule displaying *AcCYC2d* expression. **(D)** Young disc floret showing *AcCYC2a* mRNA signal in stamens and ovule but not in corolla lobes. **(E)** SEM image of disc flowers comparable to those in **(D)**.

functional evolutionary transitions. Moreover, the orthologous *CYC2* genes in *Helianthus* and *Gerbera* do not necessarily share the same function (Broholm et al., 2014), which might allow similar gene repertories shaping different inflorescence morphologies.

The maintenance of each *CYC2* paralog in different species of Asteraceae suggests that they existed before the species diversification had taken place in this family. This recruitment and maintenance of CYC2 function along the history of Asteraceae results in independent evolution of an adaptive trait, the heterogamous capitula (Chapman et al., 2012; Hileman, 2014). The convergent headed inflorescence in Asteraceae and Dipsacaceae could be the consequence of similar diversification patterns of *CYC1*- and *CYC2*-like genes in both families (Howarth and Donoghue, 2005; Carlson et al., 2011; Specht and Howarth, 2014). Although the *CYC/TB1* phylogenetic pattern is not a consequence of differential rate shifts between the inferred CYC lineages according to the BAMM analyses performed here, whole genome duplication events in Asteraceae occurring in the last 40 million years (Barker et al., 2008, 2015) could be responsible for the observed diversity of *CYC/TB1* genes.

Now that several *CYC/TB1* genes isolated from separate studies consistently fit into specific CYC lineages in a comprehensive phylogeny (**Figure 1A**), it will be convenient from here onto add and assign newly isolated genes to this framework. It will encourage an appropriate association of new *CYC/TB1* isolated genes with its phylogenetic origin and will allow a more consistent classification.

CYC/TB1 Proteins Evolve under Purifying and Episodic Positive Selection

Analysis of our *CYC/TB1* gene dataset suggests a pervasive purifying selection with bursts of episodic positive selection, where a very small proportion of sites evolve at unconstrained non-synonymous rate ($q^+ < 38\%$; **Table S12**). Chapman et al. (2008) found that the per-site frequency of synonymous substitutions was saturated on many internal branches of *Helianthus CYC/TB1* genes and that TCP and R domain were evolving under strong purifying selection. Similarly, *CYC/TB1* genes from Antirrhineae were also subjected to strong purifying selection (Hileman and Baum, 2003). Even in recent duplicates of *RAY2* in *Senecio vulgaris*, there is no evidence of positive selection that justifies their divergence (Chapman and Abbott, 2009). In terms of events of episodic codon selection associated with the stem/crown clades of the main CYC lineages identified in this study, only one codon has episodic positive selection at the stem node of the lineage CYC3b (**Figure 1**). For the remaining main CYC lineages, there were no identified episodes of positive selection associated with their origin or diversification, suggesting that selection changes are not important performers of the *CYC/TB1* main phylogenetic patterns. Although some minor clades seem to be diversified after the positive selection (**Figure 1**), it might be affecting the quaternary rather than the primary/secondary protein structure used for phylogenetic reconstruction. With the lack of a significant difference between the diversification rates in the CYC/TB1 gene lineages suggested by our BAMM analysis, it seems that it is a more uniform and recent pattern of evolution of these genes.

In cases, such as the bird toll-like receptors (Grueber et al., 2014) and the eudicot X-intrinsic proteins (Venkatesh et al., 2015), there is a similar pattern of predominant purifying selection with rounds of episodic positive selection as inferred here for the *CYC/TB1* genes. The high constraints imposed by the purifying selection of *CYC/TB1* proteins could be maintaining their general patterns/functions conserved along the eudicots, whereas the episodic positive selection might allow a subtle modulation of protein-protein interactions, such as binding regulation or protein differential heteromeric combinations (see e.g., differential capacity of dimerization of *CYC/TB1* proteins in *Gerbera* and *Helianthus* in Tähtiharju et al., 2012).

Expression Patterns of *CYC/TB1* Are Similar in Asteraceae Orthologs

Comparing the *CYC/TB1* genes of *Anacyclus clavatus* with their identified orthologs, the expression patterns are similar. For example, *AcCYC2b*, detected in young capitula and highly expressed in ray flowers (**Figure 2D**), lies in the orthologous set of *Senecio RAY1* and *HaCYC2d* (lineage CYC 2b; **Figure 1**). *RAY1* is expressed in young inflorescences in the peripheral area of the ray floral meristem in radiate and non-radiate capitula of *Senecio* (see Figure 2 in Kim et al., 2008) and *HaCYC2d* is one of the strongest candidates for conferring ray flower identity in *Helianthus* (Tähtiharju et al., 2012). *AcCYC2d* has a similar expression profile to *AcCYC2b* in qPCR analyses (**Figures 2F**), and is orthologous to *HaCYC2a* and *GhCYC7* (lineage CYC 2d; **Figure 1**). *GhCYC7* and *HaCYC2a* are expressed in different tissues, but *GhCYC7* appears in earlier stages of ray and trans flowers, similar to *HaCYC2a* in ray flowers (Chapman et al., 2008; Tähtiharju et al., 2012; Juntheikki-Palovaara et al., 2014). Also, *AcCYC2d* (**Figure 3A**) and *GhCYC7* are expressed in early stamen primordia of disc and ray flowers (Juntheikki-Palovaara et al., 2014).

Nevertheless, in the lineages CYC2c and CYC2a (**Figure 1A, Figure S4**), the expression pattern of orthologous genes is not as similar as in *CYC2b* and *CYC2d*. In CYC2c, the orthologous *AcCYC2c, RAY2* and *HaCYC2e* genes are highly expressed in young and mature ray flowers (Kim et al., 2008; Tähtiharju et al., 2012) but *HaCYC2e* appears widely expressed in several tissues in PCR assays (Chapman et al., 2008). In the lineage CYC2a, the genes *AcCYC2a, GhCYC4, GhCYC9*, and *HaCYC2b* are expressed in different tissues, but expression is nonetheless higher in ray flowers of *Gerbera* and *Helianthus* (**Figure 3C**; Chapman et al., 2008; Tähtiharju et al., 2012). The expression of *AcCYC2a* and *HaCYC2b* seems not affected when actinomorphic tubular ray flowers are formed in the trumpet individual (**Figure 2C**) and the tubular mutants of *Helianthus*, respectively (see Figure 2B in Chapman et al., 2012). Therefore, despite the fact that in Asteraceae the *CYC2b* and *CYC2d* orthologous genes display similar expression patterns, it is not always possible to predict a particular CYC/*TB1* gene expression pattern from the phylogenetic framework.

In the case of genes with unstable phylogenetic positions, such as *HaCYC2c, GhCYC2, GhCYC3*, and *GhCYC5* lying outside the CYC2 lineages 2a–2d (**Figure 1, Figure S1**), there are redundant expression patterns and multiple functions (**Figure S4**). Whereas, *GhCYC5* seems to be involved in the control of the flower density and in floral organ fusion, *GhCYC2* is expressed in the dorsal part of the ray flowers, reproductive whorls, the ligule and the perianth throat of ray flowers (Broholm et al., 2008; Tähtiharju et al., 2012; Juntheikki-Palovaara et al., 2014). *GhCYC3* and *HaCYC2c* are crucial for the ray flower identity and are expressed in meristem, perianth and ovules of ray flowers (Tähtiharju et al., 2012; Chapman et al., 2008). Juntheikki-Palovaara et al. (2014) suggest that redundancy of the *CYC2* genes in *Gerbera* reflect a functional specificity for the CYC2 proteins obtained by the formation of specific protein complexes. In Asteraceae, the maintenance of the inflorescence unit may require a cross regulation between the *CYC2* genes from different lineages, analogous to the interactions of *CYC2* genes identified in *Primulina heterotricha* from Gesneriaceae (Gao et al., 2008; Yang et al., 2012).

A gradient of expression of the *CYC/TB1* genes occurs in pseudanthial structures bearing different morphologies (e.g., in the Myrtacean *Actinodium cunninghamii* where a cluster of fertile actinomorphic flowers are surrounded by ray-shaped branched shoots; Claßen-Bockhoff et al., 2013). This pattern is coincident with the centripetal gradient of floral morphology of the Asteracean inflorescence (Harris, 1999; Citerne et al., 2010). Aside from *HaCYC2b* in wild type *Helianthus* (Chapman et al., 2012), the *CYC2* genes of wild *Anacyclus* (*AcCYC*), *Gerbera* (*GhCYC*), and *Helianthus* (*HaCYC*) are usually highly expressed in the zygomorphic ray flowers relative to the disc flowers (**Figure 3**; Chapman et al., 2012; Tähtiharju et al., 2012). Expression patterns in *Helianthus* mutants agree with this general Asteraceae profile (Berti et al., 2005; Fambrini et al., 2006, 2011; Chapman et al., 2012). The double flowered mutant (*dbl*), with disc flowers displaying bilateral ray-like corollas, expresses *HaCYC2c* ectopically, an important loci for the establishment of ray flower identity. On the other hand, in the tubular-rayed (*tub*) mutants with ray flowers displaying tubular actinomorphic corollas (similar to the trumpet phenotype in *Anacyclus*; **Figure 2B**), *HaCYC2c* is expressed at lower levels due to the presence of transposable elements. Although we cannot suggest a direct ortholog gene of *HaCYC2c* in Anthemideae due its phylogenetic unstable position (**Figure 1, Figure S1**), qPCR analysis in *Anacyclus* suggests a lower expression of the *CYC2* genes *AcCYC2b, AcCYC2c* and *AcCYC2d* in the actinomorphic ray flowers of the trumpets (**Figures 2B–D**).

Our results support the role of CYC 2 genes in the evolution of Asteraceae flower morphological diversity and illustrate their evolution, diversification, and expression patterns in *Anacyclus*. From an evolutionary perspective, the phylogenetic analyses show that CYC2 gene family has diversified in Asteraceae into four main paralogs, which has been accompanied by an increased structural and functional complexity in inflorescences across the different lineages (Chapman et al., 2008; Tähtiharju et al., 2012). The comparison of gene expression analyses in CYC paralogs and their phylogenetic relationship suggests that different Asteraceae lineages have mostly conserved their roles in determining floral symmetry (Garcês et al., 2016). However, this work also confirms previous evidence proposed by Fambrini and Pugliesi (2017) for a consistent functional recruitment of CYC2 genes in the development of microspores (pollen) and macrospores (ovule) in female and bisexual flowers of the capitulum. This observation opens a new field for the study of the involvement of CYC2 genes in the evolution of sexual systems in Asteraceae.

AUTHOR CONTRIBUTIONS

JF, MB, IA, and PC conceived the study. IA did the fieldwork, MB and IA maintained *Anacyclus* living collections, MB performed phylogenetic analyses, MB and PC conducted the *in situ* hybridization, MB and GS carried out the qPCR analysis, JF performed the speciation rate shifts analysis. MB, JF, IA, and PC discussed the results and wrote the manuscript.

ACKNOWLEDGMENTS

We are grateful to F. Durán, A. Gallego, A. Herrero, B. Ríos, Y. Ruiz (RJB-CSIC) and R. Torices (Université de Lausanne) for their contribution during the collection of plant material and CYC sequences of *Anacyclus* and *Matricaria*. R. Riina photographed the trumpet phenotypes. L. Barrios (CTI-CSIC) provided helpful advice on qPCR statistical analysis. We particularly acknowledge support from to F. Chevalier, I. Domínguez, E. Gonzalez-Grandío and M. Nicolas (CNB-CSIC) for help with *in situ* hybridization and their valuable comments. We thank K. McCreath for revision of the English language. This work was supported by grants from the Spanish Ministry of Economy and Competitivness, Plan Estatal de I+D+I and European Social Fund to JF, IA, and PC (CGL2007-66516, CGL2013-49097-C2-2-P and BIO2014-57011-R), "Juan de la Cierva" program co-financed by the European Social Fund to MB (JCI-2010-07374) and Proyecto Intramural CSIC 201430E023 to JF.

SUPPLEMENTARY MATERIAL

The Supplementary Material for this article can be found online at: http://journal.frontiersin.org/article/10.3389/fpls.2017.00589/full#supplementary-material

Figure S1 | Comparison of the CYC/TB1 summary trees based on nucleotide (A,C,D) and amino acid (B) data sets.

Figure S2 | Maximum clade credibility (MCC) tree from the Bayesian Inference analysis of *CYC/TB1*genes.

Figure S3 | Longitudinal sections of floral tissues of *A. clavatus* hybridized with the sense probe.

Figure S4 | Summary of expression patterns of selected CYC2 genes in the Asteraceae/Calyceraceae clade.

Table S1 | CYC-like genes/clones isolated from *Anacyclus* and *Matricaria* (Asteraceae, Anthemideae).

Table S2 | Primers used for inverse PCR.

Table S3 | Eudicot CYC-like genes included in the phylogenetic analyses.

Table S4 | Datasets, trees and main statistics of the Bayesian analyses.

Table S5 | Annealing temperature, product size and efficiency of the CYC2 gene specific primers used for qPCR in *Anacyclus clavatus* tissues.

Table S6 | Main results of the qPCR essays performed on wild and trumpet individuals of *A. clavatus* following the Pfaffl method.

Table S7 | Kolmogorov-Smirnoff test for probability distribution. $\alpha = 0.05$.

Table S8 | Ratio of expression (target EexpΔCt/ control EexpΔCt) of AcCYC2 genes in trumpet and wild individuals.

Table S9 | *t*-test paired two sample for means of trumpet and wild expression of AcCYC2 genes. $\alpha = 0.05$.

Table S10 | Primers used to amplify the CYC2 gene specific probes for *in situ* hybridization in *Anacyclus clavatus*.

Table S11 | Main lineages reconstructed from the different Bayesian and ML analyses performed and their supports.

Table S12 | Forty two *CYC/TB1* codons with significant evidence of positive selection according to the Mixed Effects Model of Episodic Selection (MEME) method.

REFERENCES

Aguilar-Martínez, J. A., Poza-Carrión, C., and Cubas, P. (2007). *Arabidopsis BRANCHED1* acts as an integrator of branching signals within axillary buds. *Plant Cell* 19, 458–472. doi: 10.1105/tpc.106.048934

Barker, M. S., Kane, N. C., Matvienko, M., Kozik, A., Michelmore, R. W., Knapp, S. J., et al. (2008). Multiple paleopolyploidizations during the evolution of the Compositae reveal parallel patterns of duplicate gene retention after millions of years. *Mol. Biol. Evol.* 25, 2445–2455. doi: 10.1093/molbev/msn187

Barker, M. S., Li, Z., Kidder, T. I., Reardon, C. R., Lai, Z., Oliveira, L. O., et al. (2015). Most Compositae (Asteraceae) are descendants of a paleohexaploid and all share a paleotetraploid ancestor with the Calyceraceae. *Am. J. Bot.* 103, 1203–1211. doi: 10.3732/ajb.1600113

Bazinet, A. L., Zwickl, D. J., and Cummings, M. P. (2014). A gateway for phylogenetic analysis powered by grid computing featuring GARLI 2.0. *Syst. Biol.* 63, 812–818. doi: 10.1093/sysbio/syu031

Bello, M. A., Álvarez, I., Torices, R., and Fuertes-Aguilar, J. (2013). Floral development and evolution of capitulum structure in Anacyclus (Anthemideae, Asteraceae). *Ann. Bot.* 112, 1597–1612. doi: 10.1093/aob/mcs301

Berti, F., Fambrini, M., Turi, M., Bertini, D., and Pugliesi, C. (2005). Mutations of corolla symmetry affect carpel and stamen development in *Helianthus annuus*. *Can. J. Bot.* 83, 1065–1072. doi: 10.1139/b05-047

Bremer, K. (1994). *Asteraceae, Cladistics and Classification*. Portland, OR: Timber Press.

Broholm, S. K., Tähtiharju, S., Laitinen, R. A. E., Albert, V. A., Teeri, T. H., and Elomaa, P. (2008). A TCP domain trascript factor controls flower type specification along the radial axis of the *Gerbera* (Asteraceae) inflorescence. *Proc. Natl. Acad. Sci. U.S.A.* 105, 9117–9122. doi: 10.1073/pnas.0801359105

Broholm, S. K., Teeri, T. H., and Elomaa, P. (2014). "Molecular control of inflorescence development in Asteraceae," in *Advances in Botanical Research: The Molecular Genetics of Floral Transition and Flower Development*, ed F. Fornara (Oxford: Academic Press), 297–333.

Busch, A., and Zachgo, S. (2007). Control of corolla monosymmetry in the *Brassicaceae Iberis* amara. *Proc. Natl. Acad. Sci. U.S.A.* 16, 16714–16719. doi: 10.1073/pnas.0705338104

Busch, A., and Zachgo, S. (2009). Flower symmetry evolution: towards understanding the abominable mystery of angiosperm radiation. *Bioessays* 31, 1181–1190. doi: 10.1002/bies.200900081

Carlson, S. E., Howarth, D. G., and Donoghue, M. J. (2011). Diversification of CYCLOIDEA-like genes in *Dipsacaceae* (Dipsacales): implications for the evolution of capitulum inflorescences. *BMC Evol. Biol.* 11:325. doi: 10.1186/1471-2148-11-325

Carolin, R. C. (2007). "Goodeniaceae," in *The Families and Genera of Vascular Plants*, eds J. W. Kadereit, and C. Jeffrey (Berlin; Heidelberg: Springer-Verlag), 589–598.

Chapman, M. A., and Abbott, R. J. (2009). Introgression of fitness genes across a ploidy barrier. *New Phytol.* 186, 63–71. doi: 10.1111/j.1469-8137.2009.03091.x

Chapman, M. A., Leebens-Mack, J. H., and Burke, J. M. (2008). Positive selection and expression divergence following gene duplication in the sunflower CYCLOIDEA gene family. *Mol. Biol. Evol.* 25, 1260–1273. doi: 10.1093/molbev/msn001

Chapman, M. A., Tang, S., Draeger, D., Nambeesan, S., Shaffer, H., Barb, J. G., et al. (2012). Genetic analysis of floral symmetry in Van Gogh's sunflowers reveals independent recruitment of CYCLOIDEA genes in the Asteraceae. *PLoS Genet.* 8:e1002628. doi: 10.1371/journal.pgen.1002628

Citerne, H., Jabbour, F., Nadot, S., and Damerval, C. (2010). The evolution of floral symmetry. *Adv. Bot. Res.* 54, 85–137. doi: 10.1016/S0065-2296(10)54003-5

Citerne, H., Le Guilloux, M., Sannier, J., Nadot, S., and Damerval, C. (2013). Combining phylogenetic and syntenic analyses for understanding the evolution of TCP ECE genes in eudicots. *PLoS ONE* 8:e74803. doi: 10.1371/journal.pone.0074803

Claßen-Bockhoff, R., Ruonala, R., Bull-Hereñu, K., Marchant, N., and Albert, V. A. (2013). The unique pseudanthium of Actinodium (Myrtaceae)-morphological reinvestigation and possible regulation by CYCLOIDEA-like genes. *Evodevo* 4:8. doi: 10.1186/2041-9139-4-8

Coen, E. S., Romero, J. M., Doyle, S., Elliott, R., Murphy, G., and Carpenter, R. (1990). Floricaula: a homeotic gene required for flower development in *Antirrhinum majus*. *Cell* 63, 1311–1322. doi: 10.1016/0092-8674(90)90426-F

Costa, M. M., Fox, S., Hanna, A. I., Baxter, C., and Coen, E. (2005). Evolution of regulatory interactions controlling floral asymmetry. *Development* 132, 5093–5101. doi: 10.1242/dev.02085

Cubas, P., Lauter, N., Doebley, J., and Coen, E. (1999). The TCP domain: a motif found in proteins regulating plant growth and development. *Plant J.* 18, 215–222. doi: 10.1046/j.1365-313X.1999.00444.x

Darriba, D., Taboada, G. L., Doallo, R., and Posada, D. (2011). ProtTest 3: fast selection of best-fit models of protein evolution. *Bioinformatics* 27, 1164–1165. doi: 10.1093/bioinformatics/btr088

Darriba, D., Taboada, G.-L., Doallo, R., and Posada, D. (2012). jModelTest 2: more models, new heuristics and parallel computing. *Nat. Methods* 9, 772. doi: 10.1038/nmeth.2109

Delport, W., Poon, A. F., Frost, S. D., and Kosakovsky Pond, S. L. (2010). Datamonkey 2010: a suite of phylogenetic analysis tools for evolutionary biology. *Bioinformatics* 26, 2455–2457. doi: 10.1093/bioinformatics/btq429

Doyle, J., and Doyle, J. (1987). A rapid DNA isolation procedure for small quantities of fresh leaf tissue. *Phytochem. Bull.* 19, 11–15.

Ehrendorfer, F., Schweizer, D., Greger, H., and Humphries, C. (1977). Chromosome banding and synthetic systematics in *Anacyclus* (Asteraceae-Anthemideae). *Taxon* 26, 387–394. doi: 10.2307/1220037

Fambrini, M., Bertini, D., Cionini, G., Michelotti, V., and Pugliesi, C. (2006). "Mutations affecting corolla symmetry in sunflower," in *Floriculture, Ornamental and Plant Biotechnology*, ed J. A. Texeira da Silva (London: Global Science Books, Ltd.), 61–70.

Fambrini, M., and Pugliesi, C. (2017). CYCLOIDEA 2 clade genes: key players in the control of floral symmetry, inflorescence architecture, and reproductive organ development. *Plant Mol. Biol. Rep.* 35, 20–36. doi: 10.1007/s11105-016-1005-z

Fambrini, M., Salvini, M., and Pugliesi, C. (2011). A transposon-mediate inactivation of a CYCLOIDEA-like gene originates polysymmetric and androgynous ray flowers in *Helianthus annuus*. *Genetica* 139, 1521–1529. doi: 10.1007/s10709-012-9652-y

Finlayson, S. A. (2007). *Arabidopsis TEOSINTE BRANCHED1-LIKE 1* regulates axillary bud outgrowth and is homologous to monocot *TEOSINTE BRANCHED1*. *Plant Cell Physiol.* 48, 667–677. doi: 10.1093/pcp/pcm044

Fobert, P. R., Coen, E. S., Murphy, G. J., and Doonan, J. H. (1994). Patterns of cell divisions revealed by transcriptional regulation of genes during the cell cycle in plants. *EMBO J.* 13, 616–624.

Gao, Q., Tao, J. H., Yan, D., Wang, Y. Z., and Li, Z. Y. (2008). Expression differentiation of CYC-like floral symmetry genes correlated with their protein sequence divergence in *Chirita heterotricha* (Gesneriaceae). *Dev. Genes Evol.* 218, 341–351. doi: 10.1007/s00427-008-0227-y

Garcês, H. M. P., Spencer, V. M. R., and Kim, M. (2016). Control of floret symmetry by RAY3, SvDIV1B, and SvRAD in the capitulum of *Senecio vulgaris*. *Plant Physiol.* 171, 2055–2068. doi: 10.1104/pp.16.00395

Grueber, C. E., Wallis, G. P., and Jamieson, I. G. (2014). Episodic positive selection in the evolution of avian Toll-like receptor innate immunity genes. *PLoS ONE* 9:e89632. doi: 10.1371/journal.pone.0089632

Harris, E. (1999). Capitula in the Asteridae: a widespread and varied phenomenon. *Bot. Rev.* 65, 348–369. doi: 10.1007/BF02857754

Heath, L., van der Walt, E., Varsani, A., and Martin, D. P. (2006). Recombination patterns in aphthoviruses mirror those found in other picornaviruses. *J. Virol.* 80, 11827–11832. doi: 10.1128/JVI.01100-06

Hileman, L. C. (2014). Trends in flower symmetry evolution revealed through phylogenetic and developmental genetic advances. *Philos. Trans. R. Soc. Lond. B Biol. Sci.* 369, 20130348. doi: 10.1098/rstb.2013.0348

Hileman, L. C., and Baum, D. A. (2003). Why Do Paralogs Persist? Molecular Evolution of *CYCLOIDEA* and related floral symmetry genes in Antirrhineae (Veronicaceae). *Mol. Biol. Evol.* 20, 591–600. doi: 10.1093/molbev/msg063

Howarth, D. G., and Donoghue, M. J. (2005). Duplications in *CYC*-like genes form Dispsacales correlate with floral form. *Int. J. Plant Sci.* 166, 357–370. doi: 10.1086/428634

Howarth, D. G., and Donoghue, M. J. (2006). Phylogenetic analysis of the "ECE"' (CYC/TB1) clade reveals duplications predating the core eudicots. *Proc. Natl. Acad. Sci. U.S.A.* 103, 9101–9106. doi: 10.1073/pnas.0602827103

Humphries, C. J. (1981). Cytogenetic and cladistic studies in *Anacyclus* (Compositae: Anthemidae). *Nord. J. Bot.* 1, 83–96. doi: 10.1111/j.1756-1051.1981.tb01038.x

Jackson, D. (1991). "*In-situ* hybridisation in plants," in *Molecular Plant Pathology: A Practical Approach*, eds D. J. Bowles, S. J. Gurr, and M. McPherson (Oxford: Oxford University Press), 163–174.

Jeffrey, C. (1977). "Corolla forms in Compositae-some evolutionary and taxonomic speculations," in *The Biology and Chemistry of the Compositae*, eds V. H. Heywood, J. B. Harborne, and B. L. Turner (London: Academic Press), 111–118.

Juntheikki-Palovaara, I., Tähtiharju, S., Lan, T., Broholm, S. K., Rijpkema, A. S., Ruonala, R., et al. (2014). Functional diversification of duplicated CYC2 clade genes in regulation of inflorescence development in *Gerbera hybrida* (Asteraceae). *Plant J.* 79, 783–796. doi: 10.1111/tpj.12583

Kearse, M., Moir, R., Wilson, A., Stones-Havas, S., Cheung, M., Sturrock, S., et al. (2012). Geneious basic: an integrated and extendable desktop software platform for the organization and analysis of sequence data. *Bioinformatics* 28, 1647–1649. doi: 10.1093/bioinformatics/bts199

Kim, M., Cui, M. L., Cubas,P., Gillies, A., Lee, K., Chapman, M. A. et al. (2008). Regulatory genes control a key morphological and ecological trait transferred between species. *Science* 322, 1116–1119. doi: 10.1126/science.11 64371

Kosakovsky Pond, S. L., Frost, S. D. W., and Muse, S. V. (2005). HyPhy: hypothesis testing using phylogenies *Bioinformatics* 21, 676–679. doi: 10.1093/bioinformatics/bti079

Laitinen, R. A., Broholm, S., Albert, V. A., Teeri, T. H., and Elomaa, P. (2006). Patterns of MADS-box gene expression mark flower-type development in *Gerbera hybrida* (Asteraceae). *BMC Plant Biol.* 6:11. doi: 10.1186/1471-2229-6-11

Luo, D., Carpenter, R., Vincent, C., Copsey, L., and Coen, E. (1996). Origin of floral asymmetry in Antirrhinum. *Nature* 383, 794–799. doi: 10.1038/383794a0

Maddison, W. P., and Maddison, D. R. (2011). *Mesquite: A Modular System for Evolutionary Analysis.* Version 2.75. Available online at: http://mesquiteproject.org.

Martin, D., and Rybicki, E. (2000). RDP: detection of recombination amongst aligned sequences. *Bioinformatics* 16, 562–563. doi: 10.1093/bioinformatics/16.6.562

Martín-Trillo, M., and Cubas, P. (2010). TCP genes: a family snapshot ten years later. *Trends Plant Sci.* 15, 31–39. doi: 10.1016/j.tplants.2009.11.003

Miller, M. A., Pfeiffer, W., and Schwartz, T. (2010). "Creating the CIPRES Science Gateway for inference of large phylogenetic trees," in *Proceedings of the Gateway Computing Environments Workshop (GCE), 14 Nov. 2010.* (New Orleans, LA), 1–8.

Murrel, B., Wertheim, J. O., Moola, S., Weighill, T., Scheffler, K., and Pond, S. L. K. (2012). Detecting individual sites subject to episodic diversifying selection. *PLoS Genet.* 8:e1002764. doi: 10.1371/journal.pgen.1002764

Navaud, O., Dabos, P., Carnus, E., Tremousaygue, D., and Hervé, C. (2007). TCP Transcription factors predate the emergence of land plants. *J. Mol. Evol.* 65, 23–33. doi: 10.1007/s00239-006-0174-z

Nicolas, M., and Cubas, P. (2015). "The role of TCP transcription factors in shaping flower structure, leaf morphology, and plant architecture," in *Plant Transcription Factors: Evolutionary, Structural and Functional Aspects*, ed D. González (London: Elsevier Inc.), 250–267.

Panero, J. L., Freire, S. E., Espinar, L. A., Crozier, B. S., Barboza, G. E., and Cantero, J. J. (2014). Resolution of deep nodes yields an improved backbone phylogeny and a new basal lineage to study early evolution of Asteraceae. *Mol. Phylogenet. Evol.* 80, 43–53. doi: 10.1016/j.ympev.2014.07.012

Panero, J. L., and Funk, V. A. (2008). The value of sampling anomalous taxa in phylogenetic studies: major clades of the Asteraceae revealed. *Mol. Phylogenet. Evol.* 47, 757–782. doi: 10.1016/j.ympev.2008.02.011

Pfaffl, M. W. (2001). A new mathematical model for relative quantification in real-time RT-PCR. *Nucleic Acids Res.* 29, e45. doi: 10.1093/nar/29.9.e45

Posada, D., and Crandall, K. A. (1998). Modeltest: testing the model of DNA substitution. *Bioinformatics* 14, 817–818. doi: 10.1093/bioinformatics/14.9.817

Pozner, R., Zanotti, C., and Johnson, L. A. (2012). Evolutionary origin of the *Asteraceae capitulum*: insights from the Calyceraceae. *Am. J. Bot.* 99, 1–13. doi: 10.3732/ajb.1100256

Preston, J. C., Hileman, L. C., and Cubas, P. (2011). Reduce, reuse, and recycle: developmental evolution of trait diversification. *Am. J. Bot.* 98, 397–403. doi: 10.3732/ajb.1000279

Rabosky, D. L. (2014). Automatic detection of key innovations, rate shifts, and diversity-dependence on phylogenetic trees. *PLoS ONE* 9:e89543, doi: 10.1371/journal.pone.0089543

Rabosky, D. L., Donnellan, S. C., Grundler, M., and Lovette, I. J. (2014). Analysis and visualization of complex macroevolutionary dynamics: an example from Australian scincid lizards. *Syst. Biol.* 63, 610–627. doi: 10.1093/sysbio/syu025

Ronquist, F., Teslenko, M., van der Mark, P., Ayres, D. L., Darling, A., Hohna, S., et al. (2012). Mr Bayes 3.2: efficient bayesian phylogenetic inference and model choice across a large model space. *Syst. Biol.* 61, 539–542. doi: 10.1093/sysbio/sys029

Smith, M. (1992). Analyzing the mosaic structure of genes. *J. Mol. Evol.* 34, 126–129. doi: 10.1007/BF00182389

Specht, C. D., and Howarth, D. G. (2014). Adaptation in flower form: a comparative evodevo approach. *New Phytol.* 206, 74–90. doi: 10.1111/nph.13198

Stamatakis, A., Hoover, P., and Rougemont, J. (2008). A rapid bootstrap algorithm for the RAxML web servers. *Syst. Biol.* 57, 758–771. doi: 10.1080/10635150802429642

Tähtiharju, S., Rijpkema, A. S., Vetterli, A., Albert, V. A., Teeri, T. H., and Elomaa, P. (2012). Evolution and diversification of the *CYC/TB1* gene family in Asteraceae—a comparative study in *Gerbera* (Mutisieae) and Sunflower (Heliantheae). *Mol. Biol. Evol.* 29, 1155–1166. doi: 10.1093/molbev/msr283

Torices, R., Agudo, A., and Álvarez, I. (2013). Not only size matters: achene morphology affects time of seedling emergence in three heterocarpic species of *Anacyclus* (Anthemideae, Asteraceae). Anales Jard. *Bot. Madrid* 70, 48–55. doi: 10.3989/ajbm.2351

Venkatesh, J., Yu, J. W., Gaston, D., and Park, S. W. (2015). Molecular evolution and functional divergence of X-intrinsic protein genes in plants. *Mol. Genet. Genomics* 290, 443. doi: 10.1007/s00438-014-0927-0

Yang, X., Pang, H. B., Liu, B. L., Qiu, Z. J., Gao, Q., Wei, L., et al. (2012). Evolution of double positive autoregulatory feedback loops in *CYCLOIDEA2* clade genes is associated with the origin of floral zygomorphy. *Plant Cell* 24, 1834–1847. doi: 10.1105/tpc.112.099457

Zhou, X. R., Wang, Y. Z., Smith, J. F., and Chen, R. J. (2008). Altered expression patterns of *TCP* and *MYB* genes relating to the floral developmental transition from initial zygomorphy to actinomorphy in *Bournea* (Gesneriaceae). *New Phytol.* 178, 532–543. doi: 10.1111/j.1469-8137.2008.02384.x

Conflict of Interest Statement: The authors declare that the research was conducted in the absence of any commercial or financial relationships that could be construed as a potential conflict of interest.

4

Transcriptomic Analysis Implies that GA Regulates Sex Expression via Ethylene-Dependent and Ethylene-Independent Pathways in Cucumber (*Cucumis sativus* L.)

*Yan Zhang[1,2], Guiye Zhao[1,2], Yushun Li[1,2], Ning Mo[1,2], Jie Zhang[1,2] and Yan Liang[1,2]**

[1] College of Horticulture, Northwest A&F University, Yangling, China, [2] State Key Laboratory of Crop Stress Biology in Arid Region, Northwest A&F University, Yangling, China

Edited by:
José M. Romero,
University of Seville, Spain

Reviewed by:
Miguel Blazquez,
Spanish National Research Council, Spain
Xiaolan Zhang,
China Agricultural University, China

Correspondence:
Yan Liang
liangyan@nwsuaf.edu.cn

Sex differentiation of flower buds is an important developmental process that directly affects fruit yield of cucumber (*Cucumis sativus* L.). Plant hormones, such as gibberellins (GAs) and ethylene can promote development of male and female flowers, respectively, however, the regulatory mechanisms of GA-induced male flower formation and potential involvement of ethylene in this process still remain unknown. In this study, to unravel the genes and gene networks involved in GA-regulated cucumber sexual development, we performed high throughout RNA-Seq analyses that compared the transcriptomes of shoot tips between GA_3 treated and untreated gynoecious cucumber plants. Results showed that GA_3 application markedly induced male flowers but decreased ethylene production in shoot tips. Furthermore, the transcript levels of *M* (*CsACS2*) gene, ethylene receptor *CsETR1* and some ethylene-responsive transcription factors were dramatically changed after GA_3 treatment, suggesting a potential involvement of ethylene in GA-regulated sex expression of cucumber. Interestingly, GA_3 down-regulated transcript of a C-class floral homeotic gene, *CAG2*, indicating that GA may also influence cucumber sex determination through an ethylene-independent process. These results suggest a novel model for hormone-mediated sex differentiation and provide a theoretical basis for further dissection of the regulatory mechanism of male flower formation in cucumber.

Statement: We reveal that GA can regulate sex expression of cucumber via an ethylene-dependent manner, and the *M* (*CsACS2*), *CsETR1*, and *ERFs* are probably involved in this process. Moreover, *CAG2*, a C-class floral homeotic gene, may also participate in GA-modulated cucumber sex determination, but this pathway is ethylene-independent.

Keywords: cucumber, ethylene, gibberellin, sex expression, transcriptome

---

Abbreviations: ACC, 1-aminocyclopropane-1-carboxylate; DEG, differentially expressed gene; DGE, digital gene expression; FDR, false discovery rate; FID, flame ionization detector; FPKM, fragments per kilobase of transcript sequence per millions base pairs sequenced; GA, gibberellin; GO, gene ontology; qRT-PCR, quantitative real-time PCR.

INTRODUCTION

Cucumber (*Cucumis sativus* L.) is a typical monoecious plant with distinct male and female flowers, and has been served as a model system for studying physiological and molecular aspects of sex determination in plants (Malepszy and Niemirowicz-Szczytt, 1991; Bai and Xu, 2013). During the early stages of cucumber flower development, both stamen primordia and carpel primordia are initiated, however, sex differentiation occurs just after the hermaphroditic stage, subsequently, female or male flower is formed and developed through the selective developmental arrest of stamen or carpel, respectively (Bai et al., 2004).

Sex differentiation in cucumber is mainly determined by *F*, *M*, and *A* genes. Among them, *F* (*CsACS1G*) and *M* (*CsACS2*) genes encoding two ACC synthases (key enzymes in ethylene biosynthetic pathway) govern female sex expression in cucumber, and the *F* gene promotes female flower development (Trebitsh et al., 1997; Mibus and Tatlioglu, 2004; Knopf and Trebitsh, 2006), while the *M* gene inhibits stamen development in flower buds (Yamasaki et al., 2001, 2003; Saito et al., 2007; Li et al., 2009, 2012). In contrast, the *A* gene inhibits female flower development and facilitates male flower formation (Pierce and Wehner, 1990). The interaction of *F*, *M*, and *A* genes eventually determines various sexual phenotypes of cucumber.

In addition to genetic control, sex expression of cucumber can be affected by phytohormones, such as ethylene and GAs. Particularly, ethylene is considered as a potent sex hormone in cucumber that can induce formation of female flowers (Malepszy and Niemirowicz-Szczytt, 1991). Ethylene content in shoot tip of gynoecious cucumber is higher than that of monoecious plant (Rudich et al., 1972; Fujita and Fujieda, 1981; Trebitsh et al., 1987). Treatment with exogenous ethylene or ethylene-releasing reagent can increase the numbers of female and bisexual flowers in monoecious and andromonoecious lines, respectively (MacMurray and Miller, 1968; Iwahori et al., 1969). Until now, the molecular mechanism of ethylene-regulated sex determination of cucumber has been well understood. Except for the *F* and *M* genes, other ethylene biosynthetic genes, such as *CSACO2* and *CSACO3*, which encode ACC oxidases, are also involved in sex expression of cucumber, but the transcript levels of *CSACO2* and *CSACO3* in the shoot tips show a negative correlation with femaleness, indicating an existence of a feedback inhibition mechanism underlying such correlation (Kahana et al., 1999). Overexpression of *CsACO2*, driven by the *AP3* promoter, can arrest the stamen development by inducing chromatin condensation in *Arabidopsis* (Hao et al., 2003; Duan et al., 2008). Moreover, an ethylene receptor, *CsETR1*, has been demonstrated to play a key role in stamen arrest in female cucumber flowers through induction of DNA damage (Wang et al., 2010).

Gibberellins, one class of tetracyclic diterpenoid phytohormones, can promote the male tendency in cucumber. GA production in andromonoecious cucumber in higher than that in gynoecious and monoecious plants (Hemphill et al., 1972). Exogenous GA_3 application can increase the ratio of maleness to femaleness in monoecious cucumber and induce the formation of male flowers in gynoecious plants (Wittwer and Bukovac, 1962; Pike and Peterson, 1969). In addition, GA signaling pathway is involved in stamen and anther development in hermaphroditic plants, such as *Arabidopsis* and rice (*Oryza sativa*) (Cheng et al., 2004; Fleet and Sun, 2005; Aya et al., 2009; Sun, 2010, 2011; Plackett et al., 2011; Song et al., 2013). In this pathway, GA first binds with *GID1* receptor and promotes the interaction between GID1 and DELLA proteins (repressors of GA signaling), leading to a rapid degradation of DELLA proteins by an ubiquitin-proteasome pathway, and the proteolysis of DELLA proteins releases their inhibitory effect on GA action and allows plant growth and development (Fleet and Sun, 2005; Murase et al., 2008; Harberd et al., 2009; Sun, 2010, 2011; Plackett et al., 2014). *GAMYB* is a positive regulator in GA signaling pathway and acts as an important downstream gene of DELLA proteins (Olszewski et al., 2002; Achard et al., 2004; Fleet and Sun, 2005). GA can induce *GAMYB* transcript through degradation of DELLA proteins, resulting in an enhanced flowering and anther development (Achard et al., 2004). In our previous studies, we identified two GA signaling genes, *CsGAIP* and *CsGAMYB1*, which belong to *DELLA* and *GAMYB* family, respectively. Both of them were predominantly expressed in the male specific organs during cucumber flower development. *CsGAIP* can inhibit stamen development through transcriptional repression of B class floral homeotic genes *APETALA3* (*AP3*) and *PISTILLATA* (*PI*) in *Arabidopsis* (Zhang et al., 2014a). However, whether *CsGAIP* is involved in GA-regulated sex determination in cucumber flowers is still unknown. Notably, *CsGAMYB1* can also mediate sex expression of cucumber. Knockdown of *CsGAMYB1* in cucumber results in decreased ratio of nodes with male to female flowers (Zhang et al., 2014b). Despite the current knowledge of GA-regulated sex expression of cucumber, the precise regulatory pathway in this complex process remains elusive.

Although both ethylene and GA can mediate sex expression of cucumber, their regulatory functions appear to be opposite. Atsmon and Tabbak (1979) interpreted that the GA-regulated sex differentiation has no effect on ethylene production, and there is a balance in the content of ethylene and GA in controlling the sex expression of cucumber. However, Yin and Quinn (1995) proposed a "one-hormone hypothesis" which posited that ethylene plays a dominant role in cucumber sex determination and GA may regulate the maleness through inhibiting ethylene production. However, our previous studies demonstrated that GA-*CsGAMYB1* signaling could regulate sex differentiation in cucumber through an ethylene-independent process (Zhang et al., 2014b). Therefore, a potential crosstalk between GA and ethylene pathways that determine sex expression in cucumber still remains unclear.

Besides, members of the MADS-box gene family can also regulate the sexual development in cucumber. *AGAMOUS* (*AG*), the C-class floral homeotic gene, specifies stamen and carpel identity (Lohmann and Weigel, 2002). There are three *AG* homologs in cucumber, *CAG1*, *CAG2*, and *CAG3*, in which *CAG1* and *CAG3* are expressed in both stamen and carpel, while *CAG2* is particularly restricted to the carpel. However, the expression levels of these three genes do not appear to be mediated by ethylene and GA (Kater et al., 1998; Perl-Treves et al., 1998).

Moreover, *ERAF17*, an another MADS-box gene, can be induced by ethylene and may be involved in female flowers formation in cucumber (Ando et al., 2001).

In our study, in order to understand the genes and gene networks that may be involved in GA-modulated cucumber sex determination, we performed RNA-Seq analyses to compare the transcriptomes of shoot apices between GA_3-treated and control gynoecious cucumber plants. GA_3 application induced male flowers but reduced ethylene production in the shoot apices. Notably, GA-regulated sex differentiation was associated with the changes in transcript levels of *M* (*CsACS2*) gene, ethylene receptor *ETR1*, ethylene-responsive transcription factors, and *CAG2* (a C-class floral homeotic gene), suggesting a potential involvement of both ethylene-dependent and -independent processes in GA-mediated cucumber sexual development. Thus, our results built a foundation for dissecting the molecular mechanism of male flower development in cucumber.

MATERIALS AND METHODS

Plant Materials and Growth Conditions

A gynoecious cucumber (*C. sativus* L.) line 13-3B was used in this study. The seeds were germinated on wet filter paper in a Petri dish at 28°C in dark overnight. Then the resulting seedlings were grown in a growth chamber under 16 h/8 h with 25°C/18°C in day/night, respectively. Upon two true-leaf stage, plants were transferred to a greenhouse in the experimental field of the Northwest A&F University. Pest control and water management were carried out according to standard practices.

Exogenous GA_3 Treatment

For male flowers induction in the gynoecious cucumber 13-3B, 1000 ppm GA_3 (dissolved in 0.1% ethanol) or deionized water with 0.1% ethanol (Control) were applied by foliar spray for three times at 7 day intervals, starting when the first true leaf was approximately 2.5 cm in diameter. The sex of the flowers on each node of the main stem was recorded until anthesis of flowers on node 25.

In addition, ethylene production in shoot apices was measured after 7 days of the third GA_3 treatment. And the RNA-Seq analyses were performed in shoot apices from the cucumber plants firstly treated with GA_3 for 6 h, 12 h and the Control, respectively. GA_3 was acquired from Sigma-Aldrich Chemical Co. (Shanghai, China).

Quantification of Ethylene

The ethylene production was measured by gas chromatography as described previously with some modifications (Zhang et al., 2014b). In brief, the excised shoot apices from cucumber plants treated with exogenous GA_3 and the Control were enclosed in 10 mL vessels after weighing and sealed with rubber stoppers. After incubation at 25°C for 16 h, 1 mL of head gas was withdrawn from each vessel using a syringe and injected into a gas chromatograph (GC-9A, Shimadzu, Japan) equipped with a hydrogen FID and an activated alumina column for the measurement of ethylene production. Standard ethylene gas was used for calibrating the instrument. Amount of ethylene was calculated per 1 g fresh weight and per hour.

RNA Extraction and Quality Test

The shoot apices from the gynoecious cucumber plants were collected at 6 and 12 h after the first treatment with GA_3 and the Control. Samples were immediately frozen in liquid nitrogen and stored at −80°C for RNA-Seq analyses. Total RNA was isolated using the RNA extraction kit (Promega, USA). RNA was checked by RNase-free agarose gel electrophoresis to avoid possible degradation and contamination, and then examined using the NanoPhotometer spectrophotometer (IMPLEN, Westlake Village, CA, USA) for RNA purity. RNA concentration and integrity were measured and assessed using the Qubit RNA Assay Kit in Qubit 2.0 Flurometer (Life Technologies, Carlsbad, CA, USA) and RNA Nano 6000 Assay Kit of the Bioanalyzer 2100 system (Agilent Technologies, Santa Clara, CA, USA), respectively.

Digital Gene Expression (DGE) Library Construction and Sequencing

Digital gene expression libraries were constructed using the NEBNext Ultra Directional RNA Library Prep Kit for Illumina (NEB, Ispawich, USA) following instructions of manufacturer and six index codes were added to attribute sequences to various samples (Wang et al., 2009). Briefly, poly (A) mRNA was isolated from 3 μg total RNA using oligo-dT magnetic beads (Life Technologies, Carlsbad, CA, USA), and then broken into short fragments by adding fragmentation buffer. First-strand cDNA was synthesized using random hexamer-primed reverse transcription, followed by the synthesis of the second-strand cDNA using RNase H and DNA polymerase I. After adenylation of the 3′ ends of cDNA fragments, NEBNext adapter oligonucleotides were ligated to prepare for hybridization, and then the cDNA fragments were purified using AMPure XP system (Beckman Coulter, Beverly, MA, USA) to select the fragments of preferentially 150–200 bp in length. The size-selected, adaptor-ligated cDNA fragments were enriched using Phusion High-Fidelity DNA polymerase, Universal PCR primers and Index Primer in the PCR reaction. PCR products were purified with AMPure XP system and library quality was assessed using the Agilent Bioanalyzer 2100 system. At last, the cDNA libraries were sequenced on an Illumina HiSeq 4000 platform using the paired-end technology by Novogene Co. (Beijing, China).

Bioinformatics Analysis of DGE Data

Raw reads were pre-processed to remove low quality sequences (there were more than 50% bases with quality lower than 20 in one sequence), reads with more than 5% N bases (bases unknown) and reads containing adaptor sequences. Then the clean reads were mapped to the cucumber genome (Chinese long) v2[1] using TopHat (Huang et al., 2009; Trapnell et al., 2009), allowing up to one mismatch. Unigenes mapped by at

[1] http://www.icugi.org/

least one read, in at least one sample, were identified for further analysis.

In this study, samples from three treatments (GA 6 h, GA 12 h, and Control) were prepared for genome-wide expression analyses. Two biological replicates were performed for each treatment, and thus six DGE libraries were sequenced. 44.28–60.29 million raw reads from each library were generated. After removal of low-quality tags and adapter sequences, 42.31–57.63 million high-quality clean reads with a total of 6.35–8.64G bases were obtained. Among these clean reads, the percentage of Q20 (base quality more than 20) and GC was 97.28–97.52% and 43.46–43.63%, respectively (**Table 1**). Furthermore, we clustered these clean reads into unique tags, which were mapped to the cucumber genome using TopHat (Huang et al., 2009; Trapnell et al., 2009). About 36.28–49.30 million clean reads (85.21–85.74% of total clean reads) from RNA-Seq data in the six libraries were mapped uniquely to the cucumber genome (**Table 1**).

The R package edgeR used to identify the DGEs (Robinson et al., 2010). The expression level of each gene was calculated and normalized to FPKM. The FDR was used to determine the threshold of the P-value in multiple tests. In our study, the FDR < 0.05 and fold change > 2 were used as significance cut-offs of the expression differences.

Furthermore, GO enrichment analysis of DGEs was performed using the GOseq R package (Young et al., 2010). GO terms with corrected P-value < 0.05 were considered significantly enriched by differential expressed genes.

Quantitative Real-Time PCR (qRT-PCR) Validation

Quantitative Real-Time PCR analyses were performed using the independent cucumber shoot apices in the same time point of GA_3 application as those used for RNA-Seq. Total RNA was isolated using the RNA extraction kit (Promega, USA), and cDNA was synthesized using the PrimeScript RT reagent Kit (TaKaRa, China). qRT-PCR was carried out using SYBR Premix Ex Taq (TaKaRa, China) on an ABI 7500 Real-Time PCR System (Applied Biosystems, USA). The cucumber α-*TUBULIN* (*TUA*) gene was used as an internal control in analyzing gene expression. Three biological replicates were performed for each experiment. The gene specific primers for qRT-PCR are listed in Supplementary Table S5.

RESULTS

Exogenous GA_3 Induces Male Flowers Formation and Inhibits Ethylene Production in Gynoecious Cucumber

To verify the effect of GA on sex expression of cucumber, a gynoecious cucumber line 13-3B was treated with GA_3 and the sex of flowers on each node was recorded until anthesis of the flowers on node 25 of the main stems. As shown in **Figure 1**, there were no male flower nodes in the control plants, however, GA_3 treatment induced male flowers in the gynoecious cucumber line (**Figure 1A**), accounting for 51.2% of nodes with male flowers (**Figure 1B**). Interestingly, formation of male flowers occurred mainly at the lower node positions as compared with the location of female flowers on the main stems (**Figure 1A**). These observations suggested that exogenous GA can promote male flowers formation in cucumber.

It is well known that ethylene can also control sex determination of cucumber (Malepszy and Niemirowicz-Szczytt, 1991; Bai and Xu, 2013). To assess potential involvement of ethylene in GA-regulated sex expression of cucumber, ethylene production in the shoot apices was measured in GA_3-treated and control plants. As shown in **Figure 1C**, ethylene production was significantly decreased after GA_3 treatment, suggesting that ethylene might function as a negative factor in GA-regulated male flower formation in cucumber.

Identification of Differentially Expressed Genes (DEGs) in Shoot Apices from the Gynoecious Cucumber Plants Treated with GA_3 and the Control

To identify the genes and gene networks that are involved in GA-regulated cucumber sex expression, the genome-wide expression analyses were performed to compare the transcriptome profiles of the shoot apices between the gynoecious cucumber plants treated with GA_3 for different time points (6 and 12 h) and the Control through the digital gene expression (DGE) approach (Eveland et al., 2010). Based on deep sequencing, 23,911 unigenes were collected in six libraries. Using fold change > 2 and FDR < 0.05 as the significance cut-offs, we identified 1073 DEGs including 727 up-regulated genes and 346 down-regulated genes after GA_3 treatment for 6 h compared with the Control. And we also found that 1590 genes were differentially expressed, in which 765 genes were up-regulated and 825 genes were down-regulated after GA_3 treatment for 12 h (**Figure 2**; Supplementary Tables S1 and S2). Moreover, 594 DEGs containing 303 up-regulated genes (**Figure 2A**) and 291 down-regulated genes (**Figure 2B**) were shared in the two sets of transcriptome comparisons.

Verification of RNA-Seq Data by Quantitative Real Time RT-PCR Analyses

To validate the DEGs identified by RNA-Seq, we performed quantitative real time RT-PCR (qRT-PCR) assays using the independent cucumber shoot apices in the same time point after GA_3 treatment as those used for RNA-Seq analysis. Twenty DEGs were randomly chosen for qRT-PCR analyses, in which 10 genes including five up-regulated genes and five down-regulated genes were from the set of GA 6 h vs. Control, and the other 10 genes containing five up-regulated genes and five down-regulated genes were from the set of GA 12 h vs. Control. As shown in **Figure 3**, all the 20 genes showed the similar expression patterns in the qRT-PCR analyses as those in the RNA-Seq data, although the particular values of fold-change were different. The Pearson correlation coefficient between qRT-PCR and RNA-Seq data was 0.975 ($P = 3.5E{-}13$), indicating that the RNA-Seq results were highly reliable.

TABLE 1 | Summary of the transcriptome assembly.

Samples	Control_rep1	Control _rep2	GA 6 h_rep1	GA 6 h_rep2	GA 12 h_rep1	GA 12 h_rep2
Raw reads	50,991,998	56,788,784	60,288,014	58,393,488	44,280,996	52,072,848
Clean reads (%)	48,705,848 (95.52)	54,275,188 (95.57)	57,631,340 (95.59)	55,720,738 (95.42)	42,311,862 (95.55)	49,993,570 (96.01)
Clean bases	7.31G	8.14G	8.64G	8.36G	6.35G	7.50G
Q20 (%)	97.52	97.42	97.38	97.43	97.28	97.36
GC (%)	43.63	43.60	43.51	43.53	43.46	43.54
Mapped clean reads (%)	42,119,358 (86.48)	46,783,558 (86.20)	49,881,555 (86.55)	48,281,412 (86.65)	36,728,637 (86.80)	43,283,227 (86.58)
Unique mapped clean reads (%)	41,622,535 (85.46)	46,245,871 (85.21)	49,302,733 (85.55)	47,674,994 (85.56)	36,276,717 (85.74)	42,766,514 (85.54)

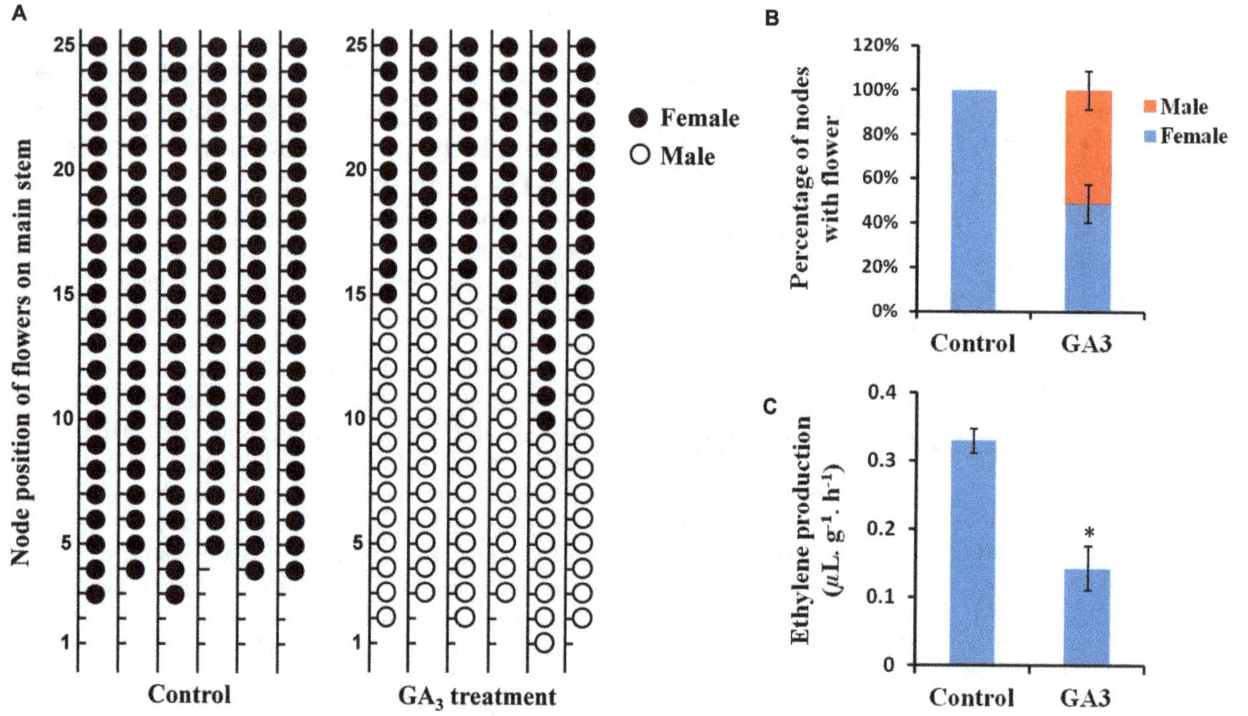

FIGURE 1 | Effects of exogenous GA_3 on sex expression and ethylene production in gynoecious cucumber. (A) Sex expression of the flowers on the first 25 nodes of the main stems in gynoecious cucumber plants treated with deionized water (with 0.1% ethanol) (Control) or 1000 ppm GA_3. The black and white circles represent female and male flowers, respectively. Lower nodes without circle indicate the vegetative nodes. **(B)** The percentage of the nodes with female or male flowers up to the 25th node on the main stems of the Control and GA_3 treatment lines. Values are the means ± SE from six independent plants. **(C)** Quantification of ethylene released from shoot tips in gynoecious cucumber plants at 7 days after treatment with GA_3 and the Control. Six biological replicates were performed for this experiment. Vertical bars represent the standard errors. The asterisk indicates the significant difference ($P < 0.01$) between the Control and GA_3 treatment lines by Duncan's test.

The *M* Gene Is Involved in GA-Regulated Sex Differentiation of Cucumber

Given that ethylene production was significantly decreased in the cucumber plants treated with GA_3 (**Figure 1C**), we screened the ethylene biosynthetic genes in the DGEs. We found that the transcript of *M* (*CsACS2*) gene encoding an ACC synthase was inhibited in both sets of transcriptome comparisons (GA 6 h vs. Control and GA 12 h vs. Control) (**Table 2**), and the qRT-PCR verification displayed the same expression pattern (**Figure 3**). Since the *M* gene is believed to inhibit stamen development in flower buds (Saito et al., 2007; Li et al., 2009, 2012), we speculated that GA might release the inhibitory effect of the *M* gene and allow male flowers to develop in cucumber.

In addition, another two ethylene biosynthetic genes, *CsACO1* and *CsACO3* which encode two ACC oxidases, were also differently expressed in the shoot apices after GA_3 treatment. *CsACO3* expression was dramatically down-regulated in the set of GA 12 h vs. Control, but not changed in GA 6 h vs. Control. However, there was an increase in the transcript level of *CsACO1* in both GA 6 h vs. Control and GA 12 h vs. Control, but the

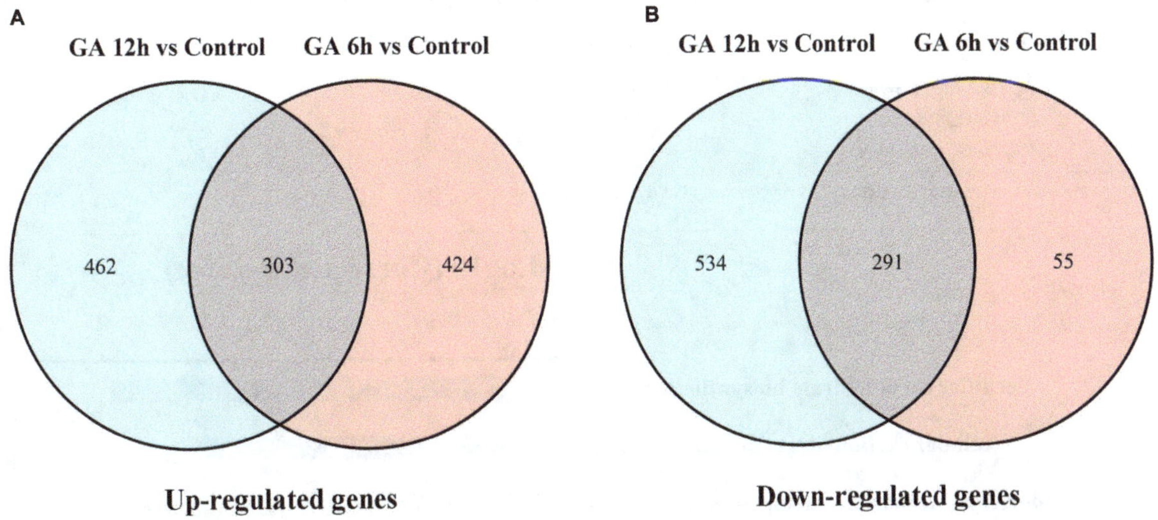

FIGURE 2 | Venn diagrams of DEGs that were significantly upregulated (A) or downregulated **(B)** after 6 or 12 h of GA_3 treatment.

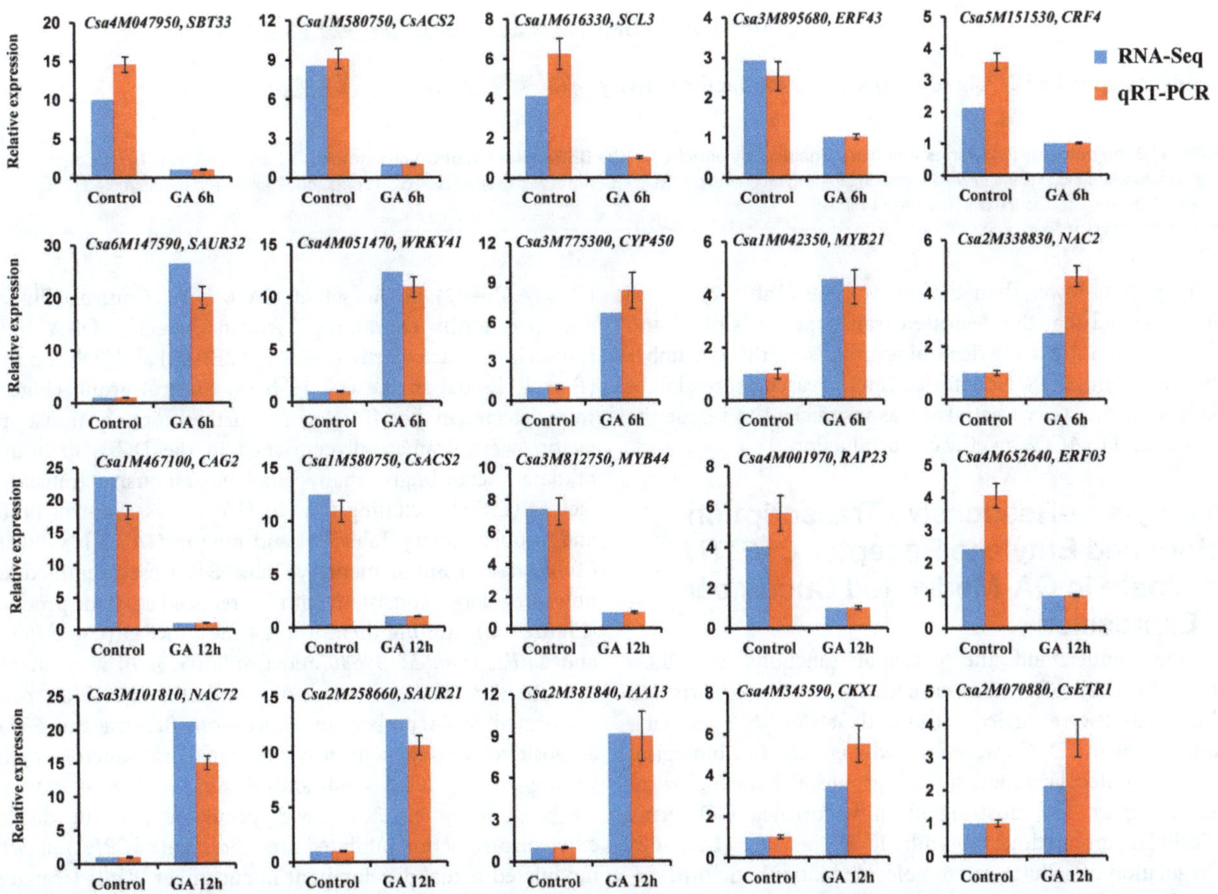

FIGURE 3 | qRT-PCR validation of DEGs identified by RNA-Seq. Twenty DEGs including ten up-regulated genes and ten down-regulated genes from the two sets of transcriptome comparisons (GA 6 h vs. Control and GA 12 h vs. Control) were randomly selected for qRT-PCR confirmation. The blue and red bars represent RNA-Seq and qRT-PCR data, respectively. The cucumber *TUA* gene was used as an internal control, and these experiments were repeated with three biological samples. Error bars indicate the standard errors.

TABLE 2 | List of differentially expressed ethylene biosynthetic genes identified by RNA-Seq in the shoot apices of GA_3 treatment plants and the Control.

Gene ID	Gene annotation	Fold change (GA 6 h/Control)	FDR	Fold change (GA 12 h/Control)	FDR
Csa1M580750	*M (CsACS2, ACC Synthase 2)*	−8.53	5.33E−23	−12.47	9.94E−22
Csa6M160180	*CsACO1 (ACC Oxidase 1)*	2.24	1.25E−06	2.56	5.59E−05
Csa6M421630	*CsACO3 (ACC Oxidase 2)*	/	/	−2.07	6.92E−04

The oblique line represents that the gene expression has no change in GA 6 h vs. Control.

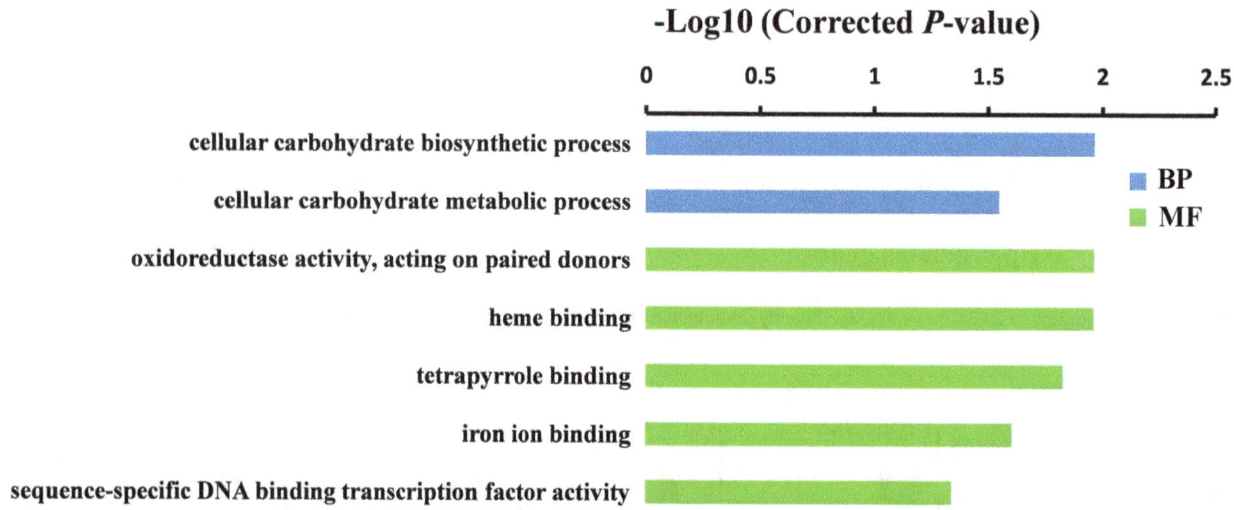

FIGURE 4 | Gene ontology (GO) terms that were significantly enriched in the DEGs after 6 h of GA_3 treatment. GO terms belong to biological processes (BP) and molecular functions (MFs) were shown in blue and green, respectively. GO terms were sorted based on corrected *P*-value, and the corrected *P*-value < 0.05 was used as the significance cut-off.

fold change was lower than that of *M* gene (**Table 2**). These results suggested that the decreased transcript levels of *M* and *CsACO3* might inhibit ethylene biosynthesis in the cucumber plants treated with GA_3. Nonetheless, an increased expression of *CsACO1* following GA_3 treatment was insufficient to rescue the effect of *M* and *CsACO3* on ethylene production.

The Ethylene-Responsive Transcription Factors and Ethylene Receptor *CsETR1* Participate in GA-Modulated Cucumber Sex Expression

To further understand the potential functions of DEGs identified by DGE, GO term enrichment analyses (Corrected *P*-value < 0.05) were carried out in both sets of RNA-Seq data. We found that the DEGs were markedly enriched in biological process and molecular function (MF) groups. For the biological process category, the most significantly enriched GO terms were "cellular carbohydrate biosynthetic process" ($P = 1.1E-02$) and "regulation of cellular macromolecule biosynthetic process" ($P = 5.8E-04$) in GA 6 h vs. Control and GA 12 h vs. Control groups, respectively (**Figures 4** and **5**). While five GO terms including "oxidoreductase activity, acting on paired donors" ($P = 1.1E-02$), heme binding ($P = 1.1E-02$), tetrapyrrole binding ($P = 1.5E-02$), iron ion binding ($P = 2.5E-02$), and "sequence-specific DNA binding transcription factor activity" ($P = 4.6E-02$) in the set of GA 6 h vs. Control (**Figure 4**) and two terms containing "sequence-specific DNA binding transcription factor activity" ($P = 2.2E-03$) and "DNA binding" ($P = 1.1E-02$) in the GA 12 h vs. Control group (**Figure 5**) were detected in the MF category. Furthermore, the transcription factors were dramatically enriched in the DGEs in both sets of data. Accordingly, many ethylene-responsive transcription factors (ERFs) including four in GA 6 h vs. Control (**Table 3** and Supplementary Table S3) and nine in GA 12 h vs. Control (**Table 4** and Supplementary Table S4) were identified to be down-regulated, consistent with the reduced ethylene production (**Figure 1C**). Among them, three genes, *ERF43* (*Csa3M895680*) and *CRF2s* (*Csa5M139630* and *Csa4M051360*), were shared in the two sets. These observations indicated that *ERFs* may be implicated in cucumber sex expression. Because the *ERFs* act as positive regulators in ethylene signal transduction pathway (Wang et al., 2002; Guo and Ecker, 2004; Prescott et al., 2016; Zhang et al., 2016), we speculated that the decreased ethylene production inhibited the expression of *ERFs*, followed by modulated sexual development in cucumber plants treated with GA_3.

Interestingly, we also noticed that an ethylene receptor, *CsETR1*, was enriched in the GO term of "DNA binding" (**Figure 5**), and its expression was increased by 2.12-fold in shoot apices after GA_3 treatment for 12 h (**Table 4** and Supplementary Table S4), but not changed in GA 6 h vs. Control

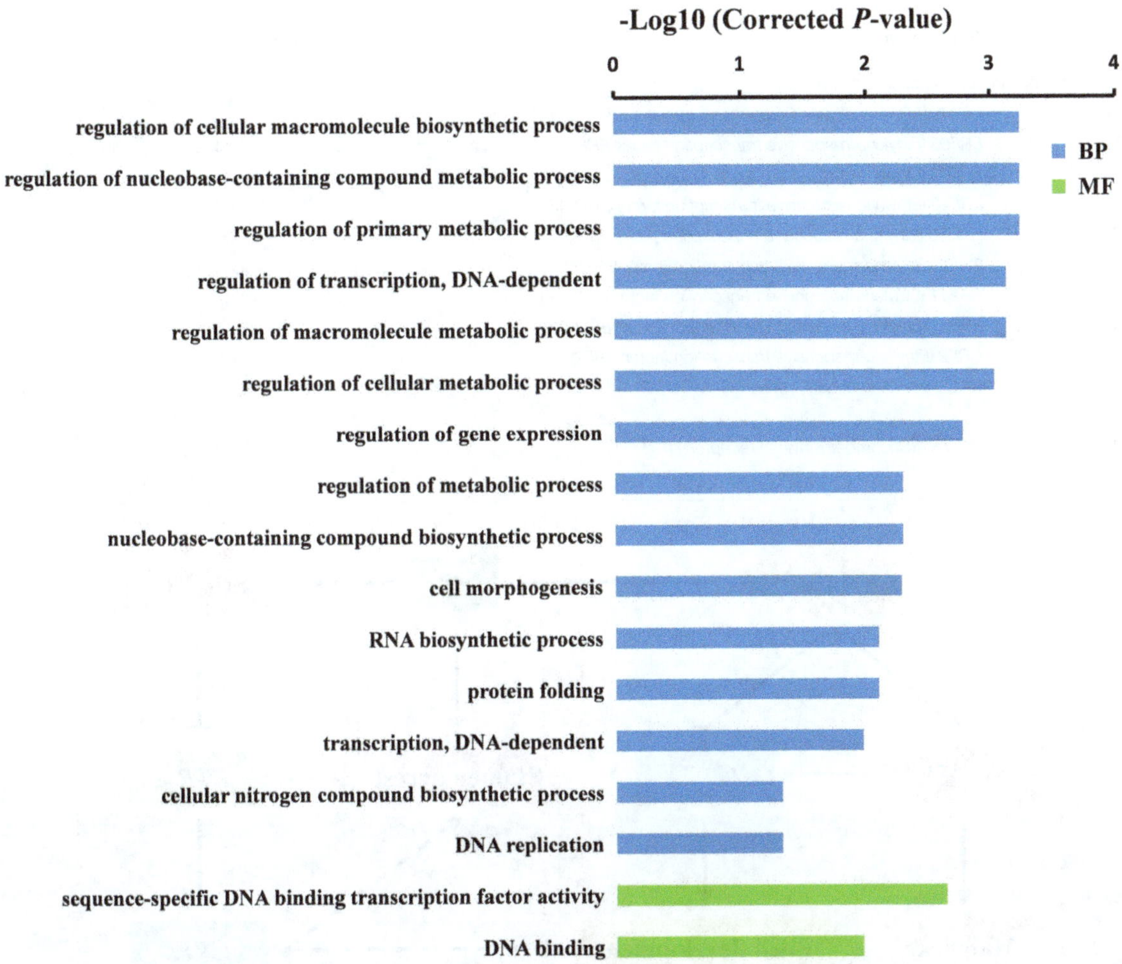

FIGURE 5 | Gene ontology terms that were significantly enriched in the DEGs after 12 h of GA_3 treatment. GO terms belong to BP and MFs were shown in blue and green, respectively. GO terms were sorted based on corrected *P*-value, and the corrected *P*-value < 0.05 was used as the significance cut-off.

TABLE 3 | List of selected ethylene signaling factors in the DEGs with enriched GO terms after GA_3 treatment for 6 h.

Gene ID	Gene annotation	Fold change (GA 6 h/Control)	FDR
Sequence-specific DNA binding transcription factor activity			
Csa3M895680	*ERF43 (Ethylene-responsive transcription factor ERF043)*	−2.93	1.08E−16
Csa5M139630	*CRF2 (Ethylene-responsive transcription factor CRF2)*	−2.51	1.89E−02
Csa4M051360	*CRF2 (Ethylene-responsive transcription factor CRF2)*	−2.23	2.66E−02
Csa5M151530	*CRF4 (Ethylene-responsive transcription factor CRF4)*	−2.12	1.14E−03

group. While the qRT-PCR verification revealed the same expression pattern (**Figure 3**). *ETR1* is an important member of ethylene receptors family that acts as negative regulator in the ethylene signaling pathway (Wang et al., 2002; Guo and Ecker, 2004; Light et al., 2016; Prescott et al., 2016). Previous studied have confirmed that *CsETR1* plays a negative role in stamen arrest during development of flower buds in cucumber (Wang et al., 2010). In accordance with these findings, we speculated that up-regulated *CsETR1* may promote male flowers formation by alleviating stamen arrest in cucumber after GA_3 treatment.

GA May Restrain the Female Tendency via Transcriptional Inhibition on *CAG2* in Cucumber

Through GO term enrichment analyses, we further identified an *AG* (C-class floral homeotic gene) homolog *CAG2*, which was enriched in the "sequence-specific DNA binding transcription factor activity" group. *CAG2* is one of the three *AG* genes in cucumber that controls pistil development due to its specific expression in the carpel (Kater et al., 1998; Perl-Treves et al., 1998). We found that the transcript level of *CAG2* was

TABLE 4 | List of selected ethylene signaling factors and *AGAMOUS* (*AG*) homolog in the DEGs with enriched GO terms after GA_3 treatment for 12 h.

Gene ID	Gene annotation	Fold change (GA 12 h/Control)	FDR
equence-specific DNA binding transcription factor activity			
Csa2M092800	*BBM2 (AP2-like ethylene-responsive transcription factor BBM2)*	$-\infty$	1.86E−02
Csa2M382540	*ERF03* (*Ethylene-responsive transcription factor ERF003*)	−3.44	3.98E−02
Csa2M382550	*ERF03* (*Ethylene-responsive transcription factor ERF003*)	−2.03	8.55E−05
Csa3M895680	*ERF43* (*Ethylene-responsive transcription factor ERF043*)	−3.06	1.93E−13
Csa4M001970	*RAP23* (*Ethylene-responsive transcription factor RAP2-3*)	−4.03	5.42E−10
Csa4M051360	*CRF2* (*Ethylene-responsive transcription factor CRF2*)	−2.44	3.40E−02
Csa4M192030	*RA211* (Ethylene-responsive transcription factor RAP2-11)	−3.87	4.03E−02
Csa4M652640	*ERF03* (*Ethylene-responsive transcription factor ERF003*)	−2.42	5.66E−03
Csa5M139630	*CRF2* (*Ethylene-responsive transcription factor CRF2*)	−3.02	9.74E−03
Csa1M467100	*CAG2* (*Floral homeotic protein AGAMOUS 2*)	−23.48	7.84E−03
DNA binding			
Csa2M070880	*CsETR1* (*Cucumber Ethylene receptor 1*)	2.12	6.99E−06

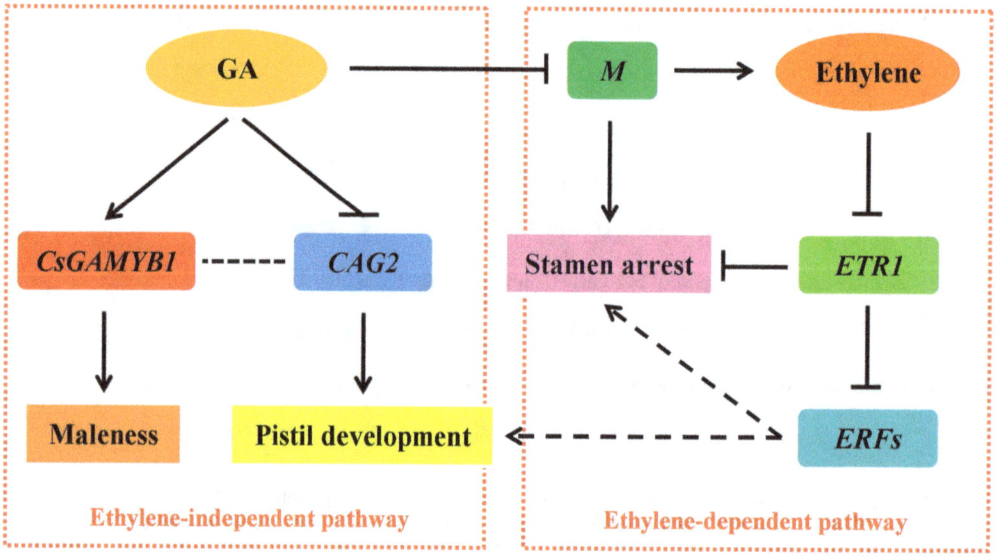

FIGURE 6 | A proposed model showing GA-regulated sex expression in cucumber. GA can promote male flowers formation via ethylene-dependent (right) and ethylene-independent (left) pathways. Arrows and T-bars indicate positive and negative effects, respectively. The solid lines define proved regulations, whereas the dotted lines represent proposed regulations.

significantly decreased by 23.48-fold in shoot apices after 12 h of GA_3 treatment (**Table 4** and Supplementary Table S4), and the qRT-PCR assay showed the same expression pattern (**Figure 3**). Our data implied that GA may restrain the femaleness via inhibiting the *CAG2* expression in cucumber.

DISCUSSION

Sex differentiation of flower buds is an important developmental process that directly affects the product yield in cucumber. In addition to genetic control, sex expression can be modified by plant hormones and environmental conditions. (Malepszy and Niemirowicz-Szczytt, 1991). Among various plant hormones, ethylene can induce female flowers formation (MacMurray and Miller, 1968; Iwahori et al., 1969), and the underlying molecular mechanism has been widely documented (Yamasaki et al., 2001, 2003; Mibus and Tatlioglu, 2004; Knopf and Trebitsh, 2006; Saito et al., 2007; Li et al., 2009, 2012; Wang et al., 2010). GA can promote male flowers development (Wittwer and Bukovac, 1962; Pike and Peterson, 1969), but the regulatory pathway remains elusive. In addition, a potential crosstalk between GA and ethylene in controlling sex determination of cucumber still remains disputed. In this study, through genome-wide expression analyses, we showed that GA may promote cucumber maleness via an ethylene-dependent pathway by altering expression of the *M* (*CsACS2*) gene, ethylene receptor *CsETR1* and ethylene-responsive transcription factors. Nevertheless, we also found that GA may also restrain femaleness through an ethylene-independent pathway regulating *CAG2*, a C-class floral homeotic gene (**Figure 6**).

GA May Promote Male Tendency via an Ethylene-Dependent Pathway in Cucumber

Upon exogenous GA_3 treatment in the gynoecious cucumber line 13-3B, the male flowers were markedly induced, meanwhile, ethylene production in the shoot apices was significantly decreased (**Figure 1**) due to collaborative regulation by three ethylene biosynthetic genes such as *M* (*CsACS2*), *CsACO1*, and *CsACO3*. Notably, *M* and *CsACO1* were significantly down-regulated and up-regulated in both sets of transcriptome comparisons, respectively, however, *CsACO3* transcript was only decreased in GA 12 h vs. Control group (**Figure 3**; **Table 2**). Since both *CsACO1* and *CsACO3* encode ACC oxidases, and they showed opposite expression patterns and similar fold changes in RNA-Seq data (**Table 2**), we speculated that interplay between up-regulation of *CsACO1* and down-regulation of *CsACO3* offset the ACC oxidase activity in shoot apices after 12 h of GA_3 treatment. This implied that the reduced *M* transcript might play major role in inhibiting ethylene biosynthesis. Given that the *M* gene can directly inhibit stamen development in cucumber flower buds (Saito et al., 2007; Li et al., 2009, 2012), our data revealed that GA might release the inhibitory effect of the *M* gene on stamen arrest and restrain ethylene production, by down-regulating the *M* gene expression (**Figure 6**, right panel).

In ethylene signal transduction pathway, the receptors such as ETRs function as negative regulators, while ERFs, downstream components of receptors, act as positive transcription factors (Wang et al., 2002; Guo and Ecker, 2004; Light et al., 2016; Prescott et al., 2016; Zhang et al., 2016). RNA-Seq data displayed that the ethylene receptor *CsETR1* in cucumber, which is thought to be a negative regulatory factor in stamen arrest of flower buds (Wang et al., 2010), was markedly up-regulated after 12 h of GA_3 treatment (**Figure 3**; **Table 4** and Supplementary Table S4). Up-regulation of *CsETR1* occurred at relatively later time point than down-regulation of the *M* gene (**Table 2**), suggesting that the increased *CsETR1* transcript was not directly caused by GA_3 rather by reduced ethylene production. Moreover, the expression levels of some *ERFs* were dramatically decreased after GA_3 treatment (**Tables 3** and **4**, Supplementary Tables S3 and S4), and they were probably involved in cucumber sex expression through inhibiting maleness or promoting femaleness. However, this understanding was based on bioinformatics analysis, the precise roles of *ERFs* on cucumber flower development remained unclear and should be verified in future studies using advanced physiological and molecular techniques. These observations indicated that increased *CsETR1* expression may stimulate male tendency through direct inhibition in stamen arrest or down-regulation on the *ERFs* transcript in cucumber after GA_3 treatment (**Figure 6**, right panel).

Furthermore, despite we have shown that GA can regulate cucumber sex expression in cooperation with ethylene, how GA modulated ethylene and which genes participated in this process were unknown. Given that the DELLA proteins are central repressors of GA responses (Sun, 2010, 2011; Plackett et al., 2014), and accumulating evidence suggested that DELLA proteins play important roles in ethylene-mediated plant growth and development processes through interactions with some regulatory factors in ethylene signaling pathway, such as CTR1 (CONSTITUTIVE TRIPLE RESPONSE1), EIN3/EIL1 (ETHYLENE INSENSITIVE 3/EIN3-LIKE 1), RAP 2.3 (RELATED TO APETALA 2.3), and ERF11 (ETHYLENE RESPONSE FACTOR 11) (Achard et al., 2003, 2007; Pierik et al., 2009; An et al., 2012; Luo et al., 2013; Marín-de la Rosa et al., 2014; Zhou et al., 2016). Therefore, we proposed that DELLA proteins may be involved in the sex differentiation of cucumber coupled with GA and ethylene in a collaborative regulation at the protein level, since the expressions of four *DELLA* homologs in cucumber, *CsGAIP*, *CsGAI1*, *CsGAI2*, and *CsGAI3* (Zhang et al., 2014a), had no change after GA_3 treatment (Supplementary Tables S1 and S2).

In addition, GA and ethylene can also cooperatively regulate other aspects of plant growth and development. For example, GA promoted apical hook development of *Arabidopsis*, in part through transcriptional regulation of several genes in ethylene biosynthetic pathway mediated by DELLA proteins (Gallego-Bartolome et al., 2011). In this process, GA induced ethylene production, that was opposite to our results in cucumber sex expression. We speculated that this distinction may be due to different roles of hormones in various plant developmental processes. In fact, GA and ethylene showed similar functions in hook development, but they may play opposite roles in sex determination, thus, the regulatory mechanisms were different.

GA May Inhibit Femaleness or Induce Maleness of Cucumber via an Ethylene-Independent Pathway

AGAMOUS, the C-class floral homeotic gene, belongs to the MADS-box family. In *Arabidopsis*, GA can induce the expression of *AG* gene (Yu et al., 2004). However, our data provided a novel point that an *AG* homolog *CAG2* in cucumber was down-regulated upon GA_3 treatment (**Figure 3**; **Table 4** and Supplementary Table S4). This distinction may be due to different abilities of these two genes to induce reproductive organ fate, in which, *AG* in *Arabidopsis* controls stamen and carpel development (Lohmann and Weigel, 2002), but *CAG2* is particularly restricted to the carpel (Kater et al., 1998; Perl-Treves et al., 1998). Previous studies showed that *CAG2* transcripts were not mediated by ethylene (Perl-Treves et al., 1998), hence, we speculated that GA probably suppressed pistil development through inhibiting the *CAG2* expression, thereby allowing male flowers to develop, and ethylene was not involved in this process. Moreover, in our previous study, we demonstrated that transcript of *CsGAMYB1* was upregulated by GA_3 treatment in male flower buds and silencing of *CsGAMYB1* could suppress masculinization of cucumber, but the ethylene production and expression of *F* and *M* genes were not changed in the *CsGAMYB1*-RNAi lines (Zhang et al., 2014b). These observations suggested that GA can also regulate sex expression of cucumber via an ethylene-independent pathway (**Figure 6**, left panel). However, the relationship between *CAG2* and *CsGAMYB1* regulations remains obscure and that should be verified in further studies.

Besides, DELLA proteins can down-regulate the expression of floral homeotic genes, *AP3*, *PI* and *AG*, subsequently inhibit flower development in *Arabidopsis* (Yu et al., 2004). *CsGAIP*, a DELLA homolog in cucumber, may restrain staminate development through transcriptional repression of *AP3* and *PI* in *Arabidopsis* (Zhang et al., 2014a). However, the expression levels of *AP3* and *PI* were not changed after GA_3 treatment (Supplementary Tables S1 and S2), indicating that they did not participate in GA-regulated male tendency. Accordingly, there may be a regulatory relationship between DELLA proteins and floral homeotic gene *CAG2* during GA-modulated sexual development in cucumber, however, the mechanism would be possibly different from that of *Arabidopsis*.

In summary, our data revealed a novel viewpoint that GA might control sex differentiation of cucumber via both ethylene-dependent and ethylene-independent pathways, and DELLA proteins were likely to be involved in both processes. However, this model was proposed by bioinformatics data, therefore, elucidation of the critical roles of DELLA proteins in flower development by cucumber transformation, and identification of the relationships among DELLAs and ethylene regulatory factors, GA-DELLA-*CsGAMYB1* signaling and *CAG2* gene, will shed new light on the molecular details of GA-regulated sex expression in cucumber.

Evolution of Unisexual Flower in Cucumber and Potential Involvement of Hormones

Generally, typical unisexual flowers have two morphological types. The type I is unisexual by abortion. Initiation of stamen and pistil occurs in all flowers, followed by the developmental arrest in one or another organ. The type II is unisexual from inception. Only stamen or pistil is initiated and it does not go through a hermaphroditic stage (Lebel-Hardenack and Grant, 1997; Ainsworth, 2000; Mitchell and Diggle, 2005). Now, it is believed that the morphology of cucumber flowers belongs to the type I (Bai et al., 2004), but its evolutionary mechanism is largely unknown. Bai and Xu proposed a "miR initiative" hypothesis (Bai and Xu, 2013), where they speculated that unisexual cucumber flowers are evolved from a hermaphrodite ancestor. The first step in the evolutionary process might be the miRNA-regulated arrest of ovary development, and this predication is based on the altered expression of miRNAs, such as miR396a, 156b, 159a, 171b, and 166a, in male flowers (Bai et al., 2004; Sun et al., 2010). And this event leads to environment-dependent andromonoecy which has no progeny. Then, the *M* gene is recruited. On the one hand, the *M* gene promotes ethylene biosynthesis, resulting in the rescue of ovary development for seed set, because the ethylene might regulate the miRNA production. On the other hand, the *M* gene inhibits stamen development to avoid self-pollination and maintain cross-pollination. So, the monoecious genotype is generated through the cooption of the *M* gene. The andromonoecious genotype is produced by the loss-of-function *m* gene, which is regarded as a reverted point mutation. Further, the *F* gene is coopted and generate the gynoecious genotype (Sun et al., 2010).

Until now, a potential role of GA in unisexual flower evolution of cucumber has not been reported. But based on the possible function of *M* gene on evolutionary development of cucumber flower and the effect of GA on the transcript of *M* gene (**Table 2**), we speculated that GA might be involved in the process of cucumber flower evolution through interaction with ethylene. In addition, previous studies showed that GA signaling system can regulate anther development by modulation of miR159/GAMYB (Achard et al., 2004). In this pathway, miR159 acts as a post-transcriptional regulator of *GAMYB* transcript levels. GA relieves the DELLA repression of GAMYB, which is mediated by the GA activation of miR159. As mentioned above, miR159 is likely to participate in the arrest of ovary development in the evolutionary process of cucumber flower. And the *GAMYB* homolog *CsGAMYB1* can regulate cucumber sex expression via an ethylene-independent pathway (Zhang et al., 2014b). These observations further revealed the possible involvement of GA in unisexual flower evolution of cucumber, but this process might be dependent on miR159 and GAMYB, and have no relationship with ethylene. Finally, it is worth noting that these viewpoints are built on the basis of the "miR initiative" hypothesis and needed to be tested in further work.

AUTHOR CONTRIBUTIONS

YZ and YaL designed the experiments. YZ, GZ, and NM performed the experiments. YZ, YuL, and JZ analyzed the data. YZ wrote the paper along with YaL. All authors reviewed the manuscript.

FUNDING

This work was supported by National Natural Science Foundation of China (31601770), Science and Technology Research and Development Program of Shaanxi Province (2015NY098), Fundamental Research Funds for Northwest A&F University (2452015024), and Doctoral Scientific Research Foundation of Northwest A&F University (Z109021504) to YZ.

ACKNOWLEDGMENTS

We thank Dr. Huazhong Ren (China Agricultural University) for providing the cucumber 13-3B seeds, and members of the Liang Laboratory for helpful discussions and technical assistance.

SUPPLEMENTARY MATERIAL

The Supplementary Material for this article can be found online at: http://journal.frontiersin.org/article/10.3389/fpls.2017.00010/full#supplementary-material

REFERENCES

Achard, P., Baghour, M., Chapple, A., Hedden, P., Van Der Straeten, D., Genschik, P., et al. (2007). The plant stress hormone ethylene controls floral transition via DELLA-dependent regulation of floral meristem-identity genes. *Proc. Natl. Acad. Sci. U.S.A.* 104, 6484–6489. doi: 10.1073/pnas.0610717104

Achard, P., Herr, A., Baulcombe, D. C., and Harberd, N. P. (2004). Modulation of floral development by a gibberellin-regulated microRNA. *Development* 131, 3357–3365. doi: 10.1242/dev.01206

Achard, P., Vriezen, W. H., Van Der Straeten, D., and Harberd, N. P. (2003). Ethylene regulates *arabidopsis* development via the modulation of DELLA protein growth repressor function. *Plant Cell* 15, 2816–2825. doi: 10.1105/tpc.015685

An, F., Zhang, X., Zhu, Z., Ji, Y., He, W., Jiang, Z., et al. (2012). Coordinated regulation of apical hook development by gibberellins and ethylene in etiolated *Arabidopsis* seedlings. *Cell Res.* 22, 915–927. doi: 10.1038/cr.2012.29

Ando, S., Sato, Y., Kamachi, S., and Sakai, S. (2001). Isolation of a MADS-box gene (ERAF17) and correlation of its expression with the induction of formation of female flowers by ethylene in cucumber plants (*Cucumis sativus* L.). *Planta* 213, 943–952. doi: 10.1007/s004250100571

Ainsworth, C. (2000). Boys and girls come out to play: the molecular biology of dioecious plants. *Ann. Bot.* 86, 211–221. doi: 10.1006/anbo.2000.1201

Atsmon, D., and Tabbak, C. (1979). Comparative effects of gibberellin, silver nitrate and aminoethoxyvinyl glycine on sexual tendency and ethylene evolution in the cucumber plant (*Cucumis sativus* L.). *Plant Cell Physiol.* 20, 1547–1555.

Aya, K., Ueguchi-Tanaka, M., Kondo, M., Hamada, K., Yano, K., Nishimura, M., et al. (2009). Gibberellin modulates anther development in rice via the transcriptional regulation of GAMYB. *Plant Cell* 21, 1453–1472. doi: 10.1105/tpc.108.062935

Bai, S. L., Peng, Y. B., Cui, J. X., Gu, H. T., Xu, L. Y., Li, Y. Q., et al. (2004). Developmental analyses reveal early arrests of the spore-bearing parts of reproductive organs in unisexual flowers of cucumber (*Cucumis sativus* L.). *Planta* 220, 230–240. doi: 10.1007/s00425-004-1342-2

Bai, S. N., and Xu, Z. H. (2013). Unisexual cucumber flowers, sex and sex differentiation. *Int. Rev. Cell Mol. Biol.* 304, 1–55. doi: 10.1016/B978-0-12-407696-9.00001-4

Cheng, H., Qin, L., Lee, S., Fu, X., Richards, D. E., Cao, D., et al. (2004). Gibberellin regulates *Arabidopsis* floral development via suppression of DELLA protein function. *Development* 131, 1055–1064. doi: 10.1242/dev.00992

Duan, Q. H., Wang, D. H., Xu, Z. H., and Bai, S. N. (2008). Stamen development in *Arabidopsis* is arrested by organ-specific overexpression of a cucumber ethylene synthesis gene CsACO2. *Planta* 228, 537–543. doi: 10.1007/s00425-008-0756-7

Eveland, A. L., Satoh-Nagasawa, N., Goldshmidt, A., Meyer, S., Beatty, M., Sakai, H., et al. (2010). Digital gene expression signatures for maize development. *Plant Physiol.* 154, 1024–1039. doi: 10.1104/pp.110.159673

Fleet, C. M., and Sun, T. P. (2005). A DELLAcate balance: the role of gibberellin in plant morphogenesis. *Curr. Opin. Plant Biol.* 8, 77–85. doi: 10.1016/j.pbi.2004.11.015

Fujita, Y., and Fujieda, K. (1981). Relation between sex expression types and cotyledon etiolation of cucumber in vitro. I. on the role of ethylene evolved from seedlings. *Plant Cell Physiol.* 22, 667–674.

Gallego-Bartolome, J., Arana, M. V., Vandenbussche, F., Zadnikova, P., Minguet, E. G., Guardiola, V., et al. (2011). Hierarchy of hormone action controlling apical hook development in *Arabidopsis*. *Plant J.* 67, 622–634. doi: 10.1111/j.1365-313X.2011.04621.x

Guo, H., and Ecker, J. R. (2004). The ethylene signaling pathway: new insights. *Curr. Opin. Plant Biol.* 7, 40–49. doi: 10.1016/j.pbi.2003.11.011

Hao, Y. J., Wang, D. H., Peng, Y. B., Bai, S. L., Xu, L. Y., Li, Y. Q., et al. (2003). DNA damage in the early primordial anther is closely correlated with stamen arrest in the female flower of cucumber (*Cucumis sativus* L.). *Planta* 217, 888–895. doi: 10.1007/s00425-003-1064-x

Harberd, N. P., Belfield, E., and Yasumura, Y. (2009). The angiosperm gibberellin-GID1-DELLA growth regulatory mechanism: how an "inhibitor of an inhibitor" enables flexible response to fluctuating environments. *Plant Cell* 21, 1328–1339. doi: 10.1105/tpc.109.066969

Hemphill, D. D., Baker, L. R., and Sell, H. M. (1972). Different sex phenotypes of *Cucumis sativus* L. and C. melo L. and their endogenous gibberellin activity. *Euphytica* 21, 285–291. doi: 10.1007/BF00036769

Huang, S., Li, R., Zhang, Z., Li, L., Gu, X., Fan, W., et al. (2009). The genome of the cucumber, *Cucumis sativus* L. *Nat. Genet.* 41, 1275–1281. doi: 10.1038/ng.475

Iwahori, S., Lyons, J. M., and William, L. S. (1969). Induced femaleness in cucumber by 2-chloroethanephosphonic acid. *Nature* 222, 271–272. doi: 10.1038/222271a0

Kahana, A., Silberstein, L., Kessler, N., Goldstein, R. S., and Perl-Treves, R. (1999). Expression of ACC oxidase genes differs among sex genotypes and sex phases in cucumber. *Plant Mol. Biol.* 41, 517–528. doi: 10.1023/A:1006343707567

Kater, M. M., Colombo, L., Franken, J., Busscher, M., Masiero, S., Van Lookeren Campagne, M. M., et al. (1998). Multiple AGAMOUS homologs from cucumber and petunia differ in their ability to induce reproductive organ fate. *Plant Cell* 10, 171–182. doi: 10.1105/tpc.10.2.171

Knopf, R. R., and Trebitsh, T. (2006). The female-specific Cs-ACS1G gene of cucumber. A case of gene duplication and recombination between the non-sex-specific 1-aminocyclopropane-1-carboxylate synthase gene and a branched-chain amino acid transaminase gene. *Plant Cell Physiol.* 47, 1217–1228. doi: 10.1093/pcp/pcj092

Lebel-Hardenack, S., and Grant, S. R. (1997). Genetics of sex determination in flowering plants. *Trends Plant Sci.* 2, 130–136. doi: 10.1016/S1360-1385(97)01012-1

Li, Z., Huang, S., Liu, S., Pan, J., Zhang, Z., Tao, Q., et al. (2009). Molecular isolation of the M gene suggests that a conserved-residue conversion induces the formation of bisexual flowers in cucumber plants. *Genetics* 182, 1381–1385. doi: 10.1534/genetics.109.104737

Li, Z., Wang, S., Tao, Q., Pan, J., Si, L., Gong, Z., et al. (2012). A putative positive feedback regulation mechanism in CsACS2 expression suggests a modified model for sex determination in cucumber (*Cucumis sativus* L.). *J. Exp. Bot.* 63, 4475–4484. doi: 10.1093/jxb/ers123

Light, K. M., Wisniewski, J. A., Vinyard, W. A., and Kieber-Emmons, M. T. (2016). Perception of the plant hormone ethylene: known-knowns and known-unknowns. *J. Biol. Inorg. Chem.* 21, 715–728. doi: 10.1007/s00775-016-1378-3

Lohmann, J. U., and Weigel, D. (2002). Building beauty: the genetic control of floral patterning. *Dev. Cell* 2, 135–142. doi: 10.1016/S1534-5807(02)00122-3

Luo, J., Ma, N., Pei, H., Chen, J., Li, J., and Gao, J. (2013). A DELLA gene, RhGAI1, is a direct target of EIN3 and mediates ethylene-regulated rose petal cell expansion via repressing the expression of RhCesA2. *J. Exp. Bot.* 64, 5075–5084. doi: 10.1093/jxb/ert296

MacMurray, A. L., and Miller, C. M. (1968). Cucumber sex expression modified by 2-chloroethanephosphonic acid. *Science* 162, 1397–1398. doi: 10.1126/science.162.3860.1397

Malepszy, S., and Niemirowicz-Szczytt, K. (1991). Sex determination in cucumber (*Cucumis sativus*) as a model system for molecular biology. *Plant Sci.* 80, 39–47. doi: 10.1016/0168-9452(91)90271-9

Marín-de la Rosa, N., Sotillo, B., Miskolczi, P., Gibbs, D. J., Vicente, J., Carbonero, P., et al. (2014). Large-scale identification of gibberellin-related transcription factors defines group VII ETHYLENE RESPONSE FACTORS as functional DELLA partners. *Plant Physiol.* 166, 1022–1032. doi: 10.1104/pp.114.244723

Mibus, H., and Tatlioglu, T. (2004). Molecular characterization and isolation of the F/f gene for femaleness in cucumber (*Cucumis sativus* L.). *Theor. Appl. Genet.* 109, 1669–1676. doi: 10.1007/s00122-004-1793-7

Mitchell, C. H., and Diggle, P. K. (2005). The evolution of unisexual flowers: morphological and functional convergence results from diverse developmental transitions. *Am. J. Bot.* 92, 1068–1076. doi: 10.3732/ajb.92.7.1068

Murase, K., Hirano, Y., Sun, T. P., and Hakoshima, T. (2008). Gibberellin-induced DELLA recognition by the gibberellin receptor GID1. *Nature* 456, 459–463. doi: 10.1038/nature07519

Olszewski, N., Sun, T. P., and Gubler, F. (2002). Gibberellin signaling: biosynthesis, catabolism, and response pathways. *Plant Cell* 14(Suppl.), S61–S80.

Perl-Treves, R., Kahana, A., Rosenmann, N., Xiang, Y., and Silberstein, L. (1998). Expression of multiple AGAMOUS-like genes in male and female flowers of cucumber (*Cucumis sativus* L.). *Plant Cell Physiol.* 39, 701–710. doi: 10.1093/oxfordjournals.pcp.a029424

Pierce, L. K., and Wehner, T. C. (1990). Review of genes and linkage groups in cucumber. *HortScience* 25, 605–615.

Pierik, R., Djakovic-Petrovic, T., Keuskamp, D. H., de Wit, M., and Voesenek, L. A. (2009). Auxin and ethylene regulate elongation responses to neighbor proximity signals independent of gibberellin and della proteins in *Arabidopsis*. *Plant Physiol.* 149, 1701–1712. doi: 10.1104/pp.108.133496

Pike, L. M., and Peterson, C. E. (1969). Gibberellin A4/A7, for induction of staminate flowers on the gynoecious cucumber (*Cucumis sativus* L.). *Euphytica* 18, 106–109.

Plackett, A. R., Ferguson, A. C., Powers, S. J., Wanchoo-Kohli, A., Phillips, A. L., Wilson, Z. A., et al. (2014). DELLA activity is required for successful pollen development in the Columbia ecotype of *Arabidopsis*. *New Phytol.* 201, 825–836. doi: 10.1111/nph.12571

Plackett, A. R., Thomas, S. G., Wilson, Z. A., and Hedden, P. (2011). Gibberellin control of stamen development: a fertile field. *Trends Plant Sci.* 16, 568–578. doi: 10.1016/j.tplants.2011.06.007

Prescott, A. M., McCollough, F. W., Eldreth, B. L., Binder, B. M., and Abel, S. M. (2016). Analysis of network topologies underlying ethylene growth response kinetics. *Front. Plant Sci.* 7:1308. doi: 10.3389/fpls.2016.01308

Robinson, M. D., McCarthy, D. J., and Smyth, G. K. (2010). edgeR: a Bioconductor package for differential expression analysis of digital gene expression data. *Bioinformatics* 26, 139–140. doi: 10.1093/bioinformatics/btp616

Rudich, J., Halevy, A. H., and Kedar, N. (1972). The level of phytohormones in monoecious and gynoecious cucumbers as affected by photoperiod and ethephon. *Plant Physiol.* 50, 585–590. doi: 10.1104/pp.50.5.585

Saito, S., Fujii, N., Miyazawa, Y., Yamasaki, S., Matsuura, S., Mizusawa, H., et al. (2007). Correlation between development of female flower buds and expression of the CS-ACS2 gene in cucumber plants. *J. Exp. Bot.* 58, 2897–2907. doi: 10.1093/jxb/erm141

Song, S., Qi, T., Huang, H., and Xie, D. (2013). Regulation of stamen development by coordinated actions of jasmonate, auxin, and gibberellin in *Arabidopsis*. *Mol. Plant* 6, 1065–1073. doi: 10.1093/mp/sst054

Sun, J. J., Li, F., Li, X., Liu, X. C., Rao, G. Y., Luo, J. C., et al. (2010). Why is ethylene involved in selective promotion of female flower development in cucumber? *Plant Signal. Behav.* 5, 1052–1056. doi: 10.4161/psb.5.8.12411

Sun, T. P. (2010). Gibberellin-GID1-DELLA: a pivotal regulatory module for plant growth and development. *Plant Physiol.* 154, 567–570. doi: 10.1104/pp.110.161554

Sun, T. P. (2011). The molecular mechanism and evolution of the GA-GID1-DELLA signaling module in plants. *Curr. Biol.* 21, R338–R345. doi: 10.1016/j.cub.2011.02.036

Trapnell, C., Pachter, L., and Salzberg, S. L. (2009). TopHat: discovering splice junctions with RNA-Seq. *Bioinformatics* 25, 1105–1111. doi: 10.1093/bioinformatics/btp120

Trebitsh, T., Rudich, J., and Riov, J. (1987). Auxin, biosynthesis of ethylene and sex expression in cucumber (*Cucumis sativus*). *Plant Growth Regul.* 5, 105–113. doi: 10.1007/BF00024738

Trebitsh, T., Staub, J. E., and O'Neill, S. D. (1997). Identification of a 1-aminocyclopropane-1-carboxylic acid synthase gene linked to the female (F) locus that enhances female sex expression in cucumber. *Plant Physiol.* 113, 987–995. doi: 10.1104/pp.113.3.987

Wang, D. H., Li, F., Duan, Q. H., Han, T., Xu, Z. H., and Bai, S. N. (2010). Ethylene perception is involved in female cucumber flower development. *Plant J.* 61, 862–872. doi: 10.1111/j.1365-313X.2009.04114.x

Wang, L. C., Li, H., and Joseph, R. E. (2002). Ethylene biosynthesis and signaling networks. *Plant Cell* 14(Suppl. 14), S131–S151.

Wang, Z., Gerstein, M., and Snyder, M. (2009). RNA-Seq: a revolutionary tool for transcriptomics. *Nat. Rev. Genet.* 10, 57–63. doi: 10.1038/nrg2484

Wittwer, S. H., and Bukovac, M. I. (1962). Staminate flower formation on gynoecious cucumber as influenced by the various gibberellins. *Naturwissenshaften* 49, 305–306. doi: 10.1007/BF00622719

Yamasaki, S., Fujii, N., Matsuura, S., Mizusawa, H., and Takahashi, H. (2001). The M locus and ethylene-controlled sex determination in andromonoecious cucumber plants. *Plant Cell Physiol.* 42, 608–619. doi: 10.1093/pcp/pce076

Yamasaki, S., Fujii, N., and Takahashi, H. (2003). Characterization of ethylene effects on sex determination in cucumber plants. *Sex. Plant Reprod.* 16, 103–111. doi: 10.1007/s00497-003-0183-7

Yin, T., and Quinn, J. A. (1995). Tests of a mechanistic model of one hormone regulating both sexes in *Cucumis sativus* (Cucurbitaceae). *Am. J. Bot.* 82, 1537–1546. doi: 10.2307/2446182

Young, M. D., Wakefield, M. J., Smyth, G. K., and Oshlack, A. (2010). Gene ontology analysis for RNA-seq: accounting for selection bias. *Genome Biol.* 11:R14. doi: 10.1186/gb-2010-11-2-r14

Yu, H., Ito, T., Zhao, Y., Peng, J., Kumar, P., and Meyerowitz, E. M. (2004). Floral homeotic genes are targets of gibberellin signaling in flower development. *Proc. Natl. Acad. Sci. U.S.A.* 101, 7827–7832. doi: 10.1073/pnas.0402377101

Zhang, H., Li, A., Zhang, Z., Huang, Z., Lu, P., Zhang, D., et al. (2016). Ethylene response factor TERF1, regulated by ETHYLENE-INSENSITIVE3-like factors, functions in reactive oxygen species (ROS) scavenging in tobacco (*Nicotiana tabacum* L.). *Sci. Rep.* 6:29948. doi: 10.1038/srep29948

Zhang, Y., Liu, B., Yang, S., An, J., Chen, C., Zhang, X., et al. (2014a). A cucumber DELLA homolog CsGAIP may inhibit staminate development through transcriptional repression of B class floral homeotic genes. *PLoS ONE* 9:e91804. doi: 10.1371/journal.pone.0091804

Zhang, Y., Zhang, X., Liu, B., Wang, W., Liu, X., Chen, C., et al. (2014b). A GAMYB homologue CsGAMYB1 regulates sex expression of cucumber via an ethylene-independent pathway. *J. Exp. Bot.* 65, 3201–3213. doi: 10.1093/jxb/eru1764

Zhou, X., Zhang, Z. L., Park, J., Tyler, L., Yusuke, J., Qiu, K., et al. (2016). The ERF11 transcription factor promotes internode elongation by activating gibberellin biosynthesis and signaling. *Plant Physiol.* 171, 2760–2770.

Conflict of Interest Statement: The authors declare that the research was conducted in the absence of any commercial or financial relationships that could be construed as a potential conflict of interest.

5

Genome-Wide Characterization of the MADS-Box Gene Family in Radish (*Raphanus sativus* L.) and Assessment of its Roles in Flowering and Floral Organogenesis

Chao Li[1†], Yan Wang[1†], Liang Xu[1], Shanshan Nie[1], Yinglong Chen[2], Dongyi Liang[1], Xiaochuan Sun[1], Benard K. Karanja[1], Xiaobo Luo[1] and Liwang Liu[1]*

[1] National Key Laboratory of Crop Genetics and Germplasm Enhancement, College of Horticulture, Nanjing Agricultural University, Nanjing, China, [2] School of Earth and Environment, The UWA Institute of Agriculture, The University of Western Australia, Perth, WA, Australia

Edited by:
Federico Valverde,
Spanish National Research Council,
Spain

Reviewed by:
Marie Monniaux,
Max Planck Society, Germany
Marcos Egea-Cortines,
Universidad Politécnica de Cartagena,
Spain

***Correspondence:**
Liwang Liu
nauliulw@njau.edu.cn

† These authors have contributed equally to this work.

The MADS-box gene family is an important transcription factor (TF) family that is involved in various aspects of plant growth and development, especially flowering time and floral organogenesis. Although it has been reported in many plant species, the systematic identification and characterization of MADS-box TF family is still limited in radish (*Raphanus sativus* L.). In the present study, a comprehensive analysis of MADS-box genes was performed, and a total of 144 MADS-box family members were identified from the whole radish genome. Meanwhile, a detailed list of MADS-box genes from other 28 plant species was also investigated. Through the phylogenetic analysis between radish and *Arabidopsis thaliana*, all the *RsMADS* genes were classified into two groups including 68 type I (31 Mα, 12 Mβ and 25Mγ) and 76 type II (70 MIKC^{C} and 6 MIKC*). Among them, 41 (28.47%) *RsMADS* genes were located in nine linkage groups of radish from R1 to R9. Moreover, the homologous MADS-box gene pairs were identified among radish, *A. thaliana*, Chinese cabbage and rice. Additionally, the expression profiles of *RsMADS* genes were systematically investigated in different tissues and growth stages. Furthermore, quantitative real-time PCR analysis was employed to validate expression patterns of some crucial *RsMADS* genes. These results could provide a valuable resource to explore the potential functions of *RsMADS* genes in radish, and facilitate dissecting MADS-box gene-mediated molecular mechanisms underlying flowering and floral organogenesis in root vegetable crops.

Keywords: radish, MADS-box genes, flowering, floral organogenesis, RT-qPCR

INTRODUCTION

MADS-box genes, which were primarily identified as floral homeotic genes, encode a family of transcription factors (TFs) containing a highly conserved MADS domain of approximately 60-amino-acid sequences in the N-terminal region (Norman et al., 1988), which bind to (CC[A/T]$_6$GG) that is also known as CArG boxes (Pellegrini et al., 1995; Shore and Sharrocks, 1995; Sasaki et al., 2010). Based on phylogenetic relationships, MADS-box genes have been classified into two broad groups, type I and type II genes, which were generated by single gene duplication (Alvarez-Buylla et al., 2000; Liu et al., 2013). Among them, type I proteins are further divided into three subgroups including Mα, Mβ and Mγ, while type II can be classified into the subgroups MIKC^C and MIKC* according to the sequence divergence at I domain (De Bodt et al., 2003; Wells et al., 2015). The MIKC* type proteins have a longer I domain and a less conserved K domain than the MIKC^C type (Henschel et al., 2002; Gramzow and Theißen, 2013). Previous reports revealed the type I MADS-box genes encode SRF-like domain proteins, while type II genes encode MEF2-like proteins and MIKC-type proteins (De Bodt et al., 2003; Wells et al., 2015). Intriguingly, the most well-known MADS-box proteins belong to MIKC-type proteins which contains four common domains including MADS (M), Intervening (I), Keratin (K) and the C-terminal (C) domain (Kaufmann et al., 2005). Compared with type II, the type I proteins lack the K domain and show a relatively simple gene structure that usually only have one or two exons (Smaczniak et al., 2012; Kaufmann et al., 2005). At present, 62 Type I and 46 Type II genes have been identified and characterized in *A. thaliana* (Parenicová et al., 2003). Among the 46 Type II genes, 39 MIKC^C type genes were further classified into 12 groups based on their phylogenetic relationships, nevertheless, there were only seven genes belonging to the MIKC* type (Duan et al., 2015).

In plants, increasing evidences from genetic and molecular analyses have revealed that MADS-box genes could play critical roles in regulating diverse developmental processes, such as flower organogenesis (Zahn et al., 2006), determination of flowering time (Moon et al., 2003; Adamczyk et al., 2007; Lee et al., 2007; Liu et al., 2008; Hu et al., 2014), regulation of fruit ripening (Liljegren et al., 2000), development of vegetative organs (Tapia-López et al., 2008), seed pigmentation and embryo development (Nesi et al., 2002). MIKC^C-type MADS-box genes play fundamental roles especially in flowering time control and floral organ identity. Based on the proposed ABC model (Haughn and Somerville, 1988), the ABCDE model that determines the identity of floral organs has been presented. Different floral organs identities are controlled by various combinations of types of genes, sepals (A+E), petals (A+B+E), stamens (B+C+E), carpels (C+E) and ovules (D+E) (Zahn et al., 2006). A series of correlative functional genes were found to be involved in this process, such as Class A, *APETALA1* (*AP1*); Class B, *PISTILATA* (*PI*) and *AP3*; Class C, *AGAMOUS* (*AG*); Class D, *SEEDSTICK* (*STK*); and Class E, *SEPALLATA* (*SEP1*, *SEP2*, *SEP3* and *SEP4*) (Parenicová et al., 2003).

In recent decades, several crucial MIKC^C-type genes have been suggested to modulate flowering time in *A. thaliana*. For instance, *FLOWERING LOCUS C* (*FLC*) gene has been found to inhibit flowering by encoding a specific MADS domain protein (Michaels and Amasino, 1999). *SUPPRESSOR OF OVEREXPRESSION OF CO1* (*SOC1*) gene plays a critical role in vernalization and gibberellin signal integration for flowering (Moon et al., 2003). *SHORT VEGETATIVE PHASE* (*SVP*) is considered as an important control factor of flowering time by ambient temperature (Lee et al., 2007). Moreover, *AGAMOUS-LIKE16 (AGL16)* gene targeted by miR824 contributes to the repression of plant flowering time (Hu et al., 2014).In addition, several other MIKC^C-type genes have also been proven to be involved in flowering time, such as *AGAMOUS-LIKE 24* (*AGL24*) (Liu et al., 2008), *MADS AFFECTING FLOWERING* (*MAF1/FLM*) (Ratcliffe et al., 2003) and *AGAMOUS-LIKE 15/18* (*AGL15/AGL18*) (Adamczyk et al., 2007). More intriguingly, compared to MIKC^C-type genes, relatively less study has been conducted on the functions of MIKC*-type and Type I genes. To date, MIKC*-type and Type I genes only have been shown to participate in the *A. thaliana* male and female gametophyte, respectively (Zobell et al., 2010; Masiero et al., 2011). Furthermore, recent studies have revealed that Type I genes are primarily expressed in developing seed of *A. thaliana* (Barker and Ashton, 2013).

Radish (*Raphanus sativus* L., $2n = 2x = 18$) is an important root vegetable crop of Brassicaceae family worldwide (Xu et al., 2013). In the complete life cycle of radish, bolting and flowering are some of the critical factors which affect the yield and quality. Premature bolting seriously decreases the production of vegetable crops which ultimately lead to the reduction of economic benefits (Nie et al., 2015). Consequently, it is extremely essential to explore the MADS-box gene family whose primary function is to regulate flowering time and floral organ development. Recently, genome-wide identification and characterization of MADS-box genes were reported in some plant species including *A. thaliana* (Parenicová et al., 2003), rice (Arora et al., 2007), Chinese cabbage (Duan et al., 2015), cucumber (Hu and Liu, 2012), soybean (Fan et al., 2013) and maize (Zhao et al., 2011). However, the genome-wide analysis and characterization of MADS-box genes in radish remain lacking. Especially, it is ambiguous how MADS-box genes control flowering time and floral organ development in radish. Fortunately, the completion of the radish genome sequencing makes it possible to analyze MADS-box genes (Mitsui et al., 2015). In the present study, MADS-box members from radish genome were firstly identified and divided into different classes, and the gene structures, conserved motifs and phylogenetic relationships between these members were systematically analyzed. Additionally, linkage group locations and primary prediction of gene functions were also investigated, and the expression patterns of all MIKC^C genes in radish were carried out with RT-qPCR. These results would greatly contribute to gain insight into functional analysis of MADS-box genes and facilitate dissecting MADS-box gene-mediated molecular mechanisms underlying flowering and floral organogenesis in radish and other root vegetable crops.

MATERIALS AND METHODS

Identification of MADS-Box Genes

All radish genome sequences used to identify the MADS-box genes were available from the NODAI Radish genome database[1]

[1]http://www.nodai-genome-d.org/

(Mitsui et al., 2015). To confirm the candidates of radish MADS-box genes, the proteins with SRF-TF domain (Pfam accession number:PF00319)[2] were searched against the genome protein sequences using HMM search tool with an *E*-value cut-off 1.0 (Finn et al., 2011; Finn et al., 2015). Each sequence predicted was subsequently verified through the public databases including NCBI[3], Pfam and SMART[4] to confirm its reliability (Letunic et al., 2012).

Sequence Collection from Various Plant Species

The MADS-box protein sequences of *A. thaliana* and Chinese cabbage were downloaded from TAIR database[5] and *Brassica* database (BRAD[6]) (Wang et al., 2011), respectively. *Capsicum annuum* and *Brassica oleracea* genome protein sequences were retrieved from pepper genome platform[7] (Kim et al., 2014) and *Brassica* database, respectively. The genome data of *Beta vulgaris, Fragaria vesca, Phaseolus vulgaris, Ricinus communis, Brachypodium distachyon, Setaria italica, Amborella trichopoda* and *Chlamydomonas reinhardtii* were downloaded from the genome browser phytozome[8]. All these collected genome sequences were used to screen MADS-box genes from various plant species through the Pfam database. All the sequences of the other species used in this study were collected from previous reports (Parenicová et al., 2003; Leseberg et al., 2006; Arora et al., 2007; Díaz-Riquelme et al., 2009; Zhao et al., 2011; Gramzow et al., 2012; Barker and Ashton, 2013; Fan et al., 2013; Duan et al., 2015).

Linkage Group Localization and Identification of Orthologous and Paralogous Genes

The sequences of *RsMADS* genes were searched against the genomic sequences of the scaffolds that were anchored to the integrated high-density linkage map (Kitashiba et al., 2014). The gene sequences with identity ≥98% and length difference ≤5 bp were considered to be the same genes between the two genomes, and localized to the linkage groups according to the corresponding location parameters using MapInspect Software[9].

To gain insight into the homology relationship between MADS-box genes of radish and other species, we investigated the orthologous and paralogous MADS-box genes in radish, *A. thaliana,* Chinese cabbage and rice using OrthoMCL program[10] (Li et al., 2003). Subsequently, the relationship networks of homologous genes in radish and *A. thaliana* was visualized using Cytoscape software (Shannon et al., 2003).

[2]http://pfam.xfam.org/
[3]http://www.ncbi.nlm.nih.gov/
[4]http://smart.embl-heidelberg.de/
[5]http://www.arabidopsis.org/
[6]http://brassicadb.org/brad/
[7]http://peppergenome.snu.ac.kr/
[8]http://www.phytozome.net/
[9]http://mapinspect.software.informer.com/
[10]http://www.orthomcl.org/cgi-bin/OrthoMclWeb.cgi

Identification of Protein Properties, Gene Structure and Conserved Motifs and Phylogenetic Analysis

ProtParam tool of ExPASy[11] was employed to analyze series of RsMADS protein properties like molecular weight, theoretical pI and instability index. The Pfam database and SMART were employed to determine conserved domains of proteins. After that, the GSDS[12] was adopted to reveal intron-exon structure of *RsMADS* genes. Conserved motifs were identified using Motif Elicitation (MEME) software[13], and the parameters settings as follows: (1) 10 ≤ optimum motif width ≤100; and (2) maximum number of motifs = 15. In addition, multiple alignments of MADS-box gene sequences were performed using ClustalX 2.0 with default parameters. MEGA 5.1 (Tamura et al., 2011) was then used to construct the phylogenetic tree based on neighbor-joining (NJ) method, and bootstrap values were set to 1,000 replications.

Prediction of miRNAs Targeting the *RsMADS* Genes

To identify potential miRNAs targeting the *RsMADS* genes, all *RsMADS* genes were searched against a comprehensive miRNA library on psRNATarget Server[14] with default parameters (Dai and Zhao, 2011), which was constructed according to the previously established five miRNA libraries (Xu et al., 2013; Nie et al., 2015; Sun X. et al., 2015; Sun Y. et al., 2015; Yu et al., 2015). After that, Cytoscape software was utilized to visualize the targeted relationship between predicted miRNA and corresponding *RsMADS* genes.

Expression Analysis Using Radish RNA-seq Data

Illumina RNA sequencing showed gene expression of radish varied in the different tissues and developmental stages (Mitsui et al., 2015). In this study, the Illumina RNA-Seq data, which were downloaded from NODAI radish genome database, were used for the transcriptional profiling of *RsMADS* genes in five tissues (cortical, cambium, xylem, root tip and leaf) and six stages of leaf [7, 14, 20, 40, 60 and 90 days after sowing(DAS)]. The expression level for each *RsMADS* gene was presented by the RPKM (Reads Per kb per Million reads) method (Mitsui et al., 2015). Lastly, heat maps were generated by Cluster 3.0[15] (de Hoon et al., 2004) and Tree View[16] (Saldanha, 2004).

Plant Material and Treatments

The radish advanced inbred line, 'NAU-DY13,' was used in the current study. Germinated seeds were vernalized and sown in plastic pots and cultivated in controlled-environment growth chamber with day/night temperature of 28/18°C. For

[11]http://web.expasy.org/protparam/
[12]http://gsds.cbi.pku.edu.cn/
[13]http://meme-suite.org/tools/meme
[14]http://plantgrn.noble.org/psRNATarget/
[15]http://bonsai.hgc.jp/~mdehoon/software/cluster/software.htm
[16]http://jtreeview.sourceforge.net/

vernalization treatments, germinated seeds were vernalized at 2–4°C for 0, 10 and 30 days, respectively, and grow under the middle-day (12 h light/12 h dark). For photoperiodic treatments, unvernalized seedlings were cultured under long-day (16 h light/8 h dark) and short-day (8 h light/16 h dark) treatments, respectively. Furthermore, the rest of unvernalized seedlings were treated with 200 mg/L and 800 mg/L GA_3 every other day for a week from 2-week-old seedlings under the middle-day condition. Unvernalized seedlings without any treatment grown under the middle-day were set as control (CK). Leaf samples were collected when treated seedlings were grown to three weeks old. Different flower tissues from control plants, including sepal, petal, stamen and carpel, were collected separately at reproductive stage. All the samples were collected from three randomly selected individuals and immediately frozen in liquid nitrogen and stored at −80°C for further use.

RNA Isolation and RT-qPCR

Total RNA of each sample was isolated using Trizol reagent according to the manufacturer's instructions (Invitrogen, Carlsbad, CA, USA). Then, the first-strand cDNA was synthesized using the Superscript III First-Strand Synthesis System (Invitrogen). The specific primers of *RsMADS* genes for RT-qPCR were designed using Beacon Designer 7.7 (Premier Biosoft International, Palo Alto, CA, USA). To confirm results reliability, three biological and three technical replicates were adopted. RT-qPCR reaction system and cycling profile were carried out on Bio-Rad iQ5 Real-Time PCR System. *RsActin* gene was selected as the reference gene (Xu et al., 2013). The primers used for RT-qPCR were shown in **Supplementary Table S1**. Finally, formula $-\Delta\Delta C_T$ and $2^{-\Delta\Delta C_T}$ were applied to calculate the relative expression ratio. The data were statistically analyzed with Duncan's multiple range test at the $P < 0.05$ level of significance using SPSS 20 software (SPSS Inc., USA).

RESULTS

Identification and Analysis of MADS-Box Proteins in Radish

To define the candidate MADS-box proteins in radish, a profile hidden Markov model (HMM) search against NODAI radish genome protein sequences was carried out using the SRF-TF domain (PF00319), and totally 157 putative MADS-box protein genes were obtained. The low quality sequences without start and/or stop codons were removed to ensure the reliability of these sequences, and finally a total of 146 sequences were retained. Subsequently, all remaining sequences were verified through the public databases including NCBI, Pfam and SMART. All these radish MADS-box proteins were named as RsMADS001 to RsMADS146, respectively (**Supplementary Table S2**). After searching these protein sequences against *A. thaliana* on TAIR database by BLASTP, RsMADS040 and RsMADS091 were removed, because they contained other functional domains and their homologous proteins were non-MADS-box proteins (**Supplementary Figure S1**). To study the comparative evolution among various plant species, MADS-box genes from 28 other plant species were also collected by searching for SRF-TF domain (PF00319) in their genomes (**Figure 1A**; **Supplementary Tables S3** and **S4**). Compared with other species, radish had a relatively large MADS-box gene family of 144 members, and the members of MADS-box gene family subgroups were also identified (**Figure 1A**).

Comparative Phylogenetic Analysis of *RsMADS* Genes

To better understand the phylogenetic relationships of the MADS-box genes between radish and *A. thaliana*, the classification of 108 MADS-box genes from *A. thaliana* was performed (**Supplementary Figure S2A**). An unrooted phylogenetic tree of MADS-box genes between radish and *A. thaliana* was constructed by the NJ method (**Supplementary Figure S2B**). It is quite obvious that *RsMADS* genes were divided into five clades according to the classification of the *A. thaliana*, namely subfamilies $MIKC^C$ (70), MIKC* (6), Mα (31), Mβ (12) and Mγ (25) (**Supplementary Figures S2B,D**). Additionally, an unrooted phylogenetic tree was produced using MADS-box proteins from radish, *A. thaliana* and Chinese cabbage to further confirm the phylogenetic relationships and classification of RsMADS proteins (**Supplementary Figure S2C**).

Phylogenetic trees for type I and type II MADS-box genes were separately generated using *A. thaliana* and radish proteins (**Figures 2A,B**). Totally 49% (70) of the 144 *RsMADS* genes belongs to the $MIKC^C$-type genes, which could be further divided into 12 subfamilies (**Figures 1B** and **2A**). Subgroup SOC1 and subgroup Bs, respectively, showed the largest (~21%) and smallest (~3%) number of *RsMADS* genes. However, in *A. thaliana* the largest and smallest proportion of subgroup is ~15% (subgroup SOC1 or FLC) and ~3% (subgroup AGL12), respectively (Parenicová et al., 2003). The number of *RsMADS* genes from other subgroups ranged from four to 13, interestingly, subgroup C/D and E had seven members, while subgroup B, FLC and AGL15 consisted of five members (**Figures 1B** and **2A**).

Linkage Group Localization and Orthologous Relationship Analysis

In total, 41 *RsMADS* genes (20 type I and 21 type II) accounting for 28.5% of the total MADS-box gene number were separately anchored onto the approximate location of linkage group (LG) R1-R9 (**Figure 3**; **Supplementary Table S5**). On the whole, the distribution of 41 *RsMADS* genes was relatively dispersed, but there were also some *RsMADS* gene cluster, for example, six genes clustered in the front of LG R2. Among the nine LGs, LG R4 contained the most *RsMADS* genes (9 members, ~22%), while LG R7 and R8 presented the least member (1 member, ~2%) (**Figure 3**). Moreover, there were two MIKC*-type genes that were successfully anchored on the LG R9.

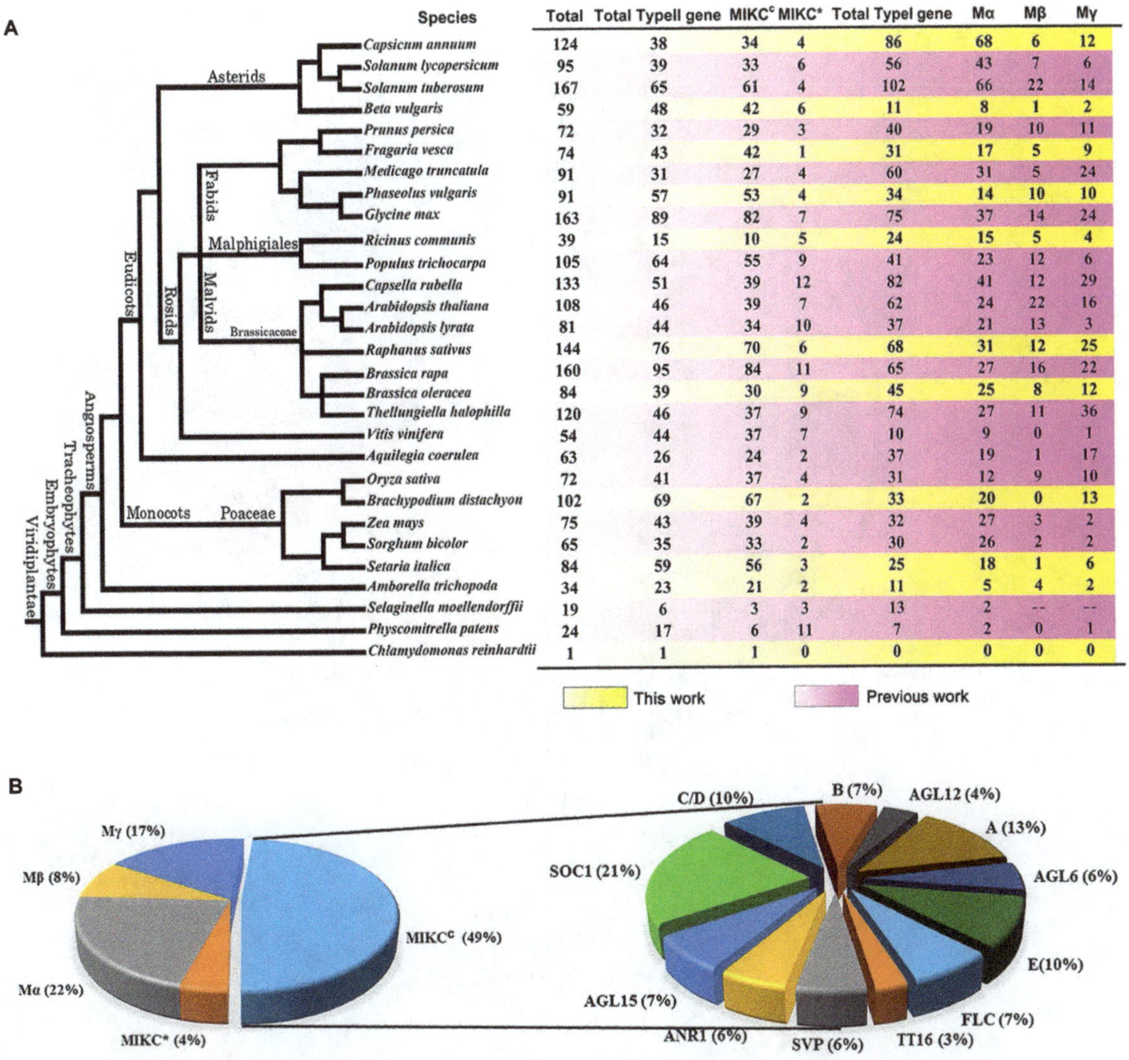

Species	Total	Total TypeII gene	MIKC^c	MIKC*	Total TypeI gene	Mα	Mβ	Mγ
Capsicum annuum	124	38	34	4	86	68	6	12
Solanum lycopersicum	95	39	33	6	56	43	7	6
Solanum tuberosum	167	65	61	4	102	66	22	14
Beta vulgaris	59	48	42	6	11	8	1	2
Prunus persica	72	32	29	3	40	19	10	11
Fragaria vesca	74	43	42	1	31	17	5	9
Medicago truncatula	91	31	27	4	60	31	5	24
Phaseolus vulgaris	91	57	53	4	34	14	10	10
Glycine max	163	89	82	7	75	37	14	24
Ricinus communis	39	15	10	5	24	15	5	4
Populus trichocarpa	105	64	55	9	41	23	12	6
Capsella rubella	133	51	39	12	82	41	12	29
Arabidopsis thaliana	108	46	39	7	62	24	22	16
Arabidopsis lyrata	81	44	34	10	37	21	13	3
Raphanus sativus	144	76	70	6	68	31	12	25
Brassica rapa	160	95	84	11	65	27	16	22
Brassica oleracea	84	39	30	9	45	25	8	12
Thellungiella halophilla	120	46	37	9	74	27	11	36
Vitis vinifera	54	44	37	7	10	9	0	1
Aquilegia coerulea	63	26	24	2	37	19	1	17
Oryza sativa	72	41	37	4	31	12	9	10
Brachypodium distachyon	102	69	67	2	33	20	0	13
Zea mays	75	43	39	4	32	27	3	2
Sorghum bicolor	65	35	33	2	30	26	2	2
Setaria italica	84	59	56	3	25	18	1	6
Amborella trichopoda	34	23	21	2	11	5	4	2
Selaginella moellendorffii	19	6	3	3	13	2	--	--
Physcomitrella patens	24	17	6	11	7	2	0	1
Chlamydomonas reinhardtii	1	1	1	0	0	0	0	0

This work / Previous work

FIGURE 1 | The classification of MADS-box family genes. (A) The number of the MADS-box family genes of 29 plant species. **(B)** The classification and proportions of *RsMADS* genes.

In the present study, orthologous and paralogous MADS-box genes between radish and other three plant species (Chinese cabbage, *A. thaliana* and rice) were comparatively analyzed by OrthoMCL Software. Among all the MADS-box genes, 60 orthologous and 38 co-orthologous gene pairs were found between radish and *A. thaliana*. Nevertheless, only 19 orthologous and 22 co-orthologous gene pairs were detected between radish and rice, and 16 orthologous and 18 co-orthologous gene pairs were found between *A. thaliana* and rice (**Figure 4**; **Supplementary Table S6**). Furthermore, a relational graph was used to visualize all the relationships among the orthologous, co-orthologous and paralogous MADS-box genes between radish and *A. thaliana* (**Supplementary Figure S3**). 29 and 50 *AtMADS* genes were determined to have no and only one orthologous gene in radish, respectively. While, 68 and 70 paralogous MADS-box gene pairs were detected in radish and *A. thaliana*, respectively (**Supplementary Figure S3**).

Characterization of RsMADS Proteins, Conserved Motif Distribution and Intron–Exon Structure

To gain insight into the molecular characterization of 144 RsMADS proteins, their physical and chemical properties including molecular weights (MWs), theoretical isoelectric points (pI), instability index and aliphatic index were analyzed, and all RsMADS proteins were hydrophilic (**Supplementary Table S2, Supplementary Figure S4**). To analyze the features of RsMADS protein sequences, the 15 conserved motifs of 144 RsMADS proteins within the different groups were predicted by performing MEME motif search tool, and the LOGO of 15 amino acid motifs were generated (**Supplementary Figure S5**). Moreover, all of the RsMADS proteins contained motif 1 and motif 3, which indicated that this highly conserved domain was MADS domain. Nevertheless, motif 4 and motif 6 were present

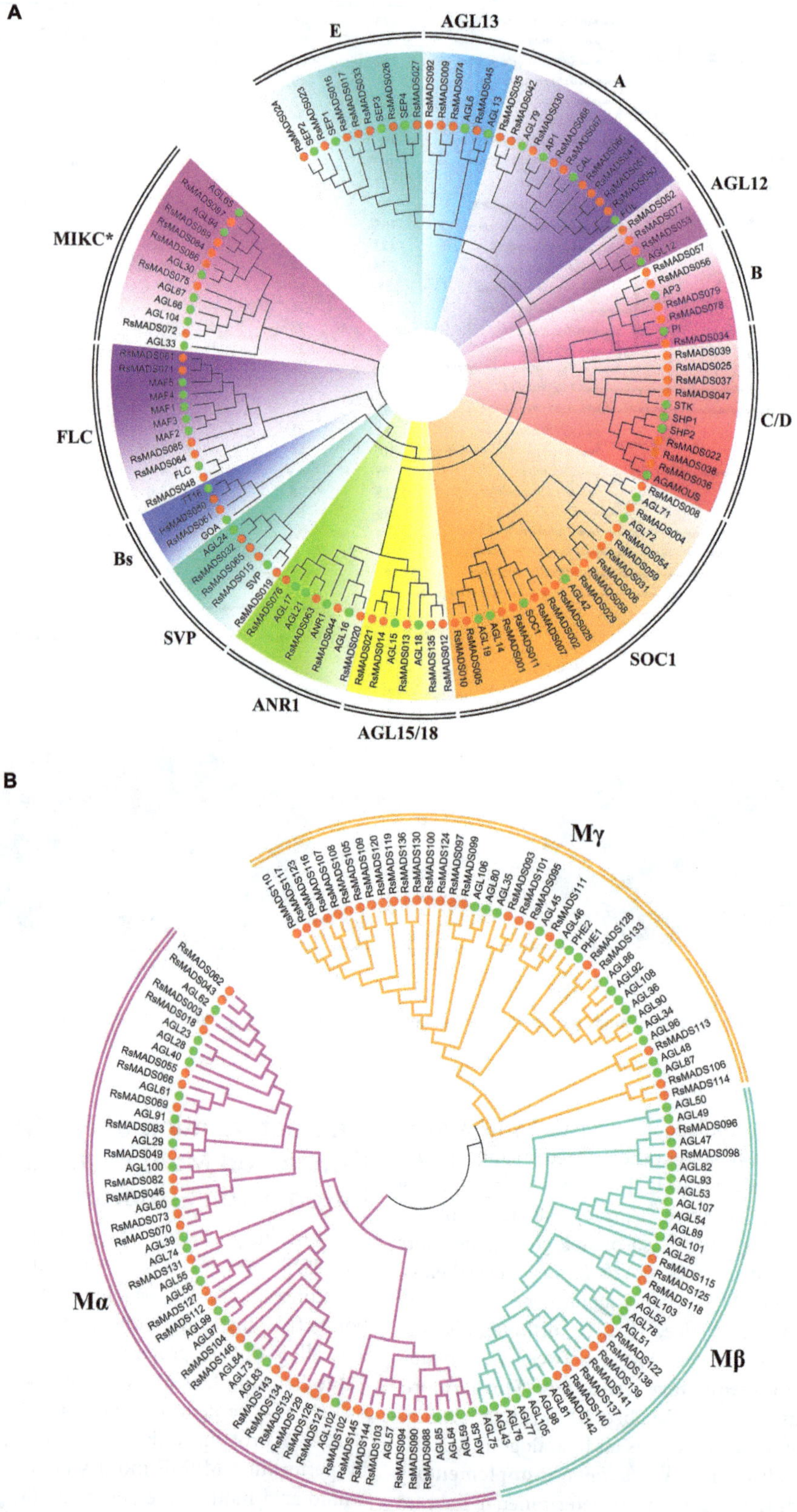

FIGURE 2 | Phylogenetic relationship of radish and *A. thaliana* type II (A) and type I (B) MADS-box proteins. Subgroups are marked in different colors.

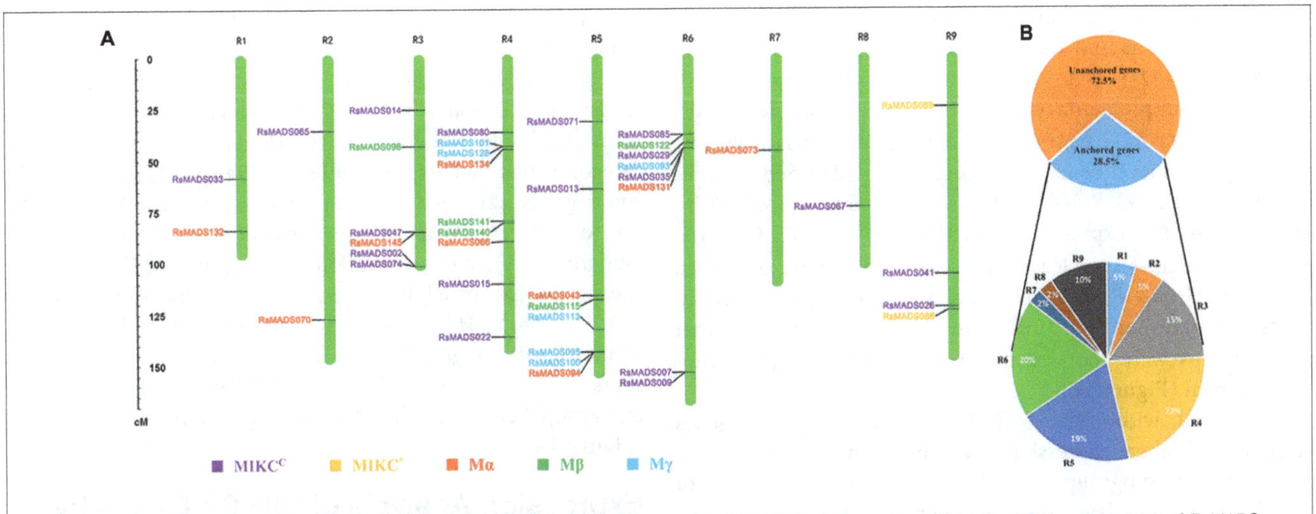

FIGURE 3 | Linkage group (LG) localization of *RsMADS* genes. (A) Distribution of the *RsMADS* genes on nine radish LGs. **(B)** The percentage of *RsMADS* gene anchored onto each LG.

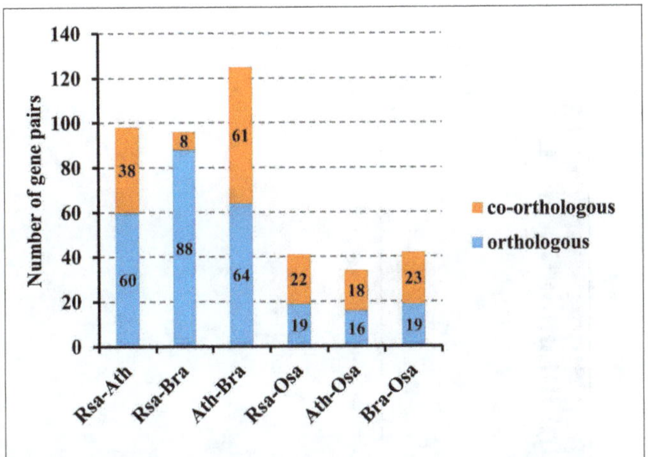

FIGURE 4 | The number of orthologous, co-orthologous gene pairs among radish (Rsa), *A. thaliana* (Ath), Chinese cabbage (Bra) and rice (Osa).

in most of the MIKC-type genes, and thus were predicted to be K-box domain. In addition, 15 motifs were submitted to the Pfam and SMART website for further identification, and provided strong evidences supporting our predictions (**Supplementary Figure S6**). Moreover, according to previous reports (Saha et al., 2015; Rameneni et al., 2014; Shu et al., 2013) and the conservative characteristics of motifs, motif 8 was predicted to represent the I domain, and motif 10 and motif 14 specified the C-terminal domain. It should be emphasized that type I group had more distinct motifs at their C-terminal regions except the MADS domain, which were more divergent than those in the type II group, and these motifs were identified as unknown by Pfam and SMART (**Supplementary Figure S6D**). Motif analysis showed that the majority of RsMADS proteins in the same subgroup shared similar motif distribution, suggesting that the proteins from the same subgroup probably had similar functions (Parenicová et al., 2003; Song et al., 2015).

Additionally, the intron–exon patterns were analyzed to investigate the structural diversity of *RsMADS* genes. Comparison of genomic DNA and cDNA showed that type I *RsMADS* genes had no or only one intron except *RsMADS133* containing five introns (**Supplementary Figure S6C, Supplementary Table S2**). Compared with type I, type II *RsMADS* genes had more complex structures. The intron numbers of type II *RsMADS* varied from 0 to 16 with an average of 5.6, and 60 (78.9%) members were consisted of at least five introns (**Supplementary Figure S6C, Supplementary Table S2**).

Analysis of miRNAs Targeting *RsMADS* Genes

To have a better understanding of the function of MADS-box gene family in radish, a comprehensive miRNA library consisting of five miRNA libraries reported from our previous studies was used to determine miRNAs targeting *RsMADS* genes by psRNATarget program. Totally, 19 known miRNAs and six potential novel miRNAs (named Rsa-miR1-Rsa-miR6) belonging to 25 miRNA families were identified as putative miRNAs which could target 25 *RsMADS* target transcripts (**Supplementary Table S7**). The regulatory relationship between putative miRNAs and their targets were presented in **Supplementary Figure S7**. *RsMADS027* was the target transcript of miR8154, miR5293 and miR831-5p, *RsMADS084* was targeted by miR5174e-5p, Rsa-miR4 and Rsa-miR3, while four transcripts (*RsMADS087*, *RsMADS125*, *RsMADS138* and *RsMADS140*) were targeted by miR5021 (**Supplementary Figure S7**). It is worth noting that miR824 was predicted to target *RsMADS020* and *RsMADS044*, whose sequences showed high similarity with *AGL16* in *A. thaliana* (**Supplementary Table S2**).

Differential Expression Analysis of *RsMADS* Genes

To estimate the expression levels of *RsMADS* genes, RPKM of 144 *RsMADS* genes in leaves from seven different development stages and in five different tissues was obtained (Mitsui et al., 2015). The results showed that the transcript abundances of different *RsMADS* genes were extremely diverse in radish (**Figure 5**; **Supplementary Table S8**). On the whole, almost all Type I *RsMADS* genes either maintained a relatively low transcriptional level or had no expression in RNA-Seq libraries except *RsMADS093*, *RsMADS097*, *RsMADS106* and *RsMADS111* (**Figure 5B**). The expression of *RsMADS097* and *RsMADS106* were downregulated in leaves with the development of radish. *RsMADS093* and *RsMADS111* have high expression levels in roots but were hardly expressed in the leaves, indicating that they perhaps were root-specific and play a vital role in root development (**Figure 5B**).

Compared with Type I *RsMADS* genes, Type II genes showed a higher expression level both in the roots and leaves except subgroup B and C/D. With the growth of radish, expression levels of some genes increased gradually, including *RsMADS042*, *RsMADS050* (subgroup A); *RsMADS001*, *RsMADS002*, *RsMADS010*, *RsMADS029* (subgroup SOC1); *RsMADS135* (subgroup AGL15/18); *RsMADS032*, *RsMADS065* (subgroup SVP), while others decreased such as *RsMADS33* (subgroup E), *RsMADS13* (subgroup AGL15/18), *RsMADS44* (subgroup ANR1) and *RsMADS36* (subgroup C/D) (**Figure 5A**). In leaves and roots, the ABCDE model genes showed low transcript levels, while SOC1, AGL15/18, ANR1, SVP and FLC categories had high expression levels (**Figure 5A**). Interestingly, some genes exhibited tissue-specific expression (**Figure 5A**). For example, *RsMADS074* (subgroup AGL13), *RsMADS050* (subgroup A) and *RsMADS089* (subgroup MIKC*) were specifically expressed in leaves, whereas some genes such as *RsMADS004*, *RsMADS008* (subgroup SOC1) and *RsMADS043*, *RsMADS044* (subgroup ANR1), were specifically expressed in roots. In addition, *RsMADS052*, *RsMADS053* and *RsMADS077* (subgroup AGL12) displayed a high expression level in root tips (**Figure 5A**).

Expression Analysis of MIKC^{C} Genes by RT-qPCR

To further reveal the function of 12 subgroups of radish MIKC^{C} genes, the relative expression levels of genes in A, B,

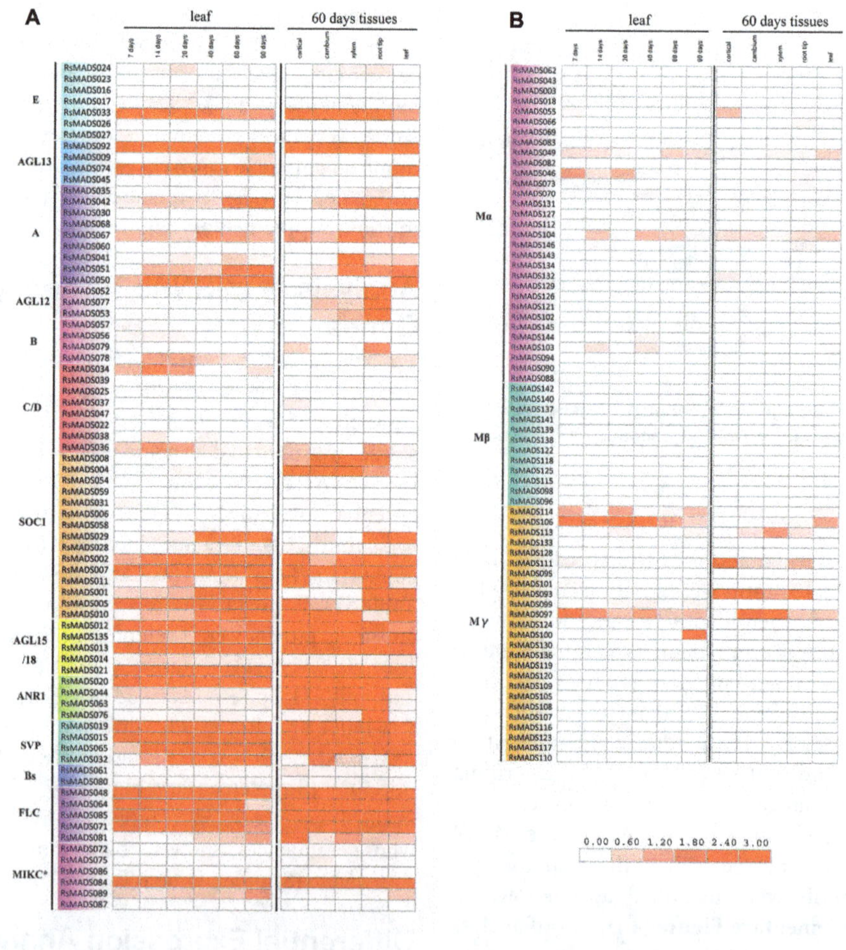

FIGURE 5 | Expression heat map of *RsMADS* genes in different stages and tissues. (A,B) Represent Type II and Type I *RsMADS* expression profiles in six stages (7, 14, 20, 40, 60 and 90 DAS) and five tissues (cortical, cambium, xylem, root tip and leaf). The expression value was calculated by reads per kilobase per million reads (RPKM). The subgroup is marked on the left side of the gene list. The scale represents relative expression value.

C/D, E, AGL6/13 and AGL12 subgroups were comprehensively investigated in various parts of floral organs (sepal, petal, stamen and carpel), and the genes in SOC1, AGL15/18, AGL16, SVP, Bs and FLC subgroups were validated under different GA concentrations, light length and vernalization time using RT-qPCR (**Supplementary Figure S8**).

All *RsMADS* genes showed differential expression patterns in different parts of floral organs (**Supplementary Figure S8**). The orthologous *RsMADS* genes with *A. thaliana* ABCDE model genes were selected for further analysis. *RsMADS68* (*AP1*) exhibited high expression level in sepal and petal, as compared with that in stamen and carpel, whereas *RsMADS057* (*AP3*) and *RsMADS078* (*PI*) were significantly expressed in petal and stamen (**Figure 6**). More interestingly, *RsMADS036* (*AG*) tended to be expressed in stamen and carpel, while *RsMADS047* (*STK*) was significantly up-regulated only in carpel (**Figure 6**). Moreover, the expression patterns of E subgroup genes were more diverse. The expression levels of *RsMADS017* (*SEP1*) and *RsMADS023* (*SEP2*) were relatively steady in the four tissues, whereas *RsMADS033* (*SEP3*) and *RsMADS026* (*SEP4*) maintained relatively high expression levels in one (petal) and two (petal and carpel) specific tissues, respectively (**Figure 6**).

Additionally, it is apparent that three different treatments (GA, light and vernalization) resulted in a wide variety of expression profiles among *RsMADS* genes (**Supplementary Figure S8**). The orthologs of seven representative genes including *RsMADS002*, *RsMADS012*, *RsMADS020*, *RsMADS021*, *RsMADS015*, *RsMADS064* and *RsMADS065* were reported to be crucial in flowering control in *A. thaliana*. The results showed that *RsMADS002* (*SOC1*) and *RsMADS065* (*AGL24*) were up-regulated under different concentrations of GA treatments, but *RsMADS015* (*SVP*), *RsMADS020* (*AGL16*), *RsMADS021* (*AGL15*) and *RsMADS064* (*FLC*) were obviously down-regulated (**Figure 7A**). *RsMADS015* (*SVP*) were down-regulated following the decrease of light lengths, whereas the transcript accumulation of *RsMADS020* (*AGL16*) and *RsMADS065* (*AGL24*) were relative lower at short-day (SD) and peaked at long-day (LD) (**Figure 7B**). Intriguingly, most members showed strong sensitivity toward vernalization treatment. Along with prolonging vernalization time, *RsMADS002* (*SOC1*) and *RsMADS065* (*AGL24*) were evidently induced, by contrast, the other five genes were inhibited inordinately (**Figure 7C**).

The MADS-Box Gene-Mediated Regulation Associated with Flowering and Floral Organogenesis

In the present study, 16 critical *RsMADS* genes including *RsMADS002* (*SOC1*), *RsMADS012* (*AGL18*), *RsMADS015* (*SVP*), *RsMADS020* (*AGL16*), *RsMADS021* (*AGL15*), *RsMADS064* (*FLC*), *RsMADS065* (*AGL24*), *RsMADS68* (*AP1*), *RsMADS057* (*AP3*), *RsMADS078* (*PI*), *RsMADS036* (*AG*), *RsMADS047* (*STK*), *RsMADS017* (*SEP1*), *RsMADS033* (*SEP3*), *RsMADS023* (*SEP2*) and *RsMADS026* (*SEP4*) were identified to be involved radish flowering and floral organ formation. According to the reported *A. thaliana* and radish flowering and floral organogenesis regulatory network (Posé et al., 2012; Sun et al., 2013; Zhou et al., 2013; Khan et al., 2014; Nie et al., 2016), some critical floral pathway integrator genes, such as *FLC*, *SVP*, *AGL16*, *SOC1*, *AGL24* and *AGL19*, were considered to respond to environmental and endogenous factors directly or indirectly through interacting with other genes, and then these genes further regulate the expression of downstream floral organ identity genes including *AP1*, *AG*, *AP3* and *SEP3*.

DISCUSSION

Higher plants routinely go through various phase transitions from germination to death mainly including juvenile phase, vegetative growth and reproductive development. The vegetative

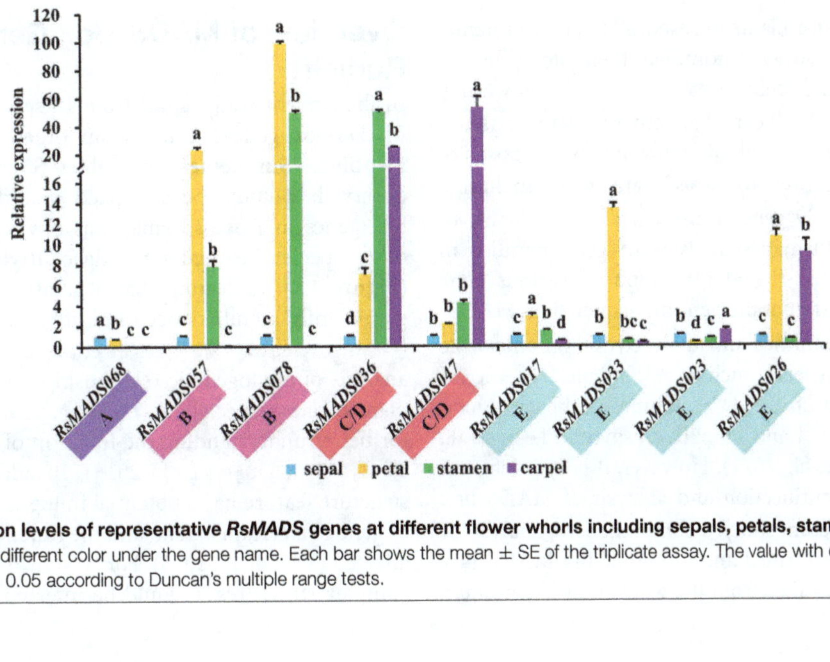

FIGURE 6 | The expression levels of representative *RsMADS* genes at different flower whorls including sepals, petals, stamens, carpels and ovules. The subgroup is marked in different color under the gene name. Each bar shows the mean ± SE of the triplicate assay. The value with different letter indicates significant difference at $P < 0.05$ according to Duncan's multiple range tests.

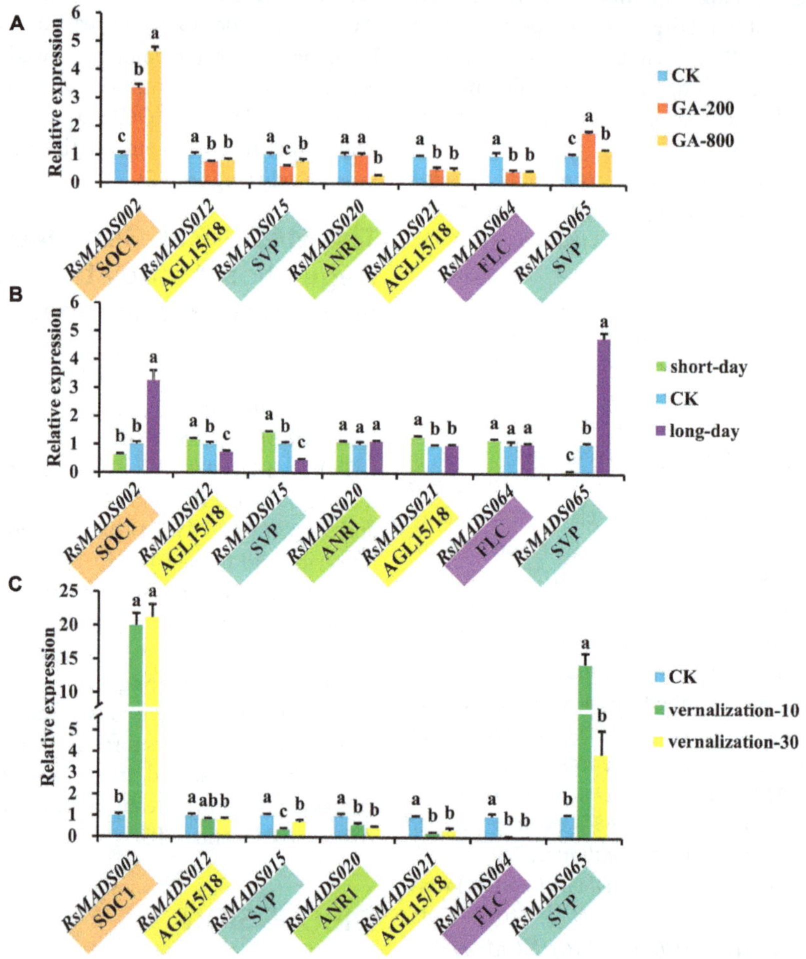

FIGURE 7 | The expression levels of representative *RsMADS* genes under GA (A), photoperiod (B) and vernalization (C) treatments. The subgroup is marked in different colors under the gene name. Each bar shows the mean ± SE of the triplicate assay. The value with different letter indicates significant difference at $P < 0.05$ according to Duncan's multiple range tests.

phase change is essential for plants in response to environmental and endogenous factors, so as to complete their life cycle and achieve reproduction successfully (Khan et al., 2014). Flowering, as a symbol of plants into the reproductive growth phase, is determined by a complex gene interaction which is composed of a crowd of flowering and organogenesis related genes including most of MADS-box family genes (Smaczniak et al., 2012). In recent years, the bioinformatics analysis of gene families in different species facilitated the identification of various gene families with the completion of the genome sequencing. MADS-box gene families were identified and analyzed at a genome-wide scale in a series of plant species such as *A. thaliana* (Parenicová et al., 2003), rice (Arora et al., 2007), Chinese cabbage (Duan et al., 2015), cucumber (Hu and Liu, 2012), soybean (Fan et al., 2013) and maize (Zhao et al., 2011). However, it is still deficient for the genome-wide identification and analysis of MADS-box genes in radish. In the present study, a comprehensive analysis of MADS-box genes was performed and a total of 144 MADS-box family members were identified from the whole radish genome.

Overview of MADS-Box Gene Family in Radish

In the current study, apart from 18 species reported previously, MADS-box gene families from other 10 species were firstly identified (**Supplementary Tables S3** and **S4**). As previously observed, it could be suggested that the number of MADS-box genes in Angiospermae was obviously larger than that in other species belonging to Algae, Bryophyta and Lycophytes (**Figure 1A**), indicating that a great expansion of MADS-box gene family members occurred after the angiosperm evolution (Gramzow et al., 2014; Duan et al., 2015). Simultaneously, the analysis of phylogenetic relationships between radish and other plant species, especially *A. thaliana*, provided a solid foundation for better understanding the function of *RsMADS* genes (Wang et al., 2015; Duan et al., 2015). In addition, the intron–exon structure feature has a potential influence on alternative splicing of gene to a certain extent, and the function of the protein will be affected (Tian et al., 2015). For type II *RsMADS* genes with more complex structures, it could be inferred that type II genes had more variable and intricate function than type I genes, which was in accordance with previous results in *A. thaliana* (Parenicová et al., 2003), Chinese cabbage (Duan et al., 2015) and soybean (Fan et al., 2013).

Previous evidence has shown that known miR5227 and novel Rsa-miR4 played a role in the bolting and flowering process of radish by high-throughput sequencing technology (Nie et al., 2015). Therefore, their target genes *RsMADS115* (*AGL103*) and *RsMADS084* (*AGL30*) might be associated with regulation of bolting and flowering time in radish. Furthermore, previous observations confirmed that miR824-regulated *AGL16* inhibited flowering in *A. thaliana* (Hu et al., 2014). In this study, miR824 was identified to target *RsMADS020* and *RsMADS044* (*AGL16*), revealing that these target genes may contribute to flowering time repression in radish.

Characterization of Critical *RsMADS* Genes in Flowering and Floral Organ Development

Biochemical and genetic studies have indicated that flowering and floral organogenesis can be modulated by MADS-box genes especially MIKC^{C} type in higher plants (Lee et al., 2013; Ó'Maoiléidigh et al., 2014). Meanwhile, a growing number of key MADS-box genes including *FLC, SOC1, SVP, AGL24, AGL16, AGL15* and *AGL18,* and ABCDE model genes involved in this process have been widely recognized (Ferrario et al., 2004; Khan et al., 2014).

Control of flowering time is an intricate genetic circuitry in response to various endogenous and exogenous cues (Wullschleger and Weston, 2012). In *A. thaliana,* molecular genetics and physiological studies revealed that five main pathways of vernalization, photoperiod gibberellin, autonomy and age controlled flowering time (Kim et al., 2009; Mutasa-Göttgens and Hedden, 2009; Srikanth and Schmid, 2011; Wang, 2014). In this study, expression profiles of seven representative genes were investigated and the results suggested that *RsMADS015* (*SVP*), *RsMADS020* (*AGL16*), *RsMADS021* (*AGL15*) and *RsMADS064* (*FLC*) might act as flowering repressor, while *RsMADS002* (*SOC1*) and *065* (*AGL24*) contributed to the flowering promotion in radish (**Figure 7**).

Flower meristem and floral identity had been explained perfectly by five kinds of genetic function genes (A-B-C-D-E), which were important in regulating different flower whorls from sepals to carpels (Ferrario et al., 2004; Li et al., 2015). In the present study, the regulatory relationships between ABCDE genes and floral organ development were analyzed in radish, and a schematic ABCDE model was proposed (**Supplementary Figure S9**). RNA-Seq and RT-qPCR analysis revealed the expression patterns of ABCDE model orthologous genes in different tissues and at different stages of flower development. These genes exhibited relatively low abundant transcripts in the leaf and root (**Figure 5A**) and a regular expression patterns in different flower whorls (**Figure 6; Supplementary Figure S8A**), which were consistent with previous studies (Su et al., 2013; Ó'Maoiléidigh et al., 2014; Xie et al., 2015), suggesting that ABCDE model genes worked in a combinatorial manner to regulate the floral morphogenesis in radish.

The Roles of MADS-Box Genes in Flowering and Flower Formation in Radish

Flowering is a coherent and sophisticated development process (Nie et al., 2015). Flowering-related genes were affected by multiple flowering signals converging on the regulation of floral organ identity genes including *SEP3, AP1, AG* and *AP3,* leading to flower formation eventually (Posé et al., 2012; Sun et al., 2013; Zhou et al., 2013; Khan et al., 2014). Considerable reports have indicated that *FLC* and *SOC1* as floral pathway integrators which were regulated by numerous genes and flowering pathways, played important roles in the flowering process (Lee et al., 2007; Franks et al., 2015). Genetic studies showed that *FLC* could block the transcriptional activation of *SOC1* and required *SVP* and *FRI* to delay flowering strongly (Helliwell et al., 2006; Lee et al., 2007; Geraldo et al., 2009; Franks et al., 2015). *AGL16,* a target gene of miR824, can help to repress flowering time by interacting indirectly with *FLC* and directly with *SVP* in *A. thaliana* (Hu et al., 2014). In this study, RT-qPCR validation showed that *FLC, SVP* and *AGL16* orthologous genes were down-expressed with the increase of GA concentration, light length and vernalization time (**Figure 7**), indicating that they may be repressors of flowering in radish.

Moreover, two other critical MADS-box genes, *SOC1* and *AGL24,* could promote flowering by responding to GA signaling (Moon et al., 2003). Additionally, *AGL15* and *AGL18* acted as the floral repressors via controlling the regulation of *SOC1* and *FT,* and *agl15 agl18* mutations presented a quick increase in *SOC1* and *FT* levels, leading to early flowering (Adamczyk et al., 2007; Fernandez et al., 2014). In the current study, *AGL24, SOC1, AGL15,* and *AGL18* orthologous genes were identified in radish, and RT-qPCR profiling showed that *AGL24* and *SOC1* orthologous genes were up-regulated, while *AGL15* and *AGL18* orthologous genes were obviously down-regulated when treated with different flowering-induced factors (**Figure 7**). These results suggested that *AGL24* and *SOC1* promoted flowering, whereas *AGL15* and *AGL18* inhibited flowering in radish. Therefore, it could be suggested that MADS-box gene family play a major role in regulating flowering time and floral meristem identity in radish.

CONCLUSION

In conclusion, a total of 144 genes encoding MADS-box TF including 68 type I and 76 type II genes were identified in the whole radish genome. Among them, 41 genes were localized on the nine linkage groups of radish. A comparative phylogenetic analysis of the MADS-box genes was carried out between radish and *A. thaliana* to classify the MADS-box proteins. Furthermore, identification of miRNAs targeting *RsMADS* transcripts shed a novel insight into the functions of

RsMADS genes at transcriptional and post-transcriptional level. In addition, RT-qPCR analysis provided a better understanding of critical functions of candidate *RsMADS* genes involved in flowering and floral organ identity in radish. Taken together, in this study, radish MADS-box gene family was comprehensively characterized, which facilitated dissecting *RsMADS* gene-mediated molecular mechanism underlying flowering and floral organogenesis in radish.

AUTHOR CONTRIBUTIONS

All authors listed, have made substantial, direct and intellectual contribution to the work, and approved it for publication.

ACKNOWLEDGMENTS

This work was in part supported by grants from the NSFC (31372064, 31501759, 31601766), National Key Technologies R&D Program of P. R China (2016YFD0100204-25), Key Technology R&D Program of Jiangsu Province (BE2016379), JASTIF(CX(16)2012) and the PAPD.

SUPPLEMENTARY MATERIAL

The Supplementary Material for this article can be found online at: http://journal.frontiersin.org/article/10.3389/fpls.2016.01390

FIGURE S1 | The protein structure and multiple sequence alignment of the (A) *RsMADS040* and (B) *RsMADS091*.

FIGURE S2 | Phylogenetic tree of radish and other species MADS-box proteins. (A) Phylogenetic tree of AtMADS proteins. **(B)** The phylogenetic tree of radish and *A. thaliana* MADS proteins. **(C)** Phylogenetic tree of radish, *A. thaliana* and Chinese cabbage MADS proteins. **(D)** Phylogenetic tree of RsMADS proteins.

FIGURE S3 | The networks of MADS-box genes in radish and *A. thaliana*. This interrelation network was constructed using radish and *A. thaliana* orthologous, co-orthologous and paralogous gene pairs. **(A)** One orthologous gene pair between radish and *A. thaliana*. **(B)** A complex network of orthologous, co-orthologous and paralogous gene pairs. **(C)** Paralogous gene pairs in radish and *A. thaliana*, respectively. **(D)** Statistics of the number of orthologous, co-orthologous and paralogous gene pairs between radish and *A. thaliana*. Orthologous gene pairs are linked in red lines; co-orthologous gene pairs are linked in yellow lines; paralogous gene pairs in radish are linked in black lines; paralogous gene pairs in *A. thaliana* are linked in blue lines.

FIGURE S4 | The physical and chemical properties of RsMADS proteins. (A) The distribution of putative isoelectric points and molecular weights. **(B)** The distribution of Grand Average of hydropathicity (GRAVY), instability index and aliphatic index.

FIGURE S5 | Sequence logos of MADS domains in radish. The overall height of the stack indicates the level of sequence conservation. The height of residues within the stack indicates the relative frequency of each residue at that position.

FIGURE S6 | The analysis of RsMADS proteins and *RsMADS* genes structure. (A) The classification of *RsMADS* genes. **(B)** Phylogenetic tree of RsMADS proteins. **(C)** Intron–exon structure distribution of 144 *RsMADS* genes. **(D)** Conserved motif distribution of 144 RsMADS proteins.

FIGURE S7 | Predicted targeted regulatory network between *RsMADS* genes and miRNAs.

FIGURE S8 | Expression analysis of *RsMADS* at different flower whorls and different treatments. Heat map representation and hierarchical clustering of *RsMADS* genes during sepals, petals, stamens, carpels and ovules **(A)**; and under vernalization, photoperiod and GA treatments **(B)**. The scale represents relative expression value. The subgroup is marked in different color on the right side of the gene list.

FIGURE S9 | Putative schematic ABCDE model of floral organ development in radish.

TABLE S1 | The primer sequences used for RT-qPCR of actin and MIKC^C genes.

TABLE S2 | The information of 144 RsMADS genes in radish.

TABLE S3 | The Information of MADS-box genes of various species from the previous reports.

TABLE S4 | The Information of MADS-box genes from different species identified in this study.

TABLE S5 | The information of linkage group localization of *RsMADS* genes.

TABLE S6 | The orthologous gene pairs and co-orthologous gene pairs in MADS-box proteins of radish, *Arabidopsis*, Chinese cabbage and rice, and the paralogous gene pairs among these species.

TABLE S7 | The information of miRNAs targeting *RsMADS* genes identified from previous five libraries.

TABLE S8 | The RPKM values of *RsMADS* genes.

REFERENCES

Adamczyk, B. J., Lehti-Shiu, M. D., and Fernandez, D. E. (2007). The MADS domain factors AGL15 and AGL18 act redundantly as repressors of the floral transition in *Arabidopsis*. *Plant J.* 50, 1007–1019. doi: 10.1111/j.1365-313X.2007.03105.x

Alvarez-Buylla, E. R., Pelaz, S., Liljegren, S. J., Gold, S. E., Burgeff, C., Ditta, G. S., et al. (2000). An ancestral MADS-box gene duplication occurred before the divergence of plants and animals. *Proc. Natl. Acad. Sci. U.S.A.* 97, 5328–5333. doi: 10.1073/pnas.97.10.5328

Arora, R., Agarwal, P., Ray, S., Singh, A. K., Singh, V. P., Tyagi, A. K., et al. (2007). MADS-box gene family in rice: genome-wide identification, organization and expression profiling during reproductive development and stress. *BMC Genomics* 8:242. doi: 10.1186/1471-2164-8-242

Barker, E., and Ashton, N. (2013). A parsimonious model of lineage-specific expansion of MADS-box genes in *Physcomitrella patens*. *Plant Cell Rep.* 2013, 1161–1177. doi: 10.1007/s00299-013-1411-8

Dai, X., and Zhao, P. X. (2011). psRNATarget: a plant small RNA target analysis server. *Nucleic Acids Res.* 39, W155–W159. doi: 10.1093/nar/gkr319

De Bodt, S., Raes, J., Florquin, K., Rombauts, S., Rouze, P., Theißen, G., et al. (2003). Genomewide structural annotation and evolutionary analysis of the type I MADS-box genes in plants. *J. Mol. Evol.* 56, 573–586. doi: 10.1007/s00239-002-2426-x

de Hoon, M. J. L., Imoto, S., Nolan, J., and Miyano, S. (2004). Open source clustering software. *Bioinformatics* 20, 1453–1454. doi: 10.1093/bioinformatics/bth078

Díaz-Riquelme, J., Lijavetzky, D., Martínez-Zapater, J. M., and Carmona, M. J. (2009). Genome-wide analysis of MIKCC-type MADS-box genes in grapevine. *Plant Physiol.* 149, 354–369. doi: 10.1104/pp.108.131052

Duan, W., Song, X., Liu, T., Huang, Z., Ren, J., Hou, X., et al. (2015). Genome-wide analysis of the MADS-box gene family in *Brassica rapa* (Chinese cabbage). *Mol. Genet. Genomics* 290, 239–255. doi: 10.1007/s00438-014-0912-7

Fan, C. M., Wang, X., Wang, Y. W., Hu, R. B., Zhang, X. M., Chen, J. X., et al. (2013). Genome-wide expression analysis of soybean MADS genes showing potential function in the seed development. *PLoS ONE* 8:e62288. doi: 10.1371/journal.pone.0062288

Fernandez, D. E., Wang, C. T., Zheng, Y., Adamczyk, B., Singhal, R., Hall, P. K., et al. (2014). The MADS-domain factors AGL15 and AGL18, along with SVP and AGL24, are necessary to block floral gene expression during the vegetative phase. *Plant Physiol.* 165, 1591–1603. doi: 10.1104/pp.114.242990

Ferrario, S., Immink, R. G., and Angenent, G. C. (2004). Conservation and diversity in flower land. *Curr. Opin. Plant Biol.* 7, 84–91. doi: 10.1016/j.pbi.2003.11.003

Finn, R. D., Clements, J., and Eddy, S. R. (2011). HMMER web server: interactive sequence similarity searching. *Nucleic Acids Res.* 39, W29–W37. doi: 10.1093/nar/gkr367

Finn, R. D., Coggill, P., Eberhardt, R. Y., Eddy, S. R., Mistry, J., Mitchell, A. L., et al. (2015). The Pfam protein families database: towards a more sustainable future. *Nucleic Acids Res.* 44, D279–D285. doi: 10.1093/nar/gkv1344

Franks, S. J., Perez-Sweeney, B., Strahl, M., Nowogrodzki, A., Weber, J. J., Lalchan, R., et al. (2015). Variation in the flowering time orthologs BrFLC and BrSOC1 in a natural population of *Brassica rapa. PeerJ* 3, 182–182. doi: 10.7717/peerj.1339

Geraldo, N., Bäurle, I., Kidou, S., Hu, X., and Dean, C. (2009). FRIGIDA delays flowering in *Arabidopsis* via a cotranscriptional mechanism involving direct interaction with the nuclear cap-binding complex. *Plant Physiol.* 150, 1611–1618. doi: 10.1104/pp.109.137448

Gramzow, L., Barker, E., Schulz, C., Ambrose, B., Ashton, N., Theißen, G., et al. (2012). Selaginella genome analysis-Entering the "homoplasy heaven" of the MADS world. *Front. Plant Sci.* 3:214. doi: 10.3389/fpls.2012.00214

Gramzow, L., and Theißen, G. (2013). Phylogenomics of MADS-box genes in plants—two opposing life styles in one gene family. *Biology* 2, 1150–1164. doi: 10.3390/biology2031150

Gramzow, L., Weilandt, L., and Theißen, G. (2014). MADS goes genomic in conifers: towards determining the ancestral set of MADS-box genes in seed plants. *Ann. Bot-Lond.* 114, 1407–1429. doi: 10.1093/aob/mcu066

Haughn, G. W., and Somerville, C. R. (1988). Genetic control of morphogenesis in *Arabidopsis. Dev. Genet.* 9, 73–89. doi: 10.1002/dvg.1020090202

Helliwell, C. A., Wood, C. C., Robertson, M., James Peacock, W., and Dennis, E. S. (2006). The *Arabidopsis* FLC protein interacts directly in vivo with SOC1 and FT chromatin and is part of a high-molecular-weight protein complex. *Plant J.* 46, 183–192. doi: 10.1111/j.1365-313X.2006.02686.x

Henschel, K., Kofuji, R., Hasebe, M., Saedler, H., Münster, T., and Theißen, G. (2002). Two ancient classes of MIKC-type MADS-box genes are present in the moss *Physcomitrella patens. Mol. Biol. Evol.* 19, 801–814. doi: 10.1093/oxfordjournals.molbev.a004137

Hu, J. Y., Zhou, Y., He, F., Dong, X., Liu, L. Y., Coupland, G., et al. (2014). miR824-regulated AGAMOUS-LIKE16 contributes to flowering time repression in *Arabidopsis. Plant Cell* 26, 2024–2037. doi: 10.1105/tpc.114.124685

Hu, L., and Liu, S. (2012). Genome-wide analysis of the MADS-box gene family in cucumber. *Genome* 55, 245–256. doi: 10.1139/g2012-009

Kaufmann, K., Melzer, R., and Theißen, G. (2005). MIKC-type MADS-domain proteins: structural modularity, protein interactions and network evolution in land plants. *Gene* 347, 183–198. doi: 10.1016/j.gene.2004.12.014

Khan, M. R. G., Ai, X. Y., and Zhang, J. Z. (2014). Genetic regulation of flowering time in annual and perennial plants. *WIREs. RNA* 5, 347–359. doi: 10.1002/wrna.1215

Kim, D. H., Doyle, M. R., Sung, S., and Amasino, R. M. (2009). Vernalization: winter and the timing of flowering in plants. *Annu. Rev. Cell Dev. Biol.* 25, 277–299. doi: 10.1146/annurev.cellbio.042308.113411

Kim, S., Park, M., Yeom, S. I., Kim, Y. M., Lee, J. M., Lee, H. A., et al. (2014). Genome sequence of the hot pepper provides insights into the evolution of pungency in *Capsicum* species. *Nat. Genet.* 46, 270–278. doi: 10.1038/ng.2877

Kitashiba, H., Li, F., Hirakawa, H., Kawanabe, T., Zou, Z., Hasegawa, Y., et al. (2014). Draft sequences of the radish (*Raphanus sativus* L.) genome. *DNA Res.* 21, 481–490. doi: 10.1093/dnares/dsu014

Lee, J. H., Ryu, H. S., Chung, K. S., Posé, D., Kim, S., Schmid, M., et al. (2013). Regulation of temperature-responsive flowering by MADS-box transcription factor repressors. *Science* 342, 628–632. doi: 10.1126/science.1241097

Lee, J. H., Yoo, S. J., Park, S. H., Hwang, I., Lee, J. S., and Ahn, J. H. (2007). Role of SVP in the control of flowering time by ambient temperature in *Arabidopsis. Gene Dev.* 21, 397–402. doi: 10.1101/gad.1518407

Leseberg, C. H., Li, A., Kang, H., Duvall, M., and Mao, L. (2006). Genome-wide analysis of the MADS-box gene family in *Populus trichocarpa. Gene* 378, 84–94. doi: 10.1016/j.gene.2006.05.022

Letunic, I., Doerks, T., and Bork, P. (2012). SMART 7: recent updates to the protein domain annotation resource. *Nucleic Acids Res.* 40, D302–D305. doi: 10.1093/nar/gkr931

Li, L., Stoeckert, C. J., and Roos, D. S. (2003). OrthoMCL: identification of ortholog groups for eukaryotic genomes. *Genome Res.* 13, 2178–2189. doi: 10.1101/gr.1224503

Li, N., Liu, Y., Zhong, M., and Li, H. (2015). Thinking out of the box: MADS-box genes and maize spikelet development. *Afr. J. Biotechnol.* 13, 4673–4679. doi: 10.5897/AJB11.3885

Liljegren, S. J., Ditta, G. S., Eshed, Y., Savidge, B., Bowman, J. L., and Yanofsky, M. F. (2000). SHATTERPROOF MADS-box genes control seed dispersal in *Arabidopsis. Nature* 404, 766–770. doi: 10.1038/35008089

Liu, C., Chen, H., Er, H. L., Soo, H. M., Kumar, P. P., Han, J. H., et al. (2008). Direct interaction of *AGL24* and *SOC1* integrates flowering signals in *Arabidopsis. Development* 135, 1481–1491. doi: 10.1242/dev.020255

Liu, Y., Cui, S., Wu, F., Yan, S., Lin, X., Du, X., et al. (2013). Functional conservation of MIKC*-Type MADS-box genes in *Arabidopsis* and rice pollen maturation. *Plant Cell* 25, 1288–1303. doi: 10.1105/tpc.113.110049

Masiero, S., Colombo, L., Grini, P. E., Schnittger, A., and Kater, M. M. (2011). The emerging importance of Type I MADS-box transcription factors for plant reproduction. *Plant Cell* 23, 865–872. doi: 10.1105/tpc.110.081737

Michaels, S. D., and Amasino, R. M. (1999). FLOWERING LOCUS C encodes a novel MADS domain protein that acts as a repressor of flowering. *Plant Cell* 11, 949–956. doi: 10.1105/tpc.11.5.949

Mitsui, Y., Shimomura, M., Komatsu, K., Namiki, N., Shibata-Hatta, M., Imai, M., et al. (2015). The radish genome and comprehensive gene expression profile of tuberous root formation and development. *Sci. Rep.* 5:10835. doi: 10.1038/srep10835

Moon, J., Suh, S. S., Lee, H., Choi, K. R., Hong, C. B., Paek, N. C., et al. (2003). The SOC1 MADS-box gene integrates vernalization and gibberellin signals for flowering in *Arabidopsis. Plant J.* 35, 613–623. doi: 10.1046/j.1365-313X.2003.01833.x

Mutasa-Göttgens, E., and Hedden, P. (2009). Gibberellin as a factor in floral regulatory networks. *J. Exp. Bot.* 60, 1979–1989. doi: 10.1093/jxb/erp040

Nesi, N., Debeaujon, I., Jond, C., Stewart, A. J., Jenkins, G. I., Caboche, M., et al. (2002). The TRANSPARENT TESTA16 locus encodes the *ARABIDOPSIS* BSISTER MADS domain protein and is required for proper development and pigmentation of the seed coat. *Plant Cell* 14, 2463–2479. doi: 10.1105/tpc.004127

Nie, S., Li, C., Xu, L., Wang, Y., Huang, D., Muleke, E. M., et al. (2016). De novo transcriptome analysis in radish (*Raphanus sativus* L.) and identification of critical genes involved in bolting and flowering. *BMC Genomics* 17:389. doi: 10.1186/s12864-016-2633-2

Nie, S., Xu, L., Wang, Y., Huang, D., Muleke, E. M., Sun, X., et al. (2015). Identification of bolting-related microRNAs and their targets reveals complex miRNA-mediated flowering-time regulatory networks in radish (*Raphanus sativus* L.). *Sci. Rep.* 5:14034. doi: 10.1038/srep14034

Norman, C., Runswick, M., Pollock, R., and Treisman, R. (1988). Isolation and properties of cDNA clones encoding SRF, a transcription factor that binds to the c-fos serum response element. *Cell* 55, 989–1003. doi: 10.1016/0092-8674(88)90244-9

Ó'Maoiléidigh, D. S., Graciet, E., and Wellmer, F. (2014). Gene networks controlling *Arabidopsis thaliana* flower development. *New Phytol.* 201, 16–30. doi: 10.1111/nph.12444

Parenicová, L., de Folter, S., Kieffer, M., Horner, D. S., Favalli, C., Busscher, J., et al. (2003). Molecular and phylogenetic analyses of the complete MADS-box transcription factor family in *Arabidopsis* new openings to the MADS world. *Plant Cell* 15, 1538–1551. doi: 10.1105/tpc.011544

Pellegrini, L., Tan, S., and Richmond, T. J. (1995). Structure of serum response factor core bound to DNA. *Nature* 376, 490–498. doi: 10.1038/376490a0

Posé, D., Yant, L., and Schmid, M. (2012). The end of innocence: flowering networks explode in complexity. *Curr. Opin. Plant Biol.* 15, 45–50. doi: 10.1016/j.pbi.2011.09.002

Rameneni, J. J., Dhandapani, V., Paul, P., Im, S., Oh, M. H., Choi, S. R., et al. (2014). Genome-wide identification, characterization, and comparative phylogeny analysis of MADS-box transcription factors in *Brassica rapa*. *Genes Genom.* 2014, 509–525. doi: 10.1007/s13258-014-0187-8

Ratcliffe, O. J., Kumimoto, R. W., Wong, B. J., and Riechmann, J. L. (2003). Analysis of the *Arabidopsis* MADS AFFECTING FLOWERING gene family: MAF2 prevents vernalization by short periods of cold. *Plant Cell* 15, 1159–1169. doi: 10.1105/tpc.009506

Saha, G., Park, J. I., Jung, H. J., Ahmed, N. U., Kayum, M. A., Chung, M. Y., et al. (2015). Genome-wide identification and characterization of MADS-box family genes related to organ development and stress resistance in *Brassica rapa*. *BMC Genomics* 16:178. doi: 10.1186/s12864-015-1349-z

Saldanha, A. J. (2004). Java Treeview—extensible visualization of microarray data. *Bioinformatics* 20, 3246–3248. doi: 10.1093/bioinformatics/bth349

Sasaki, K., Aida, R., Yamaguchi, H., Shikata, M., Niki, T., Nishijima, T., et al. (2010). Functional divergence within class B MADS-box genes TfGLO and TfDEF in *Torenia fournieri* Lind. *Mol. Genet. Genomics* 284, 399–414. doi: 10.1007/s00438-010-0574-z

Shannon, P., Markiel, A., Ozier, O., Baliga, N. S., Wang, J. T., Ramage, D., et al. (2003). Cytoscape: a software environment for integrated models of biomolecular interaction networks. *Genome Res.* 13, 2498–2504. doi: 10.1101/gr.1239303

Shore, P., and Sharrocks, A. D. (1995). The MADS-box family of transcription factors. *Eur. J. Biochem.* 229, 1–13. doi: 10.1111/j.1432-1033.1995.0001l.x

Shu, Y., Yu, D., Wang, D., Guo, D., and Guo, C. (2013). Genome-wide survey and expression analysis of the MADS-box gene family in soybean. *Mol. Biol. Rep.* 40, 3901–3911. doi: 10.1007/s11033-012-2438-6

Smaczniak, C., Immink, R. G., Angenent, G. C., and Kaufmann, K. (2012). Developmental and evolutionary diversity of plant MADS-domain factors: insights from recent studies. *Development* 139, 3081–3098. doi: 10.1242/dev.074674

Song, X., Duan, W., Huang, Z., Liu, G., Wu, P., Liu, T., et al. (2015). Comprehensive analysis of the flowering genes in Chinese cabbage and examination of evolutionary pattern of CO-like genes in plant kingdom. *Sci. Rep.* 5:14631. doi: 10.1038/srep14631

Srikanth, A., and Schmid, M. (2011). Regulation of flowering time: all roads lead to Rome. *Cell. Mol. Life Sci.* 68, 2013–2037. doi: 10.1007/s11103-013-0044-1

Su, C. L., Chen, W. C., Lee, A. Y., Chen, C. Y., Chang, Y. C. A., Chao, Y. T., et al. (2013). A modified ABCDE model of flowering in orchids based on gene expression profiling studies of the moth orchid *Phalaenopsis aphrodite*. *PLoS ONE* 8:e80462. doi: 10.1371/journal.pone.0080462

Sun, Q., Csorba, T., Skourti-Stathaki, K., Proudfoot, N. J., and Dean, C. (2013). R-loop stabilization represses antisense transcription at the *Arabidopsis* FLC locus. *Science* 340, 619–621. doi: 10.1126/science.1234848

Sun, X., Xu, L., Wang, Y., Yu, R., Zhu, X., Luo, X., et al. (2015). Identification of novel and salt-responsive miRNAs to explore miRNA-mediated regulatory network of salt stress response in radish (*Raphanus sativus* L.). *BMC Genomics* 16:197. doi: 10.1186/s12864-015-1416-5

Sun, Y., Qiu, Y., Zhang, X., Chen, X., Shen, D., Wang, H., et al. (2015). Genome-wide identification of microRNAs associated with taproot development in radish (*Raphanus sativus* L.). *Gene* 569, 118–126. doi: 10.1016/j.gene.2015.05.044

Tamura, K., Peterson, D., Peterson, N., Stecher, G., Nei, M., and Kumar, S. (2011). MEGA5: molecular evolutionary genetics analysis using maximum likelihood, evolutionary distance, and maximum parsimony methods. *Mol. Biol. Evol.* 28, 2731–2739. doi: 10.1093/molbev/msr121

Tapia-López, R., García-Ponce, B., Dubrovsky, J. G., Garay-Arroyo, A., Pérez-Ruíz, R. V., Kim, S. H., et al. (2008). An AGAMOUS-related MADS-box gene, XAL1 (AGL12), regulates root meristem cell proliferation and flowering transition in *Arabidopsis*. *Plant Physiol.* 146, 1182–1192. doi: 10.1104/pp.107.108647

Tian, Y., Dong, Q., Ji, Z., Chi, F., Cong, P., and Zhou, Z. (2015). Genome-wide identification and analysis of the MADS-box gene family in apple. *Gene* 555, 277–290. doi: 10.1016/j.gene.2014.11.018

Wang, J. W. (2014). Regulation of flowering time by the miR156-mediated age pathway. *J. Exp. Bot.* 65, 4723–4730. doi: 10.1093/jxb/eru246

Wang, W., Wu, P., Li, Y., and Hou, X. (2015). Genome-wide analysis and expression patterns of ZF-HD transcription factors under different developmental tissues and abiotic stresses in Chinese cabbage. *Mol. Genet. Genomics* 291, 1451–1464. doi: 10.1007/s00438-015-1136-1

Wang, X., Wang, H., Wang, J., Sun, R., Wu, J., Liu, S., et al. (2011). The genome of the mesopolyploid crop species *Brassica rapa*. *Nat. Genet.* 43, 1035–1039. doi: 10.1038/ng.919

Wells, C. E., Vendramin, E., Tarodo, S. J., Verde, I., and Bielenberg, D. G. (2015). A genome-wide analysis of MADS-box genes in peach [*Prunus persica* (L.) Batsch]. *BMC Plant Biol.* 15:41. doi: 10.1186/s12870-015-0436-2

Wullschleger, S. D., and Weston, D. J. (2012). Modeling the molecular and climatic controls on flowering. *New Phytol.* 194, 599–601. doi: 10.1111/j.1469-8137.2012.04142.x

Xie, W., Huang, J., Liu, Y., Rao, J., Luo, D., and He, M. (2015). Exploring potential new floral organ morphogenesis genes of *Arabidopsis thaliana* using systems biology approach. *Front. Plant Sci.* 6:829. doi: 10.3389/fpls.2015.00829

Xu, L., Wang, Y., Xu, Y., Wang, L., Zhai, L., Zhu, X., et al. (2013). Identification and characterization of novel and conserved microRNAs in radish (*Raphanus sativus* L.) using high-throughput sequencing. *Plant Sci.* 201, 108–114. doi: 10.1016/j.plantsci.2012.11.010

Yu, R., Wang, Y., Xu, L., Zhu, X., Zhang, W., Wang, R., et al. (2015). Transcriptome profiling of root microRNAs reveals novel insights into taproot thickening in radish (*Raphanus sativus* L.). *BMC Plant Biol.* 15:30. doi: 10.1186/s12870-015-0427-3

Zahn, L. M., Feng, B., and Ma, H. (2006). Beyond the ABC-Model: regulation of floral homeotic genes. *Adv. Bot. Res.* 44, 163–207. doi: 10.1016/S0065-2296(06)44004-0

Zhao, Y., Li, X., Chen, W., Peng, X., Cheng, X., Zhu, S., et al. (2011). Whole-genome survey and characterization of MADS-box gene family in maize and sorghum. *Plant Cell Tiss. Org.* 105, 159–173. doi: 10.1007/s11240-010-9848-8

Zhou, C. M., Zhang, T. Q., Wang, X., Yu, S., Lian, H., Tang, H., et al. (2013). Molecular basis of age-dependent vernalization in *Cardamine flexuosa*. *Science* 340, 1097–1100. doi: 10.1126/science.1234340

Zobell, O., Faigl, W., Saedler, H., and Münster, T. (2010). MIKC* MADS-box proteins: conserved regulators of the gametophytic generation of land plants. *Mol. Biol. Evol.* 27, 1201–1211. doi: 10.1093/molbev/msq005

Conflict of Interest Statement: The authors declare that the research was conducted in the absence of any commercial or financial relationships that could be construed as a potential conflict of interest.

6

Perspectives for a Framework to Understand Aril Initiation and Development

*Sylvia R. Silveira[1], Marcelo C. Dornelas[2] and Adriana P. Martinelli[1]**

[1] *Laboratório de Biotecnologia Vegetal, Centro de Energia Nuclear na Agricultura, Universidade de São Paulo, Piracicaba, Brazil,* [2] *Departamento de Biologia Vegetal, Instituto de Biologia, Universidade Estadual de Campinas, Campinas, Brazil*

A differentiated structure called "aril" has been described in seeds of several plant species during the course of evolution and might be considered as a supernumerary integument. Besides its ecological function in seed dispersal, the structure also represents a relevant character for systematic classification and exhibits important properties that impart agronomic value in certain species. Little is known about the molecular pathways underlying this morphological innovation because it is absent in currently used model species. A remarkable feature of the seeds of *Passiflora* species is the presence of a conspicuous aril. This genus is known for the ornamental, medicinal, and food values of its species. In view of the molecular resources and tools available for some *Passiflora* species, we highlight the potential of these species as models for developmental studies of the aril.

Keywords: aril development, integument, model species, ovule, *Passiflora*, seed

Edited by:
Federico Valverde,
Spanish National Research Council, Spain

Reviewed by:
David G. Oppenheimer,
University of Florida, USA
Simona Masiero,
University of Milan, Italy

Correspondence:
Adriana P. Martinelli
adriana.martinelli@usp.br

INTRODUCTION

The morphological diversity among plant species results from differential gene expression controlling the development of novel features that ensure the adaptation and reproductive success of a species. An important question in plant biology is when and how these features emerged during evolution. One of such novel features is the aril. The aril is a differentiated structure present in seeds of several gymnosperm and angiosperm species, forming seed dispersal units. In many species, the aril accumulates several nutritional compounds attracting and rewarding frugivorous animals. There is a great amount of information available about morphological and molecular development of plant ovules and seeds and they can be used as initial clues to investigate aril development. These appendages are often used in systematics classification, since its presence, absence, form, and function vary among taxa. The well known model species do not exhibit this feature evidencing the need for novel models to study this specific structure. A better understanding of the processes involved in aril origin and development is interesting and necessary due to its economical, ecological and phylogenetic importance.

ARIL ORIGIN AND IMPORTANCE

Several plant species develop differentiated structures associated with their seeds, often constituting diaspores, which are plant dispersal units mostly related to their dispersion syndrome (Corner, 1976). Some authors also believe that these structures originated as a protection mechanism for seeds and embryos, regardless of their role in dispersion (Mack, 2000). Also associated to the ovule/seed, either one or more integuments are found. The current theory of the evolution of integuments states that there are different evolutionary origins for the outer and inner integuments in flowering plants (Endress, 2011). In angiosperms, the inner integument is considered homologous to the single integument of extant and fossil gymnosperms (Reinheimer and Kellogg, 2009), and the outer integument may have been derived from a cupule/leaf-like structure found in several gymnosperms (Gasser et al., 1998). The integuments may or may not originate appendages that perform a defined role in seed dispersion. Such seed appendages may be wings, spines, hairs, plumes, fibers, or fleshy tissues, receiving different denominations in the literature.

Both gymnosperms and angiosperms evolved the habit of enveloping the seeds with a fleshy tissue (Lovisetto et al., 2012). Such tissue, called "aril," generally accumulate sugars and other substances that will confer biological roles similar to those of fruits (Herrera, 1989).

The use of the term "aril" is quite controversial in the literature. It has been used both in a broader sense, referring to any fleshy structure associated with the seed, but also to designate structures with a specific anatomical origin. According to Corner (1976), the term defines a structure varying from a fleshy to a more-or-less hard consistency, which develops from part of the ovule after fertilization and envelopes the seed partially or completely. Van der Pijl (1972) preferred to distinguish these structures according to their anatomical origin, the aril being originated from the funiculus. Therefore, a structure developing from other parts of the ovule are usually called arillode, false aril, or aril-like structure (**Figure 1**). Both, aril and arillode, are somehow associated with integuments. In fact, some authors consider the "true" aril as a supernumerary integument (Maheshwari, 1950; Kapil and Vasil, 1963; Endress, 2011).

As arils are generally fleshy structures, they are of extreme importance because during their development and ripening they accumulate substances that confer properties that not only attract dispersion agents, but also arouse interest for human consumption. Arils are very common in tropical and subtropical species and might accumulate oils (e.g., *Ricinus communis*), flavor- and aroma-rich compounds (*Myristica fragrans*), nutrients, and sugars (*Passiflora edulis*), among other substances.

ARIL ONTOGENY

Few studies describe the ontogeny and/or morphological aspects of aril formation and associate these with ovule development; this lack of information probably led to its controversial nomenclature. Additionally, the current model plant species do not exhibit this unique structure, making it difficult to characterize its development, especially at the molecular level.

Aril developmental stages were observed in some species of *Passiflora* (Raju, 1956; Singh, 1962; Dathan and Singh, 1973), and described in greater detail in *P. suberosa* and *Turnera ulmifolia* (Kloos and Bouman, 1980). It has also been described in Leguminosae, such as *Eriosema glaziovii* (Grear and Dengler, 1976), *Cytisus striatus*, and *C. multiflorus* (Rodriguez-Riaño et al., 2006). More recently, the development of an aril was described in Celastraceae, however, the authors showed that the origin of the aril-like structure was not from the funiculus, calling it "caruncula" (Zhang et al., 2011).

Aril development has been divided into stages by some authors, and ontogenetic descriptions suggest that it is a pre-anthesis event originating during megagametogenesis from periclinal divisions of epidermal cells of the funiculus, followed by anticlinal divisions, forming a ring or collar-like structure surrounding the ovule (Kloos and Bouman, 1980; Rodriguez-Riaño et al., 2006). The specific stage of ovule development in which the aril initiates is not very clear in most of the reports. The first divisions might be observed between the tetrad formation stage, when integuments are elongating toward the nucellus, and the beginning of megagametogenesis, when the outer integument has already enveloped the inner integument and the nucellus, forming the micropyle (Raju, 1956; Singh, 1962; Dathan and Singh, 1973; Grear and Dengler, 1976; Kloos and Bouman, 1980; Rodriguez-Riaño et al., 2006).

MOLECULAR MECHANISMS CONTROLLING INTEGUMENT INITIATION AND GROWTH

As mentioned, the aril initiates during ovule development after the emergence and growth of integuments, resembling its development and exhibiting similar patterns of polarity. Thus, to speculate on whether the aril is an extra integument, and which molecular mechanisms might be involved in its identity and development, one should look closely to the molecular basis at the integument initiation and growth.

The development of the ovule in plants has been well characterized in model species, such as *Arabidopsis* and *Petunia*, through molecular genetic studies. Several genes involved in different events of ovule development where identified through mutant screening, as reviewed by Angenent and Colombo (1996), Gasser et al. (1998), and Schneitz (1999). The results obtained from mutant characterization, patterns of gene expression, and transcriptomic analyses in the last two decades allowed for the elucidation of regulatory networks controlling the initiation and development of integuments. Most of the genes characterized encode transcription factors, and molecular studies have been performed to better understand the means by which these factors act, and how they interact regulating integument morphogenesis.

Integument formation marks the transition from the earlier established proximal-distal axis of the ovule primordia to an additional adaxial/abaxial polarity axis. Integument initiation is characterized by epidermal cell proliferation in a region between the nucellus and the funiculus. The putative transcriptional regulator NOZZLE/SPOROCYTELESS (NZZ) is required for maintaining the homeobox gene *WUSCHEL* (*WUS*) expression limited to the nucellus (**Figure 2**) (Sieber et al., 2004). Another factor restraining *WUS* in the nucellus is the interaction of BEL1 (a homeodomain protein) with an integument identity

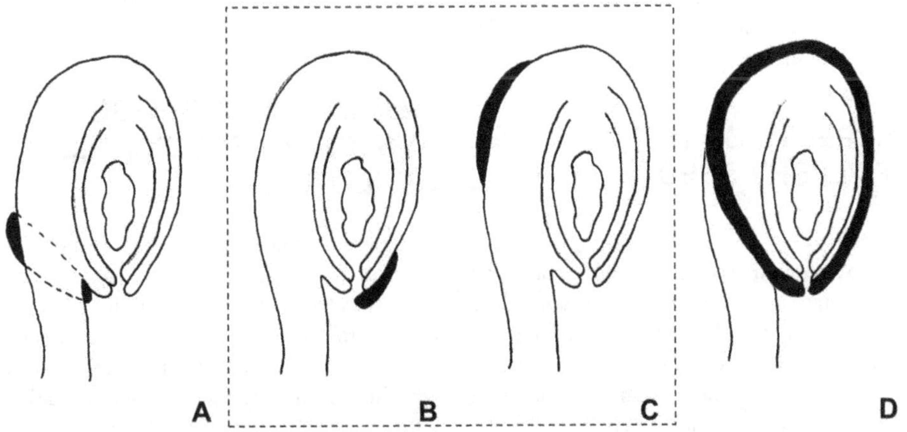

FIGURE 1 | Anatomical origin of different fleshy seed appendages or aril-like structures. Areas in black represent the fleshy structures. **(A)** True aril – develops from the funiculus as a ring around it. **(B)** Caruncle – develops from the micropilar region of the integument. **(C)** Strophiole – develops from the raphe. **(D)** Sarcotesta – involves the differentiation of the outermost layer of the outer integument after fertilization. In the dashed box **(B,C)**, structures considered as "arillodes."

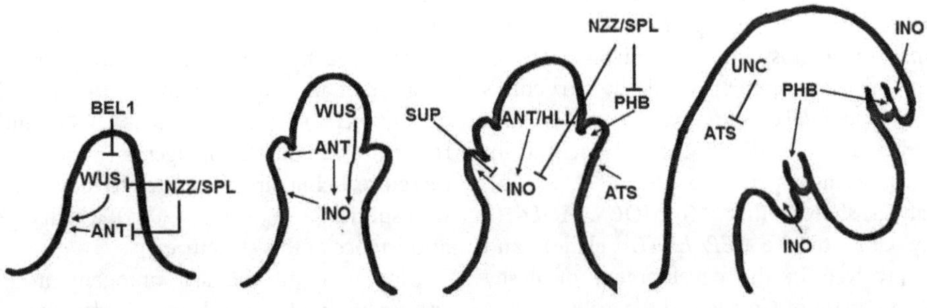

FIGURE 2 | Summary of gene interactions during early integument development (adapted from Skinner et al., 2004). Modulation of these interactions might cause the formation of supernumerary integument and/or aril-like structures.

protein complex that represses *WUS* in the chalaza, and activates *INNER NO OUTER* (*INO*) for outer integument development (**Figure 2**) (Brambilla et al., 2008). *WUS*, in turn, is sufficient to induce integument formation from the underlying chalazal tissue, since it generates downstream signals inducing meristematic activity even where it is not expressed (Gross-Hardt et al., 2002). An evidence for this is the induction of ectopic structures resembling integuments at the flanks of the funiculus, when *WUS* is ectopically expressed in the chalaza, under the control of the *AINTEGUMENTA* (*ANT*) promoter (Gross-Hardt et al., 2002). Thus, ectopic WUS expression caused by natural gain-of-function mutation(s) might be involved in the evolutionary origin of supernumerary integuments and, therefore, in structures resembling arils.

Additionally, NZZ is known to restrict both the homeodomain-leucine zipper gene *PHABULOSA* (*PHB*) in the abaxial domain of the chalazal region where the inner integument initiates (Sieber et al., 2004), and INO, which is responsible for outer integument differentiation (**Figure 2**) (Schneitz et al., 1997; Villanueva et al., 1999). *INO* expression, in turn, is restricted to the outer integument by *WUS* and, more specifically, to the abaxial side, where it is repressed by *SUPERMAN* (*SUP*). Thus, INO and SUP are responsible for the asymmetric growth of the outer integument (**Figure 2**) (Meister et al., 2002). *BEL1*, *ANT*, and *HUELLENLOS* (*HLL*) also participate directly or indirectly in *INO* negative spatial regulation (Villanueva et al., 1999). These antagonistic relations control integument polarity.

An additional mutant in which both integuments are present, but exhibits aberrant features is worth mentioning. The *unicorn* (*unc*) mutation results in excrescences emerging from the outer integument (Schneitz et al., 1997). Later on, *UNC* was found to encode an AGC VIII kinase that directly interacts with and represses the activity of ABERRANT TESTA SHAPE (ATS), a transcriptional regulator belonging to the *KANADI* family (**Figure 2**) (Enugutti et al., 2012; Enugutti and Schneitz, 2013). Thus, ectopic expression of *ATS* would provide another mechanism by which additional initiation and growth of integument-derived tissue may occur, therefore indicating an alternative possible molecular mechanism underlying the evolutionary origin of aril or aril-like structures.

Considering the amount of information on regulatory networks for integument initiation and growth, along with the fact that most of these mechanisms are conserved among different taxa, and the known morphoanatomy of arils, it becomes possible to identify the initial cues on the molecular basis of aril origin and development.

MOLECULAR ASPECTS OF THE "RIPENING" OF FLESHY SEED STRUCTURES

The development of fleshy seed structures such as the aril can be divided in three main stages: (1) initiation, which includes cell proliferation; (2) growth, with cell expansion, mainly; and (3) accumulation of storage products, which would be equivalent to a "ripening" stage. As we are assuming a similarity between integument and aril development, we considered the first two stages in the previous section, and we will now consider the third stage.

Since gymnosperms do not form ovaries that will develop into fruits after fertilization, many species developed fruit-like fleshy structures around their seeds to attract frugivorous animals that act as seed dispersers (Herrera, 1989). Because of its importance in the formation of reproductive structures in both gymnosperms and angiosperms, the involvement of MADS-Box genes in the development of fleshy structures was investigated in *Ginkgo biloba* and *Taxus baccata*, both gymnosperms (Lovisetto et al., 2012, 2013, 2015a), and in *Magnolia grandiflora*, a basal angiosperm (Lovisetto et al., 2015b). Gene expression analyses showed that *AGAMOUS*, *AGL6* (a gene phylogenetically close to the *SEPALLATA* clade), and *TM8-like* genes, are involved in the development of fleshy structures in both the sarcotesta of *Ginkgo* and the aril of *Taxus*, regardless of their anatomical origin (Lovisetto et al., 2012). Moreover, activated forms of AGL6 (AGL6::VP16) triggered ectopic outgrowths on the surface of reproductive structures in *Arabidopsis* (Koo et al., 2010). A subfamily of MADS-Box, the *B-sister* genes, is believed to be required for the correct development of ovule and seed, with their expression analyzed in the gymnosperms mentioned above (Lovisetto et al., 2013). The pattern of gene expression differed between these two species, being weaker throughout aril development in *Taxus*, indicating that the involvement of B-sisters in the formation of fleshy fruit-like structures might be dependent of their origin. In *Magnolia*, with a fleshy tissue also originating from the seed tegument, *AGAMOUS*, *AGL6*, *SEPALLATA*, and *B-sister* were also detected during the sarcotesta formation and growth (Lovisetto et al., 2015b).

There is evidence that a common set of genes was recruited independently in distantly related taxa, regulating the development of all fleshy structures, regardless of their anatomic origin in, both, gymnosperms and angiosperms. Accordingly, a group of tomato MADS-box genes have been implicated in fruit ripening, including members of the *SEPALLATA* and B-sister clades (Vrebalov et al., 2002; Yasuhiro, 2016). Altogether, these observations suggest that fleshy tissues that undergo physiological changes that involve tissue softening, pigmentation, and accumulation of sugars, aroma, and flavor (or "ripening syndrome," in general), appeared independently in fruits and seeds but are likely to be regulated, at the molecular level, by conserved pathways.

Passiflora AS A SUGGESTED MODEL SYSTEM TO STUDY ARIL DEVELOPMENT

Among the angiosperms, species belonging to *Passiflora* are noteworthy regarding their aril, which are often cited in anatomical and morphological literature as an example of a true aril. *Passiflora* is the largest genus of the family Passifloraceae with over 500 species, mostly originated in neotropical regions, with hundreds of species throughout Latin America (Kugler and King, 2004; Ulmer and MacDougal, 2004). *Passiflora* also include commercial species, such as *P. edulis*, *P. alata*, and *P. incarnata*, which are important for their ornamental, medicinal, and food values, the latter given specifically by the aril in which juice is produced and accumulated. Typically, the fruits of *Passiflora* are indehiscent berries, rarely a dehiscent capsule, very variable in shape, size, and color, and in general produces a mucilaginous or aqueous acidic pulp, forming a cupuliform or saccate aril, covering each of numerous seeds (**Figure 3**) (Cervi, 1997; Dhawan et al., 2004). Passionfruit propagation is mainly carried out by seeds (Pereira and Dias, 2000), and the aril works as a reward for its dispersing agents (Ulmer and MacDougal, 2004), therefore, being directly related to the reproductive success of the wild species (Fenster et al., 2004), which highlights the ecological importance of this structure.

Studies about the aril ontogeny in *Passiflora* are scarce, although it has been addressed in descriptions of the embryology or seed coat structure of Passifloraceae (Raju, 1956; Singh, 1962; Dathan and Singh, 1973; Kloos and Bouman, 1980). The first mention of aril initiation in *Passiflora* is from *P. suberosa* and describes it as a ring around the distal area of the funiculus (Kratzer, 1918 in Kloos and Bouman, 1980). Later studies also refer to the aril primordium as a rim, collar or ring around the funiculus in several species of the family, *P. calcarata* (Raju, 1956; Singh, 1962), *P. foetida* (Singh, 1962), *P. caerulea*, *P. molissima* (Dathan and Singh, 1973), and *P. edulis* (Dathan and Singh, 1973; Corner, 1976). These authors describe the origin of the aril as dermal, epidermal or hypodermal, and are not precise whether it develops from the funiculus, exostome, hilum, micropyle or raphe. A more detailed case study of aril development was performed using *P. suberosa* and *T. ulmifolia*, (Kloos and Bouman, 1980). According to this description, the aril is initiated during megagametogenesis by periclinal and anticlinal divisions of dermal cells, forming a rim around the funiculus from the raphe to the outer integument region at the micropyle (Kloos and Bouman, 1980). Differences among species occur mainly after fertilization. The aril of *P. suberosa* continues to grow, covers the micropyle, and by division of its apical cells, equally envelopes the developing seed, while the aril of *T. ulmifolia* grows unilaterally leaving the exostome exposed.

In spite of these descriptions of the initiation and development of the aril in *Passiflora* species, the molecular mechanisms implicated in these processes have not been described yet. Few

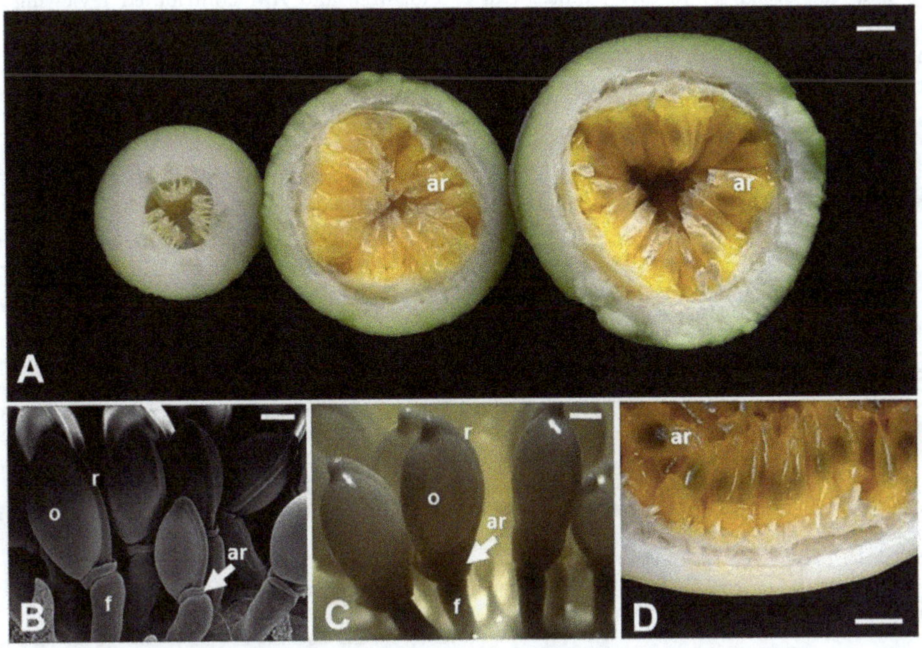

FIGURE 3 | Aril development in *Passiflora edulis*. (A) Longitudinal sections of fruits at 10, 42, or 49 days after pollination, showing arils at different developmental stages. The arils (yellow/orange) completely cover the seeds, which are observed as darker structures inside the arils. **(B)** Scanning electron microscopy of seeds at 2 days after pollination. **(C)** Seeds at 12 days after pollination. **(D)** Detail of a mature fruit in cross section, where the arils cover the whole seed. ar, aril; f, funiculus; o, ovule; r, raphe. Bars: **A,D** = 1 cm; **B** = 0.5 mm; **C** = 1 mm.

studies addressed gene expression in *Passiflora* arils, such as the analysis of differential expression among *PeETR1*, *PeERS1*, and *PeERS*. These genes encode proteins involved in ethylene perception in *Passiflora* fruit tissues, with higher levels of mRNA in arils than in seeds during fruit ripening (Mita et al., 1998; Mita et al., 2002). Nevertheless, these focused mainly on fruit ripening and, therefore, in genes involved in later aril developmental stages, and not in the identity and differentiation of this specialized structure.

Although in recent decades there has been a breakthrough in genome sequencing and genomic data analysis from crop species, efforts for entire genome sequencing were not done in *Passiflora* species, and very little is known about the genomics of this genus. The currently available sequence data in public databases are molecular markers used in phylogenetic and genetic diversity studies, such as microsatellites (Oliveira et al., 2005, 2008; Pádua et al., 2005; Cazé et al., 2012; Cerqueira-Silva et al., 2012, 2014), and internal transcribed spacers (Muschner et al., 2003; Yockteng and Nadot, 2004). On the other hand, specific transcript and genomic libraries for *Passiflora* have been constructed: a database of expressed sequence tags (ESTs) from libraries derived from *P. edulis* and *P. suberosa* reproductive tissues (Cutri and Dornelas, 2012), and a large-insert bacterial artificial chromosome (BAC) library of *P. edulis* (Santos et al., 2014). These are very resourceful for genomic studies allowing a greater understanding of gene structure and function, and the process of differentiation of complex morphological characters, which provide the diversity found among plants, such as the aril. Another useful resource that should aid these functional and developmental studies is the availability of genetic transformation and *in vitro* regeneration protocols for *Passiflora* species. Such protocols where generated by the large number of studies aiming at the genetic improvement of passion fruit, that have been carried out since the 1990s (Cerqueira-Silva et al., 2014), mainly to obtain transgenic plants resistant to the woodiness virus in *P. edulis* (Manders et al., 1994; Alfenas et al., 2005; Trevisan et al., 2006; Monteiro-Hara et al., 2011), and *P. alata* (Correa et al., 2015). Several protocols for *in vitro* regeneration via organogenesis or somatic embryogenesis for a large number of *Passiflora* species where established aiming at germplasm preservation, and recovery of transgenic plants, as reviewed by Vieira and Carneiro (2004) and Otoni et al. (2013). Although designed for breeding purposes, these methodologies are important tools to study the molecular basis of aril development. Novel genomic editing tools, such as the CRISPR/Cas9 technology, will also help in the genetic and molecular analysis of aril development.

CONCLUSION

Arils are accessory seed structures present in both gymnosperms and angiosperms, being important for seed dispersal, and might possess economic importance. Nonetheless, aril evolutionary origin and ontogenesis are largely unknown, with, both, structural and molecular information lacking and needed. Here we established parallels between ovule integuments and arils that might help the design of further studies. Our testable statements need a novel model species, since the traditional plant models do not develop arils. We postulate that *Passiflora* species are good candidates for such needed model.

AUTHOR CONTRIBUTIONS

SRS, MD, and AM designed the initial manuscript. SRS wrote the initial draft of the manuscript and conceived the figures. SRS, MD, and AM contributed reviewing and discussing the manuscript to produce its final version.

ACKNOWLEDGMENTS

The authors acknowledge financial support from Coordenação de Aperfeiçoamento de Pessoal de Nível Superior (CAPES, Brazil), Fundação de Amparo à Pesquisa do Estado de São Paulo (FAPESP, São Paulo, Brazil), and Conselho Nacional de Desenvolvimento Científico e Tecnológico (CNPq, Brazil).

REFERENCES

Alfenas, P. F., Braz, A. S. K., Torres, B., Santana, E. N., do, Nascimento, A. V. S., Carvalho, M. G., et al. (2005). Transgenic passionfruit expressing RNA derived from Cowpea aphid-borne mosaic virus is resistant to passionfruit woodiness disease. *Fitopatol. Bras.* 30, 33–38. doi: 10.1590/S0100-41582005000100006

Angenent, G. C., and Colombo, L. (1996). Molecular control of ovule development. *Trends Plant Sci.* 1, 228–232. doi: 10.1016/1360-1385(96)86900-7

Brambilla, V., Kater, M., and Colombo, L. (2008). Ovule integument identity determination in *Arabidopsis*. *Plant Signal. Behav.* 34, 246–247. doi: 10.1105/tpc.107.051797

Cazé, A. L. R., Kriedt, R. A., Beheregaray, L. B., Bonatto, S. L., and Freitas, L. B. (2012). Isolation and characterization of microsatellite markers for *Passiflora* contracta. *Int. J. Mol. Sci.* 13, 11343–11348. doi: 10.3390/ijms130911343

Cerqueira-Silva, C. B. M., Santos, E. S. L., Souza, A. M., Mori, G. M., Oliveira, E. J., Correa, R. X., et al. (2012). Development and characterization of microsatellite markers for the wild South American *Passiflora* cincinnata (Passifloraceae). *Am. J. Bot.* 99, 170–172. doi: 10.3732/ajb.1100477

Cerqueira-Silva, C. B. M., Santos, E. S. L., Vieira, J. G. P., Mori, G. M., Jesus, O. N., Corrêa, R., et al. (2014). New microsatellite markers for wild and commercial species of *Passiflora* (Passifloraceae) and cross-amplification. *Appl. Plant. Sci.* 2:1300061. doi: 10.3732/apps.1300061

Cervi, A. C. (1997). *Passifloraceæ* do Brasil. Estudo do gênero *Passiflora* L., subgênero *Passiflora*. *Fontqueria* 45, 1–92.

Corner, E. J. H. (1976). *The Seeds of Dycotyledons. 1-2.* Cambridge: Cambridge University Press.

Correa, M. F., Pinto, A. P. C., Rezende, J. A. M., Harakava, R., and Mendes, B. M. J. (2015). Genetic transformation of sweet passion fruit (*Passiflora alata*) and reactions of the transgenic plants to Cowpea aphid borne mosaic virus. *Eur. J. Plant Pathol.* 143, 813–821. doi: 10.1007/s10658-015-0733-5

Cutri, L., and Dornelas, M. C. (2012). PASSIOMA: Exploring expressed sequence tags during flower development in *Passiflora* spp. *Comp. Funct. Genomics* 2012, 1–11. doi: 10.1155/2012/510549

Dathan, A. S. R., and Singh, D. (1973). Development and structure of seed in Tacsonia Juss. and *Passiflora* L. *Proc. Indian Acad. Sci.* 77B, 5–18.

Dhawan, K., Dhawan, S., and Sharma, A. (2004). *Passiflora*: a review update. *J. Ethnopharmacol.* 94, 1–23. doi: 10.1016/j.jep.2004.02.023

Endress, P. K. (2011). Angiosperm ovules: diversity, development, evolution. *Ann. Bot.* 107, 1465–1489. doi: 10.1093/aob/mcr120

Enugutti, B., Kirchhelle, C., Oelschner, M., Torres Ruiz, R. A., Schliebner, I., Leister, D., et al. (2012). Regulation of planar growth by the *Arabidopsis* AGC protein kinase UNICORN. *Proc. Natl. Acad. Sci. U.S.A.* 109, 15060–15065. doi: 10.1073/pnas.1205089109

Enugutti, B., and Schneitz, K. (2013). Genetic analysis of ectopic growth suppression during planar growth of integuments mediated by the *Arabidopsis* AGC protein kinase UNICORN. *BMC Plant Biol.* 13:2. doi: 10.1186/1471-2229-13-2

Fenster, C. B., Armbruster, W. S., Wilson, P., Dudash, M. R., and Thomson, J. D. (2004). Pollination syndromes and floral specialization. *Ann. Rev. Ecol. Evol. Syst.* 35, 375–403. doi: 10.1146/annurev.ecolsys.34.011802.132347

Gasser, C. S., Broadhvest, J., and Hauser, B. A. (1998). Genetic analysis of ovule development. *Annu. Rev. Plant Physiol. Plant Mol. Biol.* 49, 1–24. doi: 10.1146/annurev.arplant.49.1.1

Grear, J. W., and Dengler, N. G. (1976). The seed appendage of *Eriosema* (Fabaceae). *Brittonia* 28, 281–288. doi: 10.2307/2805789

Gross-Hardt, R., Lenhard, M., and Laux, T. (2002). WUSCHEL signaling functions in interregional communication during *Arabidopsis* ovule development. *Genes Dev.* 16, 1129–1138. doi: 10.1101/gad.225202

Herrera, C. M. (1989). Seed dispersal by animals: a role in Angiosperm diversification? *Am. Nat.* 133, 309–322. doi: 10.1086/284921

Kapil, R. N., and Vasil, I. K. (1963). "Ovule," in *Recent Advances in the Embryology of Angiosperms*, ed. P. Maheshwari (Delhi: International Society of Plant Morphologists), 41–67.

Kloos, A., and Bouman, F. (1980). Case studies in aril development *Passiflora suberosa* L. and *Turnera ulmifolia* L. *Beitrage Biologie der Pflanzen* 55, 49–66.

Koo, S. C., Bracko, O., Park, M. S., Schwab, R., Chun, H. J., Park, K. M., et al. (2010). Control of lateral organ development and flowering time by the *Arabidopsis thaliana* MADS-box Gene AGAMOUS-LIKE6. *Plant J.* 62, 807–816. doi: 10.1111/j.1365-313X.2010.04192.x

Kratzer, J. (1918). Die samenentwicklung der passifloraceen. *Flora* 110, 274–343.

Kugler, E. E., and King, L. A. (2004). "A brief history of the passionflower," in *Passiflora: Passinflowers of the World*, eds T. Ulmer, J. M. MacDougal, and B. Ulmer (Portland: Timber Press), 15–26.

Lovisetto, A., Baldan, B., Pavanello, A., and Casadoro, G. (2015a). Characterization of an AGAMOUS gene expressed throughout development of the fleshy fruit-like structure produced by *Ginkgo biloba* around its seeds. *BMC Evol. Biol.* 15:139. doi: 10.1186/s12862-015-0418-x

Lovisetto, A., Guzzo, F., Busatto, N., and Casadoro, G. (2013). Gymnosperm B-sister genes may be involved in ovule/seed development and, in some species, in the growth of fleshy fruit-like structures. *Ann. Bot.* 112, 535–544. doi: 10.1093/aob/mct124

Lovisetto, A., Guzzo, F., Tadiello, A., Toffali, K., Favretto, A., and Casadoro, G. (2012). Molecular analyses of MADS-box genes trace back to gymnosperms the invention of fleshy fruits. *Mol. Biol. Evol.* 29, 409–419. doi: 10.1093/molbev/msr244

Lovisetto, A., Masiero, S., Rahim, M. A., Mendes, M. A. M., and Casadoro, G. (2015b). Fleshy seeds form in the basal Angiosperm *Magnolia grandiflora* and several MADS-box genes are expressed as fleshy seed tissues develop. *Evol. Dev.* 17, 82–91. doi: 10.1111/ede.12106

Mack, A. L. (2000). Did fleshy fruit pulp evolve as a defense against seed loss rather than as a dispersal mechanism? *J. Biosci.* 25, 93–97. doi: 10.1007/BF02985186

Maheshwari, P. (1950). *An Introduction to the Embryology of Angiosperms.* New York, NY: McGraw-Hill.

Manders, G., Otoni, W. C., d'Utra Vaz, F. B., Blackhall, N. W., Power, J. B., and Davey, M. R. (1994). Transformation of passionfruit (*Passiflora edulis* fv flavicarpa Degener.) using *Agrobacterium tumefaciens*. *Plant Cell Rep.* 13, 697–702. doi: 10.1007/BF00231627

Meister, R. J., Kotow, L. M., and Gasser, C. S. (2002). SUPERMAN attenuates positive INNER NO OUTER autoregulation to maintain polar development of *Arabidopsis* ovule outer integuments. *Development* 129, 4281–4289.

Mita, S., Kawamura, S., and Asai, T. (2002). Regulation of the expression of a putative ethylene receptor, PeERS2, during the development of passion fruit (*Passiflora edulis*). *Physiol. Plant.* 114, 271–280. doi: 10.1034/j.1399-3054.2002.1140213.x

Mita, S., Kawamura, S., Yamawaki, K., Nakamura, K., and Hyodo, H. (1998). Differential expression of genes involved in the biosynthesis and perception of ethylene during ripening of passion fruit (*Passiflora edulis* Sims). *Plant Cell Physiol.* 39, 1209–1217. doi: 10.1093/oxfordjournals.pcp.a029322

Monteiro-Hara, A. C. B. A., Jadão, A. S., Mendes, B. M. J., Rezende, J. A. M., Trevisan, F., Mello, A. P. O. A., et al. (2011). Genetic transformation of passionflower and evaluation of R1 and R2 generations for resistance to Cowpea aphid borne mosaic virus. *Plant Dis.* 95, 1021–1025. doi: 10.1094/PDIS-12-10-0873

Muschner, V., Lorenz, A., Cervi, A. C., Bonatto, S., Souza-Chies, T., Salzano, F., et al. (2003). A first molecular analysis of *Passiflora* (Passifloraceae). *Am. J. Bot.* 90, 1229–1238. doi: 10.3732/ajb.90.8.1229

Oliveira, E. J., Padua, J. G., Zucchi, M. I., Camargo, L. E. A., Fungaro, M. H. P., and Vieira, M. L. C. (2005). Development and characterization of microsatellite markers from the yellow passion fruit (*Passiflora edulis* f. flavicarpa). *Mol. Ecol. Notes* 5, 331–333. doi: 10.1111/j.1471-8286.2005.00917.x

Oliveira, E. J., Vieira, M. L. C., Garcia, A. A. F., Munhoz, C. F., Margarido, G. R. A., Consoli, L., et al. (2008). An integrated molecular map of yellow passion fruit based on simultaneous maximum-likelihood estimation of linkage and linkage phases. *J. Am. Soc. Hortic. Sci.* 133, 35–41.

Otoni, W. C., Pinto, D. L. P., Rocha, D. I., Vieira, L. M., Dias, L. L. C., Silva, M. L., et al. (2013). "Organogenesis and somatic embryogenesis in passionfruit (*Passiflora* sp.)," in *Somatic Embryogenesis and Gene Expression*, 1st Edn, eds J. Aslam, P. S. Srivastava, and M. P. Sharma (New Delhi: Narosa Publishing House), 1–17.

Pádua, J. G., Oliveira, E. J., Zucchi, M. I., Oliveira, G. C. X., Camargo, L. E. A., and Vieira, M. L. C. (2005). Isolation and characterization of microsatellite markers from the sweet passion fruit (*Passiflora alata* Curtis: Passifloraceae). *Mol. Ecol. Notes* 5, 863–865. doi: 10.1111/j.1471-8286.2005.01090.x

Pereira, K. J. C., and Dias, D. C. F. S. (2000). Germinação e vigor de sementes de maracujá-amarelo (*Passiflora edulis* Sims. f. flavicarpa Deg.) submetidas a diferentes métodos de remoção da mucilagem. *Rev. Bras. Sementes* 22, 288–291. doi: 10.1590/S1413-70542007000500010

Raju, M. V. S. (1956). Embryology of the Passifloraceae. I. Gametogenesis and seed development of *Passiflora* calcarata Mast. *J. Indian Bot. Soc.* 35, 126–138.

Reinheimer, R., and Kellogg, E. A. (2009). Evolution of AGL6-like MADS box genes in grasses (Poaceae): ovule expression is ancient and palea expression is new. *Plant Cell* 21, 2591–2605. doi: 10.1105/tpc.109.068239

Rodriguez-Riaño, T., Valuteña, F. J., and Ortega-Olivencia, A. (2006). Megasporogenesis. *Ann. Bot.* 98, 777–791. doi: 10.1093/aob/mcl166

Santos, A. A., Penha, H. A., Bellec, A., Munhoz, C. F., Pedrosa-Harand, A., Bergès, H., et al. (2014). Begin at the beginning: A BAC-end view of the passion fruit (*Passiflora*) genome. *BMC Genomics* 15:816. doi: 10.1186/1471-2164-15-816

Schneitz, K. (1999). The molecular and genetic control of ovule development. *Curr. Opin. Plant Biol.* 2, 13–17. doi: 10.1016/S1369-5266(99)80003-X

Schneitz, K., Hulskamp, M., Kopczak, S. D., and Pruitt, R. E. (1997). Dissection of sexual organ ontogenesis: a genetic analysis of ovule development in *Arabidopsis thaliana*. *Development* 124, 1367–1376.

Sieber, P., Gheyselinck, J., Gross-Hardt, R., Laux, T., Grossniklaus, U., and Schneitz, K. (2004). Pattern formation during early ovule development in *Arabidopsis thaliana*. *Dev. Biol.* 273, 321–334. doi: 10.1016/j.ydbio.2004.05.037

Singh, D. (1962). The structure and development of ovule and seed of *Passiflora foetida*. *Agra Univ. J. Res.* 11, 99–111.

Skinner, D. J., Hill, T. A., and Gasser, C. S. (2004). Regulation of ovule development. *Plant Cell* 16, S32–S45. doi: 10.1105/tpc.015933

Trevisan, F., Mendes, B. M. J., Maciel, S. C., Vieira, M. L. C., Meletti, L. M. M., and Rezende, J. A. M. (2006). Resistance to Passion fruit woodiness virus in transgenic passionflower expressing the virus coat protein gene. *Plant Dis.* 90, 1026–1030. doi: 10.1094/PD-90-1026

Ulmer, T., and MacDougal, J. M. (2004). *Passiflora: Passionflowers of the World.* Portland: Timber Press.

Van der Pijl, L. (1972). *Principles of Dispersal in Higher Plants.* Berlin: Springer.

Vieira, M. L. C., and Carneiro, M. S. (2004). "*Passiflora* spp., passionfruit," in *Biotechnology of Fruit and Nut Crops*, ed. R. E. Litz (Wallingford: CABI Publishing), 435–453.

Villanueva, J. M., Broadhvest, J., Hauser, B. A., Meister, R. J., Schneitz, K., and Gasser, C. S. (1999). INNER NO OUTER regulates abaxial-adaxial patterning in *Arabidopsis* ovules. *Genes Dev.* 13, 3160–3169. doi: 10.1101/gad.13.23.3160

Vrebalov, J., Ruezinsky, D., Padmanabhan, V., White, R., Medrano, D., Drake, R., et al. (2002). A MADS-box gene necessary for fruit ripening at the tomato ripening-inhibitor (rin) locus. *Science* 296, 343–346. doi: 10.1126/science.1068181

Yasuhiro, I. (2016). Regulation of tomato fruit ripening by MADS-Box transcription factors. *JARQ* 50, 33–38. doi: 10.6090/jarq.50.33

Yockteng, R. S., and Nadot, S. (2004). Phylogenetic relationships among *Passiflora* species based on the Glutamine Synthase nuclear gene expressed in the chloroplast (ncpGS). *Mol. Phylogenet. Evol.* 31, 379–396. doi: 10.1016/S1055-7903(03)00277-X

Zhang, X., Zhang, Z., and Stützel, T. (2011). Aril development in Celastraceae. *Feddes Repert.* 122, 445–455. doi: 10.1002/fedr.201200007

Conflict of Interest Statement: The authors declare that the research was conducted in the absence of any commercial or financial relationships that could be construed as a potential conflict of interest.

7

Expansion and Functional Divergence of AP2 Group Genes in Spermatophytes Determined by Molecular Evolution and *Arabidopsis* Mutant Analysis

Pengkai Wang[1,2], Tielong Cheng[1], Mengzhu Lu[3], Guangxin Liu[1], Meiping Li[1], Jisen Shi[1], Ye Lu[1], Thomas Laux[4] and Jinhui Chen[1]**

[1] Ministry of Education, Key Laboratory of Forest Genetics and Biotechnology, Nanjing Forestry University, Nanjing, China, [2] Suzhou Polytechnic Institute of Agriculture, Suzhou, China, [3] Laboratory of Biotechnology, Chinese Academy of Forestry, Beijing, China, [4] Institute of Biology III, University of Freiburg, Freiburg, Germany

Edited by:
Federico Valverde,
Spanish National Research Council,
Spain

Reviewed by:
Marcelo Carnier Dornelas,
State University of Campinas, Brazil
Seonghoe Jang,
Academia Sinica, Taiwan

***Correspondence:**
Thomas Laux
laux@biologie.uni-freiburg.de
Jinhui Chen
chenjh@njfu.edu.cn

The *APETALA2* (*AP2*) genes represent the AP2 group within a large group of DNA-binding proteins called AP2/EREBP. The *AP2* gene is functional and necessary for flower development, stem cell maintenance, and seed development, whereas the other members of AP2 group redundantly affect flowering time. Here we study the phylogeny of AP2 group genes in spermatophytes. Spermatophyte AP2 group genes can be classified into AP2 and TOE types, six clades, and we found that the AP2 group homologs in gymnosperms belong to the AP2 type, whereas TOE types are absent, which indicates the AP2 type gene are more ancient and TOE type was split out of AP2 type and losing the major function. In Brassicaceae, the expansion of AP2 and TOE type lead to the gene number of AP2 group were up to six. Purifying selection appears to have been the primary driving force of spermatophyte AP2 group evolution, although positive selection occurred in the AP2 clade. The transition from exon to intron of *AtAP2* in *Arabidopsis* mutant leads to the loss of gene function and the same situation was found in *AtTOE2*. Combining this evolutionary analysis and published research, the results suggest that typical AP2 group genes may first appear in gymnosperms and diverged in angiosperms, following expansion of group members and functional differentiation. In angiosperms, *AP2* genes (AP2 clade) inherited key functions from ancestors and other genes of AP2 group lost most function but just remained flowering time controlling in gene formation. In this study, the phylogenies of AP2 group genes in spermatophytes was analyzed, which supported the evidence for the research of gene functional evolution of AP2 group.

Keywords: AP2 group gene, spermatophyte, phyllogeny, selective pressures, arabidopsis mutants, functional divergence

INTRODUCTION

The genes in AP2/ERF family can be divided into subfamilies according to the number of AP2/ERF domains. There is a single AP2/ERF domain in each member of the EREBP subfamily, which function in the signal transduction pathways of stress responses and cambial tissue development (Mizoi et al., 2012; Licausi et al., 2013). The members of the AP2 subfamily contain two

AP2/ERF domains is further classified into two monophyletic groups: the AP2 group and the AINTEGUMENTA (ANT group, Shigyo et al., 2006). We selected the AP2 group genes for our study because they played key roles in the reproductive and vegetative organs development (Ohto et al., 2005; Huijser and Schmid, 2011).

There are seven conservative domains in a typical AP2 group gene, including one ethylene-responsive element binding factors (ERF) -associated amphiphilic repression motif (EAR) (Kagale et al., 2010) or EAR-like domain, a nuclear localization signal (NLS) domain, two AP2 (AP2-R1 and AP2-R2) domain (Kim et al., 2006), a linkage domain (connecting the AP2-R1 with R2), another EAR domain, and a *MIR172* target site (**Image 1**). However, not all AP2 group genes contain two typical complete AP2 domains. For example, there are six members in the AP2 gene group in *Arabidopsis*, *TARGET OF EARLY ACTIVATION TAGGED 1-3* (*TOE1-3*), *AP2*, *SCHLAFMUTZE* (*SMZ*), and SCHNARCHZAPFEN (*SNZ*). Among these six *Arabidopsis* genes, *AP2*, *TOE3*, and *TOE1* contain both complete AP2 domains (AP2-R1 and R2 domains) but there is only one typical AP2 domain (AP2-R1 domain) in *TOE2*, *SMZ*, and *SNZ* (**Image 1**). The AP2-R2 domain in these three genes are not the same as *AP2*, *TOE3*, and *TOE1*.

In *Arabidopsis*, AP2 regulates floral development, (Jofuku et al., 1994), stem cell maintenance, (Wurschum et al., 2005), seed developmenta, whereas the remaining five genes (*TOE1-3*, *SMZ*, and *SNZ*) act redundantly as flowering repressors (Zhu and Helliwell, 2011). mRNA abundance and translation of *TOE1-2*, *AP2*, *SMZ*, and *SNZ* in *Arabidopsis* are regulated by *miR172*, which is also important for regulating phase transition and determining floral organ identity in monocotyledons (Nair et al., 2010; Zhu and Helliwell, 2011). *TOE3* most likely acts redundantly with *TOE1* and *TOE2* to repress flowering. A good candidate for such a repressor is SMZ, which was originally identified in an activation-tagging screen because of its dominant late-flowering phenotype. Additionally, *SNZ*, a paralog of *SMZ*, represses flowering when expressed at high levels. Several other known regulators of flowering time have been identified as *SMZ* targets. Among them are *SMZ* itself, *SNZ*, *AP2*, and *TOE3*, suggesting complex feedback regulation among AP2 group members.

Several studies have examined the origin, phylogeny and evolution of the AP2/ERF family and the AP2 subfamily (Magnani et al., 2004; Kim et al., 2006; Shigyo et al., 2006). With the development of high-throughput DNA sequencing techniques, increasing information has become available in recent years for the of AP2/ERF gene family based on genomic data from species such as rice, apple, peach, *Hevea brasiliensis*, *Prunus mume*, *Vitis vinifera*, *Populus trichocarpa*, Chinese cabbage, and others (Zhuang et al., 2008; Licausi et al., 2010; Sharoni et al., 2011; Zhang et al., 2012; Du et al., 2013; Duan et al., 2013; Song et al., 2013). However, these studies have mainly focused on the identification, classification and expression of AP2/ERF family genes. To date, there have been no studies describing the molecular evolutionary history and the structural characterization of the AP2 group in spermatophytes. As such, an evolutionary and structural analysis of this group may provide both a reference for further functional studies and evidence for gene functional diversification.

Here, we examined 105 spermatophyte AP2 group genes from 56 spermatophytes by phylogenetic analyses, comparing whole gene sequences, homeodomains, and other motifs. The results revealed that, in general, the spermatophyte AP2 group experienced background purifying selection throughout evolution. However, analyses also showed that the AP2 group genes have also undergone positive selection, despite little evidence for positive pressure on these genes. In particular, the evolutionary relationships among members of the AP2 group were apparent from the divergence between the TOE and AP2 type. By analyzing the expression patterns, functional data and phylogenetic relationships among *AP2* genes, we reveal rules concerning the formation of new genes in the AP2 group and identified the pathway of functional evolution. We also find evidenc that the AP2 function in maintaining the stem-cell niche is to be conserved in spermatophytes.

RESULTS

The Orthologs of AP2 Group Genes from Spermatophytes Differ and Can Be Classified into Two Types and Six Clades

The composition of AP2 group orthologs differs among spermatophyte species. It is well known that the *Arabidopsis* AP2 group has six members, namely *AtAP2*, *AtTOE1-3*, *AtSMZ*, and *AtSNZ*. In fact, blast analysis of spermatophyte AP2 group gene sequences revealed that only some species in Brassicaceae contain five or six AP2 group members. Orthologs of *AtTOE2*, *AtTOE3*, *AtSMZ*, and *AtSNZ* were not found in the other species included in our study. Although the AP2-R2 domain in the *AtTOE1* orthologs of some species is not complete, these genes still clustered together to form the TOE1 clade in the prephylogenetic analysis. Therefore, the orthologs of *AtTOE2*, *AtTOE3*, *AtSMZ*, and *AtSNZ* are only been identified from Brassicaceae (**Data sheet 1**).

All predicted spermatophyte AP2 group protein sequences (105, **Data sheet 1**) were retrieved from the plant genome (Phytozome and NCBI) and protein databases (NCBI) and used to construct a maximum-likelihood phylogenetic tree (**Figure 1** and **Image 2**). According to the simplified phylogenetic tree (**Figure 1**) of spermatophyte AP2 group, all genes were categorized as two types: the AP2 type, which included the three clades TOE3, AP2-like and AP2, and the TOE1 type, which included the three clades TOE1, TOE2, SMZ/SNZ. The results of the phylogenetic analysis were consistent with those of the sequence search. For each ortholog, most of the spermatophyte sequences clustered together to form an independent clade, except in gymnosperms. The genes *AP2*, *TOE1*, *TOE2*, *SMZ/SNZ*, and *AP2L* from gymnosperms *Cycas revoluta* (*CyrAP2L*), *Ginkgo biloba* (*GibAP2L*), *Picea sitchensis* (*PisAP2L*), *Larix* × *marschlinsii* (*LamAP2La, b*), *Picea abies* (*PiaAP2La, b*), and *Pinus thunbergii* (*PitAP2La, b*) are well clustered into form six independent clades (**Figure 1**, bootstrap

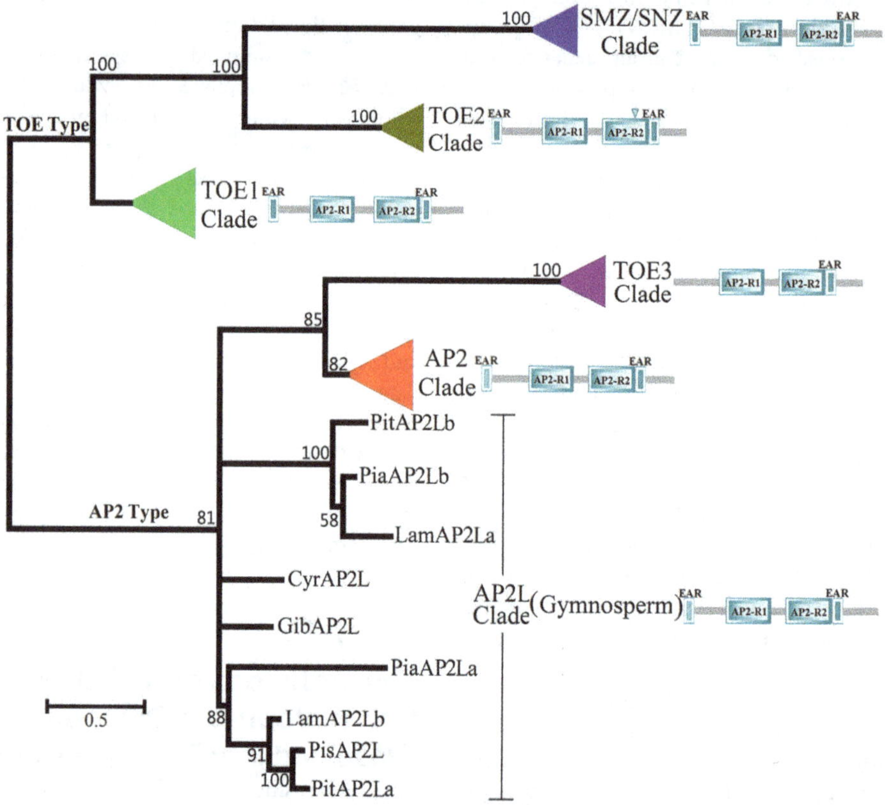

FIGURE 1 | Maximum-likelihood Phylogenetic Tree of AP2 Group Protein in Spermatophyte (simplified phylogenetic tree). The ML tree was constructed based on the whole protein sequences of spermatophyte AP2 Group gene using MEGA6.0 with 1000 bootstrap replications and Jones-Taylor-Thornton (JTT) + Gamma Distributed model with Invariant sites model (Discrete Gamma Categories = 5). There were 105 spermatophyte AP2 group genes, which were searched from 23 families, 56 spermatophytes. The different color triangles represent different clades except six gymnosperms (red, AP2 clades; green, TOE1 clade; olive, TOE2 clade; purple, TOE3 clade; blue, SMZ/SNZ clade). The detailed phylogenetic tree of every clade was shown in **Image 2**. The scale bar indicates the branch length that corresponds to 0.5 substitutions per site. The species and accession numbers are listed in **Data sheet 1**.The abbreviations used are as follows: *Pit, Pinus thunbergii*; *Pia, Picea abies*; *Lam, Larix × marschlinsii*; *Cyr, Cycas revoluta*; *Gib, Ginkgo biloba*; *Pis, Picea sitchensis*.

value >80%). *AP2L* sequences from gymnosperms were obtained from the NCBI database and clustered together with the AP2 and TOE3 clades to form a larger group, which implied that *AP2* genes might be relatively ancient in the AP2 group. The two sub-branches of Pinaceae in the AP2L cluster c adequately reflected the duplication of AP2 genes in gymnosperms. In each clade, most of the sequences from species within a single family or order clustered together well to form an independent group (**Image 2**, bootstrap value >69%). These results indicated that most of these sequences are specific at the family level. Intriguingly, in the branches of AP2 and TOE1, the phylogenetic trees are very similar to the structure of the Angiosperm Phylogeny Group system, which classifies basal angiosperms, monocots, and dicotyledons into three independent groups. In the AP2 clade, the dicots such as *Ricinus communis*, *Manihot esculenta*, *Jatropha curcas*, *Populus trichocarpa*, *Carica papaya*, *Vitis vinifera*, and *Betula platyphylla* with unisexual flowers formed a branch with lower bootstrap value (bootstrap value = 25). However, the branch of unisexual flower didn't appear in the TOE1 clade.

The Distribution of Homeodomains Varies Significantly in Different Clades and Types

There are seven common homeodomains in the typical AP2 group genes according the analytical results of MEME and Pfam (Bailey et al., 2009; Punta et al., 2011): the first EAR domain (DLNxxP or LxLxL), NLS domain, AP2-R1 domain, linkage domain, AP2-R2 domain, the second EAR domain (LxLxL), and miRNA172 target site. These homeodomains except miRNA172 target site are conserved in amino acid sequences. NLS domain, AP2-R1 domain, linkage domain, the second EAR domain and miRNA172 target site have greater sequence similarity than the first EAR domain and AP2-R2 domain.

Notably, the EAR domains differed among the clades. In the TOE3 clade (**Image 2**), there is no first EAR domain. Likewise, PtAP2, AdAP2, EgAP2, StAP2, MdTOE1, AtTOE1, GpTOE1, NhTOE1, and orthologs from the Rutaceae family also do not contain the first EAR domain. Interestingly, both EAR domains are missing in TOE1 orthologs from the Poaceae family. *LamAP2La*, *PiaAP2La*, and *PisAP2L*, three complete protein-coding genes in the AP2L clade, contain two EAR motifs. Because some N-terminal amino acid sequences of the AP2L clade were incomplete, it was not clear whether the first EAR domain exists in all AP2L clade members from gymnosperms. The EAR domains are of two types, namely DLNxxP and LxLxL. The amino acid sequence of the first EAR domain in the AP2 type (including LamAP2La, PiaAP2La, and PisAP2L) was DLNxxP, but the sequence in the TOE1 type was LxLxL, which is the same as the second EAR domain (LxLxL). Outside of the EAR domains, differences in the AP2-R2 domain were identified among AP2 group proteins. There was an incomplete AP2-R2 domain in the amino acid sequence of the TOE2 clade lacking a 15-residue insertion (**Image 1**). There are significant amino-acid sequence differences between the typical AP2-R2 domain and the SMZ/SNZ clade. The TOE2 clade has a closer phylogenetic relationship to the SMZ/SNZ clade, forming a TOE type with the TOE1 clade, indicating a common ancestry separate from the AP2 type.

The Distribution of Motifs Reflects Differences and Phylogenetic Relationships across Clades and Species

Motifs of the AP2 group were investigated, and 25 motifs including the AP2-R2 domain of the SMZ/SNZ clade, i.e., Motif 8, were then characterized relative to the homeodomains (**Figure 2**). Most of these motifs were located upstream of the NLS (7/25) or downstream of the second EAR domain (14/25). Only one motif was identified at the C-terminus (i.e., the downstream of the miR172 target site) of most AP2- and TOE-type proteins, and no C-terminal motif was seen in the proteins of clades TOE3 and SMZ/SNZ and TOE1 proteins of

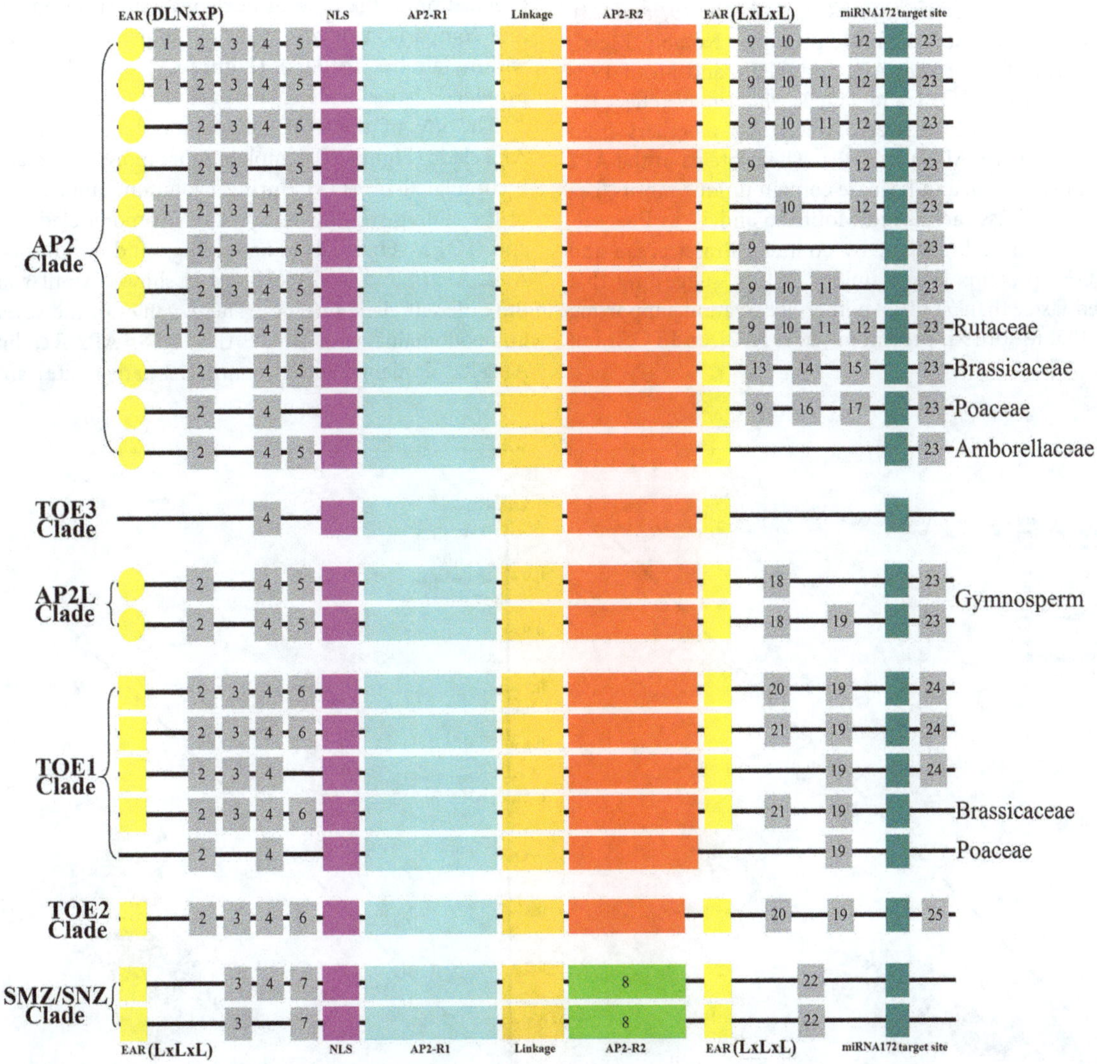

FIGURE 2 | Distribution of Homeodomains and Motifs of AP2 Group in Spermatophyte. A schematic representation of motifs obtained using MEME within the sequences is displayed. The homeodomains (EAR domain, NLS domain, AP2-R1 domain, linkage domain; AP2-R2 domain, miRNA172 target site) were showed by the same colors with **Image 1**. The different motifs were indicated by using numbers.

the Brassicaceae and Poaceae families (**Figure 2**). We also noted that most AP2 clade proteins contained more motifs (15/25), but there was only one motif (Motif 4) in TOE3 clade proteins. The differences in numbers of motifs were seen mostly in one region, i.e., between the second EAR motif and the miR172 target site. Most AP2 clade proteins had three or four motifs, whereas other proteins had only one or two. The different clades and species could be characterized based on the distribution of motifs. Every clade contained its own unique motifs: AP2 clade, Motifs 1 and 9–17; AP2L clade, Motif 18; TOE1 clade, Motifs 21 and 24; TOE2 clade, Motif 25; SMZ/SNZ clade, Motifs 7, 8, and 22). This indicated that the motifs might be related to the functional divergence of the AP2 group (**Figure 2**). Motif 4 was shared by almost all AP2 group proteins (**Figure 2**).

Interestingly, the distribution of motifs revealed a phylogenetic relationship among AP2 group proteins that agreed with the results of the phylogenetic analysis by MEGA (**Figure 1**). In the AP2L clade, all sequences were from gymnosperms and contained five or six motifs (Motifs 2, 4, 5, 18, 19, and 23), suggesting that these motifs are more primitive. Notably, Motifs 5 and 23 are specific to clades AP2 and AP2L, and Motif 19 is also specific to clades TOE1 and TOE2, which may reflect the phylogenetic relationship between AP2 group members from clade AP2L and clades AP2 and TOE1. The proteins in the AP2 clade from Brassicaceae and Poaceae contain unique motifs, i.e., Motifs 13–15 in Brassicaceae and Motifs 16 and 17 in Poaceae, forming two separate branches. By contrast, Brassicaceae and Poaceae TOE1 proteins lacked unique motifs, suggesting that *AP2* evolved faster than *TOE1*. Another observation supporting this view is that Motif 19 is present in clades AP2L and TOE1 but not in clade AP2.

Purifying Selection was the Main Driving Force in the Evolution of Spermatophyte AP2 Group Genes, but Positive Selection Still Occurred, Mostly in Clade AP2

One criterion for assessing the type of selective pressure at the protein level is to calculate ω, i.e., dN/dS, for protein-coding genes (Seo et al., 2004; Kryazhimskiy and Plotkin, 2008). The dN/dS ratio ω provides a criterion for assessing selective pressure at the protein level (Zhang et al., 2006). The ω-values of >1, 1 and <1 imply positive selection, neutral evolution and purifying selection, respectively. In the graph of the AP2 group sequences, most of the points fell between the dS axis and the diagonal, indicating $dN < dS$ (**Figure 3**) and suggesting that purifying selection dominated the selection process during evolution. Similar ω-values were obtained for each family (order) (**Figure 3B**), which also contained the different clades and the comparison between them ($p < 0.01$, *Z-test*, **Figure 3B**). Calculation of the ratio of nucleotide substitutions in a one-by-one comparison of dN to dS for individual AP2 group genes within families (orders) (**Figure 3A**) and clades (**Figure 3B**) provided further evidence for purifying selection.

The dN of the TOE1 clade was higher than that of the AP2 clade (**Figure 3B**), implying that more amino acids changes accumulated in TOE1 during evolution, similar to the results of the sliding-window analyses of ω among clades AP2, AP2L, and TOE1 (**Figure 4**). The sliding-window ω-tests on clades AP2, AP2L, and TOE1 (**Figure 4**) showed similar ω curves in the different clade lineages. The ω-values of the seven common homeodomains (the two EARs, NLS, AP2-R1, linkage, and AP2-R2 domains and the miR172 target site) in the three

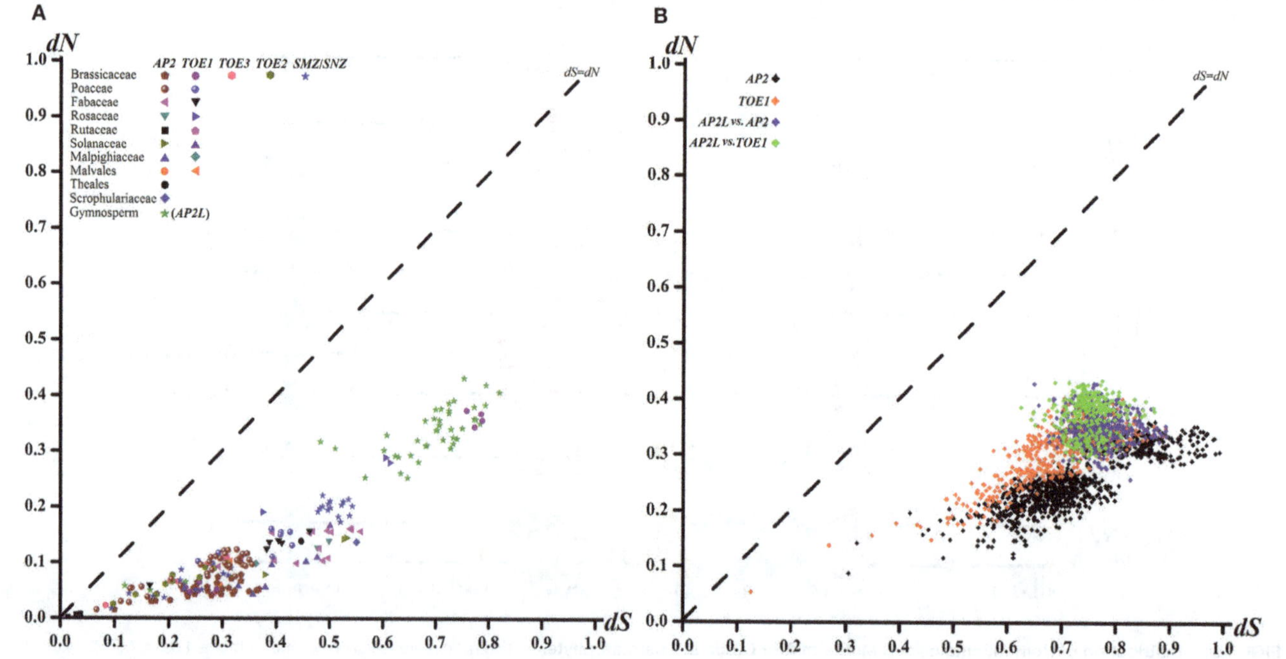

FIGURE 3 | Pairwise Comparison Plots of *dN* and *dS* Values for each family (order) AP2 group genes (A) and different clades (B).

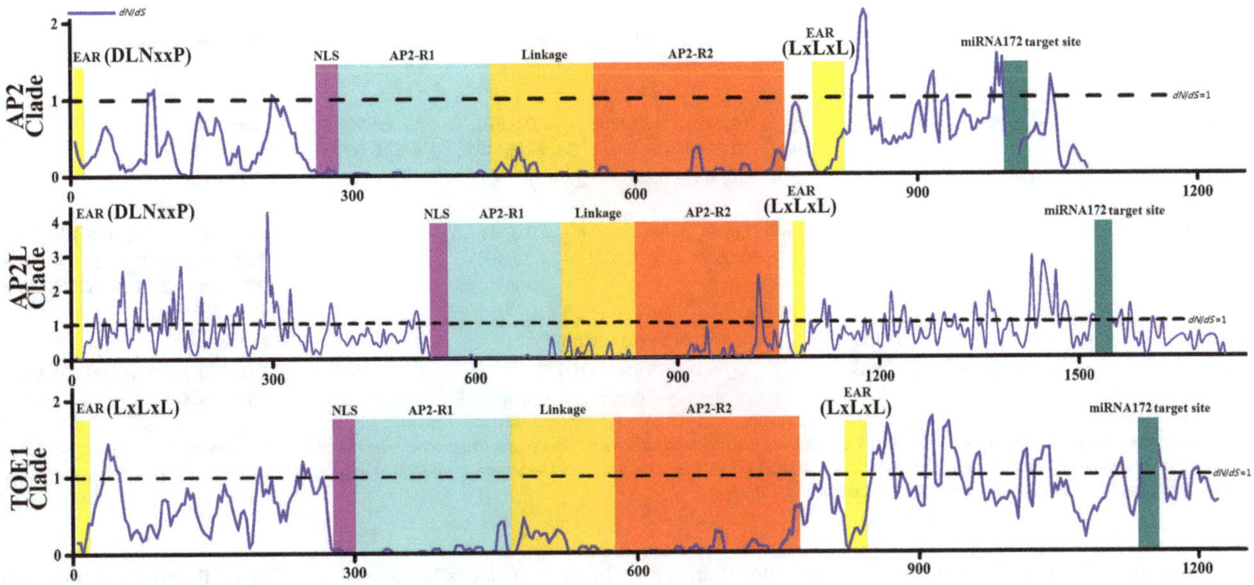

FIGURE 4 | Sliding-window Analyses of *dN/dS* among Three Clades Genes (AP2 clade,AP2L clade, and TOE1 clade). The sliding-window analyses were performed using a 30-bp sliding window moving in steps of 3-bp. Numbers on the x-axis represent the sequence positions followed below by the domain map of genes. The homeodomains (EAR domain, NLS domain, AP2-R1 domain, linkage domain; AP2-R2 domain, miRNA172 target site) were showed by the same colors with **Image 1**. The straight lines shown in the figure represent $dN = dS$.

lineages were almost all much less than 1 except for the end of the AP2-R2 domain in AP2L, which further suggested the functional importance of the common homeodomains in AP2 group proteins. The ω-values are not shown for the miR172 target site because dN was zero. The peaks above the line that marks $dN = dS$ in **Figure 4** suggest the existence of positive selection, primarily on either side of the two AP2 domains, especially downstream of domain AP2-R2. There were more $dN > dS$ peaks in the graph of the AP2L clade because fewer AP2 ortholog sequences are available for gymnosperms (only eight sequences) for analysis of evolutionary pressure, but the distribution trend of positive-selection peaks agreed with that of clades AP2 and TOE1. Compared with clade AP2, TOE1 contained more positive-selection peaks, implying that the purifying selective pressure on the TOE1 clade was relatively weak and enabled more nonsynonymous substitutions to be retained. Potentially, the ω-value differences of different regions are related to functional divergence in AP2 group genes.

To examine if ω varied among branches of each clade (**Figure 1**), the free-ratio and one-ratio models in the Codeml program of PAML 4.2 were chosen and used to detect selective pressure acting on some branches (**Data sheet 2**). The values of ω for these AP2 group genes were similar (0.2083–0.2897) and substantially less than 1. However, the free-ratio model fit the data better than the one-ratio model for the protein-coding sequences from clades AP2, AP2L, and TOE1, suggesting that the genes from these three clades possibly experienced different selective pressures. Conversely, when coding sites in genes of clades TOE3, TOE2, and SMZ/SNZ were analyzed, the codon-substitution free-ratio model, which allows for different ω-values among the branches, did not fit the data any better than the one-ratio model, which assumes a single mean ω-value for the branches. The primary reason for this result was that the genes in clades TOE3, TOE2, and SMZ/SNZ were all from Brassicaceae. All AP2 group genes were analyzed by the free-ratio and one-ratio models, and the results also suggested that ω varied among branches. Therefore, genes of different clades experienced different selective pressures.

Six codon substitution models, namely M0 (one-ratio), M1a (nearly neutral), M2a (positive selection), M3 (discrete), M7 (beta), and M8 (beta and ω), were implemented in PAML 4.2 to analyze the positive selection and identify positively selected sites in all AP2 group genes. The likelihood values and parameter estimates of all AP2 group gene sequences from the six models applied in the Codeml program are listed in **Table 1**. The average ω-values in the six models ranged from 0.2794 to 0.5952, providing evidence for purifying selection. Although the average ω was 0.2794 for all sites of the AP2 group genes by the M0 model, this model was rejected as a result of the low-likelihood value (−87829.48516) and the LRT statistic (2 delta lambda statistic, $2\Delta l$, **Table 1**). No positively selected sites were identified by the M3 model because $\omega < 1$, but two models (M2a and M8) allowed for positive selection, indicated by 38 and 17 positively selected sites with $\omega > 1$, respectively. Because of the overestimate of the number of actual positively selected sites (Anisimova et al., 2001, 2002), the results under model M3 were not used to identify positively selected sites. To reduce or avoid possible false-positive results, positively selected sites identified simultaneously by models M2a and M8 in Codeml were defined as positively selected. The LRT statistic demonstrated that the

TABLE 1 | Likelihood values and parameter estimates for AP2 group genes sequences in Spermatophyte.

Model code	lnL	*dN/dS*	Estimates of parameters	$2\Delta l$	Positive selection
M0 (one-ratio)	−87829.4852	0.2794	$\omega = 0.2794$		None
M3 (discrete)	−81346.4866	0.3541	$p_0 = 0.4298, p_1 = 0.2336$ ($p_2 = 0.3365$), $\omega_0 = 0.0223, \omega_1 = 0.3017$ $\omega_2 = 0.8143$	12965.9972 ($P = 0.0000$)	None
M1a (NearlyNeutral)	−82440.5178	0.5171	$p_0 = 0.5144$ ($p_1 = 0.4856$)		Not allowed
M2a (PositiveSelection)	−82283.7557	0.5952	$p_0 = 0.5014, p_1 = 0.4197$ ($p_2 = 0.0789$), $\omega_0 = 0.0594, \omega_1 = 1.0000$ $\omega_2 = 1.8466$	313.5242 ($P = 0.0000$)	44D **54G** 57V **74G 75S 76S 77A 78G 79K 80A** 81T **82N 83V** 276H **279Q** 284R **286N 287Q 289Q 290Q** 291L 353T
M7 (beta)	−81006.4547	0.3169	$p = 0.2595, q = 0.5591$	87.1449	Not allowed
M8 (beta & ω)	−80962.8822	0.3433	$p_0 = 0.9578$ ($p_1 = 0.0422$), $p = 0.3013, q = 0.727, \omega = 1.5014$	($P = 0.0000$)	**74G 75S 76S 77A** 79K **80A 82N** 287Q **289Q**

lnL: the log-likelihood difference between the two models; $2\Delta l$: twice the log-likelihood difference between the two models. The values in parentheses represent the significant level of 0.01 with a $\chi 2$ distribution at d.f. = 4 (M0 vs. M3) or 2 (M1a vs. M2a and M7 vs. M8). The amino acid sequence of AtAP2 was used as the sequence reference, and positive selected sites were identified with posterior probability $p > 0.95$. In boldface, $p > 0.99$.

two selection models fitted the data significantly better than the null models without positive selection, supporting the view that certain amino acids in AP2 group proteins experienced strong positive selection. At the level of posterior probability >0.95, 22, and 9 sites in the AP2 group genes were identified as being under positive selection ($\omega > 1$) by the selection models M2a and M8, respectively (**Table 1**). There were 15 sites with posterior probability >0.99 among the 22 positively selected sites in the M2a model and 7 among the 9 positively selected sites in the M8 model. All 9 positively selected sites detected by M8 were also identified by M2a at the level of posterior probability >0.99. The positively selected sites were mainly concentrated in two regions—upstream of the NLS and downstream of the second EAR domain. Both regions showed the corresponding positive-selection peaks in **Figure 4**. All of this evidence supports the existence of positive selection and positively selected sites in AP2 group genes during spermatophyte evolution.

The likelihood values and parameters of the three main clade branches (AP2, AP2L, and TOE1) of AP2 group proteins were estimated by the six models to detect whether there was positive selection in clades AP2, AP2L, and TOE1 (**Table 2**). Positive selection was only detected in clade AP2 by the two positive selection models (M2a and M8), and no such positively selected sites were discovered in AP2L and TOE1. The positively selected sites in clade AP2 were not identified by the M3 (discrete) model because ω was <1, which was similar to the results for all AP2 genes. Of the sites with posterior probability >0.95 in AP2 clade proteins, six and four sites were identified to be under positive selection ($\omega > 1$) by selection models M2a and M8, respectively, and the number of positively selective sites was four and two in M2a and M8, respectively, at posterior probability >0.99. The positively selected sites of clade AP2 were mainly concentrated upstream of the miR172 target site, and there were also corresponding positive-selection peaks in **Figure 4**. The parameter estimates from the six models were quite similar between clades AP2L and TOE1. The parameters of the M2a model for AP2L and TOE1 revealed a lack of positive selection in the two clades because $\omega = 1$. The LRT statistic demonstrated that the M8 model did not fit the data significantly better than the M7 model without positive selection, suggesting that no amino acid sites in AP2L and TOE1 underwent positive selection. Therefore, the analysis of selective pressure of the three main clade branches of spermatophyte AP2 group genes indicated that clades AP2L and TOE1 experienced similar adaptive evolutionary mechanisms and only the AP2 clade underwent positive selection. Because *TOE2*, *TOE3*, *SMZ*, and *SNZ* are only found in Brassicaceae, the analysis of positive selection of Brassicaceae AP2 group genes and five branches (*AP2*, *TOE1*, *TOE2*, *TOE3*, and SMZ/SNZ) was performed (**Data sheet 3**), with the results revealing a lack of positive selection in the AP2 group genes of Brassicaceae, which indicates that the expansion of the number of AP2 group genes in Brassicaceae was not caused by positive selection.

Certain Amino Acid Residues in Common Homeodomains Reflect the Evolutionary Relationship from *AP2L* to *AP2* and *TOE1*

Both the distribution of motifs and the analysis of selective pressure suggested that *AP2L* may have diverged to yield the two structurally and functionally distinct genes *AP2* and *TOE1*. Specific motifs (Motifs 5, 19, and 23) of clades AP2 and TOE1 were also found in AP2L proteins, and AP2L and TOE1 experienced similar adaptive evolutionary processes. By comparing all AP2, AP2L, and TOE1 proteins, 10 amino acid sites that may reflect the evolutionary relationship in common homeodomains were identified (**Figure 5**), one in the AP2-R1 domain, three in the linkage domain, five in the AP2-R2 domain and one in the second EAR domain. These sites could be divided into three categories: AP2L having the same amino acids as TOE1 (three sites), AP2L having the same amino acids as AP2 (one site) and AP2L having two amino acids from AP2 and two from TOE1 (six sites). Only one site belonging to the second category suggested that *AP2* evolved faster than *TOE1*,

TABLE 2 | Likelihood Values and Parameter Estimates for AP2, AP2L, and TOE1 Clades Genes Sequences.

Clade	Model code	lnL	*dN/dS*	Estimates of parameters	$2\Delta l$	Positive selection
AP2	M0 (one-ratio)	−32964.4428	0.2083	$\omega = 0.2083$		None
	M3 (discrete)	−30570.7018	0.2613	$p_0 = 0.5033, p_1 = 0.3059$ ($p_2 = 0.1908$), $\omega_0 = 0.0129, \omega_1 = 0.2785$ $\omega_2 = 0.8891$	4787.48207 ($P = 0.0000$)	None
	M1a (NearlyNeutral)	−31071.0231	0.4177	$p_0 = 0.6109$ ($p_1 = 0.3891$)		Not allowed
	M2a (PositiveSelection)	−31036.8254	0.4636	$p_0 = 0.6053, p_1 = 0.3574$ ($p_2 = 0.0373$), $\omega_0 = 0.0466, \omega_1 = 1.0000$ $\omega_2 = 2.0876$	68.395406 ($P = 0.0000$)	**298D** **303D** 304S **306A** **307G** 326S
	M7 (beta)	−30488.3157	0.2556	$p = 0.2052, q = 0.5973$	24.892428 ($P = 0.0000$)	Not allowed
	M8 (beta&ω)	−30475.8695	0.2654	$p_0 = 0.9713$ ($p_1 = 0.0287$) $p = 0.2195, q = 0.0466, \omega = 1.5340$		**298D** 303D 306A **307G**
AP2L	M0 (one-ratio)	−11930.6163	0.2897	$\omega = 0.2897$		None
	M3 (discrete)	−11582.3172	0.3853	$p_0 = 0.2811, p_1 = 0.5226$ ($p_2 = 0.1963$), $\omega_0 = 0.0063, \omega_1 = 0.3192$ $\omega_2 = 1.1041$	696.598213 ($P = 0.0000$)	24C 25S **50S** 66S 67M 68S 72P 111I **112V** 164A 172P 175A 356H 371D **391C** 435V **445P** 456R **462Q** 466S 467G 475D 522R 533M 534Q 538I 542P 543T 546A 547L
	M1a (NearlyNeutral)	−11675.8826	0.4763	$p_0 = 0.6021$ ($p_1 = 0.3979$)		Not allowed
	M2a (PositiveSelection)	−11675.8826	0.4763	$p_0 = 0.6021, p_1 = 0.3028$ ($p_2 = 0.0951$), $\omega_0 = 0.1302, \omega_1 = 1.0000$ $\omega_2 = 1.0000$	NA	None
	M7 (beta)	−11599.3964	0.3641	$p = 0.4323, q = 0.7545$		Not allowed
	M8 (beta&ω)	−11597.7023	0.3896	$p_0 = 0.9625$ ($p_1 = 0.0375$) $p = 0.4627$, $q = 0.9010, \omega = 1.6976$	3.388152 ($P = 0.1838$)	None
TOE1	M0 (one-ratio)	−31052.6234	0.2710	$\omega = 0.2710$		None
	M3 (discrete)	−29048.0536	0.3427	$p_0 = 0.3819, p_1 = 0.3179$ ($p_2 = 0.3002$), $\omega_0 = 0.0129, \omega_1 = 0.2758$ $\omega_2 = 0.8330$	4009.139619 ($P = 0.0000$)	None
	M1a (NearlyNeutral)	−29473.6938	0.5461	$p_0 = 0.4793$ ($p_1 = 0.5207$)		Not allowed
	M2a (PositiveSelection)	−29473.6938	0.5461	$p_0 = 0.4793, p_1 = 0.0583$ ($p_2 = 0.4624$), $\omega_0 = 0.0531, \omega_1 = 1.0000$ $\omega_2 = 1.0000$	NA	None
	M7 (beta)	−29004.4021	0.3346	$p = 0.2721, q = 0.5409$		Not allowed
	M8 (beta&ω)	−29001.2895	0.3491	$p_0 = 0.9613$ ($p_1 = 0.0387$) $p = 0.2834, q = 0.6361, \omega = 1.3657$	6.225359 ($P = 0.0445$)	None

lnL: the log-likelihood difference between the two models; $2\Delta l$: twice the log-likelihood difference between the two models. The values in parentheses represent the significant level of 0.01 with a $\chi 2$ distribution at d.f. = 4 (M0 vs. M3) or 2 (M1a vs. M2a and M7 vs. M8). The amino acid sequences of AtAP2, PiaAP2La, and AtTOE1 were respectively used as the sequences reference. Positive selected sites in AP2 and AP2L clades were identified with posterior probability $p > 0.95$, In boldface, $p > 0.99$.

in agreement with the results of the adaptive evolution analysis. The sites having two amino acids from AP2 and TOE1 also provided evidence that *AP2L* diverged into two structurally and functionally distinct genes, *AP2* and *TOE1* OR genes, *AP2* and *TOE1*, through structural and functional changes.

Alignment of AP2 Group Gene Homolog Sequences from *A. thaliana* Demonstrated the Expanded Mode of Spermatophyte AP2 Group

Like certain other Brassicaceae plants, *Arabidopsis* has six AP2 group genes, which is unique because the orthologs of *TOE2*, *TOE3*, *SMZ*, and *SNZ* were not found in the other spermatophytes analyzed in this study. Accordingly, the detailed genomic information available for *Arabidopsis* was very helpful for exploring the expansion of the AP2 group in spermatophytes. Alignment of AP2 group genes of *Arabidopsis* revealed conserved regions (**Figure 6**). Ten exons were identified in the *AtAP2* genomic DNA sequence, nine in *AtTOE1-3* and seven in *AtSMZ* and *AtSNZ*. Unsurprisingly, the exons corresponding to the first EAR, NLS, AP2-R1, and linkage domains exhibited stronger conservation in these six genes, and the conserved sequences of the AP2-R2 domain in *AtAP2*, *AtTOE1* and *AtTOE3* are also shown in **Figure 6**. Most notably, a region in intron 5 of *AtTOE2* was very similar to exon 6 of *AtAP2*, *AtTOE1*, and *AtTOE3*, strongly suggesting that exon 6 was lost in the course of *AtTOE2* evolution. There was also some sequence conservation in introns 1, 3, and 4. Overall, *AtTOE1* and *AtTOE2* were the most closely related of these six genes, and the similarity between *AtSMZ* and *AtSNZ* was highest. The conservation among other genes was mainly in the introns, such as between *AtTOE3* and *AtSMZ*

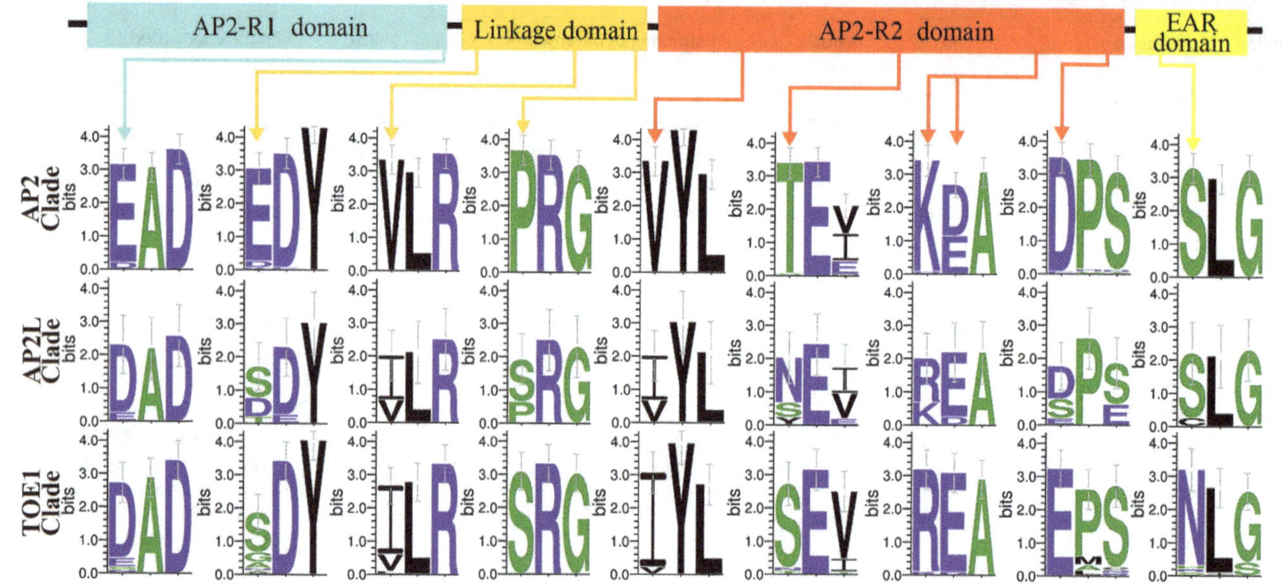

FIGURE 5 | Clade-specific sites in Homeodomains among Three Clades Genes (AP2 clade,AP2L clade, and TOE1 clade). Sequence logos of clade-specific sites in homeodomains identified in AP2Ls, AP2s, and TOE1s sequences. The arrows marked clade-specific sites and their position in homeodomains.

and between *AtTOE2* and *AtSNZ*. In the phylogenetic tree of all AP2 group genes in Brassicaceae (**Figure 1** and **Image 3**), the relationship between clades SMZ and SNZ was paralogous and so were the relationships between the AP2 clade and the TOE3, TOE2, and SMZ/SNZ clades and the AP2 and TOE1 types. These results indicated that gene duplication was an important cause of the expansion of the Brassicaceae AP2 group. Structural changes and rearrangements after gene duplication could have resulted in functional divergence.

Arabidopsis Mutant Analysis Supports Evidence for Functional Divergence after Gene Expansion of the AP2 Group

Mutant studies indicate that *AP2* is involved in the regulation of the stem-cell niche in the shoot meristem (Wurschum et al., 2005), floral development (Jofuku et al., 1994), and seed mass (Ohto et al., 2005), whereas *AtSMZ*, *AtSNZ*, *AtTOE1*, *AtTOE2*, and *AtTOE3* redundantly affect flowering time (Jung et al., 2007, 2014; Yant et al., 2010). We isolated a novel *ap2* allele from an ethyl methanesulfonate mutagenesis screen for abnormal expression patterns of the shoot meristem stem cell marker *pCLV3:YFPer* transgenic line. In contrast to wild-type plants, where *pCLV3:YFPer* is expressed in the three stem-cell layers of torpedo-stage embryos (**Image 4A**), *pCLV3:YFPer* signal was only observed in the epidermal layer of the mutant *ap2* (*2-132*) mutant at levels comparable to wild-type, whereas expression in the subjacent layers was strongly reduced (**Image 4B**). At the seedling stage, *ap2* (*2-132*) mutants failed to develop a wild-type like shoot meristem. *ap2* (*2-132*) plants also were late flowering with abnormal flower phenotype and abnormal *CLV3* expression in the stem-cell niche of shoots (**Image 4A–D,H**).

Comparing the genomic and coding sequences of *AtAP2* in the wild type and *2-132*, a single-base exchange in intron 6 of *AtAP2* was identified (**Image 4E**), which affected pre-mRNA splicing and led to loss of exon 6 of *AtAP2* in *2-132* (**Image 4F**), and consequently 15 amino acids were lost from the AP2-R2 domain (**Image 4G**). Remarkably, exon loss also occurred in *AtTOE2*, and this lost exon is homologous with exon 6 of *AtAP2* (**Figure 6** and **Image 4G**). Thus, fluctuation in the number of exons may facilitate divergence in gene function and also may be one of the ways new genes are formed.

DISCUSSION

Typical AP2 Group Genes First Appeared in Gymnosperms and Evolved into AP2 and TOE Types through Whole-Genome and Gene Duplication in Angiosperms

The AP2 domain was previously considered a plant specific core construct of the AP2 family, but it has more recently also been found in cyanobacterium, ciliates, and viruses (Magnani et al., 2004). These newly identified non-plant proteins with an AP2 domain are predicted HNH endonucleases, a kind of homing endonuclease (Chevalier and Stoddard, 2001). An HNH-AP2 homing endonuclease may have been transported into plants via endosymbiosis, horizontal transfer or other lateral gene transfer events. In the process of formation of the AP2 subfamily containing two AP2 domains, tandem duplication is likely to have played a major role (Magnani et al., 2004). A protein containing two AP2 domains has been identified in *Chlamydomonas reinhardtii*, but it does not clustered with the AP2 subfamily, though the amino acid composition of its AP2

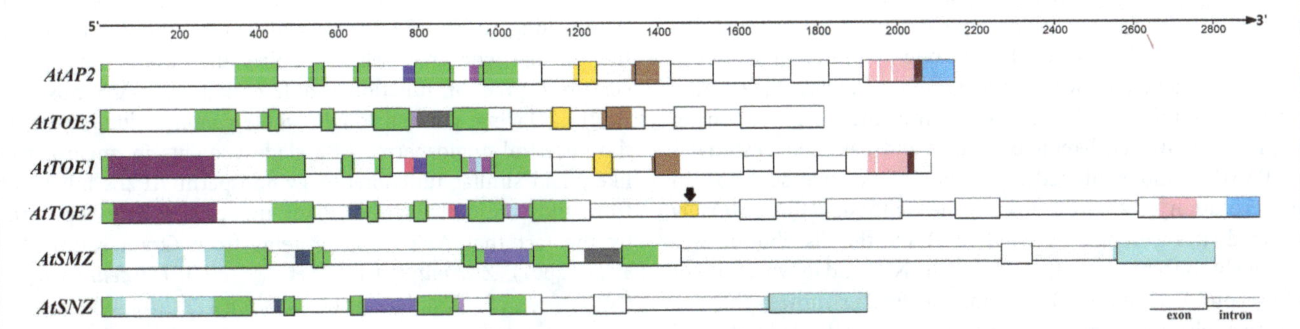

FIGURE 6 | Alignment of Genomic Sequences of AP2 Group Genes in *A. thaliana*. The sequence regions with same color in this figure have higher homology with each other. Black arrows means one intron region (gold) in *AtTOE2* has higher sequence similarity with the sixth exon in *AtAP2*, *AtTOE3,* and *AtTOE1*.

domain is very similar to the AP2 group (Shigyo et al., 2006). It is regarded as a sister to the AP2 and ANT groups in terms of phylogenetic relationships (Shigyo et al., 2006). AP2/ERF proteins with two AP2 domains from *Physcomitrella* cluster with the ANT group (Kim et al., 2006). *C. reinhardtii* belongs to the Chlorophyta lineage, which is sister to the Streptophyta lineage (Charophyceae and land plants, Karol et al., 2001). This suggests that the AP2 and EREBP subfamilies diverged before the Chlorophyta lineage diverged from the Streptophyta lineage (Shigyo et al., 2006). In addition, we found no orthologs of AP2 group genes in our database searches of alga, moss and fern. *AP2L* from gymnosperms were identified in searching the orthologs of AP2 group genes, demonstrating the ancestral polyploidy event during the formation of gymnosperms (Jiao et al., 2011). AP2 group genes were also detected in basal angiosperms (*Amborella trichopoda, AmtAP2; Gnetum parvifolium, GpTOE1; Nymphaea hybrid cultivar, NhTOE1*) and respectively clustered into the AP2 and TOE1 clades. The presence of angiosperm genes in both the AP2- and TOE-type lineages suggests the duplication that gave rise to these two lineages followed the divergence of gymnosperms and angiosperms. The whole-genome duplication in ancestral angiosperms (Jiao et al., 2011) may have led to the AP2 group genes falling into two broad categories: the AP2 and TOE types.

Gene Duplication and Motif Changes Produced New Genes in the Angiosperm AP2 Group

The distribution diagram (**Figure 2**) of motifs and homeodomains exhibited the evolutionary relationships in the AP2 group. There were ANT orthologs but no AP2 orthologs in alga, moss and fern, which supports the argument that the AP2 group first appeared in gymnosperms. There was no differentiation between the AP2 and TOE types in gymnosperms, though all *AP2Ls* from Pinaceae were divided into two sub-branches by the whole-genome duplication in ancestral gymnosperms, which indicates that the duplication did not lead to the formation of TOE-type genes. The whole-genome duplication in the ancestral angiosperm was the likely basis of AP2 group gene differentiation. In this process, there were more new motifs in AP2-type than TOE-type genes (**Figure 2**). The homeodomains have evolved little in the AP2 group gene differentiation, especially the NLS and AP2-R1 domains. Most obviously, the first EAR domain transformed from DLNxxP to LxLxL in the TOE type and disappeared in the TOE3 clade. These changes caused AP2 group genes to diverge into the AP2 and TOE types. Our analysis demonstrated there are AP2 and TOE1 orthologs in most angiosperms except Brassicaceae, which contains six AP2 group genes, suggesting that other gene or whole-genome duplication events occurred in the course of evolution. In fact, the extensive complete genome analyses in *Arabidopsis* supports the model that two recent whole-genome duplication events occurred in Brassicaceae and one triplication event occurred in eudicots (Bowers et al., 2003; Tuskan et al., 2006; Lyons et al., 2008; Barker et al., 2009). Interestingly, only in Brassicaceae, new functional AP2 group orthologs appeared. In most angiosperms, polyploidy could simply cause increased gene copy numbers of *AP2* and *TOE1*. Some changes of motifs and homeodomains have taken place in the new AP2 group genes. The AP2 group was different in AP2 domain and motifs than the ANT group (Kim et al., 2006). Like the isolation of AP2 and ANT groups, the new AP2 group genes in Brassicaceae may have been formed by a similar mechanism. For instance, the deletion and amino acid changes mainly occurred in the TOE2 and SMZ/SNZ AP2-R2 domains but TOE3 was formed by the removal of motifs. Compared with TOE2s, SMZs/SNZs evolved two new specific motifs (**Figure 2**). The analysis of mutant and genomic sequence alignment supports the view that new genes and functional differentiation could be produced by exon changes and genomic sequence rearrangements.

Different Selective Pressures Drove the Evolution of Different Clades in Spermatophyte AP2 Group

The selective pressures analysis of all and each clade of the AP2 group genes in spermatophyte suggested that AP2 group genes experienced different evolutionary patterns and each clade encountered various selective pressures, demonstrating that complex selective pressures drove the evolution of the AP2 group. As DNA-binding proteins containing two AP2 domains, AP2 group genes needed to maintain high conservation in the AP2 domains and NLS i.e., to facilitate nuclear translocation and DNA binding. All ω-values were <1 in the one-by-one comparisons of AP2 group genes, showing a background of purifying selection during evolution and coinciding with the conservation of sequence. However, likelihood values and parameter estimates in PAML demonstrated the presence of positive selection in the evolution of all AP2 group genes, although the positively selected sites were few (<3% of sites) and the distribution was relatively concentrated (upstream of the NLS and downstream of the second EAR domain). The analysis further showed that the positive selection occurred exclusively in the AP2 clade. Related to the fact that *AP2* was the main functional gene in the AP2 group, this positive selection might have lead to the protein functional changes. There was no positive selection in other clades of the AP2 group other than the AP2 clade. Accordingly, there may have been positive e selection at the time of divergence of the AP2 and TOE types. In the subsequent evolution, every clade experienced differential selective pressures, particularly the AP2 and TOE1 clades. Notably, more amino acid changes accumulated in the TOE1 clade during evolution, which was supported by the pairwise and sliding-window analysis of *dN* and *dS* in all AP2 group genes. As suggested above, different clades of the AP2 group experienced different evolutionary patterns, which might be associated with gene function. AP2 clade genes, as the main functional gene of the AP2 group, required high conservation and probably changed accordingly with angiosperm diversification.

AP2 Clade Genes Retained Ancestor Gene Function of AP2 Group

Phylogenetic analysis demonstrated that all orthologs of AP2 group genes (AP2L) in gymnosperms belong to the AP2 type and are most closely related to the AP2 clade, which suggests that gymnosperms only contain AP2 homologous genes, and that the ancestor of seed plants had AP2-like genes but no TOE1-like genes. In spermatophyte evolution, the AP2 group genes diversified but some functions might be conserved in the most recent common ancestor of extant spermatophytes. In the model plant *Arabidopsis*, the functions of the six AP2 group members (*AtAP2, AtTOE1–3, AtSMZ,* and *AtSNZ*) have been fully studied, and all are repressors of flowering, but only AtAP2 exhibits multiple functions in the development of flowers, fruit, seeds and stem cells (Wurschum et al., 2005; Mathieu et al., 2009; Huijser and Schmid, 2011; Ripoll et al., 2011). The expression patterns of AP2Ls in gymnosperms (*Larix × marschlinsii, P. abies, P. thunbergii*) have been examined in previous studies. AP2Ls are expressed in female and male cones, leaves, stems and roots (Vahala et al., 2001; Shigyo and Ito, 2004). AP2L from P. abies shows functional similarities to AtAP2 in floral patterning when overexpressed in *Arabidopsis* (Nilsson et al., 2007) and the overexpression of AtAP2 also affects floral patterning in *Nicotiana benthamiana* (Mlotshwa et al., 2006). AP2Ls are also expressed during somatic embryogenesis in gymnosperms (Guillaumot et al., 2008). Remarkably, there are multiple homologous genes of the AP2 group in gymnosperms and their expression patterns differ from each other, though they belong to theAP2 type together with the AP2 clade in the phylogenetic tree, which implies that there has been a certain degree of functional differentiation in gymnosperm AP2Ls. Having gone through whole-genome duplication in the ancestral angiosperm, AP2 clade proteins in angiosperms likely had similar functions to gymnosperm AP2Ls inheriting from the common ancestor. This model is widely supported by the fact that AP2 clade proteins from *Oryza sativa* (Lee et al., 2007), *Zea mays* (Chuck et al., 2007), *Hordeum vulgare* (Nair et al., 2010), *Solanum lycopersicon* (Karlova et al., 2011), *Solanum tuberosum* (Martin et al., 2009), *Actinidia deliciosa* (Varkonyi-Gasic et al., 2012), *Petunia hybrida* (Maes et al., 1999, 2001), and *Crocus sativus* (Tsaftaris et al., 2012) have been linked to AP2Ls and they function in floral organ identity and development, fruit development, lodicules development, branching and tuberization. To meet the complex requirements of biological and functional diversity, AP2 clades genes have been constantly evolving by structural changes and adaptive evolution, which is in good agreement with our analysis of structural (motifs and homeodomains) alignment and selective pressure.

Gene Numberic Expansion of AP2 Group Producted New Genes with Similar Functions

With gene or whole-genome duplications in spermatophyte evolution, there was expansion of the number of AP2 group genes, especially in Brassicaceae. The phylogenetic tree of the Brassicaceae AP2 group reflects the evolutionary relationship: The TOE3 clade is a sister to the AP2 clade belonging to the AP2 type; the TOE2 and SMZ/SNZ clades are paralogous and cluster together with the TOE1 clade in the TOE type. The analysis results have gained support from the functional research of AP2 group proteins in *Arabidopsis*. AP2 group proteins from *Arabidopsis* are control factors of flowering. AP2-type proteins (AtAP2 and AtTOE3) participate in floral organ identity and development (Aukerman and Sakai, 2003; Chen, 2004; Jung et al., 2014), but TOE-type proteins (AtTOE1 and AtTOE2) are involved in flowering control and developmental phases of plant (Huijser and Schmid, 2011). The single mutant and overexpression of *AtTOE3* showed no visible phenotypic effects, while the expression pattern was different from other AP2 group genes in *Arabidopsis* (Jung et al., 2007; Yant et al., 2010). Recent research has shown that overexpression of an miR172-resistant *AtTOE3* can control floral organ identity and flowering time when the miR172 target site is mutated (Jung et al., 2014). AtTOE3 binds to the second intron of *AGAMOUS* (*AtAG*) and represses its expression like AtAP2 (Yant et al., 2010; Jung et al., 2014), which means that the function of AtTOE3 is similar to AtAP2, but is strongly constrained by *miR172*. Overexpression of *AtSMZ, AtSNZ, AtTOE1,* or *AtTOE2* causes late flowering (Aukerman and Sakai, 2003; Chen, 2004; Jung et al., 2007) and quadruple (*smz snz toe1 toe2*) and sextuple (*ap2 toe3 smz snz toe1 toe2*) mutants flower earlier than any single or double mutant (Jung et al., 2007; Mathieu et al., 2009; Yant et al., 2010), which also indicates that the function of TOE-type genes is similar in

Arabidopsis. The functional studies of the AP2 group proteins in *Arabidopsis* supports the model that orthologs formed by gene or whole-genome duplications could be transformed into new genes by changes in motifs and homeodomains.

AP2 Function to Maintain the Stem Cell Niche was Conservative Function in Spermatophyte

In *Arabidopsis*, the marker genes *WUS* and *CLV3* of shoot meristem stem-cell niche are regulated by AP2 and TOE3, and TOE1 does not act redundantly with AP2 in stem-cell maintenance (Wurschum et al., 2005). Likewise, the expression of *WUS* and *CLV3* is changed in *Arabidopsis* by overexpressing *AP2L* of *P. abies* (Nilsson et al., 2007). Overexpression of *AtAP2* in *N. benthamiana* also causes expression changes of *NbWUS* (Mlotshwa et al., 2006). Accordingly, the function of maintaining the stem-cell niche was conserved in the spermatophyte AP2 and AP2L clades. *AP2Ls* from *Larix* × *marschlinsii* are expressed during somatic embryogenesis and germination (Guillaumot et al., 2008), which also provides support for this view. Mutant analysis showed that changes in AP2-R1 and R2 lead to loss of stem-cell maintenance. *AtTOE3* and *AtTOE1* do not function in stem-cell maintenance, which indicates that the changes in motifs and homeodomains could also cause the loss of function in stem-cell maintenance.

The spermatophyte AP2 group contains many orthologs because of gene or whole-genome duplications. These orthologs may have evolved different functions, which means that the AP2 group may be a very valuable research area for examining new gene formation. The problems in AP2 group research focus on two aspects: evolution and functional differentiation. In gymnosperm, there was no differentiation into AP2 type and TOE type. In angiosperm, AP2 group genes were divided into two types and finished the functional differentiation. But more gene and genome data was needed to support the conclusion, especially gymnosperm. There are only two members (*AP2s* and *TOE1s*) in angiosperm AP2 group expect Brassicaceae. Current data suggested *TOE2s*, *TOE3s*, *SMZs*, and *SNZs* only exist in Brassicaceae, which means gene expansion and functional differentiation occurred during the formation of Brassicaceae but the detailed process was still unknown. *AP2s* are the the main function genes in AP2 group. After the formation of *TOE1s*, the function of flower time was only preserved. In Brassicaceae, *SMZs* and *SNZs* that belong to the same type with *TOE1s* own some new gene function. But so far, that still needs to be further studied. The correspondence among expression patterns, function and phylogenetic relatedness requires further study. The functionally similar genes in this study suggest functional diversification within the AP2 group. Comprehensive studies of expression and function and intensive phylogenetic characterization of the AP2 group genes will give us a clearer indication of the roles these genes play in the developmental processes of different species as well as the function of AP2 group genes in the evolutionary history of spermatophytes.

MATERIALS AND METHODS

Sequence Data

We retrieved the nucleotide and amino acid sequences for the *Arabidopsis* AP2 group from the Arabidopsis Information Resource database (www.arabidopsis.org). A BLASTP search was then performed using the AtAP2, AtTOE1–3, AtSMZ, and AtSNZ sequences as the query to retrieve AP2 group gene sequences from the NCBI (www.ncbi.nlm.nih.gov) and Phytozome databases (www.phytozome.org). The identified sequences were from the 56 species of spermatophytes (**Data sheet 1**). All selected AP2 group amino acid sequences contain six or seven conserved domains (**Image 1**). Prior to the phylogenetic analysis of these AP2 group gene protein sequences with *Arabidopsis* AP2 group, we ensured that all the AP2 group sequences clustered together as well as with the AtAP2 group. For our analysis, we selected the most typical AP2 group genes from a large number of paralogs but retained all query results of gymnosperms (**Data sheet 1**). The genes that fell outside the AP2 group were not analyzed.

Phylogenetic Analysis and Tree Construction

After deletion of identical sequences, only 105 sequences were used for phylogenetic analysis (**Data sheet 1**). They were aligned together using CLUSTAL 1.83. The phylogenetic tree of AP2 group genes was obtained by using ML (maximum likelihood) (MEGA 6.0) methods, and the reliability of the trees was evaluated by the bootstrap method with 1000 replications. The dN/dS value was used to detect positive selection.

Identification of Sequence Motifs

To identify motifs shared among related proteins within the AP2 group gene, we used the MEME motif search tool was used with its default settings. The maximum number of possible motifs was set to 35, and the maximum width was 300. Identified motifs were annotated using SMART (http://smart.embl-heidelberg.de/) and Pfam (http://pfam.sanger.ac.uk/).

Analysis of Adaptive Evolution and Identification of Selective Pressures

The program Codeml implemented in the PAML 4.0 software package was used to investigate the adaptive evolution of AP2 group protein-coding sequences. A total of 105 aligned AP2 group genes sequences, isolated from the different clades, were selected to test whether they were under purifying selection. Six models of codon substitution, M0 (one-ratio), M1a (Nearly Neutral), M2a (Positive Selection), M3 (discrete), M7 (beta), and M8 (beta & ω) were used in the analysis. M0 assumes that all sites have the same ω ratio. M1a assumes two classes of sites in proteins in proportions p0 and p1 (1–p0) with $0 < \omega_0 < 1$ (purifying selection) and $\omega_1 = 1$ (neutral sites). M2a adds a proportion (p2) to account for a class of sites where ω_2 is estimated from the data and can be > 1. M3 uses a general discrete distribution with three site classes, with the proportions (p0, p1, and p2) and the ω ratios (ω_0, ω_1, and ω_2) estimated from the data. M7 assumes a beta distribution (p, q) for 10 different ω

ratios in the interval (0, 1). M8 adds an extra class of sites with positive selection ($\omega > 1$) to the beta (M7) model. Therefore, the null models M0, M1a and M7 fix the ω ratios between 0 and 1, and do not allow the presence of positively selected sites. The alternative models M2a, M3, and M8 account for positive selection by using parameters, which estimate ω greater than 1, and allow for the variable ω along codon sequence.

The likelihood ratio test (LRT) was performed to detect the presence of positively selected sites by comparing the models that do not allow for positive selection with the models that allow for positive selection. The LRT was performed by taking twice the difference in log likelihood between nested models and testing for significance using the $\chi 2$ distribution with the degrees of freedom equivalent to the difference in the number of parameters between models. If the LRT yields a statistically significant result, then positive selection is inferred. In the present study, three LRTs (M0 vs. M3, M1a vs. M2a, and M7 vs. M8) were used to detect positive selection. The Bayes empirical Bayes (BEB) approach implemented in M2a and M8 was used to determine the positively selected sites by calculating the posterior probabilities (p) of ω classes for each site. The sites with high posterior probabilities ($p > 0.95$) coming from the class with $\omega > 1$ were believed to be under positive selection.

Plant Growth, Mutant Lines, and Mapping

All plants were in the Landsberg erecta (Ler) ecotype, which was also used as the wild-type control. The *2-132* mutant was identified in the M2 generation of ethylmethanesulfonate-mutagenized *pCLV3:YFPer* plants. *ap2-1* and *ap2-2* mutants have been described.

The *2-132* mutation was mapped using plants with a wild-type-like phenotype in the F2 generation from a cross of an *2-132*/+ plant with the Columbia ecotype. The initial mapping was done using CAPS markers from The Arabidopsis Information Resource (http://www.arabidopsis.org). The dCAPS markers used for fine mapping of the *2-132* mutation were generated based on the information available from CEREON.

AUTHOR CONTRIBUTIONS

PW, JS, and JC planned and designed the research. PW, TC, and Meiping Li performed experiments. Mengzhu Lu, TL conducted experiments. PW, GL, YL, and JC collected and analyzed data. PW wrote the manuscript.

FUNDING

This work was supported by National High Technology Research and Development Program of China (863 Program, 2013AA102705), Specialized National Basic Research Program of China (973 Program, 2012CB114504), Natural Science Foundation of Jiangsu University grant 13KJA220001, Talent project by the Ministry of Science and Technology, Co-Innovation Center for Sustainable Forestry in Southern China, Nanjing Forestry University, and Priority Academic Program Development of Jiangsu Higher Education Institutions.

SUPPLEMENTARY MATERIAL

The Supplementary Material for this article can be found online at: http://journal.frontiersin.org/article/10.3389/fpls.2016.01383

Image 1 | Conservative Domains Distributions of AP2 Group and the Consensus Amino Acid Sequence of AP2-R2 Domain in Brassicaceae. Different colors represent different conservative domains (EAR domain, yellow; purple, NLS domain; aqua, AP2-R1 domain; gold, linkage domain; red and green, AP2-R2 domain; darkcyan, miRNA172 target site). The amino acid composition of AP2-R2 domain in different AP2 group genes is poor conservative, especially in TOE2s, SMZs, and SNZs.

Image 2 | Detailed Phylogenetic Trees of five clade in Figure 1. The different colorst represent different clades (red, AP2 clades; green, TOE1 clade; olive, TOE2 clade; purple, TOE3 clade; blue, SMZ/SNZ clade). The species and accession numbers are listed in **Data sheet 1**. The abbreviations used are as follows: *Ad*, *Actinidia deliciosa*; *Amt*, *Amborella trichopoda*; *Bp*, *Betula platyphylla*; *Al*, *Arabidopsis lyrata*; *At*, *Arabidopsis thaliana*; *Aa*, *Arabis alpine*; *Bn*, *Brassica napus*; *Br*, *Brassica rapa*; *Cr*, *Capsella rubella*; *La*, *Lepidium appelianum*; *Th*, *Thellungiella halophile*; *Cp*, *Carica papaya*; *Jc*, *Jatropha curcas*; *Rc*, *Ricinus communis*; *Me*, *Manihot esculenta*; *Ca*, *Cicer arietinum*; *Gm*, *Glycine max*; *Lj*, *Lotus japonicas*; *Mt*, *Medicago truncatula*; *Pv*, *Phaseolus vulgaris*; *Ps*, *Pisum sativum*; *Gp*, *Gnetum parvifolium*; *Gr*, *Gossypium raimondii*; *Eg*, *Eucalyptus grandis*; *Nh*, *Nymphaea hybrid cultivar*; *Aes*, *Aegilops speltoides*; *Aet*, *Aegilops tauschii*; *Bd*, *Brachypodium distachyon*; *Hv*, *Hordeum vulgare*; *Os*, *Oryza sativa*; *Si*, *Setaria italic*; *Sb*, *Sorghum bicolor*; *Tt*, *Triticum turanicum*; *Zm*, *Zea mays*; *Fv*, *Fragaria vesca*; *Md*, *Malus × domestica*; *Pp*, *Prunus persica*; *Cc*, *Citrus clementina*; *Cs*, *Citrus sinensis*; *Ct*, *Citrus trifoliate*; *Pt*, *Populus trichocarpa*; *Anm*, *Antirrhinum majus*; *Mg*, *Mimulus guttatus*; *Ph*, *Petunia × hybrid*; *Sl*, *Solanum lycopersicum*; *St*, *Solanum tuberosum*; *Tc*, *Theobroma cacao*; *Cas*, *Camellia sinensis*; *Vv*, *Vitis vinifera*.

Image 3 | Phylogenetic Analysis of AP2 Group Protein in Brassicaceae (simplified phylogenetic tree). The ML tree was constructed based on the whole protein sequences of spermatophyte AP2 Group gene using MEGA6.0 with 1000 bootstrap replications and Jones-Taylor-Thornton (JTT) + Gamma Distributed model (Discrete Gamma Categories = 5).

Image 4 | Phenotype, Site of Mutation and Mechanism of Mutation of *2-132*. (A,B) Fluorescence microscope of pCLV3:YFPer in wild-type **(A)** and *2-132* **(B)** homozygous mutant torpedo stage embryos. The yellow fuorescent protein indicate the location of the embryonic stem cell niche. In 100 self-crossed progeny embryos of *2-132* heterozygote, the number of abnormal and normal yellow fuorescent in embryonic stem cell niche was 29: 71, which was no significant difference with 3:1 by χ^2-test. **(C,D)** The phenotype of flower in in wild-type **(C)** and *2-132* **(D)** homozygous mutant. The sepals of *2-132* **(D)** homozygous mutant transform into leaves morphologically and the petals are like sepals. E Genomic organization of AP2. The mutant sites of the *2-132*, *l28*, *ap2-1*, *ap2-2*, and *ap2-7* mutations are shown. The exon sequences of the two AP2 domains are marked (aqua, AP2-R1 domain; red, AP2-R2 domain). The point mutation in the genomic sequence of *2-132* is highlighted. **(F,G)** The sequencing results of *AP2* (genomic DNA and mRNA) from wild-type and *2-132* homozygous mutant. The sequences of mRNA show there are 45 bases deletion in *AP2* of *2-132* homozygous mutant which happens to be the sixth exon of wild-type *AP2*. In the *AtTOE2* of wild-type, this exon also does not exist. **(H)** The phenotype of flowering and height growth (centimeter) of wild-type and *2-132* homozygous mutant. The number of rosette leaves in *2-132* **(B)** homozygous mutant during flowering time is more than wild-type but the height growth is less.

Data Sheet 1 | The List of AP2 Group Genes in this Article.

Data Sheet 2 | Branch Model Test for Each Clade Genes.

Data Sheet 3 | Likelihood Values and Parameter Estimates for *AP2s*, *TOE1s*, *TOE3s*, *TOE2s*, and *SMZSNZs* in Brassicaceae.

REFERENCES

Anisimova, M., Bielawski, J. P., and Yang, Z. (2001). Accuracy and power of the likelihood ratio test in detecting adaptive molecular evolution. *Mol. Biol. Evol.* 18, 1585–1592. doi: 10.1093/oxfordjournals.molbev.a003945

Anisimova, M., Bielawski, J. P., and Yang, Z. (2002). Accuracy and power of bayes prediction of amino acid sites under positive selection. *Mol. Biol. Evol.* 19, 950–958. doi: 10.1093/oxfordjournals.molbev.a004152

Aukerman, M. J., and Sakai, H. (2003). Regulation of flowering time and floral organ identity by a MicroRNA and its APETALA2-like target genes. *Plant Cell* 15, 2730–2741. doi: 10.1105/tpc.016238

Bailey, T. L., Boden, M., Buske, F. A., Frith, M., Grant, C. E., Clementi, L., et al. (2009). MEME SUITE: tools for motif discovery and searching. *Nucleic Acids Res.* 37(suppl. 2), W202–W208. doi: 10.1093/nar/gkp335

Barker, M. S., Vogel, H., and Schranz, M. E. (2009). Paleopolyploidy in the Brassicales: analyses of the Cleome transcriptome elucidate the history of genome duplications in Arabidopsis and other Brassicales. *Genome Biol. Evol.* 1, 391–399. doi: 10.1093/gbe/evp040

Bowers, J. E., Chapman, B. A., Rong, J., and Paterson, A. H. (2003). Unravelling angiosperm genome evolution by phylogenetic analysis of chromosomal duplication events. *Nature* 422, 433–438. doi: 10.1038/nature01521

Chen, X. (2004). A microRNA as a translational repressor of APETALA2 in Arabidopsis flower development. *Science* 303, 2022–2025. doi: 10.1126/science.1088060

Chevalier, B. S., and Stoddard, B. L. (2001). Homing endonucleases: structural and functional insight into the catalysts of intron/intein mobility. *Nucleic Acids Res.* 29, 3757–3774. doi: 10.1093/nar/29.18.3757

Chuck, G., Meeley, R., Irish, E., Sakai, H., and Hake, S. (2007). The maize *tasselseed4* microRNA controls sex determination and meristem cell fate by targeting *Tasselseed6/indeterminate* spikelet1. *Nat. Genet.* 39, 1517–1521. doi: 10.1038/ng.2007.20

Du, D., Hao, R., Cheng, T., Pan, H., Yang, W., Wang, J., et al. (2013). Genome-wide analysis of the AP2/ERF gene family in *Prunus mume*. *Plant Mol. Biol. Rep.* 31, 741–750. doi: 10.1007/s11105-012-0531-6

Duan, C., Argout, X., Gébelin, V., Summo, M., Dufayard, J.-F., Leclercq, J., et al. (2013). Identification of the *Hevea brasiliensis* AP2/ERF superfamily by RNA sequencing. *BMC Genomics* 14:30. doi: 10.1186/1471-2164-14-30

Guillaumot, D., Lelu-Walter, M. A., Germot, A., Meytraud, F., Gastinel, L., and Riou-Khamlichi, C. (2008). Expression patterns of LmAP2L1 and LmAP2L2 encoding two-APETALA2 domain proteins during somatic embryogenesis and germination of hybrid larch (*Larix* x *marschlinsii*). *J. Plant Physiol.* 165, 1003–1010. doi: 10.1016/j.jplph.2007.08.009

Huijser, P., and Schmid, M. (2011). The control of developmental phase transitions in plants. *Development* 138, 4117–4129. doi: 10.1242/dev.063511

Jiao, Y., Wickett, N. J., Ayyampalayam, S., Chanderbali, A. S., Landherr, L., Ralph, P. E., et al. (2011). Ancestral polyploidy in seed plants and angiosperms. *Nature* 473, 97–100. doi: 10.1038/nature09916

Jofuku, K. D., Den Boer, B., Van Montagu, M., and Okamuro, J. K. (1994). Control of Arabidopsis flower and seed development by the homeotic gene APETALA2. *Plant Cell* 6, 1211–1225. doi: 10.1105/tpc.6.9.1211

Jung, J.-H., Lee, S., Yun, J., Lee, M., and Park, C.-M. (2014). The miR172 target TOE3 represses AGAMOUS expression during Arabidopsis floral patterning. *Plant Sci.* 215–216, 29–38. doi: 10.1016/j.plantsci.2013.10.010

Jung, J. H., Seo, Y. H., Seo, P. J., Reyes, J. L., Yun, J., Chua, N. H., et al. (2007). The GIGANTEA-regulated microRNA172 mediates photoperiodic flowering independent of CONSTANS in Arabidopsis. *Plant Cell* 19, 2736–2748. doi: 10.1105/tpc.107.054528

Kagale, S., Links, M. G., and Rozwadowski, K. (2010). Genome-wide analysis of ethylene-responsive element binding factor-associated amphiphilic repression motif-containing transcriptional regulators in Arabidopsis. *Plant Physiol.* 152, 1109–1134. doi: 10.1104/pp.109.151704

Karlova, R., Rosin, F. M., Busscher-Lange, J., Parapunova, V., Do, P. T., Fernie, A. R., et al. (2011). Transcriptome and metabolite profiling show that APETALA2a is a major regulator of tomato fruit ripening. *Plant Cell* 23, 923–941. doi: 10.1105/tpc.110.081273

Karol, K. G., McCourt, R. M., Cimino, M. T., and Delwiche, C. F. (2001). The closest living relatives of land plants. *Science* 294, 2351–2353. doi: 10.1126/science.1065156

Kim, S., Soltis, P. S., Wall, K., and Soltis, D. E. (2006). Phylogeny and domain evolution in the APETALA2-like gene family. *Mol. Biol. Evol.* 23, 107–120. doi: 10.1093/molbev/msj014

Kryazhimskiy, S., and Plotkin, J. B. (2008). The population genetics of dN/dS. *PLoS Genet.* 4:e1000304. doi: 10.1371/journal.pgen.1000304

Lee, D. Y., Lee, J., Moon, S., Park, S. Y., and An, G. (2007). The rice heterochronic gene *SUPERNUMERARY BRACT* regulates the transition from spikelet meristem to floral meristem. *Plant J.* 49, 64–78. doi: 10.1111/j.1365-313X.2006.02941.x

Licausi, F., Giorgi, F. M., Zenoni, S., Osti, F., Pezzotti, M., and Perata, P. (2010). Genomic and transcriptomic analysis of the AP2/ERF superfamily in *Vitis vinifera*. *BMC Genomics* 11:719. doi: 10.1186/1471-2164-11-719

Licausi, F., Ohme-Takagi, M., and Perata, P. (2013). APETALA2/Ethylene Responsive Factor (AP2/ERF) transcription factors: mediators of stress responses and developmental programs. *New Phytol.* 199, 639–649. doi: 10.1111/nph.12291

Lyons, E., Pedersen, B., Kane, J., Alam, M., Ming, R., Tang, H., et al. (2008). Finding and comparing syntenic regions among Arabidopsis and the outgroups papaya, poplar, and grape: CoGe with rosids. *Plant Physiol.* 148, 1772–1781. doi: 10.1104/pp.108.124867

Maes, T., Van de Steene, N., Zethof, J., Karimi, M., D'Hauw, M., Mares, G., et al. (2001). Petunia *Ap2*-like genes and their role in flower and seed development. *Plant Cell* 13, 229–244. doi: 10.1105/tpc.13.2.229

Maes, T., Van Montagu, M., and Gerats, T. (1999). The inflorescence architecture of *Petunia hybrida* is modified by the *Arabidopsis thaliana Ap2* gene. *Dev. Genet.* 25, 199–208. doi: 10.1002/(SICI)1520-6408(1999)25:3<199::AID-DVG3>3.0.CO;2-L

Magnani, E., Sjölander, K., and Hake, S. (2004). From endonucleases to transcription factors: evolution of the AP2 DNA binding domain in plants. *Plant Cell Online* 16, 2265–2277. doi: 10.1105/tpc.104.023135

Martin, A., Adam, H., Díaz-Mendoza, M., Żurczak, M., González-Schain, N. D., and Suárez-López, P. (2009). Graft-transmissible induction of potato tuberization by the microRNA *miR172*. *Development* 136, 2873–2881. doi: 10.1242/dev.031658

Mathieu, J., Yant, L. J., Mürdter, F., Küttner, F., and Schmid, M. (2009). Repression of Flowering by the miR172 Target SMZ. *PLoS Biol* 7:e1000148. doi: 10.1371/journal.pbio.1000148

Mizoi, J., Shinozaki, K., and Yamaguchi-Shinozaki, K. (2012). AP2/ERF family transcription factors in plant abiotic stress responses. *Biochim. Biophys. Acta* 1819, 86–96. doi: 10.1016/j.bbagrm.2011.08.004

Mlotshwa, S., Yang, Z., Kim, Y., and Chen, X. (2006). Floral patterning defects induced by Arabidopsis APETALA2 and microRNA172 expression in *Nicotiana benthamiana*. *Plant Mol. Biol.* 61, 781–793. doi: 10.1007/s11103-006-0049-0

Nair, S. K., Wang, N., Turuspekov, Y., Pourkheirandish, M., Sinsuwongwat, S., Chen, G., et al. (2010). Cleistogamous flowering in barley arises from the suppression of microRNA-guided HvAP2 mRNA cleavage. *Proc. Natl Acad. Sci. U.S.A.* 107, 490–495. doi: 10.1073/pnas.0909097107

Nilsson, L., Carlsbecker, A., Sundas-Larsson, A., and Vahala, T. (2007). APETALA2 like genes from *Picea abies* show functional similarities to their Arabidopsis homologues. *Planta* 225, 589–602. doi: 10.1007/s00425-006-0374-1

Ohto, M.-A., Fischer, R. L., Goldberg, R. B., Nakamura, K., and Harada, J. J. (2005). Control of seed mass by APETALA2. *Proc. Natl. Acad. Sci. U.S.A.* 102, 3123–3128. doi: 10.1073/pnas.0409858102

Punta, M., Coggill, P. C., Eberhardt, R. Y., Mistry, J., Tate, J., Boursnell, C., et al. (2011). The Pfam protein families database. *Nucleic Acids Res.* 40, D290–D301. doi: 10.1093/nar/gkr1065

Ripoll, J. J., Roeder, A. H. K., Ditta, G. S., and Yanofsky, M. F. (2011). A novel role for the floral homeotic gene *APETALA2* during *Arabidopsis* fruit development. *Development* 138, 5167–5176. doi: 10.1242/dev.073031

Seo, T. K., Kishino, H., and Thorne, J. L. (2004). Estimating absolute rates of synonymous and nonsynonymous nucleotide substitution in order to characterize natural selection and date species divergences. *Mol. Biol. Evol.* 21, 1201–1213. doi: 10.1093/molbev/msh088

Sharoni, A. M., Nuruzzaman, M., Satoh, K., Shimizu, T., Kondoh, H., Sasaya, T., et al. (2011). Gene structures, classification and expression models of the AP2/EREBP transcription factor family in rice. *Plant Cell Physiol.* 52, 344–360. doi: 10.1093/pcp/pcq196

Shigyo, M., Hasebe, M., and Ito, M. (2006). Molecular evolution of the AP2 subfamily. *Gene* 366, 256–265. doi: 10.1016/j.gene.2005.08.009

Shigyo, M., and Ito, M. (2004). Analysis of gymnosperm two-AP2-domain-containing genes. *Dev. Genes Evol.* 214, 105–114. doi: 10.1007/s00427-004-0385-5

Song, X. M., Li, Y., and Hou, X. L. (2013). Genome-wide analysis of the AP2/ERF transcription factor superfamily in Chinese cabbage (*Brassica rapa* ssp *pekinensis*). *BMC Genomics* 14:573. doi: 10.1186/1471-2164-14-573

Tsaftaris, A., Pasentsis, K., Madesis, P., and Argiriou, A. (2012). Sequence characterization and expression analysis of three APETALA2-like genes from saffron Crocus. *Plant Mol. Biol. Rep.* 30, 443–452. doi: 10.1007/s11105-011-0355-9

Tuskan, G. A., Difazio, S., Jansson, S., Bohlmann, J., Grigoriev, I., Hellsten, U., et al. (2006). The genome of black cottonwood, *Populus trichocarpa* (Torr. & Gray). *Science* 313, 1596–1604. doi: 10.1126/science.1128691

Vahala, T., Oxelman, B., and Arnold, S. V. (2001). Two *APETALA2*-like genes of *Picea abies* are differentially expressed during development. *J. Exp. Bot.* 52, 1111–1115. doi: 10.1093/jexbot/52.358.1111

Varkonyi-Gasic, E., Lough, R., Moss, S. A., Wu, R., and Hellens, R. (2012). Kiwifruit floral gene *APETALA2* is alternatively spliced and accumulates in aberrant indeterminate flowers in the absence of miR172. *Plant Mol. Biol.* 78, 417–429. doi: 10.1007/s11103-012-9877-2

Wurschum, T., Gross-Hardt, R., and Laux, T. (2005). APETALA2 regulates the stem cell niche in the Arabidopsis shoot meristem. *Plant Cell* 18, 295–307. doi: 10.1105/tpc.105.038398

Yant, L., Mathieu, J., Dinh, T. T., Ott, F., Lanz, C., Wollmann, H., et al. (2010). Orchestration of the floral transition and floral development in Arabidopsis by the bifunctional transcription factor APETALA2. *Plant Cell Online* 22, 2156–2170. doi: 10.1105/tpc.110.075606

Zhang, C. H., Shangguan, L. F., Ma, R. J., Sun, X., Tao, R., Guo, L., et al. (2012). Genome-wide analysis of the AP2/ERF superfamily in peach (*Prunus persica*). *Genet. Mol. Res.* 11, 4789–4809. doi: 10.4238/2012.October.17.6

Zhang, C. Y., Wei, J. F., and He, S. H. (2006). Adaptive evolution of the spike gene of SARS coronavirus: changes in positively selected sites in different epidemic groups. *BMC Microbiol.* 6:88. doi: 10.1186/1471-2180-6-88

Zhu, Q.-H., and Helliwell, C. A. (2011). Regulation of flowering time and floral patterning by miR172. *J. Exp. Bot.* 62, 487–495. doi: 10.1093/jxb/erq295

Zhuang, J., Cai, B., Peng, R.-H., Zhu, B., Jin, X.-F., Xue, Y., et al. (2008). Genome-wide analysis of the AP2/ERF gene family in *Populus trichocarpa*. *Biochem. Biophys. Res. Commun.* 371, 468–474. doi: 10.1016/j.bbrc.2008.04.087

Conflict of Interest Statement: The authors declare that the research was conducted in the absence of any commercial or financial relationships that could be construed as a potential conflict of interest.

8

TCP Transcription Factors at the Interface between Environmental Challenges and the Plant's Growth Responses

*Selahattin Danisman**

Molecular Cell Physiology, Faculty of Biology, Bielefeld University, Bielefeld, Germany

Edited by:
José M. Romero,
University of Seville, Spain

Reviewed by:
Cristina Ferrandiz,
Consejo Superior de Investigaciones Cientificas – Instituto de Biologia Molecular y Celular de Plantas, Spain
Nobutaka Mitsuda,
National Institute of Advanced Industrial Science and Technology, Japan
Daniel H. Gonzalez,
National University of the Littoral, Argentina

***Correspondence:**
Selahattin Danisman
selahattin.danisman@uni-bielefeld.de

Plants are sessile and as such their reactions to environmental challenges differ from those of mobile organisms. Many adaptions involve growth responses and hence, growth regulation is one of the most crucial biological processes for plant survival and fitness. The plant-specific TEOSINTE BRANCHED 1, CYCLOIDEA, PCF1 (TCP) transcription factor family is involved in plant development from cradle to grave, i.e., from seed germination throughout vegetative development until the formation of flowers and fruits. TCP transcription factors have an evolutionary conserved role as regulators in a variety of plant species, including orchids, tomatoes, peas, poplar, cotton, rice and the model plant *Arabidopsis*. Early TCP research focused on the regulatory functions of TCPs in the development of diverse organs via the cell cycle. Later research uncovered that TCP transcription factors are not static developmental regulators but crucial growth regulators that translate diverse endogenous and environmental signals into growth responses best fitted to ensure plant fitness and health. I will recapitulate the research on TCPs in this review focusing on two topics: the discovery of TCPs and the elucidation of their evolutionarily conserved roles across the plant kingdom, and the variety of signals, both endogenous (circadian clock, plant hormones) and environmental (pathogens, light, nutrients), TCPs respond to in the course of their developmental roles.

Keywords: transcription factor, TCP, development, evolution, plant hormones, signaling

DISCOVERY OF TCPs – OF *PELORIA* AND OTHER MUTANTS

Developmental plasticity is important for plant survival because plants are sessile organisms that have to adapt to suboptimal environmental conditions. It is crucial that these developmental adaptions are balanced, which means that multiple environmental stimuli have to be perceived and weighed against each other before a plant adjusts its growth. Hence, a plethora of regulatory proteins is involved in governing developmental responses to the environment. One family of

Abbreviations: bHLH, basic helix-loop-helix; BRC, BRANCHED; CCA1, CIRCADIAN CLOCK ASSOCIATED 1; CIN, CINCINNATA; CUC, CUP-SHAPED COTYLEDON; CYC, CYCLOIDEA; FT, FLOWERING LOCUS T; GA, gibberellic acid; ICS1, ISOCHORISMATE SYNTHASE 1; IDR, intrinsically disordered region; *jaw*-D, *JAGGED AND WAVY*-D; LA, LANCEOLATE; LHY, LATE ELONGATED HYPOCOTYL; LOX, LIPOXYGENASE; NLS, nuclear localization signal; PRR, PSEUDO RESPONSE REGULATOR; TB1, TEOSINTHE BRANCHED 1; TCP, TEOSINTE BRANCHED 1, CYCLOIDEA, PCF1; TOC1, TIMING OF CAB EXPRESSION1.

transcription factors that is involved in multiple developmental processes are the TEOSINTE BRANCHED 1, CYCLOIDEA, PCF1 (TCP) proteins.

The common toadflax (*Linaria vulgaris*) is a perennial plant with bilateral, zygomorphic flowers that is native to Europe and large parts of northern Asia. When Carl Linnaeus was presented with a common toadflax that did not exhibit zygomorphic but radially symmetric flowers, he called it *peloria* from the Old Greek πέλωρ (pelór), which means *monster*. Linnaeus speculated that this monster was a hybrid between the common toadflax and a thitherto unknown plant and he was surprised to see that this hybrid was nevertheless able to propagate through seeds (Linnaeus and Rudberg, 1744). Whereas his hybrid hypothesis proved to be wrong, he used this case as evidence against immutability, the belief that all species are created at the beginning of the world and are unchanging (Smith, 1821). *Peloria* is a natural variation that occurs in toadflax, snapdragons (*Antirrhinum majus*) (Darwin, 1868) and in foxgloves (*Digitalis purpurea*) (Keeble et al., 1910), amongst other species.

About 250 years later, Luo et al. (1996) isolated the *CYCLOIDEA (CYC)* gene which is only expressed in the dorsal parts of the snapdragon flower and which is responsible for the regulation of zygomorphic flowers. A double mutant of *CYC* and its close homolog *DICHOTOMA* leads to radially symmetric snapdragon flowers (Luo et al., 1996). Cubas et al. (1999b) found that a homolog of the *CYC* gene was also responsible for floral symmetry in the common toadflax. Here, they could show that the *CYC* gene in peloric mutants was extensively methylated and silenced (Cubas et al., 1999b). At about the same time, Doebley et al. (1995) analyzed two quantitative trait loci that control morphological differences between domesticated maize (*Zea mays*) and its wild progenitor teosinte. They found the *teosinte branched 1* (*tb1*) mutation, which leads to increased side shoot outgrowth, and showed that the difference between the maize and the teosinte variant of *TB1* lies mainly in the regulatory regions of the gene, i.e., whereas the function remains the same, the expression pattern is different between domesticated maize and teosinte (Wang et al., 1999).

Kosugi et al. (1995) found that two promoter motifs that are important for the transcriptional regulation of the proliferating cell nuclear antigen (PCNA) gene in rice (*Oryza sativa*) were bound by two transcription factors that were designated PCF1 and PCF2 (Kosugi and Ohashi, 1997). Finally, Cubas et al. (1999a) determined that the above described proteins TB1, CYC and PCF1 and PCF2 share a conserved non-canonical bHLH region, the eponymous TCP domain (Kosugi and Ohashi, 1997).

FORM AND FUNCTION OF TCP TRANSCRIPTION FACTORS

Whereas, this review will mainly focus on the evolutionarily conserved roles of TCPs in the regulation of plant development and their interactions with endogenous and environmental signals, it is crucial to understand how they function. TCP transcription factors are divided into two classes, class I and class II TCPs. These classes differ in the composition of their respective NLSs, the length of the second helix in the bHLH domain, and the presence of an arginine-rich domain of unknown functionality outside the bHLH domain (Cubas et al., 1999a). This so-called R domain is not found in class I TCPs and was predicted to form a hydrophilic α-helix or a coiled-coil structure that mediates protein–protein interactions (Lupas et al., 1991; Cubas et al., 1999a).

The basic region of the TCP domain is essential for DNA binding. Replacement of a conserved glycine–proline pair in the basic region by two lysines completely abolished DNA binding activity of TCP4 in electrophoretic mobility shift studies (Aggarwal et al., 2010). Addition of the major groove binding dye methyl green reduced TCP4 binding to DNA, indicating that TCP4 binds to the major groove in double stranded DNA (Aggarwal et al., 2010).

In various experimental approaches, class I and class II TCP proteins have been shown to recognize GC-rich sequences in target gene promoters (Kosugi and Ohashi, 1997; Li et al., 2005; Viola et al., 2011; Danisman et al., 2012). The differences between class I and class II binding preferences are dependent on the presence of glycine or aspartic acid at positions 11 or 15, respectively (Viola et al., 2012). Interestingly, the class I and class II consensus binding site sequences are not mutually exclusive, indicating that at least a subset of potential target genes are targeted by both class I and class II TCP proteins. This led to speculations about a possible antagonistic relation between class I and class II TCPs, where these proteins compete for common target genes and inhibit or activate gene expression depending on which class dominates the target gene promoter (Li et al., 2005). So far, this was shown in one case only, where the *Arabidopsis* class I TCP transcription factor TCP20 binds to the same promoter as the class II TCP4 and regulates the target gene *LIPOXYGENASE2(LOX2)* in the opposite direction to TCP4 (Danisman et al., 2012). It is likely though that more cases of class I-class II TCP antagonisms will be discovered in the future, as the two classes are frequently discovered to be involved in the same biological processes.

Similar to many transcription factor families, TCPs require dimerization to bind to DNA, as addition of deoxycholate, an inhibitor of protein–protein interactions, to electrophoretic mobility shift assays leads to a reduction of TCP binding to target sequences (Trémousaygue et al., 2003). Dimerization between TCP transcription factors first has been described between PCF1 and PCF2 in rice, which form homo- and heterodimers (Kosugi and Ohashi, 1997). Whereas the homodimer of TCP20 for example does not bind to the promoter of the iron homeostasis regulator *BHLH39* in yeast one-hybrid experiments, the TCP20 heterodimer with TCP8 or TCP21 does (Andriankaja et al., 2014). A systematic yeast two-hybrid approach between *Arabidopsis* TCPs found that many protein–protein interactions are possible between TCPs and that there is a preference to bind to TCPs of the own class, i.e., class I TCPs preferably interact with class I TCPs and class II TCPs preferably interact with class II TCPs (Danisman et al., 2013). Dimerization of TCPs are facilitated by IDR (Valsecchi et al., 2013). These are characterized by low compactness, low globularity and higher structural flexibility and are typically present extensively in eukaryotic transcription factors (Liu et al., 2006). The C-terminal IDR of TCP8 is needed for self-assembly of TCP8 in dimers and higher order complexes. These IDRs potentially facilitate the flexibility of TCPs in the choice of interacting partners and thus increases the number of potential functions TCP transcription factors can be involved in Thieulin-Pardo et al. (2015). TCPs not only interact with TCPs: protein–protein interactions with a plethora of other proteins has been described, including negative regulators of effector-triggered immunity (Kim et al., 2014), components of the circadian clock (Pruneda-Paz et al., 2009, 2014; Giraud et al., 2010), and others (Trémousaygue et al., 2003; Tao et al., 2013).

EVOLUTIONARY CONSERVED ROLES OF TCPs

The three eponymous TCP proteins were characterized as regulators of branching, floral symmetry, and the cell cycle (Doebley et al., 1995; Luo et al., 1996). Later, both CYC-like and the PCF-like TCPs were shown to be involved in leaf development (Kosugi and Ohashi, 1997; Palatnik et al., 2003). TCP research since then has focused on these three developmental processes, mainly identifying evolutionarily conserved processes in a wide array of plant species and the role of cell cycle regulation in the observed phenotypes. Recently it became clear however that TCPs are not limited to branching, floral symmetry and leaf development, and neither are they limited to cell cycle mediated regulation of growth. Both will be discussed further below.

TEOSINTE BRANCHED 1, CYCLOIDEA, PCF1 transcription factors belong to an evolutionary conserved family that first appears in fresh water algae of the *Charophyta* family (Navaud et al., 2007). In the bryophyte *Physcomitrella patens,* knockout of the TCP transcription factor *Pp*TCP5 leads to increased numbers of sporangia that are attached to a single seta, reminiscent of **branching** phenotypes of *tcp* mutants in higher land plants (Ortiz-Ramírez et al., 2016). Hence, control of meristematic activity of axillary meristems with a subsequent effect on branching patterns seems to be an ancient role of TCP transcription factors (Ortiz-Ramírez et al., 2016). Consistent with this finding, branching phenotypes are apparent both in monocot and dicot plant species. Overexpression of the rice OsTB1, an ortholog of maize TB1, led to a strong decrease in tiller number. The number of axillary buds was not affected in these plants but their outgrowth was Takeda et al. (2003). This fits to the observation that it is not the formation of axillary meristems but the outgrowth of these that is affected by TCPs (Braun et al., 2012). This has been shown in peas (Braun et al., 2012), poplar (Muhr et al., 2016), *Arabidopsis* (Aguilar-Martínez et al., 2007; Poza-Carrión et al., 2007) and potato (Nicolas et al., 2015).

TCP effect on **floral development** was shown in a wide range of plant species, including *Arabidopsis*, Antirrhinum, annual candytuft (*Iberis amara*) (Busch and Zachgo, 2007; Busch et al., 2012), angiosperms like *Aristolochia arborea* and *Saruma henryi* (Horn et al., 2015), Gerbera species (Broholm et al., 2008), rice (Yuan et al., 2009), sunflowers (Fambrini et al., 2012), peas (Wang et al., 2008), ragworts (Kim et al., 2008), Morrow's honeysuckle (*Lonicera morrowii*) (Howarth and Donoghue, 2006), *Knautia macedonica* (Berger et al., 2016), and orchids (De Paolo et al., 2015).

Phylogenetic analysis revealed that the CYCLOIDEA-like TCPs underwent two major duplication events that both predate the formation of core eudicots (Howarth and Donoghue, 2006). In *Arabidopsis*, all three CYC clades are represented by TCP12, TCP1 and TCP18, respectively (Howarth and Donoghue, 2006). Especially the CYC2 clade, represented by TCP1 in *Arabidopsis*, underwent multiple additional duplications and has been studied for its effect on floral symmetry, as it contains the original *CYC* gene of Antirrhinum (Howarth and Donoghue, 2006). An interesting side note is that the duplication of the CYCLOIDEA-like TCPs nearly coincides with the major duplication events of the homeotic MADS-box transcription factors APETALA3, AGAMOUS and SEPALLATA, all three of them important factors for the definition of organ identity in flowering plants (Howarth and Donoghue, 2006). This suggests that the genetic components that are important for the definition of floral organs diversified at a similar time as the components that are important for the growth regulation of these. TCP transcription factors have been identified as targets of *Arabidopsis* APETALA1 and SEPALLATA3 (Kaufmann et al., 2009, 2010), highlighting a possible link between organ identity formation and growth regulation between MADS-box transcription factors and TCPs (Dornelas et al., 2011).

In Antirrhinum, CYC regulates symmetry via the Myb-domain transcription factor RADIALIS (Corley et al., 2005). Overexpression of CYC in *Arabidopsis* leads to larger petals containing enlarged petal cells (Costa et al., 2005). Regulation of floral growth is not restricted to the CYC-like class II TCPs. In the *jaw*-D mutant, petal development is different from wild type *Arabidopsis* (Palatnik et al., 2003). Nag et al. (2009) showed that this depends on miR319 regulation of *TCP4.* A microRNA-resistant form of *TCP4* under the control of an *APETALA3* promoter is expressed in floral organs only and leads to dramatically smaller flowers that only consist of carpels and sepals, missing any petals or stamens, whereas the seedlings of these plants look normal (Nag et al., 2009).

The zinc-finger transcriptional repressor RABBIT EARS controls the expression of the TCPs *TCP5, TCP13,* and *TCP17* and misexpression of both *RABBIT EARS* and these *TCPs* leads to aberrant petal development in *Arabidopsis* (Huang and Irish, 2015). Repression of these TCPs leads to an early stop of mitotic activity during petal development (Huang and Irish, 2015). Interestingly the opposite occurs upon downregulation of *TCP5, TCP13,* and *TCP17* in leaves, where leaf cells continue with mitotic divisions for a longer time than in wild type plants (Efroni et al., 2008). Here, the effect of TCP transcription factors on organ development is dependent on the organ-context. This underlines the importance of the regulatory interplay between TCPs and organ identity regulators. While there are hints at this interplay between TCPs and MADS box transcription factors during flower development, such an interplay remains to be shown during the development of other organs (Dornelas et al., 2011).

First indications for a role of TCPs in **leaf development** comes from work in Antirrhinum (Nath et al., 2003). The Antirrhinum class II TCP mutant *cin* displays crinkly leaves, which are the result of a change in the regulation of the cell cycle during leaf development (Nath et al., 2003). Essentially, mitotic divisions of developing leaf cells in the leaf tip are arrested first and those at the leaf base are arrested last. The result of this successive arresting behavior is a so called arrest front that moves from the leaf tip to the leaf base. The form of this arrest front is different in *cin* leaves than in wild type leaves, leading to a modified leaf curvature (Nath et al., 2003). In *Arabidopsis*, similar behavior is observed in the *jaw*-D mutant (Palatnik et al., 2003). *Jaw*-D is an overexpressor of the microRNA miR319a in which the *CIN*-like class II TCPs *TCP2, TCP3, TCP4, TCP10,* and *TCP24* are downregulated (Palatnik et al., 2003). *Jaw*-D mutants display serrated leaves, abnormal petals and delayed leaf development and senescence (Palatnik et al., 2003). This phenotype derives from delayed leaf development, in which the mitotic arrest front starts later than in wild type plants (Efroni et al., 2008). Recently, it was shown that miR319a-regulated TCP transcription factors act redundantly with NGATHA transcription factors to limit meristematic activity of leaf meristems during leaf development (Alvarez et al., 2016). This phenotype was also apparent in plants expressing an artificial microRNA against the class II TCPs *TCP5, TCP13,* and *TCP17* and the phenotype was extremely strong when these plants were crossed with *jaw*-D plants (Efroni et al., 2008).

Class II TCPs also regulate leaf development in tomato compound leaves. An ortholog of the *Arabidopsis* miR319-sensitive TCPs in tomato is *LA* and it is under the control of the tomato miR319 (Ori et al., 2007). *La* mutants exhibit simple leaves, whereas overexpression of miR319 without *LA* insensitivity to the microRNA leads to increased partitioning of the compound leaves. Also, miR319 overexpressing tomato leaves grow 3 months longer than wild type leaves and show the marks of late differentiation, which is a behavior that is identical to *Arabidopsis jaw*-D plants (Ori et al., 2007; Efroni et al., 2008). Overexpression of miR319 in the monocot *Agrostis stolonifera* (creeping bentgrass) leads to downregulation of class II TCPs and to the formation of wider and thicker leaves that are different from the wild type (Zhou et al., 2013). This phenotype stems from an increased number of cells in the transgenic bentgrass, similar to *jaw*-D in *Arabidopsis* (Efroni et al., 2008; Zhou et al., 2013). In general, expression of CIN-like genes is closely correlated with leaf shapes both in Solanaceae species and in the desert poplar (*Populus euphratica*) (Shleizer-Burko et al., 2011; Ma et al., 2016).

Expression of TCP3 with a dominant repressor domain led to severe disturbance of *Arabidopsis* development in all organs (Koyama et al., 2007), involving ectopic shoot formation, serrated leaves, modified sepals and petals, and wavy silique formation. This was due to misexpression of boundary specific genes, i.e., *CUC* and *LATERAL ORGAN BOUNDARIES* (Koyama et al., 2007). Also in Antirrhinum, an ortholog of *Arabidopsis* TCP15 was found to interact with CUPULIFORMIS, a protein that is related to *Arabidopsis* CUC proteins (Weir et al., 2004). Furthermore, the two *Arabidopsis* class I TCPs TCP14 and TCP15 were shown to be redundant in affecting cell proliferation during leaf development and in other tissues in *Arabidopsis*. The most obvious effect though was seen in internode length, which is reduced in *tcp14 tcp15* mutants and leads to shorter plants (Kieffer et al., 2011).

Whereas TCP functions have thus been very well-characterized in these branching, flower and leaf development over a wide array of plant species (**Figure 1**), there are hints that this is just a subset of TCP roles in development. TCPs were shown to be upregulated upon imbibition of dry seeds and germination of *tcp14* transposon insertion lines seemed to be lower than in wild type seeds (Tatematsu et al., 2008). Although here, expression of *TCP14* in the transposon lines was not necessarily lower than in the wild type, indicating that *TCP14* may not be the only cause of the reduced germination rate (Tatematsu et al., 2008). Downregulation of *TCP* expression in cotton led to reduced cotton hair fiber length as well as a higher of lateral shoots and a stunted growth indicative of a reduced apical dominance (Hao et al., 2012). Overexpression of miR319 in Chinese cabbage not only led to altered leaf development, also the cabbage heads were rounder than in cabbage with low *miR319* expression and higher expression of its target gene *BrpTCP4-1* (Mao et al., 2014). Heterologous expression of the rice OsTCP19 in *Arabidopsis* led to a lower number of lateral roots (Mukhopadhyay et al., 2015). In cucumber, mutation of a *TCP* gene led to a unique plant phenotype. The affected cucumber plants did not develop tendrils but shoots instead. The authors of this study hypothesize that here TCPs not only affect growth of an organ but also determine organ identity (Wang S.et al., 2015). A similar phenotype was found in melons where a single-nucleotide mutation in *CmTCP1* led to the Chiba tendril-less mutation. Also here, the tendrils were converted to shoot and leaf-like structures (Mizuno et al., 2015). This would be the first indication that TCPs can act as organ identity regulators. Further research has yet to uncover whether the function of TCPs in organ identity regulation of tendrils is a unique and novel role or whether other plant organs also need TCPs to define their identity.

TCP FUNCTIONS EFFECT ON THE CELL CYCLE – DIRECT OR INDIRECT?

Early, TCP research focused on the cell cycle as main target of TCP regulation (Kosugi and Ohashi, 1997; Li et al., 2005). Whereas, binding to cell cycle genes has been shown in certain cases (Li et al., 2005; Davière et al., 2014), close analysis of cell division patterns and transcript changes during *jaw*-D leaf development indicated that the class II-TCP dependent regulation of the cell cycle may be indirect (Efroni et al., 2008). Also binding of the class I TCP TCP20 to cell cycle genes has been shown only once and *in vitro* (Li et al., 2005), whereas direct target gene analysis indicate that hormone synthesis, especially jasmonate synthesis, is rather directly targeted by TCP20 (Danisman et al., 2012). Both TCP4 and TCP20 affect leaf development via the synthesis of methyl jasmonate, a hormone that has multiple functions in plant development and response to wounding and pathogens (Schommer et al., 2008). Jasmonate, usually known for its role in wounding and pathogen response, does also affect the cell cycle (Świątek et al., 2002).

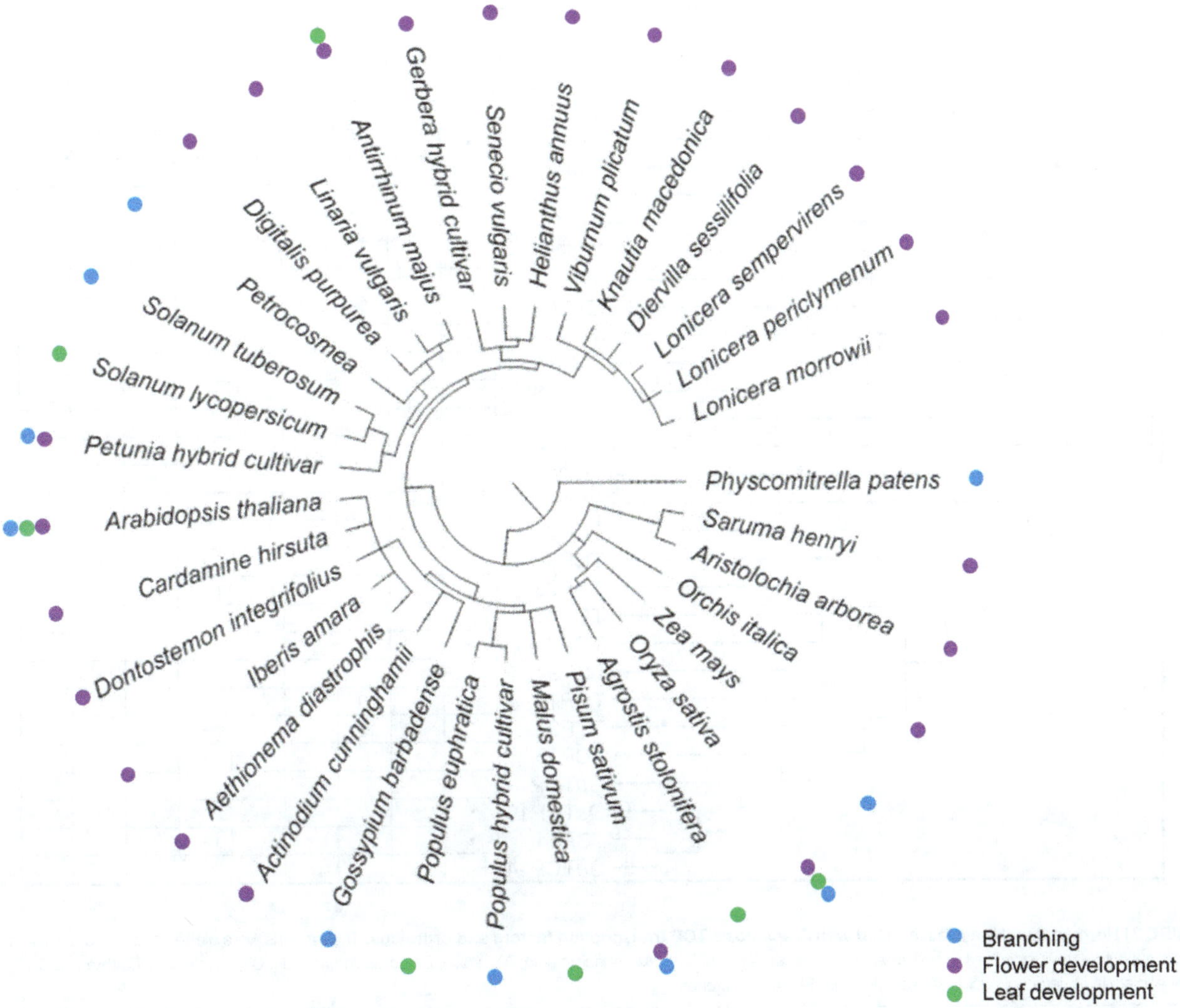

FIGURE 1 | Phylogenetic tree of plant species in which TCP transcription factors are involved in branching (Takeda et al., 2003; Aguilar-Martínez et al., 2007; Poza-Carrión et al., 2007; Bai et al., 2012; Braun et al., 2012; Drummond et al., 2015; Nicolas et al., 2015; Muhr et al., 2016) **(blue dots), flower development** (Linnaeus and Rudberg, 1744; Keeble et al., 1910; Corley et al., 2005; Costa et al., 2005; Busch and Zachgo, 2007; Broholm et al., 2008; Kim et al., 2008; Nag et al., 2009; Yuan et al., 2009; Howarth et al., 2011; Busch et al., 2012; Tähtiharju et al., 2012; Claßen-Bockhoff et al., 2013; Juntheikki-Palovaara et al., 2014; De Paolo et al., 2015; Horn et al., 2015; Lucero et al., 2015; Wang et al., 2008; Wang X.et al., 2015; Yang et al., 2015; Berger et al., 2016) **(purple dots) or leaf development** (Kosugi and Ohashi, 1997; Nath et al., 2003; Palatnik et al., 2003; Koyama et al., 2007, 2010a,b; Ori et al., 2007; Efroni et al., 2008; Kieffer et al., 2011; Mimida et al., 2011; Sarvepalli and Nath, 2011; Danisman et al., 2012, 2013; Aguilar-Martínez and Sinha, 2013; Burko et al., 2013; Tao et al., 2013; Zhou et al., 2013; Ballester et al., 2015; Huang and Irish, 2015; Ma et al., 2016) **(green dots), respectively.** The phylogenetic tree was created using Phylotree and iTOL (Letunic and Bork, 2016).

Jasmonate is not the only plant hormone that may mediate TCP regulation to the cell cycle. The evidence for hormone involvement in TCP-mediated growth regulation accumulated in the recent years (Nicolas and Cubas, 2016). TCP functions have been associated with abscisic acid (Tatematsu et al., 2008; González-Grandío et al., 2013; Mukhopadhyay et al., 2015), auxin (Kosugi et al., 1995; Ben-Gera and Ori, 2012; Uberti-Manassero et al., 2012; Das Gupta et al., 2014), brassinosteroid (Guo et al., 2010), cytokinin (Steiner et al., 2012; Efroni et al., 2013), GA (Yanai et al., 2011; Das Gupta et al., 2014; Davière et al., 2014), jasmonic acid (Schommer et al., 2008; Danisman et al., 2012), salicylic acid (Wang X.et al., 2015), and strigolactone signaling pathways (Dun et al., 2012; Hu et al., 2014) (**Figure 2**).

Apart from hormonal control of growth, TCP transcription factors are also involved in other biological processes that in turn affect growth. For example, binding sites of TCP transcription factors have been identified in the promoters of *CYTOCHROME C1* and 103 genes that are encoding components of the mitochondrial oxidative phosphorylation machinery and protein biogenesis (Welchen and Gonzalez, 2006). The authors of this study proposed that the TCP transcription factors binding these sites coordinate mitochondria genesis and function with growth in new organs (Welchen and Gonzalez, 2006). Another

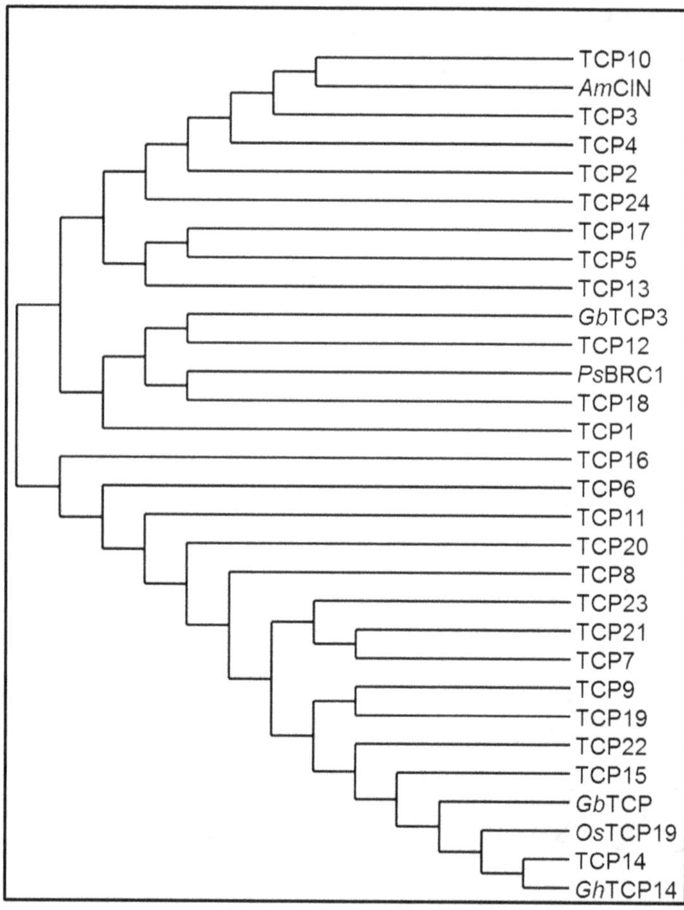

	AA	AX	BR	CK	ET	GA	JA	SA	SL
TCP10				•			•		
*Am*CIN		•		•					
TCP3		•		•			•		
TCP4	•	•		•			•		
TCP2				•			•		
TCP24				•			•		
TCP17									
TCP5									
TCP13									
*Gb*TCP3									
TCP12	•								
*Ps*BRC1				•					•
TCP18									
TCP1			•						
TCP16									
TCP6									
TCP11									
TCP20							•		
TCP8						•		•	
TCP23									
TCP21								•	
TCP7									
TCP9							•	•	
TCP19									
TCP22						•			
TCP15		•		•		•			
*Gb*TCP									
*Os*TCP19	•						•	•	
TCP14	•			•		•			
*Gh*TCP14		•							

FIGURE 2 | Hormonal pathways associated with *Arabidopsis* TCP transcription factors and orthologs. The proteins were plotted according to their phylogenetic similarity using PhyML and TreeDyne (Dereeper et al., 2008). AA, abscisic acid; AX, auxin; BR, brassinosteroids; CK, cytokinin; ET, ethylene; GA, gibberellic acid; JA, jasmonic acid; SA, salicylic acid; SL, strigolactones.

study showed these genes contain a GGGC(C/T) element in their promoters which is important for diurnal regulation of their gene expression (Giraud et al., 2010). These promoters are bound by TCP transcription factors, implying a role in diurnal regulation of transcripts of the mitochondrial oxidative phosphorylation machinery (Giraud et al., 2010). Earlier TCP21 was found to bind to the promoter of the core clock gene *CCA1* and regulate its expression (Pruneda-Paz et al., 2009). TCP21 serves as an inhibitor of *CCA1* during the day and dimerization of TOC1 with TCP21 abolishes its binding to the *CCA1* promoter. In a double mutant with the clock gene LHY, *tcp21/lhy* greatly reduces the period of *CCA1* expression (Pruneda-Paz et al., 2009). Not only TCP21, other TCPs have also been found to bind to *CCA1* in yeast based studies and co-immunoprecipitation experiments (Giraud et al., 2010; Pruneda-Paz et al., 2014). A recent study also showed that TCP20 and TCP22 act as activators of *CCA1* in the morning, fulfilling an important role in the circuity of the circadian clock (Wu et al., 2016). This means that TCP proteins bind to the promoters of clock genes, regulate their expression, dimerize with clock proteins and bind to downstream targets of the clock (Pruneda-Paz et al., 2009, 2014; Giraud et al., 2010; Wu et al., 2016) (**Figure 3**). Altogether, it becomes clear that TCPs not only affect growth via the cell cycle. Instead, they act in different biological processes that directly or indirectly affect growth.

MEDIATING ENVIRONMENTAL SIGNALS INTO GROWTH RESPONSES

This picture becomes even more complex, as TCPs also mediate environmental signals into growth responses. TCPs were found to be involved in pathogen defense. First, an extensive study showed that both *Pseudomonas syringae* and *Hyaloperonospora arabidopsidis* infection led to reduction of TCP14 protein (Mukhtar et al., 2011). Secreted proteins from pathogenic bacteria transferred by the Aster leafhopper (*Macrosteles quadrilineatus*) to *Arabidopsis* were able to dimerize with and destabilize TCP2, TCP4, and TCP7 proteins, comprising both classes of TCP transcription factors (Sugio et al., 2011, 2014). Overexpression of the responsible phytoplasma protein SECRETED ASTER YELLOWS-WITCHES BROOM PROTEIN 11 in *Arabidopsis* destabilizes TCP2, TCP3, TCP4, TCP5, TCP10,

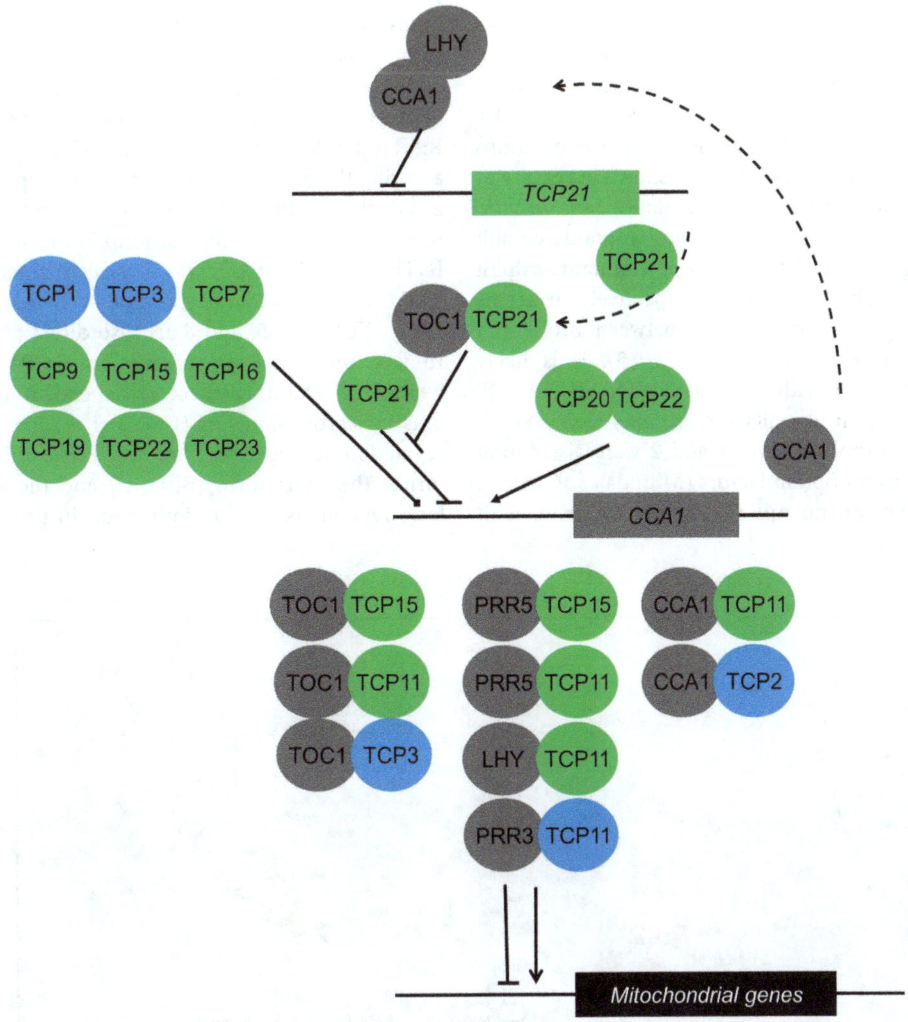

FIGURE 3 | Interactions of TCP transcription factors with components of the circadian clock both within the central clock circuitry and in downstream processes. Class I and class II TCPs are depicted in green and blue, respectively. Known clock components are depicted in gray. Proteins are represented as circles, genes in squares. Dimers are depicted as overlapping circles. *CCA1* inhibition by TCP21 is abolished by dimerization of TCP21 with TOC1. The CCA1/LHY dimer inhibits *TCP21* expression (Pruneda-Paz et al., 2009). The effect of nine TCPs that bind to the *CCA1* promoter in yeast one-hybrid studies is unknown (Pruneda-Paz et al., 2014). Downstream of the clock, TCP/clock component heterodimers regulate rhythmic expression of mitochondrial proteins depending on the number and arrangement of TCP binding sites in the mitochondrial gene promoters (Giraud et al., 2010).

TCP13, TCP17, and TCP24 and leads to *jaw*-D-like phenotypes (Sugio et al., 2011). Additionally, jasmonic acid levels in infected *Arabidopsis* leaves are significantly reduced in comparison with untreated leaves, indicating that the plant's defense mechanisms are reduced upon infection by the pathogen. A similar effect has been found in apples, where the plant pathogen *Candidatus Phytoplasma mali* binds to two TCP transcription factors and induces morphogenetic changes that co-occur with reduction of jasmonic acid, salicylic acid, and abscisic acid levels (Janik et al., 2016). Further studies identified the class I TCPs TCP8 and TCP9 as important factors for the expression of *ICS1*, which encodes for a key enzyme in salicylic acid synthesis (Wang X.et al., 2015). In another study, TCP21 has been identified to bind to the promoter of *ICS1* and induction of *ICS1* expression by salicylic acid is blocked in *tcp21* mutants (Zheng et al., 2015). Class I TCPs also interact with proteins known to regulate *ICS1* expression, i.e., the transcription factors WRKY28, NAC019 and ETHYLENE INSENSITIVE 1 and the calmodulin binding protein SYSTEMIC ACQUIRED RESISTANCE DEFICIENT 1. Consequently, the *tcp8 tcp9* double mutant shows increased sensitivity to infection with *Pseudomonas syringae* pv. *maculicola* ES4326 (Wang X.et al., 2015). TCP transcription factors partially control pathogen defense via a second pathway, i.e., by antagonizing the effect of SUPPRESSOR OF rps4-RLD1, a protein that negatively regulates effector-triggered immunity in *Arabidopsis* (Kim et al., 2014). Lack of TCPs in the triple mutant *tcp8 tcp14 tcp15* leads to increased growth of *Pseudomonas syringae* DC3000 when compared to wild type plants (Kim et al., 2014).

Recent studies showed that TCP transcription factors regulate flowering time. A knockout of the class I TCP transcription factor TCP23 led to earlier flowering than the wild type, whereas TCP23 overexpressing lines showed delayed flowering behavior (Balsemão-Pires et al., 2013). The floral transition of axillary meristems in *Arabidopsis* is controlled by an interaction between the flowering time proteins FT and TWIN SISTER OF FT and BRC1 (Niwa et al., 2013). The protein–protein interactions between these transcription factors have been shown in yeast two-hybrid, bimolecular fluorescence complementation, and *in vitro* pull-down assays (Niwa et al., 2013). As *brc1* mutants exhibit accelerated flowering and *ft* and *twin sister of ft* mutants exhibit slower flowering of axillary meristems, respectively, it seems that there is an antagonistic relationship between BRC1 and the flowering time proteins (Niwa et al., 2013). It is likely that dimerization of BRC1 with FT and TWIN SISTER OF FT represses their function in axillary meristems (Niwa et al., 2013). The apple FT orthologs *Md*FT1 and 2 were also found to interact with TCP transcription factors (Mimida et al., 2011). Overexpression of the tomato miR319 led to flowering with fewer leaves than in wild type tomato and it was shown that LA binds to the promoters of the tomato *APETALA1* and *FRUITFUL* orthologs (Burko et al., 2013).

Perception of the red to far-red light ratio (R:FR) informs a plant of shading by neighboring vegetation and a lower R:FR ratio leads to suppressed axillary meristem outgrowth, allowing the plant to invest in a longer hypocotyl and eventually avoid the shading. In *Arabidopsis*, hypocotyl elongation is regulated via the bHLH transcription factor PHYTOCHROME INTERACTING FACTOR 4, which among others activates *YUCCA8* expression to promote cell elongation (Sun et al., 2012). *YUCCA2, 5,* and *8* are also direct target genes of TCP4. In fact, induced overexpression of TCP4 leads to elongated hypocotyls and this effect is dependent on both auxin and brassinosteroid signaling (Challa et al., 2016). In potato, *BRC1a* regulation is dependent on the R:FR. BRC1a comes in two forms: the short form (BRC1a^{S}) and the alternatively spliced long version (BRC1a^{L}). Both result in proteins but the shorter

FIGURE 4 | Schematic figure depicting the diversity of environmental signals that affect TCP functions in plants (Mukhtar et al., 2011; Sugio et al., 2011; Balsemão-Pires et al., 2013; González-Grandío et al., 2013; Niwa et al., 2013; Viola et al., 2013, 2016; Guan et al., 2014; Hu et al., 2014; Kim et al., 2014; Mukhopadhyay et al., 2015; Nicolas et al., 2015; Kumar et al., 2016).

form is cytoplasmatic and does not bind to target genes in the nucleus. The ratio between these two forms changes upon decapitation of potato shoots, exposure to darkness, and under low R:FR conditions (Nicolas et al., 2015). Whereas decapitation leads to a relative increase in $BRC1a^S$, darkness and low R:FR treatments lead to a relative increase in $BRC1a^L$ content. The longer $BRC1a^L$ protein subsequently inhibits axillary branch elongation in potato shoots and stolons (Nicolas et al., 2015). *Arabidopsis brc1* and *brc2* (*tcp12* and *tcp18*) show a reduced response to R:FR and the response is abolished in the *brc1 brc2* double mutant (González-Grandío et al., 2013). TCP transcription factors are also involved in axillary bud outgrowth of Petunia. Here, *Gh*TCP3 acts in conjunction with DECREASED APICAL DOMINANCE 2, a receptor protein that normally inactivates strigolactones in response to decreased R:FR (Drummond et al., 2015). Rice OsTCP15 is involved in the mesocotyl elongation in response to darkness and responds to strigolactone and cytokinin treatments, outlining the interplay between TCPs and different plant hormones in developmental regulation that is responsive to the environment (Hu et al., 2014).

Viola et al. (2013) showed that class I TCPs contain a conserved cysteine-20 which is sensitive to treatments by oxidants in a dose-dependent manner. This redox-dependent behavior of TCP15 is important for its effect in anthocyanin biosynthesis. A mutant in which the cysteine-20 of TCP15 was replaced by a serine accumulates less anthocyanin under high light stress than wild type plants (Viola et al., 2016). Plant extracts from TCP15 overexpressing plants showed that exposure to prolonged high light conditions leads to an abolishment of TCP15 DNA-binding activity *in vivo*, mirroring the *in vitro* phenotype (Viola et al., 2013, 2016). Thus, TCP15 function is reactive to high light input. While the anthocyanin response is not a direct developmental response, further analysis may show that there is a developmental effect.

Not only light affects TCPs, also other signals are perceived and lead to TCP-mediated growth regulation. For example, Guan et al. (2014) showed that TCP20 is involved in nutrient foraging of *Arabidopsis* roots. In split-root experiments wild type *Arabidopsis* develops an increased number of lateral roots in medium containing high nitrate concentrations (i.e., 5 mM NO_3^-) and close to no lateral roots in medium containing low nitrate concentrations (i.e., 0 mM NO_3^-). *Tcp20* plants do not exhibit this behavior, indicating that the regulation of root foraging is under the control of TCP20 (Guan et al., 2014). Interestingly, *TCP20* transcript levels are not under the control of nitrate levels, indicating that TCP20 is regulated on protein level, either by forming specific protein–protein dimers in the case of nitrate deficiency or via another regulatory mechanism. In rice, the transcript of the class I TCP *OsTCP19* is upregulated during salt stress and water-deficit treatments (Mukhopadhyay et al., 2015). Heterologous *OsTCP19* overexpression in *Arabidopsis* leads to reduced numbers of lateral roots but increased abiotic stress tolerance, i.e., plants grew better on Mannitol-containing medium and recovered better after water-deficit treatments. Here, *LOX2* expression was reduced in the OsTCP19 overexpressors and ABA signaling genes were upregulated (Mukhopadhyay et al., 2015). Recent experiments revealed up- and down-regulation of several TCPs in *Arabidopsis* under osmotic stress, although a functional analysis of their role in response to osmotic stress has not been done yet (Kumar et al., 2016). In summary, these few results are first indications that TCPs are no mere static regulators of development, but that they do directly translate environmental signals into growth regulation (**Figure 4**).

CONCLUSION AND OUTLOOK

TCP transcription factors play a role in a multitude of growth processes over a wide range of plant species (**Figure 1**). They affect growth directly via the cell cycle and indirectly via influencing plant hormonal signaling and the circadian clock (**Figures 2** and **3**). Additionally, recent discoveries link TCP-controlled growth responses with environmental signals such as R:FR, high light stress, salt stress or the presence or absence of nutrients.

TCP transcription factors are involved in so many important developmental processes and interact with so many plant hormones that it is likely that future plant research will also uncover a lot more signals that TCPs react to. This will also mean that future TCP research will have to more closely elucidate how the interaction of TCPs with different signaling networks is regulated to ensure a measured response to environmental challenges. This research will have to uncover the roles of dimerization, transcriptional and post-transcriptional regulation as well as post-translational modifications in controlling and ensuring specific TCP functions in plant development.

Plant pathogens are targeting TCP transcription factors to manipulate plant architecture in their favor. If plant pathogens use TCPs in their best interests, maybe so should we. TCP transcription factors will be valuable tools in optimizing plant architecture and hardening plants in response to environmental challenges.

AUTHOR CONTRIBUTIONS

SD drafted, wrote and critically revised the article.

FUNDING

Work in the author's lab is supported by the Bielefeld Young Researcher's Fund and the core grant of Bielefeld University to D. Staiger.

ACKNOWLEDGMENTS

We acknowledge support for the Article Processing Charge by the Deutsche Forschungsgemeinschaft and the Open Access Publication Fund of Bielefeld University.

REFERENCES

Aggarwal, P., Gupta, M. D., Joseph, A. P., Chatterjee, N., Srinivasan, N., and Nath, U. (2010). Identification of specific DNA binding residues in the TCP family of transcription factors in *Arabidopsis*. *Plant Cell* 22, 1174–1189. doi: 10.1105/tpc.109.066647

Aguilar-Martínez, J. A., Poza-Carrión, C., and Cubas, P. (2007). *Arabidopsis* BRANCHED1 acts as an integrator of branching signals within axillary buds. *Plant Cell* 19, 458–472. doi: 10.1105/tpc.106.048934

Aguilar-Martínez, J. A., and Sinha, N. (2013). Analysis of the role of *Arabidopsis* class I TCP genes AtTCP7, AtTCP8, AtTCP22, and AtTCP23 in leaf development. *Front. Plant Sci.* 4:406. doi: 10.3389/fpls.2013.00406

Alvarez, J. P., Furumizu, C., Efroni, I., Eshed, Y., and Bowman, J. L. (2016). Active suppression of a leaf meristem orchestrates determinate leaf growth. *eLife* 5:e15023. doi: 10.7554/eLife.15023

Andriankaja, M. E., Danisman, S., Mignolet-Spruyt, L. F., Claeys, H., Kochanke, I., Vermeersch, M., et al. (2014). Transcriptional coordination between leaf cell differentiation and chloroplast development established by TCP20 and the subgroup Ib bHLH transcription factors. *Plant Mol. Biol.* 85, 233–245. doi: 10.1007/s11103-014-0180-2

Bai, F., Reinheimer, R., Durantini, D., Kellogg, E. A., and Schmidt, R. J. (2012). TCP transcription factor, BRANCH ANGLE DEFECTIVE 1 (BAD1), is required for normal tassel branch angle formation in maize. *Proc. Natl. Acad. Sci. U.S.A.* 109, 12225–12230. doi: 10.1073/pnas.1202439109

Ballester, P., Navarrete-Gómez, M., Carbonero, P., Oñate-Sánchez, L., and Ferrándiz, C. (2015). Leaf expansion in *Arabidopsis* is controlled by a TCP-NGA regulatory module likely conserved in distantly related species. *Physiol. Plant.* 155, 21–32. doi: 10.1111/ppl.12327

Balsemão-Pires, E., Andrade, L. R., and Sachetto-Martins, G. (2013). Functional study of TCP23 in *Arabidopsis thaliana* during plant development. *Plant Physiol. Biochem.* 67, 120–125. doi: 10.1016/j.plaphy.2013.03.009

Ben-Gera, H., and Ori, N. (2012). Auxin and LANCEOLATE affect leaf shape in tomato via different developmental processes. *Plant Signal. Behav.* 7, 1255–1257. doi: 10.4161/psb.21550

Berger, B. A., Thompson, V., Lim, A., Ricigliano, V., and Howarth, D. G. (2016). Elaboration of bilateral symmetry across Knautia macedonica capitula related to changes in ventral petal expression of CYCLOIDEA-like genes. *EvoDevo* 7, 8. doi: 10.1186/s13227-016-0045-7

Braun, N., Germain, A., de, S., Pillot, J.-P., Boutet-Mercey, S., Dalmais, M., et al. (2012). The Pea TCP transcription factor PsBRC1 acts downstream of strigolactones to control shoot branching. *Plant Physiol.* 158, 225–238. doi: 10.1104/pp.111.182725

Broholm, S. K., Tähtiharju, S., Laitinen, R. A. E., Albert, V. A., Teeri, T. H., and Elomaa, P. (2008). A TCP domain transcription factor controls flower type specification along the radial axis of the *Gerbera* (Asteraceae) inflorescence. *Proc. Natl. Acad. Sci.* 105, 9117–9122. doi: 10.1073/pnas.0801359105

Burko, Y., Shleizer-Burko, S., Yanai, O., Shwartz, I., Zelnik, I. D., Jacob-Hirsch, J., et al. (2013). A role for APETALA1/FRUITFULL transcription factors in tomato leaf development. *Plant Cell* 25, 2070–2083. doi: 10.1105/tpc.113.113035

Busch, A., Horn, S., Mühlhausen, A., Mummenhoff, K., and Zachgo, S. (2012). Corolla monosymmetry: evolution of a morphological novelty in the brassicaceae family. *Mol. Biol. Evol.* 29, 1241–1254. doi: 10.1093/molbev/msr297

Busch, A., and Zachgo, S. (2007). Control of corolla monosymmetry in the Brassicaceae *Iberis amara*. *Proc. Natl. Acad. Sci. U.S.A.* 104, 16714–16719. doi: 10.1073/pnas.0705338104

Challa, K. R., Aggarwal, P., and Nath, U. (2016). Activation of YUCCA5 by the transcription factor TCP4 integrates developmental and environmental signals to promote hypocotyl elongation in *Arabidopsis*. *Plant Cell* 28, 2117–2130. doi: 10.1105/tpc.16.00360

Claßen-Bockhoff, R., Ruonala, R., Bull-Hereñu, K., Marchant, N., and Albert, V. A. (2013). The unique pseudanthium of Actinodium (Myrtaceae) - morphological reinvestigation and possible regulation by CYCLOIDEA-like genes. *EvoDevo* 4, 8. doi: 10.1186/2041-9139-4-8

Corley, S. B., Carpenter, R., Copsey, L., and Coen, E. (2005). Floral asymmetry involves an interplay between TCP and MYB transcription factors in *Antirrhinum*. *Proc. Natl. Acad. Sci. U.S.A.* 102, 5068–5073. doi: 10.1073/pnas.0501340102

Costa, M. M. R., Fox, S., Hanna, A. I., Baxter, C., and Coen, E. (2005). Evolution of regulatory interactions controlling floral asymmetry. *Development* 132, 5093–5101. doi: 10.1242/dev.02085

Cubas, P., Lauter, N., Doebley, J., and Coen, E. (1999a). The TCP domain: a motif found in proteins regulating plant growth and development. *Plant J.* 18, 215–222. doi: 10.1046/j.1365-313X.1999.00444.x

Cubas, P., Vincent, C., and Coen, E. (1999b). An epigenetic mutation responsible for natural variation in floral symmetry. *Nature* 401, 157–161. doi: 10.1038/43657

Danisman, S., Dijk, A. D. J., van Bimbo, A., Wal, F., van der Hennig, L., Folter, S., et al. (2013). Analysis of functional redundancies within the *Arabidopsis* TCP transcription factor family. *J. Exp. Bot.* 64, 5673–5685. doi: 10.1093/jxb/ert337

Danisman, S., Wal, F., van der Dhondt, S., Waites, R., Folter, S., de Bimbo, A., et al. (2012). *Arabidopsis* Class I and Class II TCP transcription factors regulate jasmonic acid metabolism and leaf development antagonistically. *Plant Physiol.* 159, 1511–1523. doi: 10.1104/pp.112.200303

Darwin, C. R. (1868). *The Variation of Animals and Plants Under Domestication.* London: John Murray.

Das Gupta, M., Aggarwal, P., and Nath, U. (2014). CINCINNATA in *Antirrhinum majus* directly modulates genes involved in cytokinin and auxin signaling. *New Phytol.* 204, 901–912. doi: 10.1111/nph.12963

Davière, J.-M., Wild, M., Regnault, T., Baumberger, N., Eisler, H., Genschik, P., et al. (2014). Class I TCP-DELLA interactions in inflorescence shoot apex determine plant height. *Curr. Biol.* 24, 1923–1928. doi: 10.1016/j.cub.2014.07.012

De Paolo, S., Gaudio, L., and Aceto, S. (2015). Analysis of the TCP genes expressed in the inflorescence of the orchid *Orchis italica*. *Sci. Rep.* 5:16265. doi: 10.1038/srep16265

Dereeper, A., Guignon, V., Blanc, G., Audic, S., Buffet, S., Chevenet, F., et al. (2008). Phylogeny.fr: robust phylogenetic analysis for the non-specialist. *Nucleic Acids Res.* 36, W465–W469. doi: 10.1093/nar/gkn180

Doebley, J., Stec, A., and Gustus, C. (1995). teosinte branched1 and the origin of maize: evidence for epistasis and the evolution of dominance. *Genetics* 141, 333–346.

Dornelas, M. C., Patreze, C. M., Angenent, G. C., and Immink, R. G. H. (2011). MADS: the missing link between identity and growth? *Trends Plant Sci.* 16, 89–97. doi: 10.1016/j.tplants.2010.11.003

Drummond, R. S. M., Janssen, B. J., Luo, Z., Oplaat, C., Ledger, S. E., Wohlers, M. W., et al. (2015). Environmental control of branching in *Petunia*. *Plant Physiol.* 168, 735–751. doi: 10.1104/pp.15.00486

Dun, E. A., Germain, A., de, S., Rameau, C., and Beveridge, C. A. (2012). Antagonistic action of strigolactone and cytokinin in bud outgrowth control. *Plant Physiol.* 158, 487–498. doi: 10.1104/pp.111.186783

Efroni, I., Blum, E., Goldshmidt, A., and Eshed, Y. (2008). A protracted and dynamic maturation schedule underlies *Arabidopsis* leaf development. *Plant Cell* 20, 2293–2306. doi: 10.1105/tpc.107.057521

Efroni, I., Han, S.-K., Kim, H. J., Wu, M.-F., Steiner, E., Birnbaum, K. D., et al. (2013). Regulation of leaf maturation by chromatin-mediated modulation of cytokinin responses. *Dev. Cell* 24, 438–445. doi: 10.1016/j.devcel.2013.01.019

Fambrini, M., Salvini, M., and Pugliesi, C. (2012). A transposon-mediate inactivation of a CYCLOIDEA-like gene originates polysymmetric and androgynous ray flowers in *Helianthus annuus*. *Genetica* 139, 1521–1529. doi: 10.1007/s10709-012-9652-y

Giraud, E., Ng, S., Carrie, C., Duncan, O., Low, J., Lee, C. P., et al. (2010). TCP transcription factors link the regulation of genes encoding mitochondrial proteins with the circadian clock in *Arabidopsis thaliana*. *Plant Cell* 22, 3921–3934. doi: 10.1105/tpc.110.074518

González-Grandío, E., Poza-Carrión, C., Sorzano, C. O. S., and Cubas, P. (2013). BRANCHED1 promotes axillary bud dormancy in response to shade in *Arabidopsis*. *Plant Cell* 25, 834–850. doi: 10.1105/tpc.112.108480

Guan, P., Wang, R., Nacry, P., Breton, G., Kay, S. A., Pruneda-Paz, J. L., et al. (2014). Nitrate foraging by *Arabidopsis* roots is mediated by the transcription factor TCP20 through the systemic signaling pathway. *Proc. Natl. Acad. Sci. U.S.A.* 111, 15267–15272. doi: 10.1073/pnas.1411375111

Guo, Z., Fujioka, S., Blancaflor, E. B., Miao, S., Gou, X., and Li, J. (2010). TCP1 Modulates brassinosteroid biosynthesis by regulating the expression of the key biosynthetic gene DWARF4 in *Arabidopsis thaliana*. *Plant Cell* 22, 1161–1173. doi: 10.1105/tpc.109.069203

Hao, J., Tu, L., Hu, H., Tan, J., Deng, F., Tang, W., et al. (2012). GbTCP, a cotton TCP transcription factor, confers fibre elongation and root hair development by a complex regulating system. *J. Exp. Bot.* 63, 6267–6281. doi: 10.1093/jxb/ers278

Horn, S., Pabón-Mora, N., Theuß, V. S., Busch, A., and Zachgo, S. (2015). Analysis of the CYC/TB1 class of TCP transcription factors in basal angiosperms and magnoliids. *Plant J.* 81, 559–571. doi: 10.1111/tpj.12750

Howarth, D. G., and Donoghue, M. J. (2006). Phylogenetic analysis of the "ECE" (CYC/TB1) clade reveals duplications predating the core eudicots. *Proc. Natl. Acad. Sci. U.S.A.* 103, 9101–9106. doi: 10.1073/pnas.0602827103

Howarth, D. G., Martins, T., Chimney, E., and Donoghue, M. J. (2011). Diversification of CYCLOIDEA expression in the evolution of bilateral flower symmetry in Caprifoliaceae and *Lonicera* (Dipsacales). *Ann. Bot.* 107, 1521–1532. doi: 10.1093/aob/mcr049

Hu, Z., Yamauchi, T., Yang, J., Jikumaru, Y., Tsuchida-Mayama, T., Ichikawa, H., et al. (2014). Strigolactone and cytokinin act antagonistically in regulating rice mesocotyl elongation in darkness. *Plant Cell Physiol.* 55, 30–41. doi: 10.1093/pcp/pct150

Huang, T., and Irish, V. F. (2015). Temporal control of plant organ growth by TCP transcription factors. *Curr. Biol.* 25, 1765–1770. doi: 10.1016/j.cub.2015.05.024

Janik, K., Mithöfer, A., Raffeiner, M., Stellmach, H., Hause, B., and Schlink, K. (2016). An effector of apple proliferation phytoplasma targets TCP transcription factors - a generalized virulence strategy of phytoplasma? *Mol. Plant Pathol.* doi: 10.1111/mpp.12409 [Epub ahead of print].

Juntheikki-Palovaara, I., Tähtiharju, S., Lan, T., Broholm, S. K., Rijpkema, A. S., Ruonala, R., et al. (2014). Functional diversification of duplicated CYC2 clade genes in regulation of inflorescence development in *Gerbera hybrida* (Asteraceae). *Plant J.* 79, 783–796. doi: 10.1111/tpj.12583

Kaufmann, K., Muiño, J. M., Jauregui, R., Airoldi, C. A., Smaczniak, C., Krajewski, P., et al. (2009). Target genes of the MADS transcription factor SEPALLATA3: integration of developmental and hormonal pathways in the *Arabidopsis* flower. *PLoS Biol.* 7:e1000090. doi: 10.1371/journal.pbio.1000090

Kaufmann, K., Wellmer, F., Muiño, J. M., Ferrier, T., Wuest, S. E., Kumar, V., et al. (2010). Orchestration of floral initiation by APETALA1. *Science* 328, 85–89. doi: 10.1126/science.1185244

Keeble, F., Pellew, M. C., and Jones, W. N. (1910). The inheritance Op peloria and flowercolour in foxgloves (digitalis Purpurea). *New Phytol.* 9, 68–77. doi: 10.1111/j.1469-8137.1910.tb05554.x

Kieffer, M., Master, V., Waites, R., and Davies, B. (2011). TCP14 and TCP15 affect internode length and leaf shape in *Arabidopsis*. *Plant J.* 68, 147–158. doi: 10.1111/j.1365-313X.2011.04674.x

Kim, M., Cui, M.-L., Cubas, P., Gillies, A., Lee, K., Chapman, M. A., et al. (2008). Regulatory genes control a key morphological and ecological trait transferred between species. *Science* 322, 1116–1119. doi: 10.1126/science.1164371

Kim, S. H., Son, G. H., Bhattacharjee, S., Kim, H. J., Nam, J. C., Nguyen, P. D. T., et al. (2014). The *Arabidopsis* immune adaptor SRFR1 interacts with TCP transcription factors that redundantly contribute to effector-triggered immunity. *Plant J.* 78, 978–989. doi: 10.1111/tpj.12527

Kosugi, S., and Ohashi, Y. (1997). PCF1 and PCF2 specifically bind to cis elements in the rice proliferating cell nuclear antigen gene. *Plant Cell* 9, 1607–1619. doi: 10.1105/tpc.9.9.1607

Kosugi, S., Suzuka, I., and Ohashi, Y. (1995). Two of three promoter elements identified in a rice gene for proliferating cell nuclear antigen are essential for meristematic tissue-specific expression. *Plant J.* 7, 877–886. doi: 10.1046/j.1365-313X.1995.07060877.x

Koyama, T., Furutani, M., Tasaka, M., and Ohme-Takagi, M. (2007). TCP Transcription factors control the morphology of shoot lateral organs via negative regulation of the expression of boundary-specific genes in *Arabidopsis*. *Plant Cell* 19, 473–484. doi: 10.1105/tpc.106.044792

Koyama, T., Mitsuda, N., Seki, M., Shinozaki, K., and Ohme-Takagi, M. (2010a). TCP transcription factors regulate the activities of ASYMMETRIC LEAVES1 and miR164, as Well as the auxin response, during differentiation of leaves in *Arabidopsis*. *Plant Cell* 22, 3574–3588. doi: 10.1105/tpc.110.075598

Koyama, T., Sato, F., and Ohme-Takagi, M. (2010b). A role of TCP1 in the longitudinal elongation of leaves in *Arabidopsis*. *Biosci. Biotechnol. Biochem.* 74, 2145–2147. doi: 10.1271/bbb.100442

Kumar, D., Hazra, S., Datta, R., and Chattopadhyay, S. (2016). Transcriptome analysis of *Arabidopsis* mutants suggests a crosstalk between ABA, ethylene and GSH against combined cold and osmotic stress. *Sci. Rep.* 6, 36867. doi: 10.1038/srep36867

Letunic, I., and Bork, P. (2016). Interactive tree of life (iTOL) v3: an online tool for the display and annotation of phylogenetic and other trees. *Nucleic Acids Res.* 44, W242–W245. doi: 10.1093/nar/gkw290

Li, C., Potuschak, T., Colón-Carmona, A., Gutiérrez, R. A., and Doerner, P. (2005). *Arabidopsis* TCP20 links regulation of growth and cell division control pathways. *Proc. Natl. Acad. Sci. U.S.A.* 102, 12978–12983. doi: 10.1073/pnas.0504039102

Linnaeus, C., and Rudberg, D. (1744). *Dissertatio Botanica de Peloria*. Uppsala: Q-Med AB.

Liu, J., Perumal, N. B., Oldfield, C. J., Su, E. W., Uversky, V. N., and Dunker, A. K. (2006). Intrinsic disorder in transcription factors. *Biochemistry (Mosc.)* 45, 6873–6888. doi: 10.1021/bi0602718

Lucero, L. E., Uberti-Manassero, N. G., Arce, A. L., Colombatti, F., Alemano, S. G., and Gonzalez, D. H. (2015). TCP15 modulates cytokinin and auxin responses during gynoecium development in *Arabidopsis*. *Plant J.* 84, 267–282. doi: 10.1111/tpj.12992

Luo, D., Carpenter, R., Vincent, C., Copsey, L., and Coen, E. (1996). Origin of floral asymmetry in *Antirrhinum*. *Nature* 383, 794–799. doi: 10.1038/383794a0

Lupas, A., Dyke, M. V., and Stock, J. (1991). Predicting coiled coils from protein sequences. *Science* 252, 1162–1164. doi: 10.1126/science.252.5009.1162

Ma, X., Ma, J., Fan, D., Li, C., Jiang, Y., and Luo, K. (2016). Genome-wide Identification of TCP family transcription factors from populus euphratica and their involvement in leaf shape regulation. *Sci. Rep.* 6:32795. doi: 10.1038/srep32795

Mao, Y., Wu, F., Yu, X., Bai, J., Zhong, W., and He, Y. (2014). microRNA319a-targeted *Brassica rapa* ssp. pekinensis TCP genes modulate head shape in chinese cabbage by differential cell division arrest in leaf regions. *Plant Physiol.* 164, 710–720. doi: 10.1104/pp.113.228007

Mimida, N., Kidou, S.-I., Iwanami, H., Moriya, S., Abe, K., Voogd, C., et al. (2011). Apple FLOWERING LOCUS T proteins interact with transcription factors implicated in cell growth and organ development. *Tree Physiol.* 31, 555–566. doi: 10.1093/treephys/tpr028

Mizuno, S., Sonoda, M., Tamura, Y., Nishino, E., Suzuki, H., Sato, T., et al. (2015). *Chiba Tendril-Less* locus determines tendril organ identity in melon (*Cucumis melo* L.) and potentially encodes a tendril-specific TCP homolog. *J. Plant Res.* 128, 941–951. doi: 10.1007/s10265-015-0747-2

Muhr, M., Prüfer, N., Paulat, M., and Teichmann, T. (2016). Knockdown of strigolactone biosynthesis genes in Populus affects BRANCHED1 expression and shoot architecture. *New Phytol.* 212, 613–626. doi: 10.1111/nph.14076

Mukhopadhyay, P., Tyagi, A. K., and Tyagi, A. K. (2015). OsTCP19 influences developmental and abiotic stress signaling by modulating ABI4-mediated pathways. *Sci. Rep.* 5:9998. doi: 10.1038/srep09998

Mukhtar, M. S., Carvunis, A.-R., Dreze, M., Epple, P., Steinbrenner, J., Moore, J., et al. (2011). Independently evolved virulence effectors converge onto hubs in a plant immune system network. *Science* 333, 596–601. doi: 10.1126/science.1203659

Nag, A., King, S., and Jack, T. (2009). miR319a targeting of TCP4 is critical for petal growth and development in *Arabidopsis*. *Proc. Natl. Acad. Sci. U.S.A.* 106, 22534–22539. doi: 10.1073/pnas.0908718106

Nath, U., Crawford, B. C. W., Carpenter, R., and Coen, E. (2003). Genetic control of surface curvature. *Science* 299, 1404–1407. doi: 10.1126/science.1079354

Navaud, O., Dabos, P., Carnus, E., Tremousaygue, D., and Hervé, C. (2007). TCP transcription factors predate the emergence of land plants. *J. Mol. Evol.* 65, 23–33. doi: 10.1007/s00239-006-0174-z

Nicolas, M., and Cubas, P. (2016). TCP factors: new kids on the signaling block. *Curr. Opin. Plant Biol.* 33, 33–41. doi: 10.1016/j.pbi.2016.05.006

Nicolas, M., Rodríguez-Buey, M. L., Franco-Zorrilla, J. M., and Cubas, P. (2015). A recently evolved alternative splice site in the BRANCHED1a gene controls potato plant architecture. *Curr. Biol.* 25, 1799–1809. doi: 10.1016/j.cub.2015.05.053

Niwa, M., Daimon, Y., Kurotani, K., Higo, A., Pruneda-Paz, J. L., Breton, G., et al. (2013). BRANCHED1 interacts with FLOWERING LOCUS T to repress the floral transition of the axillary meristems in *Arabidopsis*. *Plant Cell* 25, 1228–1242. doi: 10.1105/tpc.112.109090

Ori, N., Cohen, A. R., Etzioni, A., Brand, A., Yanai, O., Shleizer, S., et al. (2007). Regulation of LANCEOLATE by miR319 is required for compound-leaf development in tomato. *Nat. Genet.* 39, 787–791. doi: 10.1038/ng2036

Ortiz-Ramírez, C., Hernandez-Coronado, M., Thamm, A., Catarino, B., Wang, M., Dolan, L., et al. (2016). A transcriptome atlas of physcomitrella patens provides insights into the evolution and development of land plants. *Mol. Plant.* 9, 205–220. doi: 10.1016/j.molp.2015.12.002

Palatnik, J. F., Allen, E., Wu, X., Schommer, C., Schwab, R., Carrington, J. C., et al. (2003). Control of leaf morphogenesis by microRNAs. *Nature* 425, 257–263. doi: 10.1038/nature01958

Poza-Carrión, C., Aguilar-Martínez, J. A., and Cubas, P. (2007). Role of TCP Gene BRANCHED1 in the control of shoot branching in *Arabidopsis*. *Plant Signal. Behav.* 2, 551–552. doi: 10.4161/psb.2.6.4811

Pruneda-Paz, J. L., Breton, G., Nagel, D. H., Kang, S. E., Bonaldi, K., Doherty, C. J., et al. (2014). A genome-scale resource for the functional characterization of *Arabidopsis* transcription factors. *Cell Rep.* 8, 622–632. doi: 10.1016/j.celrep.2014.06.033

Pruneda-Paz, J. L., Breton, G., Para, A., and Kay, S. A. (2009). A functional genomics approach reveals CHE as a component of the *Arabidopsis* circadian clock. *Science* 323, 1481–1485. doi: 10.1126/science.1167206

Sarvepalli, K., and Nath, U. (2011). Interaction of TCP4-mediated growth module with phytohormones. *Plant Signal. Behav.* 6, 1440–1443. doi: 10.4161/psb.6.10.17097

Schommer, C., Palatnik, J. F., Aggarwal, P., Chételat, A., Cubas, P., Farmer, E. E., et al. (2008). Control of jasmonate biosynthesis and senescence by miR319 targets. *PLoS Biol.* 6:e230. doi: 10.1371/journal.pbio.0060230

Shleizer-Burko, S., Burko, Y., Ben-Herzel, O., and Ori, N. (2011). Dynamic growth program regulated by LANCEOLATE enables flexible leaf patterning. *Development* 138, 695–704. doi: 10.1242/dev.056770

Smith, S. J. E. (1821). *A Selection of the Correspondence of Linnaeus, and other Naturalists: From the Original Manuscripts.* Harlow: Longman.

Steiner, E., Efroni, I., Gopalraj, M., Saathoff, K., Tseng, T.-S., Kieffer, M., et al. (2012). The *Arabidopsis* O-Linked N-Acetylglucosamine transferase SPINDLY interacts with class I TCPs to facilitate cytokinin responses in leaves and flowers. *Plant Cell* 24, 96–108. doi: 10.1105/tpc.111.093518

Sugio, A., Kingdom, H. N., MacLean, A. M., Grieve, V. M., and Hogenhout, S. A. (2011). Phytoplasma protein effector SAP11 enhances insect vector reproduction by manipulating plant development and defense hormone biosynthesis. *Proc. Natl. Acad. Sci. U.S.A.* 108, E1254–E1263. doi: 10.1073/pnas.1105664108

Sugio, A., MacLean, A. M., and Hogenhout, S. A. (2014). The small phytoplasma virulence effector SAP11 contains distinct domains required for nuclear targeting and CIN-TCP binding and destabilization. *New Phytol.* 202, 838–848. doi: 10.1111/nph.12721

Sun, J., Qi, L., Li, Y., Chu, J., and Li, C. (2012). PIF4–mediated activation of YUCCA8 expression integrates temperature into the auxin pathway in regulating *Arabidopsis* hypocotyl growth. *PLOS Genet* 8:e1002594. doi: 10.1371/journal.pgen.1002594

Świątek, A., Lenjou, M., Van Bockstaele, D., Inzé, D., and Van Onckelen, H. (2002). Differential effect of jasmonic acid and abscisic acid on cell cycle progression in tobacco BY-2 cells. *Plant Physiol.* 128, 201–211. doi: 10.1104/pp.010592

Tähtiharju, S., Rijpkema, A. S., Vetterli, A., Albert, V. A., Teeri, T. H., and Elomaa, P. (2012). Evolution and diversification of the CYC/TB1 gene family in asteraceae—a comparative study in *Gerbera* (Mutisieae) and sunflower (Heliantheae). *Mol. Biol. Evol.* 29, 1155–1166. doi: 10.1093/molbev/msr283

Takeda, T., Suwa, Y., Suzuki, M., Kitano, H., Ueguchi-Tanaka, M., Ashikari, M., et al. (2003). The OsTB1 gene negatively regulates lateral branching in rice. *Plant J.* 33, 513–520. doi: 10.1046/j.1365-313X.2003.01648.x

Tao, Q., Guo, D., Wei, B., Zhang, F., Pang, C., Jiang, H., et al. (2013). The TIE1 transcriptional repressor links TCP transcription factors with TOPLESS/TOPLESS-RELATED corepressors and modulates leaf development in *Arabidopsis*. *Plant Cell* 25, 421–437. doi: 10.1105/tpc.113.109223

Tatematsu, K., Nakabayashi, K., Kamiya, Y., and Nambara, E. (2008). Transcription factor AtTCP14 regulates embryonic growth potential during seed germination in *Arabidopsis thaliana*. *Plant J.* 53, 42–52. doi: 10.1111/j.1365-313X.2007.03308.x

Thieulin-Pardo, G., Avilan, L., Kojadinovic, M., and Gontero, B. (2015). Fairy "tails": flexibility and function of intrinsically disordered extensions in the photosynthetic world. *Front. Mol. Biosci.* 2:23. doi: 10.3389/fmolb.2015.00023

Trémousaygue, D., Garnier, L., Bardet, C., Dabos, P., Hervé, C., and Lescure, B. (2003). Internal telomeric repeats and "TCP domain" protein-binding sites co-operate to regulate gene expression in *Arabidopsis thaliana* cycling cells. *Plant J.* 33, 957–966. doi: 10.1046/j.1365-313X.2003.01682.x

Uberti-Manassero, N. G., Lucero, L. E., Viola, I. L., Vegetti, A. C., and Gonzalez, D. H. (2012). The class I protein AtTCP15 modulates plant development through a pathway that overlaps with the one affected by CIN-like TCP proteins. *J. Exp. Bot.* 63, 809–823. doi: 10.1093/jxb/err305

Valsecchi, I., Guittard-Crilat, E., Maldiney, R., Habricot, Y., Lignon, S., Lebrun, R., et al. (2013). The intrinsically disordered C-terminal region of *Arabidopsis thaliana* TCP8 transcription factor acts both as a transactivation and self-assembly domain. *Mol. Biosyst.* 9, 2282–2295. doi: 10.1039/c3mb70128j

Viola, I. L., Camoirano, A., and Gonzalez, D. H. (2016). Redox-dependent modulation of anthocyanin biosynthesis by the TCP transcription factor TCP15 during exposure to high light intensity conditions in *Arabidopsis*. *Plant Physiol.* 170, 74–85. doi: 10.1104/pp.15.01016

Viola, I. L., Güttlein, L. N., and Gonzalez, D. H. (2013). Redox modulation of plant developmental regulators from the class I TCP transcription factor family. *Plant Physiol.* 162, 1434–1447. doi: 10.1104/pp.113.216416

Viola, I. L., Reinheimer, R., Ripoll, R., Manassero, N. G. U., and Gonzalez, D. H. (2012). Determinants of the DNA binding specificity of class I and Class II TCP transcription factors. *J. Biol. Chem.* 287, 347–356. doi: 10.1074/jbc.M111.256271

Viola, I. L., Uberti Manassero, N. G., Ripoll, R., and Gonzalez, D. H. (2011). The *Arabidopsis* class I TCP transcription factor AtTCP11 is a developmental regulator with distinct DNA-binding properties due to the presence of a threonine residue at position 15 of the TCP domain. *Biochem. J.* 435, 143–155. doi: 10.1042/BJ20101019

Wang, R.-L., Stec, A., Hey, J., Lukens, L., and Doebley, J. (1999). The limits of selection during maize domestication. *Nature* 398, 236–239. doi: 10.1038/18435

Wang, S., Yang, X., Xu, M., Lin, X., Lin, T., Qi, J., et al. (2015). A rare SNP identified a TCP transcription factor essential for tendril development in cucumber. *Mol. Plant* 8, 1795–1808. doi: 10.1016/j.molp.2015.10.005

Wang, X., Gao, J., Zhu, Z., Dong, X., Wang, X., Ren, G., et al. (2015). TCP transcription factors are critical for the coordinated regulation of ISOCHORISMATE SYNTHASE 1 expression in *Arabidopsis thaliana*. *Plant J.* 82, 151–162. doi: 10.1111/tpj.12803

Wang, Z., Luo, Y., Li, X., Wang, L., Xu, S., Yang, J., et al. (2008). Genetic control of floral zygomorphy in pea (*Pisum sativum* L.). *Proc. Natl. Acad. Sci. U.S.A.* 105, 10414–10419. doi: 10.1073/pnas.0803291105

Weir, I., Lu, J., Cook, H., Causier, B., Schwarz-Sommer, Z., and Davies, B. (2004). CUPULIFORMIS establishes lateral organ boundaries in *Antirrhinum*. *Development* 131, 915–922. doi: 10.1242/dev.00993

Welchen, E., and Gonzalez, D. H. (2006). Overrepresentation of elements recognized by TCP-domain transcription factors in the upstream regions of nuclear genes encoding components of the mitochondrial oxidative phosphorylation machinery. *Plant Physiol.* 141, 540–545. doi: 10.1104/pp.105.075366

Wu, J.-F., Tsai, H.-L., Joanito, I., Wu, Y.-C., Chang, C.-W., Li, Y.-H., et al. (2016). LWD–TCP complex activates the morning gene CCA1 in *Arabidopsis*. *Nat. Commun.* 7:13181. doi: 10.1038/ncomms13181

Yanai, O., Shani, E., Russ, D., and Ori, N. (2011). Gibberellin partly mediates LANCEOLATE activity in tomato. *Plant J.* 68, 571–582. doi: 10.1111/j.1365-313X.2011.04716.x

Yang, X., Zhao, X.-G., Li, C.-Q., Liu, J., Qiu, Z.-J., Dong, Y., et al. (2015). Distinct regulatory changes underlying differential expression of

TEOSINTE BRANCHED1-CYCLOIDEA-PROLIFERATING CELL FACTOR genes associated with petal variations in zygomorphic flowers of *Petrocosmea* spp. of the family Gesneriaceae. *Plant Physiol.* 169, 2138–2151. doi: 10.1104/pp.15.01181

Yuan, Z., Gao, S., Xue, D.-W., Luo, D., Li, L.-T., Ding, S.-Y., et al. (2009). RETARDED PALEA1 controls palea development and floral zygomorphy in rice. *Plant Physiol.* 149, 235–244. doi: 10.1104/pp.108.128231

Zheng, X., Zhou, M., Yoo, H., Pruneda-Paz, J. L., Spivey, N. W., Kay, S. A., et al. (2015). Spatial and temporal regulation of biosynthesis of the plant immune signal salicylic acid. *Proc. Natl. Acad. Sci. U.S.A.* 112, 9166–9173. doi: 10.1073/pnas.1511182112

Zhou, M., Li, D., Li, Z., Hu, Q., Yang, C., Zhu, L., et al. (2013). Constitutive expression of a miR319 gene alters plant development and enhances salt and drought tolerance in transgenic creeping bentgrass. *Plant Physiol.* 161, 1375–1391. doi: 10.1104/pp.112.208702

Conflict of Interest Statement: The author declares that the research was conducted in the absence of any commercial or financial relationships that could be construed as a potential conflict of interest.

9

CRP1 Protein: (Dis)similarities between *Arabidopsis thaliana* and *Zea mays*

Roberto Ferrari[1†], Luca Tadini[1†], Fabio Moratti[2†], Marie-Kristin Lehniger[3], Alex Costa[1], Fabio Rossi[4], Monica Colombo[5], Simona Masiero[1], Christian Schmitz-Linneweber[3] and Paolo Pesaresi[6*]

[1] Dipartimento di Bioscienze, Università degli studi di Milano, Milano, Italy, [2] Max-Planck-Institut für Molekulare Pflanzenphysiologie, Potsdam-Golm, Germany, [3] Molecular Genetics, Institute of Biology, Humboldt University of Berlin, Berlin, Germany, [4] Dipartimento di Biotecnologie Mediche e Medicina Traslazionale, Università degli studi di Milano, Milano, Italy, [5] Centro Ricerca e Innovazione, Fondazione Edmund Mach, San Michele all'Adige, Italy, [6] Dipartimento di Scienze Agrarie e Ambientali - Produzione, Territorio, Agroenergia, Università degli studi di Milano, Milano, Italy

Edited by:
Federico Valverde,
Consejo Superior de Investigaciones Científicas (CSIC), Spain

Reviewed by:
Jean-David Rochaix,
University of Geneva, Switzerland
Alexandra-Viola Bohne,
Ludwig Maximilian University of Munich, Germany

***Correspondence:**
Paolo Pesaresi
paolo.pesaresi@unimi.it

† These authors have contributed equally to this work.

Biogenesis of chloroplasts in higher plants is initiated from proplastids, and involves a series of processes by which a plastid able to perform photosynthesis, to synthesize amino acids, lipids, and phytohormones is formed. All plastid protein complexes are composed of subunits encoded by the nucleus and chloroplast genomes, which require a coordinated gene expression to produce the correct concentrations of organellar proteins and to maintain organelle function. To achieve this, hundreds of nucleus-encoded factors are imported into the chloroplast to control plastid gene expression. Among these factors, members of the Pentatricopeptide Repeat (PPR) containing protein family have emerged as key regulators of the organellar post–transcriptional processing. PPR proteins represent a large family in plants, and the extent to which PPR functions are conserved between dicots and monocots deserves evaluation, in light of differences in photosynthetic metabolism (C3 vs. C4) and localization of chloroplast biogenesis (mesophyll vs. bundle sheath cells). In this work we investigated the role played in the process of chloroplast biogenesis by At5g42310, a member of the Arabidopsis PPR family which we here refer to as *At*CRP1 (Chloroplast RNA Processing 1), providing a comparison with the orthologous *Zm*CRP1 protein from *Zea mays*. Loss-of-function *atcrp1* mutants are characterized by yellow-albinotic cotyledons and leaves owing to defects in the accumulation of subunits of the thylakoid protein complexes. As in the case of *Zm*CRP1, *At*CRP1 associates with the 5′ UTRs of both *psaC* and, albeit very weakly, *petA* transcripts, indicating that the role of CRP1 as regulator of chloroplast protein synthesis has been conserved between maize and Arabidopsis. *At*CRP1 also interacts with the *petB-petD* intergenic region and is required for the generation of *petB* and *petD* monocistronic RNAs. A similar role has been also attributed to *Zm*CRP1, although the direct interaction of *Zm*CRP1 with the *petB-petD* intergenic region has never been reported, which could indicate that *At*CRP1 and *Zm*CRP1 differ, in part, in their plastid RNA targets.

Keywords: PPR, anterograde signaling, chloroplast, biogenesis, RNA metabolism

INTRODUCTION

In land-plants, nuclear-encoded pentatricopeptide repeat (PPR) containing proteins constitute a large family, which regulates organelle gene expression at the RNA level (Lurin et al., 2004; O'Toole et al., 2008; Barkan and Small, 2014). They are, indeed, a major constituent of the genome-coordinating anterograde signaling pathway that evolved to adapt the expression of the organellar genomes in response to endogenous and environmental stimuli that are perceived by the nucleus (Woodson and Chory, 2008).

A typical PPR motif is characterized by a degenerate 35-amino acid repeat that folds into two antiparallel alpha helices (Small and Peeters, 2000). PPR proteins contain a tandem array of 2–30 PPR motifs, which stack together to form a superhelix with a central groove that allows the protein to bind RNA (Lurin et al., 2004; Rivals et al., 2006). According to the characteristics of their repeats, PPR proteins are generally classified into P and PLS sub-families. The P-type proteins are implicated in the determination and stabilization of 5′ and/or 3′ RNA termini, RNA splicing and translation of specific RNAs in chloroplasts and mitochondria, while PLS-type proteins are generally involved in RNA editing (Barkan and Small, 2014). Higher plants harbor several hundreds of PPR proteins, which generally have distinct, non-redundant functions in organelle biogenesis, plant growth and development and adaptation to environmental cues (Barkan and Small, 2014; Manna, 2015), as revealed by the high number of *ppr* mutants with distinct phenotypes. This is due to their ability to recognize primary RNA sequences, with each protein having different target sites, thus implying that the elucidation of the primary role of each PPR protein is greatly facilitated by the identification of its RNA targets.

The detection of few native PPR-RNA interactions through RNA immunoprecipitation on microarray (RIP-Chip) analyses and *in vitro* binding assays using PPR recombinant proteins, together with PPR crystal structures indicate that PPR proteins bind their cognate RNA targets in a sequence specific manner (Meierhoff et al., 2003; Schmitz-Linneweber et al., 2005, 2006; Williams-Carrier et al., 2008; Yin et al., 2013; Okuda et al., 2014; Shen et al., 2016). The code describing how PPR proteins recognize specific nucleotides of their RNA targets relies primarily on two amino acids that are within a single PPR motif, specifically the fifth residue in the first helix and the last residue on the loop interconnecting adjacent motifs (Barkan et al., 2012; Yin et al., 2013; Cheng et al., 2016). However, the current understanding of the code does not allow accurate large-scale computational predictions of PPR targets (Takenaka et al., 2013; Kindgren et al., 2015; Hall, 2016; Harrison et al., 2016). Predictive power is constrained by the fact that the code is degenerate and by the low accuracy of current methods used for the identification of PPR domains, which in turn leads to mismatches in the amino acid/nucleotide alignments. However, a more robust annotation of PPR domains has recently been conducted and made available at the PlantPPR database[1] (Cheng et al., 2016). Furthermore, more PPR-RNA interactions as well as crystal structures of PPR-RNA complexes need to be characterized in different species in order to improve the understanding of the code. This would also help to determine if the amino acid sequences of the PPR domains coevolved with the nucleotide sequences of their RNA targets and ultimately to determine whether there is functional conservation of PPR proteins among land plants.

The function of PPR proteins, and more generally the function of the nuclear gene complement involved in organellar RNA metabolism, have been primarily studied in maize, since the large seed reserves of maize support rapid heterotrophic growth of non-photosynthetic mutants and provide ready access to non-photosynthetic tissues for molecular biology and biochemical studies (Belcher et al., 2015). However, the degree of functional conservation of PPR proteins between maize and other species, including *Arabidopsis thaliana*, has yet to be investigated. The question is of particular interest since the elaboration of the thylakoid membrane system and the biogenesis of the multi-subunit photosynthetic complexes appear to have major differences between monocotyledonous and dicotyledonous plants (Pogson et al., 2015). Indeed in maize, and more generally in monocots, the process of chloroplast development from the proplastid to functional chloroplasts can be observed as a gradient along the leaf blade, whereas in dicots, such as *Arabidopsis thaliana*, the development of chloroplasts differs between developmental stages, plant organs – i.e., chloroplast development is different in cotyledons and leaves – and plant tissues (Pogson and Albrecht, 2011; Jarvis and Lopez-Juez, 2013).

The majority of PPR proteins are conserved at sequence level between dicots (Arabidopsis) and monocots (rice) (O'Toole et al., 2008). Orthologous pairs can readily be identified and in a number of cases, primary sequence conservation can be traced back to the roots of all embryophytes (O'Toole et al., 2008). As a matter of fact, functional differences between orthologous PPR proteins of maize and Arabidopsis have been observed. For example, the molecular phenotypes resulting from loss of the orthologous PPR proteins ATP4 (maize) and SVR7 (Arabidopsis) differ substantially (Liu et al., 2010; Zoschke et al., 2012, 2013a,b), as do the molecular defects in maize and Arabidopsis mutants lacking the PGR3 protein (Yamazaki et al., 2004; Cai et al., 2011; Belcher et al., 2015). Thus, the extent to which lessons on PPR proteins learnt from maize can be extrapolated to dicots, such as Arabidopsis, and more broadly to other organisms, needs further investigation.

In this context, we investigated here the function of and identified the RNA targets of the PPR protein At5g42310 from *Arabidopsis thaliana*, that shares high similarity with the well-characterized CRP1 (Chloroplast RNA Processing 1) protein from maize (*Zm*CRP1), and which we here refer to as *At*CRP1. Our findings indicate that *At*CRP1, like the orthologous *Zm*CRP1 (Barkan et al., 1994; Fisk et al., 1999; Schmitz-Linneweber et al., 2005), is essential for plant autotrophy since it plays a direct role in the accumulation of the cytochrome b_6/f (Cyt b_6/f) complex and of the PsaC subunit

[1] http://www.plantppr.com

of photosystem I (PSI). Furthermore *At*CRP1, similarly to *Zm*CRP1, is required for the accumulation of *petB* and *petD* monocistronic RNAs, indicating that the functional roles of CRP1 proteins are highly conserved between monocots and dicots.

MATERIALS AND METHODS

Plant Material and Growth Conditions

Arabidopsis thaliana atcrp1-1 (SALK_035048) (Alonso et al., 2003) and *atcrp1-2* (SAIL_916A02) (Sessions et al., 2002) T-DNA insertion lines were identified by searching the T-DNA Express database[2]. For promoter analyses, the putative *AtCRP1* promoter region (*AtCRP1p*, −1062 to −2 upstream the translation starting codon) was cloned into pBGWFS7 destination vector and introduced into Arabidopsis wild type background, ecotype Columbia-0 (Col-0), by *Agrobacterium tumefaciens*-mediated transformation. *At*CRP1-GFP transgenic lines were obtained by transformation of *AtCRP1/atcrp1-1* heterozygous plants with either the *At*CRP1 coding sequence fused to GFP under the control of *35S-CaMV* promoter, cloned into pB7FWG2 vector, or the genomic locus fused to GFP under the control of the native promoter, cloned into a modified pGreenII vector (Gregis et al., 2009). The GUN1 coding sequence, devoid of the stop codon, was cloned into pB7RWG2 vector, carrying an RFP reporter gene. pB7FWG2, pBGWFS7, and pB7RWG2 plasmids were obtained from Flanders Interuniversity Institute for Biotechnology of Gent (Karimi et al., 2002). Primers used for amplification of the DNA fragments cloned into the vectors, reported above, are listed in **Supplementary Table S2**. Arabidopsis Col-0 and mutant plants were grown on soil under controlled growth chamber conditions with a 16 h light/8 h dark cycle at 22°C/18°C. In the case of mesophyll protoplast preparation, Arabidopsis plants were also grown on soil in a growth chamber under the above reported conditions. Moreover, phenotypic characterization and molecular biology analyses were also conducted on plants grown on Murashige and Skoog (MS) medium (Duchefa)[3], supplemented with or without 1% (w/v) sucrose. Tobacco plants, employed for transient gene expression, were cultivated for 5–6 weeks in a greenhouse under a 12 h light/12 h dark cycle at 22°C/18°C.

Protoplast Transformation

Mesophyll protoplasts of *Arabidopsis thaliana* (Col-0) were isolated and transiently transformed according to Yoo et al. (2007) and Costa et al. (2012). Briefly, well-expanded rosette leaves from 3-to-5 week-old plants were cut into strips of 0.5–1 mm with a fresh razor blade. Leaf tissue was digested using an enzyme solution containing 1.25% cellulase Onozuka R-10 (Duchefa) and 0.3% Macerozyme R-10 (Duchefa) for 3 h at 23°C in the dark. The protoplast suspension was filtered through a 50 μm nylon mesh washed three times with W5 solution (154 mM NaCl, 125 mM $CaCl_2$, 5 mM KCl, 2 mM MES, pH 5.7 adjusted with KOH) and used for PEG-mediated transformation. For each protoplast transformation 10 μg of a MidiPrep purified DNA (QIAGEN) plasmid harboring the *35S-CaMV::AtCRP1-GFP* cassette was used. Protoplasts were maintained for 16–24 h at 23°C in the dark, before performing epifluorescent microscopy.

Transient Expression in *Nicotiana benthamiana* Leaves

Tobacco leaf infiltration was performed using *A. tumefaciens* strain GV3101/pMP90 carrying the specified constructs (see results for details) together with the p19-enhanced expression system (Voinnet et al., 2003), according to the method described by Waadt and Kudla (2008). The final OD_{600} for *A. tumefaciens* strains harboring *35S-CaMV::AtCRP1-GFP* and *35S-CaMV::GUN1-RFP* was 0.2 and 0.3, respectively. After infiltration, plants were incubated for 3–5 days under the conditions described above.

Confocal Microscopy Analysis

Confocal Scanning Laser Microscopy analyses were performed using an inverted microscope, Leica DMIRE2, equipped with a Leica TCS SP2 laser scanning device (Leica). For the simultaneous detection of GFP and chlorophyll auto-fluorescence the cells were excited (Arabidopsis mesophyll protoplasts or tobacco leaf cells) with the 488 nm line of the Argon laser and the emissions were collected between 515/535 and 650/750 nm, respectively. For RFP detection the cells were excited at 561 nm from a He/Ne laser and the emission was collected between 575/625 nm. Image analyses were performed with Fiji[4]: an open-source platform for biological-image analysis (Schindelin et al., 2012).

Nucleic Acid Analyses

Arabidopsis DNA was isolated according to Ihnatowicz et al. (2004). Isolation of total RNA from homozygous *atcrp1-1* plants at four-leaf rosette stage and RNA gel blot analyses were performed as described by Meurer et al. (2002), using 10 μg of total RNA for each sample. For the RNA slot blot hybridization experiments, one-fourth of the RNA purified from each immunoprecipitation pellet and one-tenth of the RNA purified from the corresponding supernatant were applied to a nylon membrane with a slot-blot manifold and hybridized to specific radiolabeled probes (see **Supplementary Table S2**). ^{32}P-labeled DNA probes, complementary to chloroplast genes, were amplified using the primer pairs listed in **Supplementary Table S2**. Four micrograms of total RNA, treated with TURBO DNA-*free* (Ambion by Life Technologies), were employed for first-strand cDNA synthesis using GoScript Reverse Transcription System (Promega) according to the supplier's instructions. Quantitative Real-Time PCR (qRT-PCR) was carried out on an CFX96 Real-Time system (Bio-Rad), using the primer pairs

[2]http://signal.salk.edu/cgi-bin/tdnaexpress

[3]http://www.duchefa.com

[4]https://fiji.sc/

reported in **Supplementary Table S2**. The *SAND* (Remans et al., 2008) and *ubiquitin* transcripts were used as internal references. Data from three biological and three technical replicates were analyzed with Bio-Rad CFX Manager software (V3.1).

Immunoblot Analyses

For immunoblot analyses, total proteins were prepared as described by Martinez-Garcia et al. (1999). Total proteins, corresponding to 5 mg of leaf fresh-weight (100% of WT and *atcrp1-1* samples) and isolated from plants at four-leaf rosette stage, were fractionated by SDS–PAGE (12% acrylamide [w/v]; (Schagger and von Jagow, 1987). Proteins were then transferred to polyvinylidene difluoride (PVDF) membranes (Ihnatowicz et al., 2004) and replicate filters were immunodecorated with antibodies specific for PSI (PsaA, PsaC, and PsaD), PSII (D1, PsbO) Cyt b_6/f (PetA, PetB, and PetC), ATPase (ATPase-β) subunits, PSI (Lhca1, Lhca2) and PSII (Lhcb2, Lhcb3) antenna proteins, all obtained from Agrisera[5]. The GFP antibody was purchased from Life Technologies[6].

Chloroplast Stromal Preparation and Protein Immunoprecipitation

Intact chloroplasts were isolated from 11 days old Arabidopsis plants, according to Kunst (1998), and Kupsch et al. (2012) with some modifications. Chloroplasts were directly resuspended in 300–400 μl of extraction buffer [2 mM DTT, 30 mM HEPES-KOH, pH 8.0, 60 mM KOAc, 10 mM MgOAc and proteinase inhibitor cocktail (Sigma–Aldrich-P9599)]. Two independent stromal preparations were carried out and one of them was performed in the presence of 2% sodium deoxycholate in order to solubilize the membrane-attached *At*CRP1 protein fraction. Chloroplasts were then disrupted by pulling them through a syringe (0.55 mm × 40 mm) 30–40 times. The solution was centrifuged at 21,000 × *g* at 4°C to separate the stromal from the membrane fraction.

The isolated stromal fraction was diluted with one volume of coimmunoprecipitation (CoIP) buffer (150 mM NaCl, 20 mM Tris-HCl pH 7.5, 2 mM $MgCl_2$, 0.5% Nonidet P-40 and 0.5 μg/mL Aprotinin). Five microliters of mouse anti-GFP antibody (Roche, No. 11814460001) were added to the stromal fraction and incubated for 1 h at 4°C and 13 rpm on an overhead shaker. Thereafter the coimmunoprecipitation was performed as described by Kupsch et al. (2012). Successful precipitation of *At*CRP1-GFP was confirmed by immunoblot analyses, using the same GFP antibody.

RNA Extraction and Labeling for RIP-Chip Assay

RNA immunoprecipitation-chip analyses were performed using a tiling microarray covering the complete Arabidopsis chloroplast genome (Kupsch et al., 2012). The coimmunoprecipitated RNA was isolated from pellet and supernatant fractions either by phenol-chloroform extraction or using the Direct-zol™ RNA MiniPrep kit (Zymo Research). For the phenol-chloroform extraction, RNA samples were incubated in 1% SDS and 5 mM EDTA at room temperature for 5 min to dissociate RNA-protein complexes. RNA was phenol-chloroform extracted, ethanol precipitated with the addition of Glycoblue™ Coprecipitant (Thermo Fisher Scientific), washed twice with 75% ethanol, air-dried and resuspended in 20 μl RNase-free water. For the replicate, RNA was extracted using the Direct-zol™ RNA MiniPrep kit (Zymo Research) according to the manufacturer's instructions. Before the extraction 2 μg yeast RNA was added to the coimmunoprecipitated RNA pellet. The entire RNA of the pellet fraction and 2 μg RNA of the supernatant fraction were used for labeling. The pellet and supernatant RNA were labeled with 0.5 μl Cy5 and 1 μl Cy3 dye, respectively (aRNA labeling kit, Kreatech Diagnostics). Labeling reaction, microarray hybridization, scanning, and evaluation were performed as described in Kupsch et al. (2012). Only PCR products for which more than half of all replicate spots (24 per PCR product spanning two experiments) passed our quality assessment (Kupsch et al., 2012) and were used in this analysis (**Supplementary Table S1**).

In silico Prediction of AtCRP1 Binding Sites

The putative *At*CRP1 binding motif, i.e., the nucleotide preference for each of the amino acid pairs at the fifth and last position of PPR domains, was predicted *in silico* using the reported weighting schemes (Barkan et al., 2012; Barkan and Small, 2014; Harrison et al., 2016). The software FIMO[7], which analyzes sequence databases for occurrences of known motifs (Grant et al., 2011), was employed to identify the potential binding sites of *At*CRP1 within the regions enriched in our RIP-Chip experiment. Furthermore, the same regions were searched for the presence of sRNA native footprints, by consulting the JBrowse sRNA database[8] (Ruwe et al., 2016). Numbers that delimit the native footprints refer to the chloroplast genome of *Arabidopsis thaliana* (NC_000932.1).

β-Glucuronidase (GUS) Assay

For GUS histochemical detection, plant material was fixed in 90% acetone at −20°C for 1 h. Samples were then washed three times with NaPi buffer (NaH_2PO_4 50 mM, Na_2HPO_4 50 mM; pH 7.0) and stained overnight at 37°C with *X*-gluc solution [1 mM 5-bromo-4-chloro-3-indolyl-β-D-clucoronide, 2 mM $K_3/K_4Fe(CN)_6$, 0,1% Triton (v/v), 10 mM EDTA, 50 mM NaPi pH 7.0]. 70% EtOH (v/v) was used as washing solution. Stained samples were then stored at 4°C and observed using a Zeiss Axiophot D1 microscope equipped with differential interference contrast (DIC) optics. Images were recorded with an Axiocam MRc5 camera (Zeiss) using the Axiovision program (v.4.1).

[5]http://www.agrisera.com/en/artiklar/plantalgal-cell-biology/index.html

[6]http://www.thermofisher.com

[7]http://meme-suite.org/tools/fimo

[8]https://www.molgen.hu-berlin.de/projects-jbrowse-athaliana.php

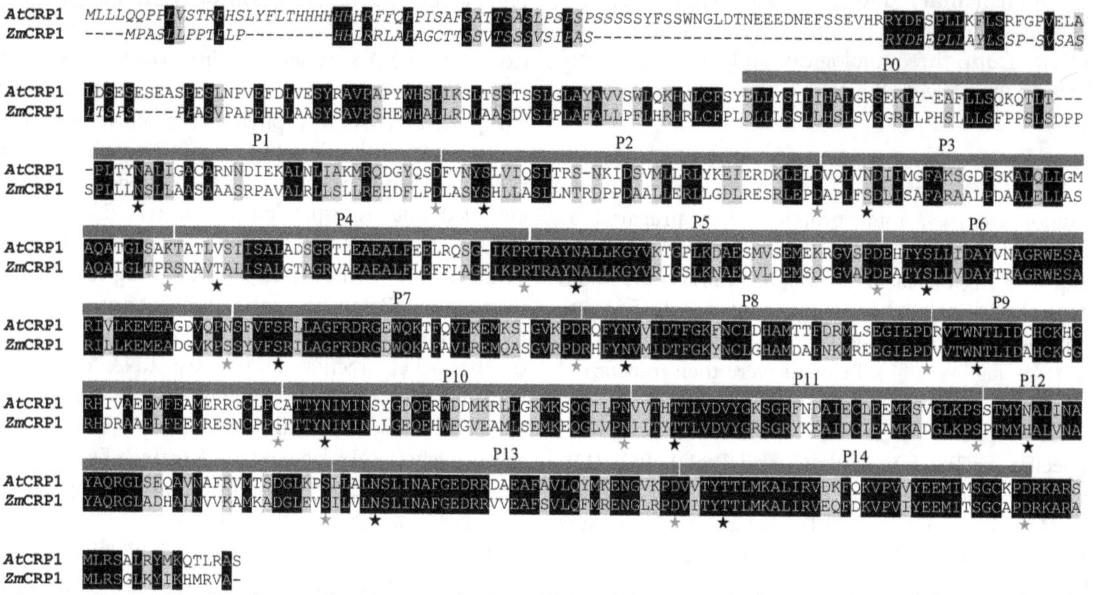

FIGURE 1 | Primary amino acid sequence alignment of *At*CRP1 and *Zm*CRP1 proteins. The amino acid sequence of the Arabidopsis CRP1 (*At*CRP1, At5g42310) was compared with CRP1 from *Zea mays* (*Zm*CRP1), using ClustalW2. Black boxes indicate strictly conserved amino acids, and gray boxes closely related ones. The predicted chloroplast transit peptides (ChloroP, http://www.cbs.dtu.dk/services/ChloroP/) are indicated in italics, and the PPR motives (P0-to-P14), identified using the PlantPPR database (http://www.plantppr.com), are marked with gray bars. The specificity determining amino acids in each PPR motif at position 5 and 35 are indicated by black and gray stars, respectively. Note that P0 motif was not considered to contribute to the identification of RNA targets, as previously reported by Barkan et al. (2012). P0 is composed of 30 aa, whereas all other P motifs are of 35 aa, with the exception of P2, wich contains 37 aa in Arabidopsis and 38 in maize.

RESULTS

*At*CRP1 Is a PPR Protein Imported into the Chloroplast

The Maize Genetics and Genomics Database (Lawrence et al., 2004)[9] was used to identify the At5g42310 gene as the Arabidopsis ortholog of *ZmCRP1* (see also Belcher et al., 2015). At5g42310 encodes a polypeptide of 709 amino acids with a calculated molecular mass of 80 kDa. Intron number (three) and position are conserved between the two genes, and BLASTP query of public Arabidopsis sequence database with *Zm*CRP1 amino acid sequence detected At5g42310 protein as the top hit with 55% sequence identity and 72% sequence similarity (**Figure 1**).

*At*CRP1 is annotated as a PPR protein and shares with *Zm*CRP1 15 PPR tandem repeats, which were predicted by using the PlantPPR database (Cheng et al., 2016). All PPR motifs are of 35 aa, with the exception of P0 which consists of 30 aa and P2 of 37 aa in Arabidopsis and 38 aa in maize. The fifth and the last residue of each PPR domain form the amino acid pairs that specify the RNA target molecules (Cheng et al., 2016), and are labeled with gray and black stars in **Figure 1**. The ChloroP server (Emanuelsson et al., 1999)[10] predicted the presence of a cTP of 54 residues (see amino acid residues in italics in **Figure 1**), indicating that *At*CRP1, like *Zm*CRP1, could be imported into the chloroplast. To corroborate the *in silico* prediction, the *At*CRP1-GFP fusion protein was expressed in transiently transformed Arabidopsis protoplasts (**Figure 2**). In agreement with the ChloroP prediction, the chimeric protein (GFP fluorescence) accumulated within the chloroplast in distinct fluorescent foci (CHL, autofluorescence of chloroplast chlorophylls, **Figure 2A**), resembling the nucleoid complexes. Indeed, *At*CRP1-GFP chimera co-localized perfectly with the GUN1-RFP fusion protein, used as a nucleoid marker in this assay (RFP fluorescence, **Figure 2B**), (Koussevitzky et al., 2007; Colombo et al., 2016; Tadini et al., 2016), in tobacco leaf cells. To further localize *At*CRP1, chloroplasts were fractionated to separate the stroma and thylakoid compartments. Immunoblot analysis, using a GFP specific antibody, allowed detection of *At*CRP1-GFP specific signal in total chloroplasts, as well as in thylakoids and in the stromal fraction, indicating that the nucleoid *At*CRP1 protein is both associated to membranes and soluble in the stroma (**Figure 2C**). These findings are in agreement with the identification of *At*CRP1 as part of Megadalton complexes in the chloroplast stroma (Olinares et al., 2010), as well as in the grana of thylakoid membranes (Tomizioli et al., 2014).

*At*CRP1 Is Essential for Plant Autotrophy

To investigate the role that *At*CRP1 plays in Arabidopsis, two lines carrying T-DNA insertions into the coding sequence of At5g42310, renamed *atcrp1-1* (Salk_035048) and *atcrp1-2*

[9]http://www.maizegdb.org/

[10]http://www.cbs.dtu.dk/services/ChloroP/

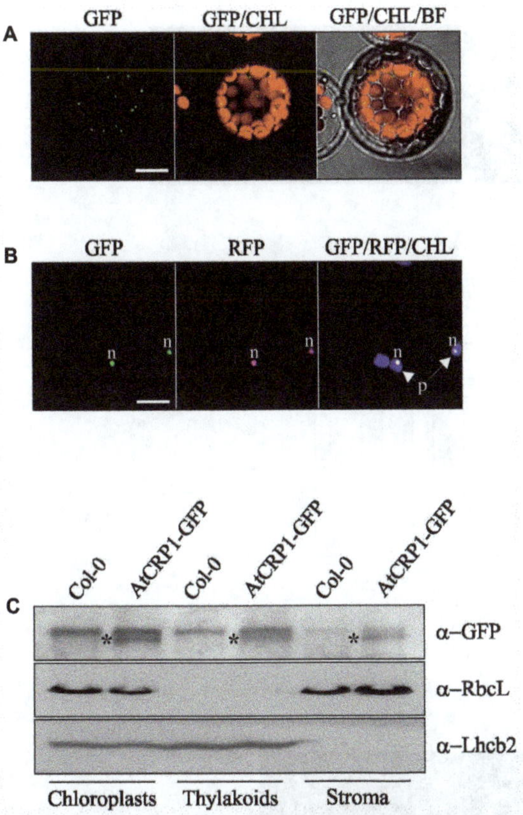

FIGURE 2 | Subcellular localization of *At*CRP1 in Arabidopsis mesophyll protoplasts and leaf cells. (A) Series of Lasers Scanning Confocal images (CLSM) of the subcellular localization of the *At*CRP1-GFP fusion protein (indicated as GFP) expressed in transiently transformed Arabidopsis (ecotype Col-0) leaf mesophyll protoplasts. The GFP signal accumulates in distinct spots within the chloroplasts, visualized by the red chlorophyll autofluorescence (CHL), resembling the pattern of chloroplast nucleoids. BF, Bright Field. **(B)** Series of CLSM images of the subcellular localization of *At*CRP1-GFP and GUN1-RFP [indicated as RFP and used as a marker of chloroplast nucleoids (Koussevitzky et al., 2007)] fusion proteins upon transient co-expression in tobacco leaf cells. The green fluorescence (GFP) co-localizes perfectly with the purple fluorescence (RFP) inside the chloroplasts (violet autofluorescence of chlorophylls, CHL), indicating that *At*CRP1 protein is part of the chloroplast nucleoids. Images are representative of three independent experiments. Bar = 10 μm; *p* = chloroplast; *n* = nucleoid. **(C)** Immunoblot analyses of proteins from Col-0 and Arabidopsis transgenic lines containing the *AtCRP1-GFP* construct under the control of *AtCRP1* native promoter (approximately 1 kb upstream of the translation start codon, see also Materials and Methods). Equal protein amounts isolated from total chloroplasts, thylakoids and stroma were loaded. Filters were immunolabeled with a GFP specific antibody to detect the localization of the *At*CRP1-GFP chimera. An antibody specific for the large subunit of RUBISCO (RbcL) was used as a marker of chloroplast stroma, whilst an Lhcb2 specific antibody was used as a marker of thylakoid membranes. Asterisks indicate the position of the AtCRP1-GFP fusion protein. One out of three immunoblots for each antibody is shown. Note that the *At*CRP1-GFP chimera is fully functional, since it was able to rescue the *atcrp1-1* mutant phenotype (see also **Figure 4**).

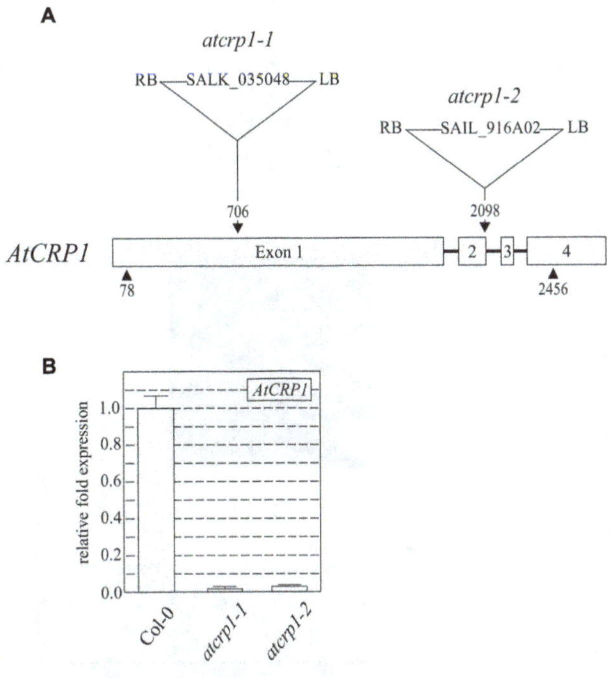

FIGURE 3 | T-DNA tagging and expression levels of *AtCRP1* gene. (A) Schematic representation of *AtCRP1* gene, where exons are indicated as numbered white boxes, while introns are shown as black lines. Arrowheads indicate the positions of translation initiation and stop codons. The locations, designations and orientations of T-DNA insertions are indicated (RB, right border; LB, left border). Note that the T-DNA insertions are not drawn to scale. **(B)** Levels of *AtCRP1* gene expression was ascertained by real-time PCR of cDNA obtained from leaves of WT (Col-0) and *atcrp1-1*, *atcrp1-2* mutant plants. Gene expression was normalized to the level of *AtCRP1* transcripts in Col-0 plants, and *SAND* and *ubiquitin* were used as internal references. The bars indicate standard deviations.

(Sail_916A02), were obtained from the T-DNA Express Arabidopsis mutant collection (**Figure 3A**; see also Materials and Methods).

Both T-DNA insertions completely suppressed the accumulation of the corresponding transcripts in homozygous mutant seedlings (**Figure 3B**), which were characterized by a paler pigmentation of cotyledons, visible even at the fully mature embryo stage (**Figure 4A**), and leaves (**Figures 4B,C**), and found to be seedling lethal under autotrophic growth conditions on soil and MS medium without sucrose, but able to develop yellow-albinotic rosette leaves and sterile inflorescence when sucrose was provided in the medium (**Figure 4C**). The mutant phenotype could be rescued by *Agrobacterium tumefaciens*-mediated transformation of heterozygous plants with either the appropriate coding sequence fused to the 35S promoter of cauliflower mosaic virus (*35S-CaMV::AtCRP1-GFP*), or the genomic sequence including a 1-Kbp fragment of the promoter region (*AtCRP1p::AtCRP1-GFP*), corroborating a direct correspondence between genotype and phenotype, and indicating that the AtCRP1-GFP chimera was fully functional, in both cases (**Figure 4D**). Interestingly, complemented plants carrying the *AtCRP1-GFP* construct under the control of the native promoter showed a fivefold increase in *AtCRP1* gene expression (**Figure 4E**), most probably as consequence of the T-DNA insertion in a highly expressed

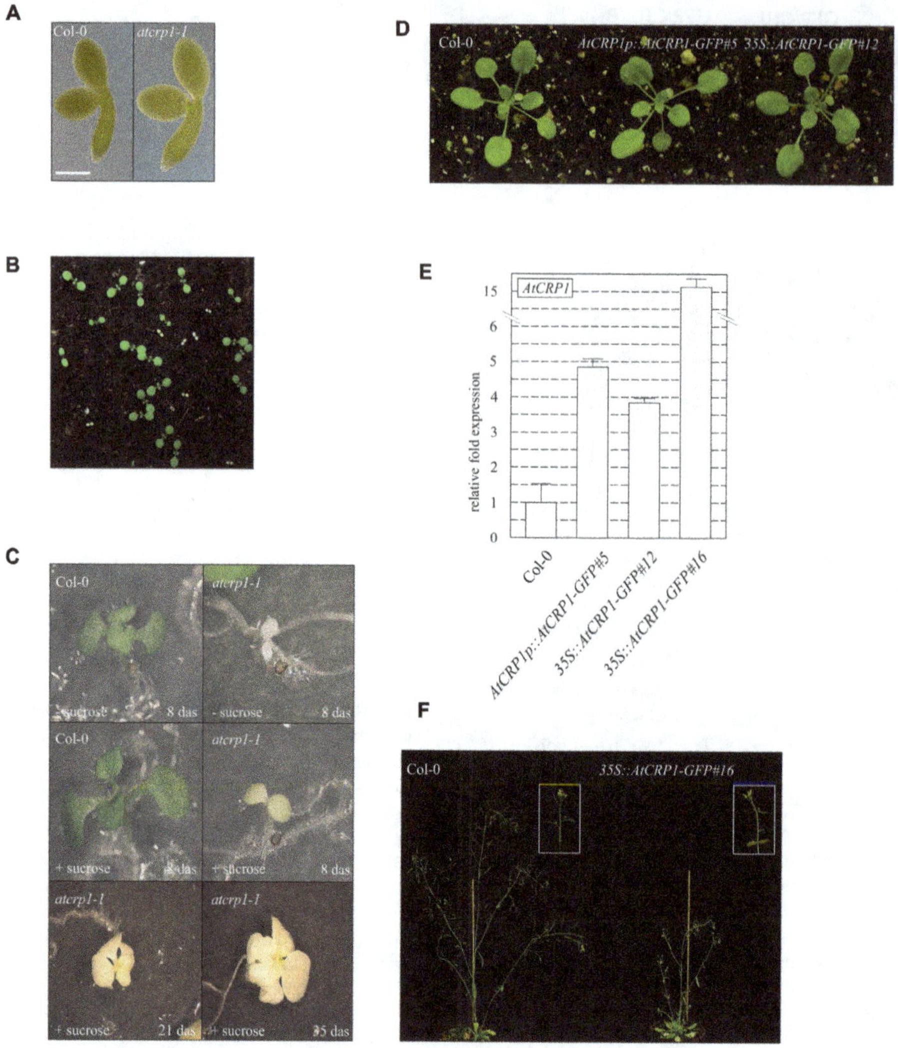

FIGURE 4 | Effects of loss of *At*CRP1 on plant development. (A) Images of isolated fully mature embryos (bent cotyledon stage) from WT (Col-0) and *atcrp1-1* seeds. The lack of *At*CRP1 protein did not alter embryo development, although mutant embryos were slight larger and paler than Col-0. **(B)** *atcrp1-1* seeds were able to germinate on soil, giving rise to yellow seedlings that accounted for about one-quarter of all seedlings, indicative of a monogenic recessive trait. Mutant seedlings did not survive past the cotyledon stage. **(C)** Mutant seedlings showed albino cotyledons when grown on MS medium without sucrose and arrested at the cotyledon stage as in **(B)**. However, when *atcrp1-1* seedlings were grown on MS medium supplement with 1% sucrose, they showed yellow-albinotic cotyledons at 8 das (days after sowing) and were able to develop up to 8–10 true leaves after 35 das. **(D)** The *atcrp1-1* seedling lethal phenotype could be fully rescued by *Agrobacterium tumefaciens*-mediated transformation of *AtCRP1/atcrp1-1* heterozygous plants with either the *AtCRP1* coding sequence fused to GFP under the control of 35S-CaMV promoter (*35S::AtCRP1-GFP#12*), or the genomic sequence fused to GFP under the control of native promoter (*AtCRP1p::AtCRP1-GFP#5*). **(E)** Real-time PCR to monitor the expression of *AtCRP1* gene in WT and complemented plants. Gene expression was normalized with respect to the level of *AtCRP1* transcripts in Col-0, and *SAND* and *ubiquitin* were used as internal references. The bars indicate standard deviations. **(F)** Col-0 and *35S::AtCRP1-GFP#16* transgenic line with about 15-folds more *AtCRP1* transcripts than WT. In this case the transgenic line shows WT-like rosette, but it is characterized by shorter and paler stems, with bleached cauline leaves and sterile flowers. A detail of the stem and inflorescence is shown in the inset. Note that the detailed molecular characterization of *At*CRP1 function was conducted on *atcrp1-1* plants, since the *atcrp1-2* seedlings showed an identical phenotype.

euchromatin region of the nuclear genome. Furthermore, a complete rescue of mutant plant phenotype could only be observed in *35S::AtCRP1-GFP* transgenic lines with a limited accumulation of *AtCRP1* transcripts (**Figures 4D,E**). Higher *AtCRP1* expression levels (around 15-folds in comparison to WT) led to transgenic plants with WT-like rosette but shorter and paler stems, bleached cauline leaves, together with sterile flowers (**Figures 4E,F**).

Temporal and spatial expression patterns of *At*CRP1, monitored by fusing the promoter region of the gene upstream of the GUS reporter gene (see also Materials and Methods), support further the key role played by *At*CRP1 during early stages of seedling and leaf development (**Figure 5**). The GUS staining could, indeed, be detected in young cotyledons and in the upper portion of the hypocotyl (**Figure 5A**). Furthermore, intense GUS signals were observable in young developing leaves

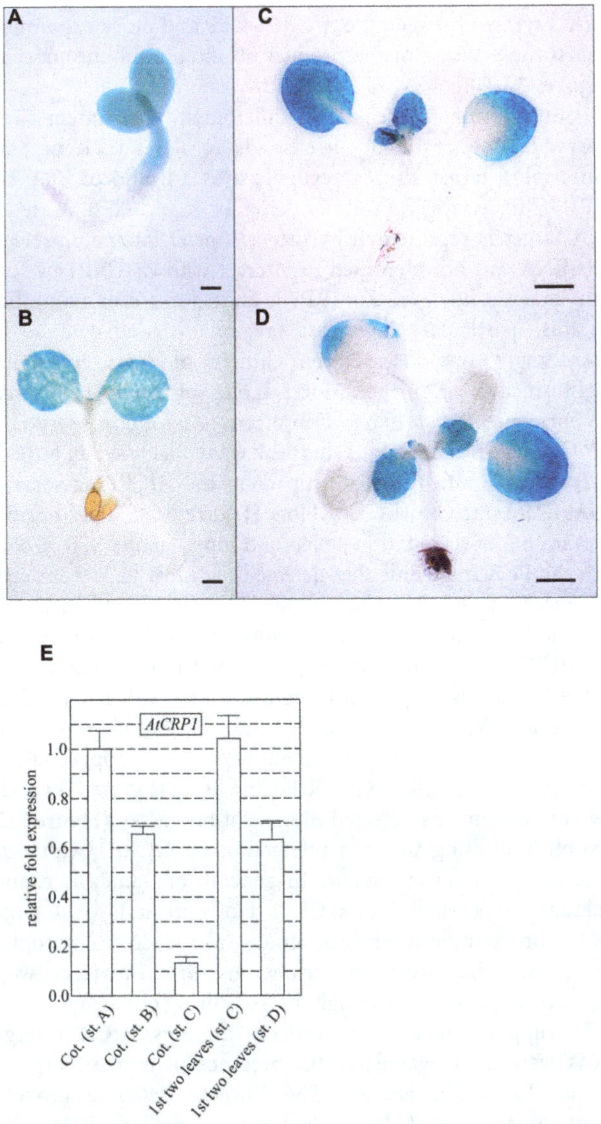

FIGURE 5 | *AtCRP1* promoter-driven β-glucuronidase (GUS) activity in cotyledons and rosette leaves. Histochemical GUS staining was conducted on seedlings at the two cotyledon stage **(A)**, at the onset of the first true leaves **(B)**, at four leaves rosette-stage **(C)**, and at the onset of the third pair of true leaves **(D)**. In general, GUS staining in younger leaves was stronger than in older leaves and the activity of *AtCRP1* promoter was below the limit of detection in cotyledons after the development of the first true leaves. **(A)** and **(B)** Bar = 1 mm, **(C)** and **(D)** bar = 1 cm. **(E)** Real-time PCR analyses were conducted with cDNA obtained from cotyledons at the developmental stages reported in **(A–C)** (Cot. st. A, Cot. st. B and Cot. st. C) and on the first pair of true leaves at stages C–D (st. C and st. D) to monitor the accumulation of *AtCRP1* transcripts. Gene expression was normalized with respect to the level of *AtCRP1* transcripts in cotyledons at stage A, and *SAND* and *ubiquitin* were used as internal references. The bars indicate standard deviations.

(**Figures 5C,D**), whereas the GUS coloration tended to decrease in old cotyledons and leaves (**Figures 5B–D**). Similar results were also obtained by monitoring the expression of *AtCRP1* in cotyledons and leaves using quantitative Real-Time PCR (qRT-PCR). In general, a high level of expression of *AtCRP1* was observed in green developing tissues, such as young cotyledons and leaves, whereas the expression decreased in older tissues (**Figure 5E**).

atcrp1 Mutant Chloroplasts Fail to Accumulate Cytochrome b_6/f Protein Complex and the PsaC Subunit of PSI

The albino pigmentation of *atcrp1* seedlings, together with their inability to grow under autotrophic conditions, indicated a defect in the thylakoid-associated photosynthetic apparatus. To verify this assumption, immunoblot analyses with antibodies specific for single subunits of the four major thylakoid protein complexes were performed on total leaf proteins. Leaf samples were harvested from *atcrp1* plants at the four-leaf rosette stage and grown on MS-medium supplemented with 1% sucrose (**Figure 6**; see also Materials and Methods). Under standard light conditions (50 μmol photons m^{-2} s^{-1}), subunits of Photosystem I (PsaA, PsaC, and PsaD), Photosystem II (D1, PsbO), Light harvesting complexes (Lhca1, Lhca2, Lhcb2, and Lhcb3) and ATPase (ATPase-β) accumulated to levels lower than 10% with respect to wild type plants. Furthermore, subunits of the Cyt b_6/f (PetA, PetB, and PetC) and PSI (PsaC) were below the limits of immunoblot detection.

In summary, these results indicate a general reduction of thylakoid protein complex subunits in *atcrp1* leaves, with a particularly severe effect on the accumulation of the Cyt b_6/f complex and PsaC.

AtCRP1 Is Associated *In vivo* with *psaC* and *petB-petD* Transcripts

*Zm*CRP1 has been previously demonstrated to associate with the *psaC* and *petA* mRNAs *in vivo* by RIP-Chip analyses (Schmitz-Linneweber et al., 2005). To investigate whether *At*CRP1 shares with *Zm*CRP1 the RNA targets, the same RIP-Chip approach employed in maize was used here. Stroma from plants expressing *At*CRP1-GFP, under the control of the native promoter (*AtCRP1p::AtCRP1-GFP*), was isolated and the fusion protein was immunoprecipitated using an anti-GFP serum. As a control, we performed mock precipitations with stroma extracted from WT plants, using the same GFP antibody. RNA was purified from the immunoprecipitation pellets and supernatants and was labeled with Cy5 (red) and Cy3 (green) fluorescent dyes, respectively. The two RNA fractions from *At*CRP1-GFP immunoprecipitations (IPs) and from mock IPs were competitively hybridized to a chloroplast genome tiling microarray (Kupsch et al., 2012). Enrichment of RNA is reflected in the ratio of red to green fluorescence for each spot on the microarray. Two biological replicate experiments were performed with stroma from *At*CRP1-GFP expressing plants and two with WT stroma. Data from the four assays were normalized and used to calculate median enrichment ratios of the red and green fluorescence signals for each PCR product among the 24 replicate spots on two arrays (**Supplementary Table S1**). To identify enrichment of RNA species specifically in the *At*CRP1-GFP immunoprecipitation,

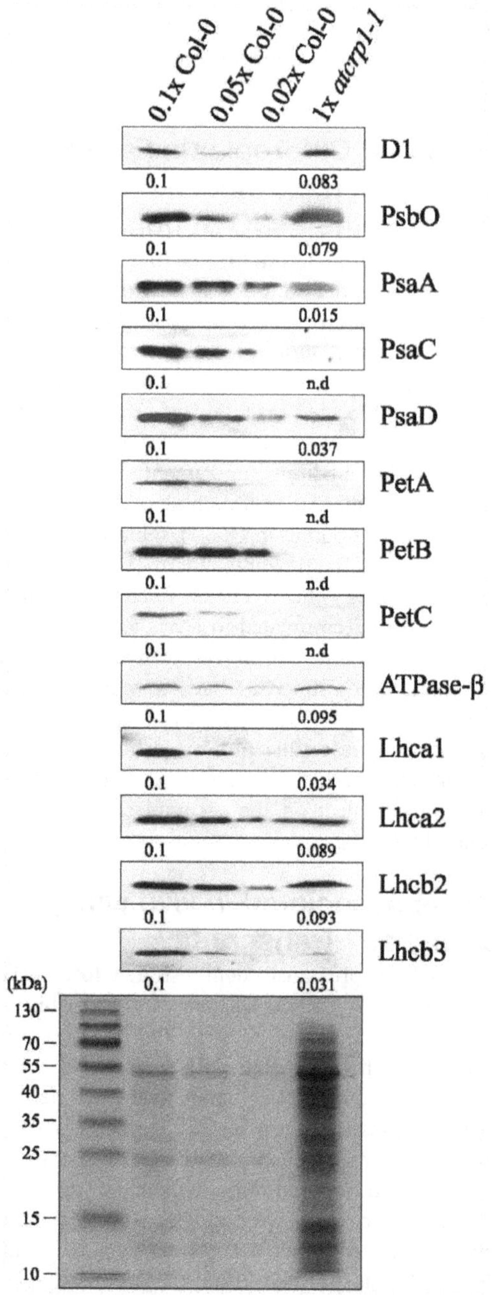

FIGURE 6 | Immunoblot analyses of thylakoid protein complexes in Col-0 and *atcrp1-1* mutant leaves. PVDF filters bearing fractionated total proteins, isolated at the four-leaf rosette stage from Col-0 and *atcrp1-1* plants grown on MS medium supplemented with 1% sucrose (see also **Figure 4**), were probed with antibodies raised against individual subunits of PSII (D1, PsbO), PSI (PsaA, PsaC, and PsaD), Cyt b_6f (PetA, PetB, and PetC), ATPase (ATPase-β), LHCI (Lhca1, Lhca2) and LHCII (Lhcb2, Lhcb3). Reduced levels of Col-0 total proteins were loaded in the lanes marked 0.1x Col-0, 0.05x Col-0, and 0.02x Col-0 in order to obtain signals from Col-0 proteins within the range of mutant protein signals (1x *atcrp1-1*). A replica SDS-PAGE stained with Coomassie-brilliant-blue is shown as loading control. Averaged relative protein abundance is given below each immunoblot and standard deviation was less than 10%. One out of three immunoblots for each antibody is shown. Note that the complete lack of Cyt b_6f and PsaC subunits was also observed in *atcrp1-2* leaves. n.d., not detected.

we plotted the difference in median enrichment ratio for each DNA fragment between the *At*CRP1-GFP and mock experiment against the position of the product on the plastid chromosome (**Figures 7A,B**).

Four prominent peaks of differential enrichment were observed. One of them corresponds to the 5′UTR of *psaC* transcript, a target already recognized as a ligand of *Zm*CRP1 in RIP-Chip assays (Schmitz-Linneweber et al., 2005). A second RNA target is represented by the *petB-petD* intergenic region. This RNA was not identified to interact with *Zm*CRP1 by RIP-Chip analysis, however, *Zm*CRP1 is known to aid in maturation of this particular intergenic region (Barkan et al., 1994). Interestingly, the observed enrichment of *rps15* transcripts might uncover a further, novel target of *At*CRP1, whereas the enrichment of *psbM/trnD* transcripts is often observed in RIP-Chip experiments, thus this peak was considered an artifact.

To corroborate the RIP-Chip data, the *At*CRP1-associated RNAs were analyzed by slot blots (**Figure 7C**). RNA purified from immunoprecipitation pellets and supernatants were probed with the PCR fragments that detected the most highly enriched sequences in the RIP-Chip assay. The data confirmed that the *psaC* and *petB-petD* transcripts are highly enriched in the *At*CRP1-GFP immunoprecipitates, but not the *rps15* RNA. *Zm*CRP1 was also reported to be associated with RNAs of the *petA* region (Schmitz-Linneweber et al., 2005; Williams-Carrier et al., 2008), however, no enrichment of *petA* transcripts could be observed in the *At*CRP1-GFP RIP-Chip assay (**Figure 7A**) and a low enrichment was detected in the slot blot assay (**Figure 7C**), possibly indicating that the interaction of *At*CRP1 with *petA* transcripts is not very stable. In general, our analysis cannot exclude the possibility that CRP1 binds to additional target RNAs, for example when interactions take place at chloroplast membranes. Since we are not using cross-linked material, weak RNA-protein interactions might be lost during our assay.

To support further the RIP-Chip findings, *At*CRP1 target RNAs were interrogated for the presence of native footprints at the JBrowse database[11]. The JBrowse database provides annotations of *Arabidopsis thaliana* organellar short RNA (sRNA), thought to be generated from protein-mediated temporary protection of target RNAs against exonucleolytic degradation (Ruwe et al., 2016; see also **Figure 8**). sRNAs were found within the 5′UTR of *psaC* (corresponding to the 117633–117597 region of chloroplast genome) and the *petB-petD* intergenic region (region 76318–76358), and an sRNA was also annotated in the 5′UTR of *petA* (region 61615–61643). Furthermore, *At*CRP1 predicted RNA binding motifs were shown to co-map with the native footprints, when the corresponding sequences were searched for the occurrence of the consensus binding motif with the FIMO program in the MEME suite[12] (**Figure 8B**; Takenaka et al., 2013). A short RNA has been also mapped upstream of *rps15*, but this region was not enriched in the RIP-Chip assay and the match with the predicted binding site of *At*CRP1 is weaker than for the *psaC*, *petB-petD,* and *petA* sRNAs.

[11]https://www.molgen.hu-berlin.de/projects-jbrowse-athaliana.php

[12]http://meme-suite.org/tools/fimo

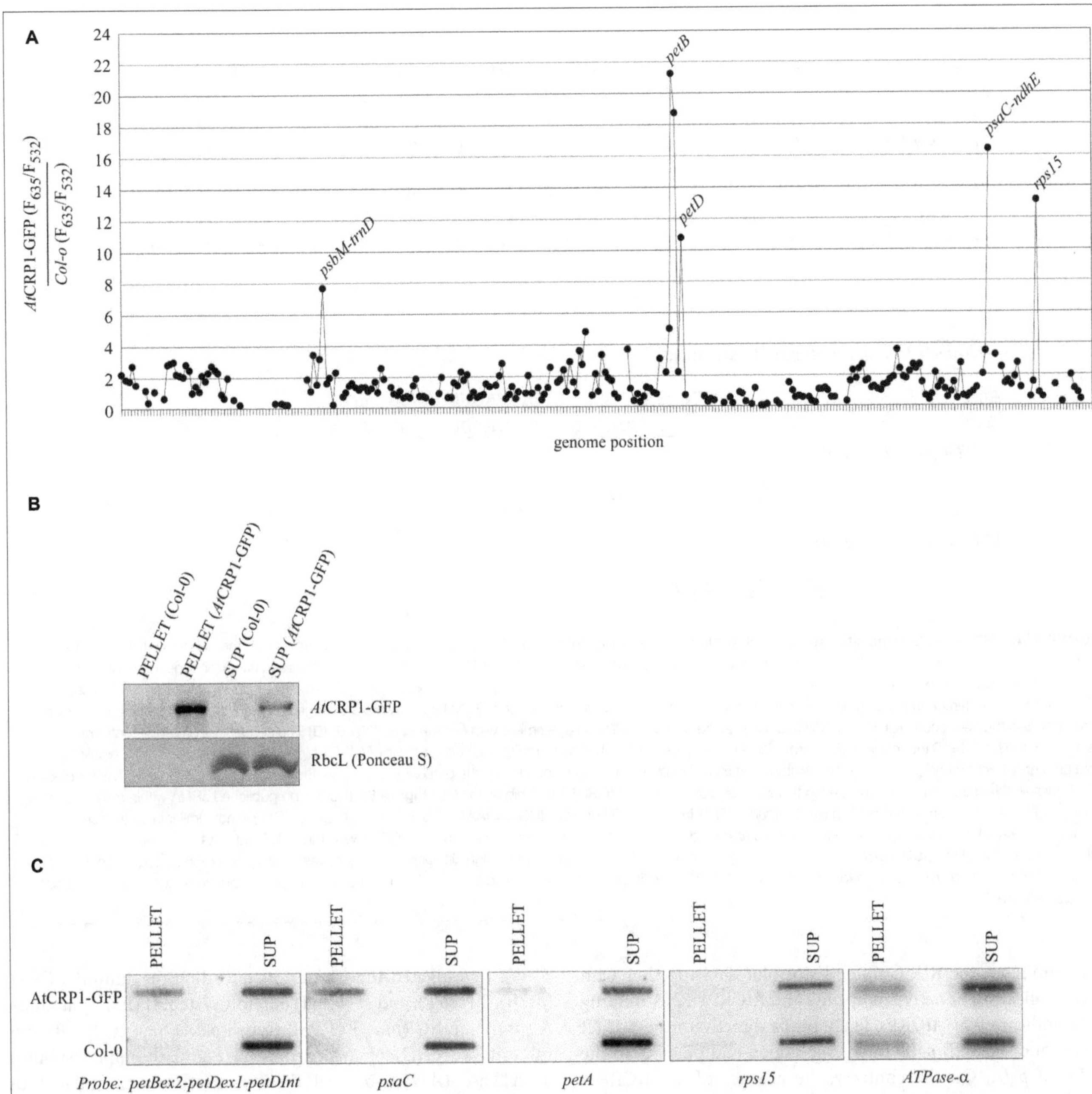

FIGURE 7 | *At*CRP1 RIP-Chip data plotted according to gene order within the plastid genome. (A) Differential enrichment ratios obtained by RNA immunoprecipitation (RIP)-Chip analysis. The enrichment ratios (F_{635}/F_{532}) obtained from an assay of *AtCRP1p::AtCRP1-GFP* chloroplast stroma extract were normalized with respect to a control assay that used WT (Col-0) chloroplast stroma extract (both assays were performed in duplicate). The median-normalized values for replicate spots were plotted according to gene order within the plastid genome. Fragments for which fewer than 13 spots per experiment (*AtCRP1p::AtCRP1-GFP/WT*) passed our manual quality control and/or yielded an F532 signal below background were excluded and appear as gap in the curve. The enrichment of *psaC* 5′UTR is in agreement with previous findings obtained by RIP-Chip analysis on CRP1 from maize (Schmitz-Linneweber et al., 2005). **(B)** Immunoblot analysis of protein fractions obtained from immunoprecipitation experiments using the anti-GFP mouse antibody and stroma material from Col-0 and *At*CRP1-GFP plants. Equal volumes of supernatant and pellet preparations were loaded onto the gel. Note that the pellet from *At*CRP1-GFP immunoprecipitation gave a stronger signal than the corresponding supernatant, implying quantitative precipitation of *At*CRP1-GFP. The fact that no signal was obtained with Col-0 extracts demonstrates the specificity of the antibody. The RbcL migration region of the Ponceau S stained nylon membrane, after transfer from SDS-PAGE, was used to verify equal loading. **(C)** Verification of *At*CRP1 RNA targets. Coimmunoprecipitations and RNA extractions from *At*CRP1-GFP and Col-0 samples were performed as for RIP-Chip assays. The RNAs were then analyzed by slot-blot hybridization with the indicated probes (see also Materials and Methods and **Supplementary Table S2**). The *ATPase-α* probe hybridization was included as a control. SUP, supernatant.

A

motif	P1	P2	P3	P4	P5	P6	P7	P8	P9	P10	P11	P12	P13	P14
AtCRP1 aa pos. 5	N	S	N	V	N	S	S	N	N	N	T	N	N	T
AtCRP1 aa pos. 35	D	D	K	R	D	N	D	D	C	N	S	S	D	D
Predicted RNA b.s.	*U*	*G*	*Y*	-	*U*	*A*	*G*	*U*	*Y*	*Y*	*R*	*Y*	*U*	*G*
ZmCRP1 aa pos. 5	N	S	S	T	N	S	S	N	N	N	T	H	N	T
ZmCRP1 aa pos. 35	D	D	R	R	D	S	D	D	G	N	S	S	D	D
Predicted RNA b.s.	*U*	*G*	*R*	*R*	*U*	*A*	*G*	*U*	*Y*	*Y*	*R*	-	*U*	*G*

B

***petB-petD Intergenic Region* (+ strand):**

At 76318-CUUACUUAUUACUUGGUGAAGGAACGAUAG**UA**U**UUUAUUG**C-76358

Zm 74582-AUAU**CGGGUAG**GUUGUGG**UA**U**UU**C**AUUG**CU-74611

***5' UTR psaC* (-strand):**

At 117633-UUUUAAUAUACCAUUCAGUUAGA**AGUUUACU**AGAUUG-117597

***5' UTR petA* (+strand):**

At 61615-GCUA**ACUUUAUUG**UAGAAAUUUUCGGGAU-61643

FIGURE 8 | *At*CRP1 RNA binding sites and the chloroplast *in vivo* footprints. (A) PPR motifs in *At*CRP1 were identified with the aid of PlantPPR database [www.plantppr.com, (Cheng et al., 2016). Amino acid residues in the 5th and last position of PPR motifs have been considered critical for sequence-specific RNA recognition, as previously reported (Barkan et al., 2012; Barkan and Small, 2014; Cheng et al., 2016; Harrison et al., 2016); see also **Figure 1**]. When the code developed for the different amino acid pairs is applied to the *At*CRP1 repeats, the sequence UGYNUAGUYYRYUG emerges as predicted RNA binding sequence (b.s.), whereas the sequence UGRRUAGUYYRNUG is predicted for *Zm*CRP1, in agreement with Barkan et al. (2012). **(B)** The sequences of *in vivo* footprints identified in the *petB-petD* intergenic region (Arabidopsis and Maize) and 5′UTR *psaC* region that co-map with *At*CRP1 binding sites (p-value < 0.01, highlighted in bold on a gray background) are shown. The Arabidopsis 5′UTR of *petA* transcripts shows also the presence of a native footprint that co-maps with *At*CRP1 binding site, however, this region was only enriched in the slot blot, but not in the AtCRP1 RIP-Chip assay (see **Figure 6**). There is no published sRNA within the *psaC* or *petA* 5′UTR of maize. A predicted binding site for maize CRP1 in the 5′-UTR of *psaC* (UGGAUAAACCAUUG; Barkan et al., 2012) is not similar in sequence to the Arabidopsis prediction shown here. Moreover, nucleic acid binding assay showed a direct interaction of *Zm*CRP1 with the 5′-UTR of *petA* (UUAGCUACCUAUCU**AAUUUAUUG**UAGAAAUU; Williams-Carrier et al., 2008), that shows high similarity with the corresponding Arabidopsis sequence (see predicted binding site highlighted in bold). Note that no *At*CRP1-specific *in vivo* footprint could be identified in the other RIP-Chip enriched regions, *rps15* and *psbM* (see also **Figure 7**).

In summary, the RIP-Chip and slot blot data together with the colocalization of native footprints and *At*CRP1 RNA binding motifs indicate that *At*CRP1 likely binds directly to the 5′UTR of *psaC* and the *petB-petD* intergenic region and possibly to the 5′UTR of *petA*. On the contrary, the absence of an *At*CRP1-specific footprint within the *rps15* RNA, together with the failure of slot blot enrichment, makes any *At*CRP1-*rps15* interaction unlikely.

*At*CRP1 Is Required for the Correct Processing of *psbB-psbT-psbH-petB-petD* Transcripts

To assess whether the lack of Cyt b_6/f complex and PsaC subunit, together with the marked reduction of all protein complex subunits observed in *atcrp1-1* thylakoids, was caused by deficiencies in transcript accumulation and *At*CRP1-dependent transcript processing, we probed the identified *At*CRP1 RNA targets and other plastid transcripts by gel blot hybridization (**Figure 9**).

We investigated the transcripts encoding the subunits CP47 (*psbB*), T (*psbT*), and H (*psbH*) of photosystem II (PSII), subunits A (*psaA*) and C (*psaC*) of PSI, Cyt *f* (*petA*), Cyt b_6 (*petB*) and subunit IV (*petD*) of cytochrome b_6/f and the alpha subunit of ATPase (*ATPase-α*). All these transcripts accumulated in *atcrp1-1* plastids to levels lower than WT, indicating that global plastid gene expression is affected by the *atcrp1-1* mutation, and explaining the marked reduction of thylakoid protein accumulation observed in *atcrp1-1* leaves.

Furthermore, the plastid polycistronic transcription unit *psbB-psbT-psbH-petB-petD* showed some striking alteration of transcript pattern in *atcrp1* samples (**Figure 9**). In particular, the monocistronic *petB* (band #4, 0.8 Kb), the dicistronic *psbH-petB* (band #3; 1.2 Kb) and the unspliced *petB* (band #2, 1.6 Kb) transcripts were barely detectable in the mutant, whereas the *petB*-unspliced *petD*-spliced dicistronic transcript (band #1, 2.2 Kb), detected with probes D, E, F, and H, accumulated to even higher levels in *atcrp1* plastids, presumably due to the failure of *At*CRP1-dependent processing between the *petB* and *petD* coding regions, as also shown in *zmcrp1* mutant plants (Barkan et al.,

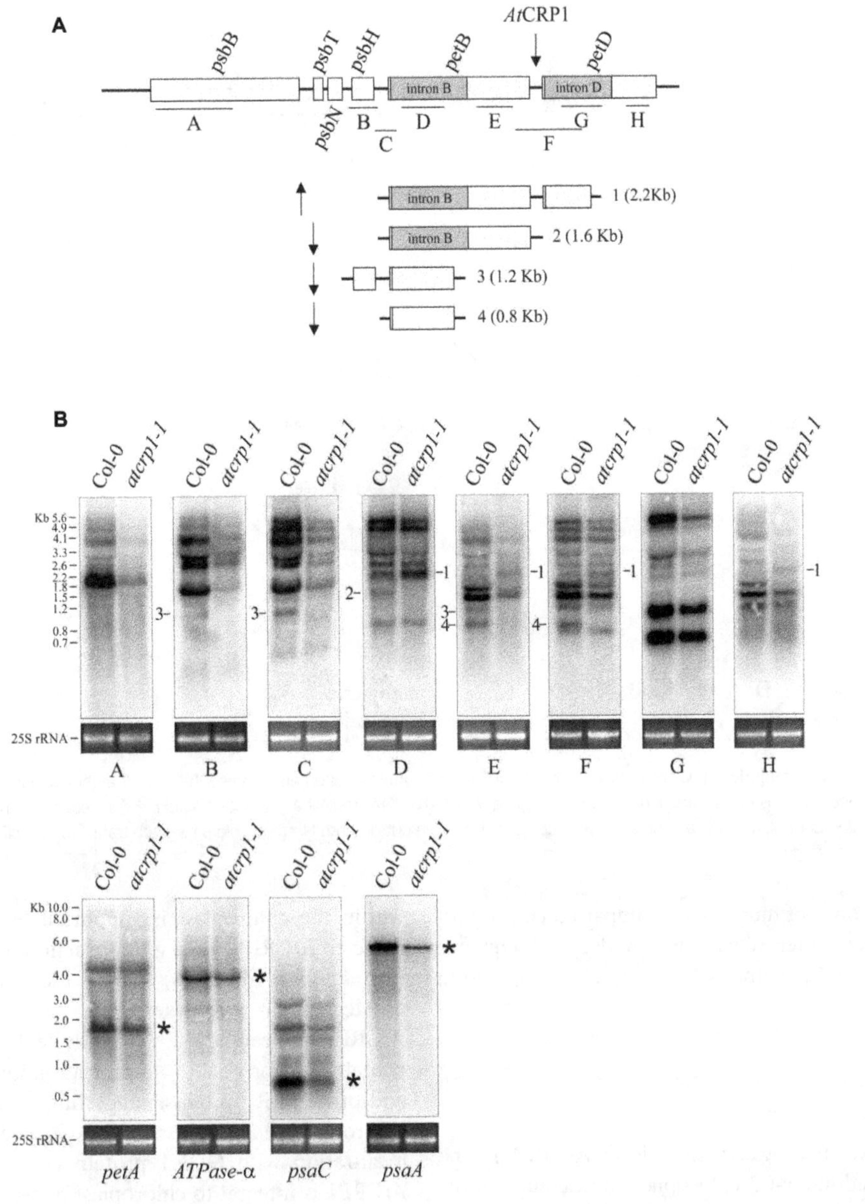

FIGURE 9 | Transcript patterns of chloroplast genes in Col-0 and *atcrp1-1* mutant leaves. (A) The structure of the *psbB* gene cluster and probes A to H used in RNA gel blots analysis in **(B)** are shown. Furthermore, processed and spliced transcripts that accumulate differentially between Col-0 and mutant chloroplasts are drawn to scale and numbered from 1 to 4. Upward arrow indicates transcripts that accumulate to higher levels in *atcrp1-1* than Col-0 chloroplasts, whilst the downward arrow is used for transcripts less abundant or absent in mutant samples. The putative binding site of *At*CRP1 within the *petB-petD* intergenic region is also indicated. **(B)** RNA gel blot analysis of the *psbB* gene cluster were performed using probes indicated as A to H, whilst *petA*, *ATPase-A*, *psaC*, and *psaA* specific probes are described in section "Materials and Methods." The identity of labeled transcripts (1–4), shown in **(A)** together with their size, was established based on the hybridization pattern, transcript size and on data reported in Meierhoff et al. (2003) and Stoppel et al. (2011). Asterisks indicate the mature transcript forms. A portion of the ethidium bromide stained Agarose gels, containing the cytosolic 25S rRNA, is included, as loading control, below each filter. One out of three Northern-blots for each transcript-specific probe is shown.

1994; Fisk et al., 1999). In contrast with maize, monocistronic and spliced *petD* transcripts of ~600 nucleotides do not accumulate to significant levels in Arabidopsis, and thus its absence was not observed in *atcrp1* plastids (Barkan et al., 1994; Barkan, 2011).

Moreover, the lack of the PsaC and PetA subunits could be the consequence of the simultaneous decrease of transcript accumulation and a possible defect in *At*CRP1-dependent activation of *psaC* and *petA* translation, as shown in *Zea mays* (Barkan et al., 1994; Schmitz-Linneweber et al., 2005). However, the specific regulatory role of *AtCRP1* in plastid protein translation is difficult to verify, owing to the general and pleiotropic decrease of mature plastid *rRNA* in *atcrp1-1* leaves, in spite of WT-like accumulation of *rrn23* and *rrn4.5* precursor forms (**Figure 10**). This rRNA accumulation pattern

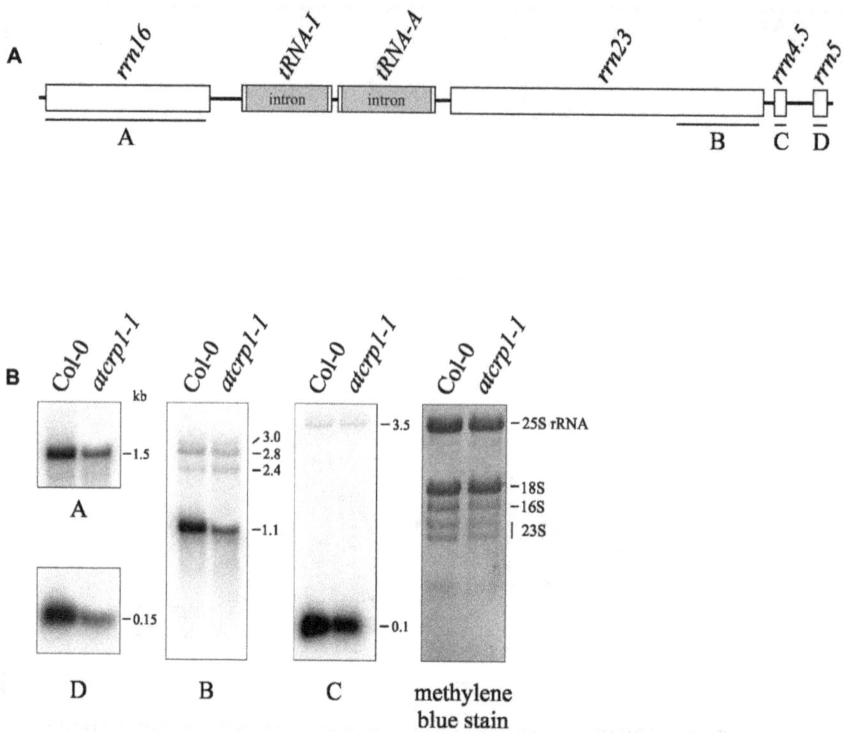

FIGURE 10 | Plastid rRNA accumulation in Colo-0 and *atcrp1-1* mutant leaves. (A) Schematic representation of the chloroplast *rrn* operon. Probes used in Northern blot analysis are indicated as black bars under each rRNA gene (A–D). **(B)** RNA gel blot analysis of plastid rRNAs were performed using the probes A-to-D described above. For loading control, a methylene blue stained filter is shown. One out of three Northern-blots for each transcript-specific probe is shown.

is very similar to the ones of mutants with impaired chloroplast translation and has been interpreted as a secondary consequence of reduced plastid protein synthesis (Tiller et al., 2012; Tadini et al., 2016).

DISCUSSION

In this study we have investigated the role of *At*CRP1 in the biogenesis of dicotyledonous-C3 chloroplasts and compared its function to the already characterized monocotyledonous-C4 chloroplast counterpart, *Zm*CRP1. Both proteins are essential for chloroplast biogenesis and photosynthetic activity, since they are required for the processing and translation of specific plastid transcripts encoding subunits of the thylakoid protein complexes. Our results indicate that *At*CRP1 and *Zm*CRP1 have very similar RNA targets and the main functional divergences are most likely due to the distinct localization of the two proteins inside the chloroplast and the partially different affinity for the RNA targets (see **Table 1**).

CRP1 Proteins Are Part of Chloroplast Nucleoids

We detected *At*CRP1 in the stroma and associated with thylakoid membranes (see **Figure 2**; **Table 1**), whereas *Zm*CRP1 is a stromal protein with no detectable association with chloroplast membranes (Fisk et al., 1999). The dual localization of *At*CRP1 within the chloroplast is supported by proteomic studies that detected *At*CRP1 in the grana-fraction of Arabidopsis thylakoids (Tomizioli et al., 2014) and in the stroma proteome, as part of Megadalton complexes (Olinares et al., 2010). In particular, *At*CRP1 appeared to be highly enriched in fractions that contained ribosomal proteins, translation factors, RNA helicases and other PPR proteins, suggesting a major role of *At*CRP1 in chloroplast gene expression. These data, together with the co-localization with GUN1 protein (see **Figure 2**), indicate that *At*CRP1 is integral to chloroplast nucleoids (Koussevitzky et al., 2007; Colombo et al., 2016; Tadini et al., 2016), i.e., the DNA-containing structures without defined boundaries that harbor the plastid gene expression machinery (Pfalz and Pfannschmidt, 2013; Melonek et al., 2016). Similarly, *Zm*CRP1 was found to be highly enriched in the nucleoid fractions of maize plastids, together with proteins involved in DNA replication, organization and repair as well as transcription, mRNA processing, splicing and editing (Majeran et al., 2012), further supporting the involvement of CRP1 proteins in plastid gene expression.

CRP1 Proteins Are Required for the Biogenesis of the Photosynthetic Apparatus

The yellow-albinotic and seedling lethal phenotype exhibited by *atcrp1* is very similar to the chlorophyll deficient and lethal phenotype of *zmcrp1* plants (Barkan et al., 1994; Fisk et al.,

TABLE 1 | Overview of the phenotypes of Arabidopsis and maize *crp1* mutants and comparison of their molecular roles in chloroplast biogenesis.

***atcrp1*[a]**	***zmcrp1*[b]**
Plant phenotype	
Seedling lethal with yellow-albinotic cotyledons and leaves. Plants are able to develop mature leaves and sterile flowers when grown on MS medium supplemented with sucrose	Seedling lethal with pale-green cotyledon and leaves. Plants are able to develop mature non-photosynthetic leaves thanks to the large reserves of maize seeds
CRP1 protein localization	
*At*CRP1 is a component of plastid nucleoids and it is found associated to thylakoid membranes and in the stroma	*Zm*CRP1 has been reported to be highly enriched in plastid nucleoids and to localize exclusively in the chloroplast stroma

Thylakoid protein content

PSI	PSII	Cyt b_6f	ATPase	PSI	PSII	Cyt b_6f	ATPase
– (/PsaC)	–	/	–	–	=	/	=

RNA targets

RIP-Chip	Slot-Blot	*In vivo* footprint	RIP-Chip	Slot-Blot	*In vivo* footprint
psaC	*psaC*	*psaC*	*psaC*	*psaC*	n.r.
petB-petD	*petB-petD*	*petB-petD*	/	/	*petB-petD*
/	*petA*	*petA*	*petA*	*petA*	n.r.
rps15	/	/	/	/	n.r.

Metabolism of chloroplast RNAs

	Accumulation	Processing defects	Accumulation	Processing defects
psaC	–	No	=	No
petB-petD	/	Yes	/	Yes
petA	–	No	=	No

[a]*Data are obtained from the present manuscript* [b]*Data are obtained from Barkan et al. (1994, 2012), Fisk et al. (1999), Schmitz-Linneweber et al. (2005). –, marked reduction; /, complete absence; =, no changes; +, increase; n.r., not reported.*

1999). Arabidopsis mutants die at the two-cotyledon stage after germination on soil, but can overcome seedling lethality on sucrose-containing media, where they develop mature leaves and sterile flowers (see **Figure 4**; **Table 1**). Similarly, non-photosynthetic *zmcrp1* plants die at about 3 weeks after germination when seed reserves are exhausted. Furthermore, the *atcrp1* phenotype appears to be typical of Arabidopsis mutants lacking components of the photosynthetic apparatus and not of the gene expression machinery or of the protein import apparatus, since the latter usually result in the premature arrest at the globular-to-heart stage of embryo development, when chloroplast biogenesis begins (Ruppel and Hangarter, 2007; Romani et al., 2012; Beeler et al., 2014). Nevertheless, the pale-green pigmentation of the mutant embryo at bent-cotyledon stage (see **Figure 4**) and the β-glucuronidase (GUS) activity observed in young developing cotyledons and rosette leaves, but not in older tissues (see **Figure 5**), indicate that *AtCRP1* gene expression and protein accumulation is required during the very early stages of the photosynthetic apparatus assembly. Immunoblot data indicate, indeed, that *At*CRP1, like *Zm*CRP1, might act as a nuclear-encoded anterograde regulatory component responsible for coordination of the accumulation of Cyt b_6/f and PSI protein complexes (see **Figure 6**). Besides their role in linear electron transport (LET), Cyt b_6/f and PSI indeed play a key role in Cyclic Electron Transport (CET), which has been reported to be enhanced in Arabidopsis green seeds and to be required for optimal seed vigor and seed germination rate (Allorent et al., 2015).

In contrast to *zmcrp1* plants (Barkan et al., 1994), the absence of *At*CRP1 destabilized the entire photosynthetic apparatus, as shown by the marked reduction of PSII core, ATPase and LHC protein levels. The general down-regulation of thylakoid complexes owing to defects in the intersystem electron transport chain appears to be a common feature of Arabidopsis photosynthetic mutants and provides clear evidence of a different adaptive response between monocot and dicot plants (Meurer et al., 1996; Varotto et al., 2000, 2002; Maiwald et al., 2003; Weigel et al., 2003; Ihnatowicz et al., 2004, 2007; Belcher et al., 2015). Furthermore, the *atcrp1-1* phenotype, both in terms of plastid transcript and plastid protein accumulation, appears to be much more drastic than the one of other *ppr* mutants required for the processing and expression of *psbB-psbT-psbH-petB-petD* operon, such as *hcf152* (Meierhoff et al., 2003), suggesting that the absence of *At*CRP1 protein might affect the activity of other factors essential for plastid gene expression. As a matter of fact, rRNA abundance is markedly reduced in *atcrp1-1* plastids, indicating a general reduction of protein synthesis, as consequence of pleiotropic effects.

RNA Targets: Commonalities and Divergences between *At*CRP1 and *Zm*CRP1 Proteins

RNA immunoprecipitation-Chip and slot blot data suggest a physical interaction between *At*CRP1 and the transcripts of *psaC*, *petB-petD* and possibly *petA*, even though it is not known whether these interactions are direct or mediated by other factors (see **Figure 7**). However, all of these RNAs harbor a region where a native footprint is annotated, raising the tempting hypothesis that *At*CRP1 is in fact the RNA-binding factor responsible for that footprint (see **Figure 8**). Furthermore, when these enriched fragments were searched for occurrences of the predicted binding motif of *At*CRP1, each of them proved to contain a hit inside the footprint region, strongly suggesting that *At*CRP1 could be the factor leaving those footprints. Nevertheless, the observation that the footprints identified in Arabidopsis *psaC*, *petA*, and *petB-petD* transcripts are larger than the 14 nucleotide size of the predicted *At*CRP1 footprint (37, 29, and 41 nucleotides in *psaC*, *petA*, and *petB-petD*, respectively) supports the view that the binding of *At*CRP1 to its targets *in vivo* could be stabilized by other protein partners. For instance, the peptide chain release factor B3 (PrfB3) has been also shown to be required for Arabidopsis autotrophic growth and for the stability of 3′ processed *petB* transcripts to adjust cytochrome b_6 levels (Stoppel et al., 2011), thus possibly being an *At*CRP1 specific protein partner. Similarly, PPR proteins involved in RNA stabilization and editing have been shown to interact with RNA Recognition Motif (RRM) proteins and other factors, indicating that larger protein complexes assembled around a PPR protein are likely to occur (Kupsch et al., 2012; Takenaka et al., 2014; Shi et al., 2015).

The interactions with the 5′UTR of *psaC* and *petA* have also been reported in the case of *Zm*CRP1 (Schmitz-Linneweber et al., 2005; Williams-Carrier et al., 2008), indicating that this feature of CRP1 function is conserved between Arabidopsis and maize. *Zm*CRP1 was also shown to bind directly to the 5′-UTR of *petA* transcripts by electrophoresis mobility shift assay (Williams-Carrier et al., 2008), favoring the possibility of a direct binding of CRP1 proteins to the corresponding RNA targets (see also **Figure 8**). Furthermore, *Zm*CRP1 has been proposed to directly control the translation of *petA* and *psaC* transcripts (Barkan et al., 1994), as shown through pulse labeling and polysome loading (in the case of *petA*), or deduced from the reduced association of *psaC* RNAs with ribosomes. Interestingly, the PsaC subunit of PSI and the PetA subunit of Cyt b_6/f could not be detected in *atcrp1* thylakoids, despite the accumulation of the corresponding transcripts with no processing defects (see also **Figure 9**), suggesting that *At*CRP1 plays a major role in translation regulation also in Arabidopsis. Unfortunately, the specific requirement of *At*CRP1 in plastid protein translation cannot be verified by comparing Col-0 and *atcrp1-1* leaves, due to the marked reduction of *rRNA* accumulation in *atcrp1-1* plastids.

In addition to the defects in *petA* translation, the complete absence of Cyt b_6/f protein complex observed in *atcrp1* thylakoids can also be attributed to processing alterations of the *psbB-psbT-psbH-petB-petD* polycistronic transcription unit. The lack of the monocistronic *petB*, the dicistronic *psbH-petB*, and the unspliced *petB* transcripts, together with the direct binding of *At*CRP1 to the *petB-petD* intergenic region, strongly support the role of *At*CRP1 in the metabolism of *petB* and *petD* transcripts. PPR protein-derived RNA-footprints are considered to arise due to exonucleolytic activity (Ruwe et al., 2016). Since sRNAs corresponding to predicted binding sites of *At*CRP1 are identified here, the most likely role for *At*CRP1 is to block exonucleases from degrading the *petB* and *petD* transcripts. A similar defect in *petB-petD* maturation has been reported in *zmcrp1* mutant plants (Barkan et al., 1994; Fisk et al., 1999; Schmitz-Linneweber et al., 2005), although no association was detected between *Zm*CRP1 and the *petB-petD* intergenic region (Schmitz-Linneweber et al., 2005), so it is still uncertain whether the role of *Zm*CRP1 is direct or indirect.

CONCLUSION

Taken together, the characterization of the functional role of *At*CRP1 in chloroplast biogenesis has highlighted several features in common with the *Zm*CRP1. Both proteins appear to control, directly or indirectly, the expression of plastid genes encoding subunits of Cyt b_6/f and PSI protein complexes. The coordination of the accumulation of these two protein complexes is fundamental to guarantee optimal photosynthesis in mature plants, but appears also to be important during seed germination, when cyclic electron transport is highly enhanced relative to LET.

Differences in RNA targets observed by immunoprecipitation and hybridization assays between *At*CRP1 and *Zm*CRP1 might be explained by a broad affinity for RNA targets, but may also have technical reasons (GFP antibody for Arabidopsis versus direct anti-ZmCRP1 antibody in maize). Evidence in favor of conservation of PPR protein activity between different species has been reported for the PLS and P subfamilies (Choury et al., 2005; Bolle and Kempken, 2006; Choury and Araya, 2006); for instance, the maize MPPR6 protein can complement loss-of-function Arabidopsis mutants lacking the orthologous protein (Manavski et al., 2012). However, functional divergence has been also observed, as in the case of orthologous PPR proteins ATP4 (maize) and SVR7 (Arabidopsis) (Liu et al., 2010; Zoschke et al., 2012, 2013a,b). Further studies aimed to verify the degree of protein activity conservation between monocots and dicots are needed to extend our knowledge of PPR protein functions and the degree of protein function conservation. The parallel characterization of PPR orthologs, including the relationship between their protein structures and the corresponding target RNA species, may represent an underestimated and powerful strategy to precisely determine the PPR code, essential for a fast and accurate large scale prediction of PPR targets, and for the functional characterization of the PPR-mediated nucleus-to-chloroplast anterograde signaling pathway.

AUTHOR CONTRIBUTIONS

RF, LT, FM, FR, SM, M-KL, CS-L, and PP participated to the organization of the manuscript. RF, LT, FM, FR, SM, MC, AC, and PP designed and carried out the experiments related to the molecular biology and biochemical characterization of atcrp1 mutants. RF, FM, LT, M-KL, and CS-L were involved in RIP-Chip and slot blot assays, as well as in the *in silico* identification of native footprints and prediction of AtCRP1 binding motif. PP wrote the manuscript.

FUNDING

This work was supported by ERA-NET Cofund FACCE SURPLUS (BarPLUS grant id. 93) to PP. Work in the lab of CS-L was supported by DFG grant SCHM 1698/5-1.

ACKNOWLEDGMENT

We thank Hannes Ruwe and Gongwei Wang for helpful comments on the sRNA analysis.

SUPPLEMENTARY MATERIAL

The Supplementary Material for this article can be found online at: http://journal.frontiersin.org/article/10.3389/fpls.2017.00163/full#supplementary-material

TABLE S1 | RNA immunoprecipitation (RIP-Chip) data.

TABLE S2 | Oligonucleotides employed for cloning, genotyping, northern blot, slot blot and qRT-PCR assays.

REFERENCES

Allorent, G., Osorio, S., Vu, J. L., Falconet, D., Jouhet, J., Kuntz, M., et al. (2015). Adjustments of embryonic photosynthetic activity modulate seed fitness in *Arabidopsis thaliana. New Phytol.* 205, 707–719. doi: 10.1111/nph.13044

Alonso, J. M., Stepanova, A. N., Leisse, T. J., Kim, C. J., Chen, H., Shinn, P., et al. (2003). Genome-wide insertional mutagenesis of *Arabidopsis thaliana. Science* 301, 653–657. doi: 10.1126/science.1086391

Barkan, A. (2011). Expression of plastid genes: organelle-specific elaborations on a prokaryotic scaffold. *Plant Physiol.* 155, 1520–1532. doi: 10.1104/pp.110.171231

Barkan, A., Rojas, M., Fujii, S., Yap, A., Chong, Y. S., Bond, C. S., et al. (2012). A combinatorial amino acid code for RNA recognition by pentatricopeptide repeat proteins. *PLoS Genet.* 8:e1002910. doi: 10.1371/journal.pgen.1002910

Barkan, A., and Small, I. (2014). Pentatricopeptide repeat proteins in plants. *Annu. Rev. Plant Biol.* 65, 415–442. doi: 10.1146/annurev-arplant-050213-040159

Barkan, A., Walker, M., Nolasco, M., and Johnson, D. (1994). A nuclear mutation in maize blocks the processing and translation of several chloroplast mRNAs and provides evidence for the differential translation of alternative mRNA forms. *EMBO J.* 13, 3170–3181.

Beeler, S., Liu, H. C., Stadler, M., Schreier, T., Eicke, S., Lue, W. L., et al. (2014). Plastidial NAD-dependent malate dehydrogenase is critical for embryo development and heterotrophic metabolism in *Arabidopsis. Plant Physiol.* 164, 1175–1190. doi: 10.1104/pp.113.233866

Belcher, S., Williams-Carrier, R., Stiffler, N., and Barkan, A. (2015). Large-scale genetic analysis of chloroplast biogenesis in maize. *Biochim. Biophys. Acta* 1847, 1004–1016. doi: 10.1016/j.bbabio.2015.02.014

Bolle, N., and Kempken, F. (2006). Mono- and dicotyledonous plant-specific RNA editing sites are correctly edited in both in organello systems. *FEBS Lett.* 580, 4443–4448. doi: 10.1016/j.febslet.2006.07.011

Cai, W., Okuda, K., Peng, L., and Shikanai, T. (2011). PROTON GRADIENT REGULATION 3 recognizes multiple targets with limited similarity and mediates translation and RNA stabilization in plastids. *Plant J.* 67, 318–327. doi: 10.1111/j.1365-313X.2011.04593.x

Cheng, S., Gutmann, B., Zhong, X., Ye, Y., Fisher, M. F., Bai, F., et al. (2016). Redefining the structural motifs that determine RNA binding and RNA editing by pentatricopeptide repeat proteins in land plants. *Plant J.* 85, 532–547. doi: 10.1111/tpj.13121

Choury, D., and Araya, A. (2006). RNA editing site recognition in heterologous plant mitochondria. *Curr. Genet.* 50, 405–416. doi: 10.1007/s00294-006-0100-3

Choury, D., Farre, J. C., Jordana, X., and Araya, A. (2005). Gene expression studies in isolated mitochondria: *Solanum tuberosum* rps10 is recognized by cognate potato but not by the transcription, splicing and editing machinery of wheat mitochondria. *Nucleic Acids Res.* 33, 7058–7065. doi: 10.1093/nar/gki1017

Colombo, M., Tadini, L., Peracchio, C., Ferrari, R., and Pesaresi, P. (2016). GUN1, a jack-of-all-trades in chloroplast protein homeostasis and signaling. *Front. Plant Sci.* 7:1427. doi: 10.3389/fpls.2016.01427

Costa, A., Gutla, P. V., Boccaccio, A., Scholz-Starke, J., Festa, M., Basso, B., et al. (2012). The *Arabidopsis* central vacuole as an expression system for intracellular transporters: functional characterization of the Cl-/H+ exchanger CLC-7. *J. Physiol.* 590, 3421–3430. doi: 10.1113/jphysiol.2012.230227

Emanuelsson, O., Nielsen, H., and von Heijne, G. (1999). ChloroP, a neural network-based method for predicting chloroplast transit peptides and their cleavage sites. *Protein Sci.* 8, 978–984. doi: 10.1110/ps.8.5.978

Fisk, D. G., Walker, M. B., and Barkan, A. (1999). Molecular cloning of the maize gene crp1 reveals similarity between regulators of mitochondrial and chloroplast gene expression. *EMBO J.* 18, 2621–2630. doi: 10.1093/emboj/18.9.2621

Grant, C. E., Bailey, T. L., and Noble, W. S. (2011). FIMO: scanning for occurrences of a given motif. *Bioinformatics* 27, 1017–1018. doi: 10.1093/bioinformatics/btr064

Gregis, V., Sessa, A., Dorca-Fornell, C., and Kater, M. M. (2009). The *Arabidopsis* floral meristem identity genes AP1, AGL24 and SVP directly repress class B and C floral homeotic genes. *Plant J.* 60, 626–637. doi: 10.1111/j.1365-313X.2009.03985.x

Hall, T. M. (2016). De-coding and re-coding RNA recognition by PUF and PPR repeat proteins. *Curr. Opin. Struct. Biol.* 36, 116–121. doi: 10.1016/j.sbi.2016.01.010

Harrison, T., Ruiz, J., Sloan, D. B., Ben-Hur, A., and Boucher, C. (2016). aPPRove: an HMM-based method for accurate prediction of RNA-pentatricopeptide repeat protein binding events. *PLoS ONE* 11:e0160645. doi: 10.1371/journal.pone.0160645

Ihnatowicz, A., Pesaresi, P., and Leister, D. (2007). The E subunit of photosystem I is not essential for linear electron flow and photoautotrophic growth in *Arabidopsis thaliana. Planta* 226, 889–895. doi: 10.1007/s00425-007-0534-y

Ihnatowicz, A., Pesaresi, P., Varotto, C., Richly, E., Schneider, A., Jahns, P., et al. (2004). Mutants for photosystem I subunit D of *Arabidopsis thaliana*: effects on photosynthesis, photosystem I stability and expression of nuclear genes for chloroplast functions. *Plant J.* 37, 839–852. doi: 10.1111/j.1365-313X.2004.02011.x

Jarvis, P., and Lopez-Juez, E. (2013). Biogenesis and homeostasis of chloroplasts and other plastids. *Nat. Rev. Mol. Cell Biol.* 14, 787–802. doi: 10.1038/nrm3702

Karimi, M., Inze, D., and Depicker, A. (2002). GATEWAY vectors for Agrobacterium-mediated plant transformation. *Trends Plant Sci.* 7, 193–195. doi: 10.1016/S1360-1385(02)02251-3

Kindgren, P., Yap, A., Bond, C. S., and Small, I. (2015). Predictable alteration of sequence recognition by RNA editing factors from *Arabidopsis. Plant Cell* 27, 403–416. doi: 10.1105/tpc.114.134189

Koussevitzky, S., Nott, A., Mockler, T. C., Hong, F., Sachetto-Martins, G., Surpin, M., et al. (2007). Signals from chloroplasts converge to regulate nuclear gene expression. *Science* 316, 715–719. doi: 10.1126/science.%201140516

Kunst, L. (1998). Preparation of physiologically active chloroplasts from *Arabidopsis. Methods Mol. Biol.* 82, 43–48.

Kupsch, C., Ruwe, H., Gusewski, S., Tillich, M., Small, I., and Schmitz-Linneweber, C. (2012). *Arabidopsis* chloroplast RNA binding proteins CP31A and CP29A associate with large transcript pools and confer cold stress tolerance by influencing multiple chloroplast RNA processing steps. *Plant Cell* 24, 4266–4280. doi: 10.1105/tpc.112.103002

Lawrence, C. J., Dong, Q., Polacco, M. L., Seigfried, T. E., and Brendel, V. (2004). MaizeGDB, the community database for maize genetics and genomics. *Nucleic Acids Res.* 32, D393–D397. doi: 10.1093/nar/gkh011

Liu, X., Yu, F., and Rodermel, S. (2010). An *Arabidopsis* pentatricopeptide repeat protein, SUPPRESSOR OF VARIEGATION7, is required for FtsH-mediated chloroplast biogenesis. *Plant Physiol.* 154, 1588–1601. doi: 10.1104/pp.110.164111

Lurin, C., Andres, C., Aubourg, S., Bellaoui, M., Bitton, F., Bruyere, C., et al. (2004). Genome-wide analysis of *Arabidopsis* pentatricopeptide repeat proteins reveals their essential role in organelle biogenesis. *Plant Cell* 16, 2089–2103. doi: 10.1105/tpc.104.022236

Maiwald, D., Dietzmann, A., Jahns, P., Pesaresi, P., Joliot, P., Joliot, A., et al. (2003). Knock-out of the genes coding for the Rieske protein and the ATP-synthase delta-subunit of *Arabidopsis*. Effects on photosynthesis, thylakoid protein composition, and nuclear chloroplast gene expression. *Plant Physiol.* 133, 191–202. doi: 10.1104/pp.103.024190

Majeran, W., Friso, G., Asakura, Y., Qu, X., Huang, M., Ponnala, L., et al. (2012). Nucleoid-enriched proteomes in developing plastids and chloroplasts from maize leaves: a new conceptual framework for nucleoid functions. *Plant Physiol.* 158, 156–189. doi: 10.1104/pp.111.188474

Manavski, N., Guyon, V., Meurer, J., Wienand, U., and Brettschneider, R. (2012). An essential pentatricopeptide repeat protein facilitates 5′ maturation and translation initiation of rps3 mRNA in maize mitochondria. *Plant Cell* 24, 3087–3105. doi: 10.1105/tpc.112.099051

Manna, S. (2015). An overview of pentatricopeptide repeat proteins and their applications. *Biochimie* 113, 93–99. doi: 10.1016/j.biochi.2015.04.004

Martinez-Garcia, J. F., Monte, E., and Quail, P. H. (1999). A simple, rapid and quantitative method for preparing *Arabidopsis* protein extracts for immunoblot analysis. *Plant J.* 20, 251–257. doi: 10.1046/j.1365-313x.1999.00579.x

Meierhoff, K., Felder, S., Nakamura, T., Bechtold, N., and Schuster, G. (2003). HCF152, an *Arabidopsis* RNA binding pentatricopeptide repeat protein involved in the processing of chloroplast psbB-psbT-psbH-petB-petD RNAs. *Plant Cell* 15, 1480–1495. doi: 10.1105/tpc.010397

Melonek, J., Oetke, S., and Krupinska, K. (2016). Multifunctionality of plastid nucleoids as revealed by proteome analyses. *Biochim. Biophys. Acta* 1864, 1016–1038. doi: 10.1016/j.bbapap.2016.03.009

Meurer, J., Lezhneva, L., Amann, K., Godel, M., Bezhani, S., Sherameti, I., et al. (2002). A peptide chain release factor 2 affects the stability of UGA-containing transcripts in *Arabidopsis* chloroplasts. *Plant Cell* 14, 3255–3269. doi: 10.1105/tpc.006809

Meurer, J., Meierhoff, K., and Westhoff, P. (1996). Isolation of high-chlorophyll-fluorescence mutants of *Arabidopsis thaliana* and their characterisation by spectroscopy, immunoblotting and northern hybridisation. *Planta* 198, 385–396. doi: 10.1007/BF00620055

Okuda, K., Shoki, H., Arai, M., Shikanai, T., Small, I., and Nakamura, T. (2014). Quantitative analysis of motifs contributing to the interaction between PLS-subfamily members and their target RNA sequences in plastid RNA editing. *Plant J.* 80, 870–882. doi: 10.1111/tpj.12687

Olinares, P. D., Ponnala, L., and Van Wijk, K. J. (2010). Megadalton complexes in the chloroplast stroma of *Arabidopsis thaliana* characterized by size exclusion chromatography, mass spectrometry, and hierarchical clustering. *Mol. Cell. Proteom.* 9, 1594–1615. doi: 10.1074/mcp.M000038-MCP201

O'Toole, N., Hattori, M., Andres, C., Iida, K., Lurin, C., Schmitz-Linneweber, C., et al. (2008). On the expansion of the pentatricopeptide repeat gene family in plants. *Mol. Biol. Evol.* 25, 1120–1128. doi: 10.1093/molbev/msn057

Pfalz, J., and Pfannschmidt, T. (2013). Essential nucleoid proteins in early chloroplast development. *Trends Plant Sci.* 18, 186–194. doi: 10.1016/j.tplants.2012.11.003

Pogson, B. J., and Albrecht, V. (2011). Genetic dissection of chloroplast biogenesis and development: an overview. *Plant Physiol.* 155, 1545–1551. doi: 10.1104/pp.110.170365

Pogson, B. J., Ganguly, D., and Albrecht-Borth, V. (2015). Insights into chloroplast biogenesis and development. *Biochim. Biophys. Acta* 1847, 1017–1024. doi: 10.1016/j.bbabio.2015.02.003

Remans, T., Smeets, K., Opdenakker, K., Mathijsen, D., Vangronsveld, J., and Cuypers, A. (2008). Normalisation of real-time RT-PCR gene expression measurements in *Arabidopsis thaliana* exposed to increased metal concentrations. *Planta* 227, 1343–1349. doi: 10.1007/s00425-008-0706-4

Rivals, E., Bruyere, C., Toffano-Nioche, C., and Lecharny, A. (2006). Formation of the *Arabidopsis* pentatricopeptide repeat family. *Plant Physiol.* 141, 825–839. doi: 10.1104/pp.106.077826

Romani, I., Tadini, L., Rossi, F., Masiero, S., Pribil, M., Jahns, P., et al. (2012). Versatile roles of *Arabidopsis* plastid ribosomal proteins in plant growth and development. *Plant J.* 72, 922–934. doi: 10.1111/tpj.12000

Ruppel, N. J., and Hangarter, R. P. (2007). Mutations in a plastid-localized elongation factor G alter early stages of plastid development in *Arabidopsis thaliana*. *BMC Plant Biol.* 7:37. doi: 10.1186/1471-2229-7-37

Ruwe, H., Wang, G., Gusewski, S., and Schmitz-Linneweber, C. (2016). Systematic analysis of plant mitochondrial and chloroplast small RNAs suggests organelle-specific mRNA stabilization mechanisms. *Nucleic Acids Res.* 44, 7406–7417. doi: 10.1093/nar/gkw466

Schagger, H., and von Jagow, G. (1987). Tricine-sodium dodecyl sulfate-polyacrylamide gel electrophoresis for the separation of proteins in the range from 1 to 100 kDa. *Anal. Biochem.* 166, 368–379. doi: 10.1016/0003-2697(87)90587-2

Schindelin, J., Arganda-Carreras, I., Frise, E., Kaynig, V., Longair, M., Pietzsch, T., et al. (2012). Fiji: an open-source platform for biological-image analysis. *Nat. Methods* 9, 676–682. doi: 10.1038/nmeth.2019

Schmitz-Linneweber, C., Williams-Carrier, R., and Barkan, A. (2005). RNA immunoprecipitation and microarray analysis show a chloroplast Pentatricopeptide repeat protein to be associated with the 5' region of mRNAs whose translation it activates. *Plant Cell* 17, 2791–2804. doi: 10.1105/tpc.105.034454

Schmitz-Linneweber, C., Williams-Carrier, R. E., Williams-Voelker, P. M., Kroeger, T. S., Vichas, A., and Barkan, A. (2006). A pentatricopeptide repeat protein facilitates the trans-splicing of the maize chloroplast rps12 pre-mRNA. *Plant Cell* 18, 2650–2663. doi: 10.1105/tpc.106.046110

Sessions, A., Burke, E., Presting, G., Aux, G., Mcelver, J., Patton, D., et al. (2002). A high-throughput *Arabidopsis* reverse genetics system. *Plant Cell* 14, 2985–2994. doi: 10.1105/tpc.004630

Shen, C., Zhang, D., Guan, Z., Liu, Y., Yang, Z., Yang, Y., et al. (2016). Structural basis for specific single-stranded RNA recognition by designer pentatricopeptide repeat proteins. *Nat. Commun.* 7:11285. doi: 10.1038/ncomms11285

Shi, X., Hanson, M. R., and Bentolila, S. (2015). Two RNA recognition motif-containing proteins are plant mitochondrial editing factors. *Nucleic Acids Res.* 43, 3814–3825. doi: 10.1093/nar/gkv245

Small, I. D., and Peeters, N. (2000). The PPR motif - a TPR-related motif prevalent in plant organellar proteins. *Trends Biochem. Sci.* 25, 46–47. doi: 10.1016/S0968-0004(99)01520-0

Stoppel, R., Lezhneva, L., Schwenkert, S., Torabi, S., Felder, S., Meierhoff, K., et al. (2011). Recruitment of a ribosomal release factor for light- and stress-dependent regulation of petB transcript stability in *Arabidopsis* chloroplasts. *Plant Cell* 23, 2680–2695. doi: 10.1105/tpc.111.085324

Tadini, L., Pesaresi, P., Kleine, T., Rossi, F., Guljamow, A., Sommer, F., et al. (2016). GUN1 controls accumulation of the plastid ribosomal protein S1 at the protein level and interacts with proteins involved in plastid protein homeostasis. *Plant Physiol.* 170, 1817–1830. doi: 10.1104/pp.15.02033

Takenaka, M., Verbitskiy, D., Zehrmann, A., Hartel, B., Bayer-Csaszar, E., Glass, F., et al. (2014). RNA editing in plant mitochondria-connecting RNA target sequences and acting proteins. *Mitochondrion* 19(Pt B), 191–197. doi: 10.1016/j.mito.2014.04.005

Takenaka, M., Zehrmann, A., Brennicke, A., and Graichen, K. (2013). Improved computational target site prediction for pentatricopeptide repeat RNA editing factors. *PLoS ONE* 8:e65343. doi: 10.1371/journal.pone.0065343

Tiller, N., Weingartner, M., Thiele, W., Maximova, E., Schöttler, M. A., and Bock, R. (2012). The plastid-specific ribosomal proteins of *Arabidopsis thaliana* can be divided into non-essential proteins and genuine ribosomal proteins. *Plant J.* 69, 302–316. doi: 10.1111/j.1365-313X.2011.04791.x

Tomizioli, M., Lazar, C., Brugiere, S., Burger, T., Salvi, D., Gatto, L., et al. (2014). Deciphering thylakoid sub-compartments using a mass spectrometry-based approach. *Mol. Cell. Proteom.* 13, 2147–2167. doi: 10.1074/mcp.M114.040923

Varotto, C., Pesaresi, P., Jahns, P., Lessnick, A., Tizzano, M., Schiavon, F., et al. (2002). Single and double knockouts of the genes for photosystem I subunits G, K, and H of *Arabidopsis*. Effects on photosystem I composition, photosynthetic electron flow, and state transitions. *Plant Physiol.* 129, 616–624. doi: 10.1104/pp.002089

Varotto, C., Pesaresi, P., Meurer, J., Oelmuller, R., Steiner-Lange, S., Salamini, F., et al. (2000). Disruption of the *Arabidopsis* photosystem I gene psaE1 affects photosynthesis and impairs growth. *Plant J.* 22, 115–124. doi: 10.1046/j.1365-313x.2000.00717.x

Voinnet, O., Rivas, S., Mestre, P., and Baulcombe, D. (2003). An enhanced transient expression system in plants based on suppression of gene silencing by the p19 protein of tomato bushy stunt virus. *Plant J.* 33, 949–956. doi: 10.1046/j.1365-313X.2003.01676.x

Waadt, R., and Kudla, J. (2008). In planta visualization of protein interactions using Bimolecular Fluorescence Complementation (BiFC). *CSH Protoc* 2008, dbrot4995.

Weigel, M., Varotto, C., Pesaresi, P., Finazzi, G., Rappaport, F., Salamini, F., et al. (2003). Plastocyanin is indispensable for photosynthetic electron flow in *Arabidopsis thaliana*. *J. Biol. Chem.* 278, 31286–31289. doi: 10.1074/jbc.M302876200

Williams-Carrier, R., Kroeger, T., and Barkan, A. (2008). Sequence-specific binding of a chloroplast pentatricopeptide repeat protein to its native group II intron ligand. *RNA* 14, 1930–1941. doi: 10.1261/rna.1077708

Woodson, J. D., and Chory, J. (2008). Coordination of gene expression between organellar and nuclear genomes. *Nat. Rev. Genet.* 9, 383–395. doi: 10.1038/nrg2348

Yamazaki, H., Tasaka, M., and Shikanai, T. (2004). PPR motifs of the nucleus-encoded factor, PGR3, function in the selective and distinct steps of chloroplast gene expression in *Arabidopsis*. *Plant J.* 38, 152–163. doi: 10.1111/j.1365-313X.2004.02035.x

Yin, P., Li, Q., Yan, C., Liu, Y., Liu, J., Yu, F., et al. (2013). Structural basis for the modular recognition of single-stranded RNA by PPR proteins. *Nature* 504, 168–171. doi: 10.1038/nature12651

Yoo, S. D., Cho, Y. H., and Sheen, J. (2007). *Arabidopsis* mesophyll protoplasts: a versatile cell system for transient gene expression analysis. *Nat. Protoc.* 2, 1565–1572. doi: 10.1038/nprot.2007.199

Zoschke, R., Kroeger, T., Belcher, S., Schottler, M. A., Barkan, A., and Schmitz-Linneweber, C. (2012). The pentatricopeptide repeat-SMR protein ATP4 promotes translation of the chloroplast atpB/E mRNA. *Plant J.* 72, 547–558. doi: 10.1111/j.1365-313X.2012.05081.x

Zoschke, R., Qu, Y., Zubo, Y. O., Borner, T., and Schmitz-Linneweber, C. (2013a). Mutation of the pentatricopeptide repeat-SMR protein SVR7 impairs accumulation and translation of chloroplast ATP synthase subunits in *Arabidopsis thaliana*. *J. Plant Res.* 126, 403–414. doi: 10.1007/s10265-012-0527-1

Zoschke, R., Watkins, K. P., and Barkan, A. (2013b). A rapid ribosome profiling method elucidates chloroplast ribosome behavior in vivo. *Plant Cell* 25, 2265–2275. doi: 10.1105/tpc.113.111567

Conflict of Interest Statement: The authors declare that the research was conducted in the absence of any commercial or financial relationships that could be construed as a potential conflict of interest.

10

Comprehensive Analysis of the CDPK-SnRK Superfamily Genes in Chinese Cabbage and its Evolutionary Implications in Plants

Peng Wu[1], Wenli Wang[1], Weike Duan[1,2], Ying Li[1] and Xilin Hou[1]*

[1] State Key Laboratory of Crop Genetics and Germplasm Enhancement, Key Laboratory of Biology and Germplasm Enhancement of Horticultural Crops in East China, Ministry of Agriculture, Nanjing Agricultural University, Nanjing, China,
[2] School of Life Science and Food Engineering, Huaiyin Institute of Technology, Huaian, China

Edited by:
José M. Romero,
University of Seville, Spain

Reviewed by:
Nigel G. Halford,
Rothamsted Research (BBSRC), UK
Jianchang Du,
Jiangsu Academy of Agricultural Sciences, China

***Correspondence:**
Xilin Hou
hxl@njau.edu.cn

The CDPK-SnRK (calcium-dependent protein kinase/Snf1-related protein kinase) gene superfamily plays important roles in signaling pathways for disease resistance and various stress responses, as indicated by emerging evidence. In this study, we constructed comparative analyses of gene structure, retention, expansion, whole-genome duplication (WGD) and expression patterns of CDPK-SnRK genes in *Brassica rapa* and their evolution in plants. A total of 49 *BrCPKs*, 14 *BrCRKs*, 3 *BrPPCKs*, 5 *BrPEPRKs*, and 56 *BrSnRKs* were identified in *B. rapa*. All BrCDPK-SnRK proteins had highly conserved kinase domains. By statistical analysis of the number of CDPK-SnRK genes in each species, we found that the expansion of the CDPK-SnRK gene family started from angiosperms. Segmental duplication played a predominant role in CDPK-SnRK gene expansion. The analysis showed that PEPRK was more preferentially retained than other subfamilies and that CPK was retained similarly to SnRK. Among the CPKs and SnRKs, CPKIII and SnRK1 genes were more preferentially retained than other groups. CRK was closest to CPK, which may share a common evolutionary origin. In addition, we identified 196 CPK genes and 252 SnRK genes in 6 species, and their different expansion and evolution types were discovered. Furthermore, the expression of *BrCDPK-SnRK* genes is dynamic in different tissues as well as in response to abiotic stresses, demonstrating their important roles in development in *B. rapa*. In summary, this study provides genome-wide insight into the evolutionary history and mechanisms of CDPK-SnRK genes following whole-genome triplication in *B. rapa*.

Keywords: CDPK-SnRK genes, *Brassica rapa*, evolutionary conservation, synteny analysis, evolutionary pattern, expression pattern

INTRODUCTION

Plants are remarkably responsive to a variety of environmental stimuli, including pathogen attack, wounding, cold, drought reception, and fluctuations in incident light (Kudla et al., 2010). Meanwhile, a variety of internal substances also affect plants growth. These external and internal signals compose a complex regulatory network that allows plants to develop in balance. Following the detection of a stress stimulus, various signal transduction pathways are switched on, resulting in physiological changes in the plant cell. As second messengers, calcium ions play an essential role in many important cellular processes, especially under stress conditions (Trewavas and Malhó, 1998; Sanders et al., 1999; Berridge et al., 2000). In plants, transient changes in calcium content in the cytosol (calcium signatures) have been observed during growth, development and stress conditions (Evans et al., 2001; Harper, 2001; Knight and Knight, 2001; Sanders et al., 2002). Intracellular Ca^{2+} signals are produced in plant cells by a variety of stimuli, such as changes in environmental conditions, interaction with microbes, and developmental programs (Bush, 1995; Ehrhardt et al., 1996; Hammond-Kosack and Jones, 1996; Knight et al., 1996; Taylor and Hepler, 1997; Pei et al., 2000; Assmann and Wang, 2001; Murata et al., 2001; McAinsh et al., 2002; Plieth and Trewavas, 2002; Ritchie et al., 2002). Plants have multiple calcium stores, including the apoplast, vacuole, nuclear envelope, endoplasmic reticulum (ER), mitochondria and chloroplasts. Therefore, each stimulus can elicit a characteristic Ca^{2+}wave by specifically altering the activity of various differentially localized Ca^{2+} channels, H/Ca^{2+} antiporters, and Ca^{2+} and H^+ ATPases (Thuleau et al., 1998; Allen et al., 2000; Harmon et al., 2000; Hwang et al., 2000). Different calcium sensors recognize specific calcium signatures and transduce them into downstream effects, including altered protein phosphorylation and gene expression patterns (Sanders et al., 1999; Rudd and Franklin-Tong, 2001).

In eukaryotes, calcium-dependent protein kinases (CDPKs) and most sucrose non-fermenting-1-related kinases (SnRKs) are involved in regulating and decoding Ca^{2+} signals (Assmann and Wang, 2001; Evans et al., 2001; Harmon et al., 2001; Cheng et al., 2002; Fasano et al., 2002; Hrabak et al., 2003; Cho et al., 2009; Kulik et al., 2011). The protein kinases also involved in stress signal transduction in plants are common to all eukaryotic organisms and include mitogen-activated protein kinases (MAPKs), glycogen synthase kinase 3 (GSK3), and S6 kinase (S6K). The CDPK-SnRK superfamily consists of seven types of protein kinases, which differ in the regulatory domains they contain (Harmon et al., 2001). CDPKs (also named CPKs) are activated by the binding of calcium to their calmodulin-like regulatory domains. The carboxyl terminal domains of CRKs (CDPK-related kinases) have sequence similarity to the regulatory domains of CPKs but do not bind calcium. PEPRKs (PEP carboxylase kinases) contain only one catalytic domain (Harmon et al., 2001). PPCKs (PEPC kinases) have a carboxyl-terminal domain that has no similarity to that of any other member of the superfamily (Hrabak et al., 2003). CCaMKs (calcium- and calmodulin-dependent protein kinases) bind both calcium ions and the calcium/calmodulin complex, whereas CaMKs (calmodulin-dependent protein kinases) bind the calcium/calmodulin complex but not calcium (Hrabak et al., 2003). In addition, there are the classical SNF1-type kinases from yeast; Halford and Hardie (1998) proposed the name SNF1-related kinase (SnRK) for this group and recognized three subgroups: SnRK1, SnRK2, and SnRK3 (Harmon et al., 2001). However, CaMK and CCaMK are absent from *Arabidopsis* (Hrabak et al., 2003). All members of the CDPK-SnRK gene superfamily have kinase domains of similar length and sequence and a similar general organization, with the kinase domains at or near the N-terminus, then the junction domains, followed by the regulatory domains (Harmon, 2003; Hrabak et al., 2003).

The plant CPKs characterized to date play substantive roles in diverse physiological processes. These processes include tolerance to salt, cold, and drought stress in rice (Saijo et al., 2000), the defense response in tobacco (Romeis et al., 2000), the accumulation of storage starch and protein in immature seeds of rice (Asano et al., 2002), the regulation and development of nodule number in *Medicago truncatula* (Gargantini et al., 2006), and the response to ABA in *Arabidopsis* (Choi et al., 2005) (38). The original, systematic report on the CPK genes family in *Arabidopsis thaliana* identified 34 CPK genes family members (Choi et al., 2005) and was followed by research in rice (*Oryza sativa*) (Ray et al., 2007) and wheat (*Triticum aestivum*) (Li et al., 2008). Recently, genome-wide analyses of the CPK gene family have been reported in maize (*Zea mays*) (Kong et al., 2013) and poplar (*Populus trichocarpa*) (Zuo et al., 2013). Meanwhile, more and more investigations of CPK genes have also involved horticultural plants, such as alfalfa (Davletova et al., 2001), potato (Raíces et al., 2003), strawberry (Llop-Tous et al., 2002), and tomato (Chico et al., 2002). Furthermore, research using transgenic plants has revealed the biological functions of a few CPK genes in higher plants. Transgenic rice constitutively overexpressing *OsCPK7* or *OsCPK13* showed enhanced tolerance to cold, salt, and drought stress (Saijo et al., 2000; Komatsu et al., 2007). In tobacco, CPK-silenced plants displayed a reduced and delayed hypersensitive response to the fungal Avr9 elicitor (Romeis et al., 2001). *GhCPK1* was the first cotton CPK gene to be identified and was considered to play a role in the calcium signaling events associated with fiber elongation (Huang et al., 2008). *Arabidopsis thaliana CPK23* (*AtCPK23*) is a positive regulator of the response to drought and salt stress (Ma and Wu, 2007), whereas *AtCPK6* may be crucial in positively regulating methyl-jasmonate signaling in guard cells (Munemasa et al., 2011). In addition, the overexpression of rice (*Oryza sativa*) *CPK7* (*OsCPK7*) significantly improves resistance to cold (Komatsu et al., 2007). Phytohormones are involved in the responses to abiotic stresses; therefore, the expression levels of members of the CPK gene family has also been shown to be regulated after treatment with various phytohormones, such as ABA, auxin and jasmonic acid. Recently, *Zea mays CPK11* was reported as a component of the jasmonic acid signaling pathway, and its concentration in cells was observed to increase in response to wounding and touch (Szczegielniak et al., 2012).

Sucrose non-fermenting-1-related protein kinase (SnRK) is homologous to SNF1 and AMP-activated protein kinase (AMPK), which is widely distributed in plants and is involved in a variety of signaling pathways. SnRK is the key switch in plant sugar signaling, stress, seed germination and seedling growth. SNF1 of yeast, AMPK of mammals and SnRK1 of plants are homologous, belonging to the SNF1 protein kinase superfamily. SNF1 was found in yeast (*Saccharomyces cerevisiae*) originally (Alderson et al., 1991). In yeast, glucose regulates the protein-protein interaction, substrate specificity and subcellular localization of the SNF1 subunit that modulates SNF1 kinase activity, resulting in the phosphorylation of activators and repressors that control transcription of multiple genes in metabolic pathways required for the utilization of alternative energy sources. In the eukaryote, SNF1 protein kinase is very strongly conserved. Many SNF1 analogs have been identified in plants. SnRK1 was discovered initially in rye (*Secale cereale* L.) (Alderson et al., 1991). At present, some members of the SnRK1 subfamily have been found in variety of model plants and some important crops, such as *Arabidopsis thaliana*, rye, barley (*Hordeum vulgare*), potatoes (*Solanum tuberosum*), tobacco, beets, etc. It may exist in all plants (Halford and Hardie, 1998; Halford et al., 2003). Studies have shown that SnRK1 is the key switch in plant sugar signaling. In addition, the regulation of glucose metabolism, hormonal regulation and sugar signaling is directly related to signal transduction (Kleinow et al., 2000; Jossier et al., 2009; Mathieu et al., 2009). SnRK2s are a plant-specific Ser/Thr protein kinase family. All of the members have a conserved N-terminal catalytic domain similar to that of SNF1/AMPK-type kinases and a short C-terminal regulatory domain that is not highly conserved. Prior to 2000, there were only a small number of studies indicating that ABA and abiotic stresses induced the expression of some *SnRK2* genes (Anderberg and Walker-Simmons, 1992; Holappa and Walker-Simmons, 1995). In 2000, SnRK2s began to be recognized as enzymes involved in abiotic stress signal transduction in plants (Li et al., 2000). By 2003, 10 SnRK2 genes had been identified and were renamed *SnRK2.1* through *SnRK2.10* (Hrabak et al., 2003). In 2009, independently, two laboratories obtained a triple *SnRK2.2/2.3/2.6* mutant. *SnRK2.2/2.3/2.6* triple-mutant plants are nearly completely insensitive to ABA, which was used to establish the role of ABA-dependent SnRK2s in the plant response to water deficit, seed maturation, and germination. These reports indicate that *SnRK2.2/2.3/2.6* function as primary positive regulators and suggest that ABA signaling is controlled by the dual modulation of *SnRK2.2/3/6* and group A PP2Cs (Fujii and Zhu, 2009; Fujii et al., 2009; Nakashima et al., 2009). SnRK3 is a protein kinase in plants, called calcineurin B-like calcium sensor-interacting protein kinase (CIPK) (Kim et al., 2000). CIPK interacts with the calcium-binding protein SOS3, SCaBPS and CBL (calcineurin B-like calcium sensor). Studies have shown that CIPK and an upstream complex of CSL interactions are involved in salt stress, sucrose and ABA signal transduction (Imamura et al., 2008). In *Arabidopsis*, PKS3, PKS18 and CIPK3 of the SnRK3 family can regulate plant growth, stomatal opening and closing and seed germination under ABA treatment (Kim et al., 2003). Arabidopsis AtCIPK1 forms complexes with AtCBL1 and AtCBL9, regulating ABA-independent and ABA-dependent pathways, respectively (D'Angelo et al., 2006). AtCIPK3 regulates ABA and cold signal transduction pathways (Kim et al., 2003). Girdhar's study showed that CBL9 interacted with CIPK3 to regulate the ABA pathway, and this finding was validated in a yeast two-hybrid experiment (Pandey et al., 2008).

During their evolution, plants have substantially altered their phenotypes to adapt to environmental changes by transforming the form and function of genes. Gene duplication, even a whole-genome duplication (WGD), offers the chance for genes to change (Rensing, 2014). Angiosperm genome evolution is characterized by polyploidization through WGD followed by diploidization, which is typically accompanied by considerable homoeologous gene loss (Stebbins, 1950). After duplication, one copy of the gene might either becomes non-functional (pseudogenized or silenced, also called gene death) or acquire a novel function (neofunctionalization). Alternatively, the two duplicates might divide the original function of the gene (Innan and Kondrashov, 2010). Preliminary analyses revealed that gene duplication and subsequent divergence are the main contributors to evolutionary momentum (Ohno et al., 1968; Chothia et al., 2003). The genome of *A. thaliana* has experienced a paleohexaploidy (β) duplication shared with most dicots and two subsequent genome duplications (α and γ) since its divergence from *Carica papaya*, along with rapid DNA sequence divergence and extensive gene loss (Bowers et al., 2003). In *A. thaliana*, some duplicated regions found in CDPK-SnRK protein kinases indicated that CDPK-SnRK protein kinases are paralogs that arose by divergence after genome duplication events (Hrabak et al., 2003). The CPK genes of *Arabidopsis* and maize have undergone both segmental and tandem duplication, contributing to the expansion of the CPK family. In *Populus*, however, segmental duplication played a predominant role in the expansion of CPK genes (Zuo et al., 2013). In addition, tandem duplication of CPK genes has not occurred in the rice genome (Asano et al., 2005).

In this study, we constructed a comprehensive comparative analysis of CDPK-SnRK genes, including phylogenetic relationships, gene structures, chromosome distribution, gene retentions, gene expansions, gene duplication and gene expression patterns, in different tissues to characterize the divergences in composition, expansion, and expression. First, we identified 555 *CPKs*, 120 *CRKs*, 5 *PPCKs*, 14 *PEPRKs*, and 697 *SnRKs* in 16 plant species. Second, we conducted a comparative genomic analysis of these genes with 16 other plant and species found that the expansion of the CDPK-SnRK family from angiosperms mainly relied on WGDs. Third, PEPRK genes were more preferentially retained than other subfamilies and CPK genes were retained similarly to SnRK genes during diploidization following WGT in *B. rapa*. Fourth, during the course of evolution, CPK appeared most recently and expanded most rapidly. Fifth, the expressions of CDPK-SnRK genes are dynamic in different tissues as well as in response to abiotic stresses, demonstrating their important roles in development in *B. rapa*. This study is the first report on CDPK-SnRK genes in *B. rapa*. and extends our understanding of the roles of the CDPK-SnRK gene superfamily in evolution and stress responses.

RESULTS

Identification and Classification of the CDPK-SnRK Superfamily of Protein Kinases in *Brassica rapa* and Comparative Analyses

In this study, genome-wide analysis of CDPK-SnRK gene family has been performed on the basis of the completed *B. rapa* genome sequence (Wang et al., 2011). Based on previously reported methods (Harmon et al., 2001; Hrabak et al., 2003), the homogeneous candidate CDPK-SnRK genes between

Brassica rapa and other species were identified by BLASTP (Supplementary Table 1). Subsequently, all candidate protein sequences were subjected to Pfam and SMART analyses. Finally, we identified 49 *BrCPKs*, 14 *BrCRKs*, 3 *BrPPCKs*, 5 *BrPEPRKs*, and 56 *BrSnRKs* named according to nomenclature proposed for CDPK-SnRK genes (Supplementary Table 2).

To better understand the expansion and evolutionary history of CDPK-SnRK genes in *B. rapa*, genes were also identified in 16 other species representing the major clades of plants. The evolutionary relationships of the species and the number of CDPK-SnRK genes are shown in **Figure 1A**. The data show that *Glycine max* contained the highest number of CDPK-SnRK genes (200), followed by *Z. mays* (193) and *M. truncatula* (188) (**Figure 1A**). However, *A. trichopoda*, a basal angiosperm species that was the single living representative of the sister lineage to all other extant flowering plants, contained the lowest number of CDPK-SnRK genes (28) in *Angiospermae*. The reason is that it originated prior to the split of eudicots and monocots and has not experienced any whole genome duplication (WGD), while the other 12 angiosperms had several rounds of WGDs/triplications after their split from *A. trichopoda*. Furthermore, the number of CDPK-SnRK genes in algae, Bryophyta and Pteridophyta was less than that in *Angiospermae*. This phenomenon was also caused by several WGD events that occurred during angiosperm evolution (**Figure 1B**). These results indicated that the expansion of the CDPK-SnRK family from angiosperms mainly relied on large-scale DNA rearrangements, namely, WGDs. The elevated duplication frequency and increased retention of CDPK-SnRK genes also contributed to neofunctionalization and caused them to gain important functions in angiosperm development.

Characteristics of Structure, and Expansion Analysis of BrCDPK-SnRK Proteins

To investigate the extent of lineage-specific expansion of the CDPK-SnRK genes in *B. rapa*, phylogenetic trees were constructed using the maximum likelihood method (**Figure 2A**). The phylogenetic tree showed that all the CDPK-SnRK genes were clustered into five distinct gene classes (CPK, CRK, PEPRK, PPCK, SnRK) (**Figure 2A**), while the CPK family was divided into four groups (I, II, III, and IV) and the SnRK family was classified into three groups (SnRK1, SnRK2, and SnRK3), consistent with the reports in *A. thaliana* (Hrabak et al., 2003). In *B. rapa*, the CPK, CRK, PEPRK, PPCK, and SnRK gene families contained 49 members, 14 members, 3 members, 5 members, and 56 members, respectively, whereas in *A. thaliana*, the CPK, CRK, PEPRK, PPCK, and SnRK families contained 34 members, 8 members, 2 members, 2 members, and 39 members, respectively. Next, the synteny of CDPK-SnRK genes between *A. thaliana* and three subgenomes in *B. rapa* was analyzed. There were 34 *CPK*, 8 *CRK*, 2 *PEPRK*, 2 *PPCK*, and 38 *SnRK* genes on the conserved collinear block (Supplementary Table 5). Meanwhile, 2 *CPKs*, 2 *CRKs*, 1 *PEPRK*, 1 *PPCK*, and 3 *SnRK*s were retained completely; conversely, 4 *CPKs* and 3 *SnRKs* from *B. rapa* were lost. Due to the *Brassica*-specific WGT event, the gene numbers of these classes in *B. rapa* were greater than those in *A. thaliana*.

Furthermore, the different domain architectures, motif compositions and gene structures of CDPK-SnRK were analyzed (**Figure 2B**, Figure S1). All members of the CDPK-SnRK superfamily have a kinase domain of similar length and sequence, with the kinase domains at or near the N-terminus, then the junction domains, followed by the regulatory domains (**Figure 2B**). Although CPK proteins have a functional kinase domain coupled with regulatory calcium-binding EF-hands, the C-terminal domains (EF-hands) of CRK proteins contain apparently degenerate calcium-binding sites with no function. Meanwhile, 10 conserved motifs were detected in BrCPK, BrCRK, BrPEPRK, BrPPCK, and BrSnRK, respectively (Figure S3). All BrCDPK-SnRK proteins had highly conserved kinase domains, which corresponded to motifs 1-4,7,9 in BrCDPK and motifs 1-4,6,7,9 in BrSnRK, whereas motif 8 was found in *BrSnRK3.1*, corresponding to the NAF/FISL domains (Figures S1A,B). The amino acid sequence of BrCDPK-SnRK was aligned with AtCDPK-SnRK protein sequences from five gene classes. In CPK, CRK, PEPRK, PPCK, and SnRK, higher sequence similarities were identified in the N-terminus, which corresponded to the conserved kinase domain (Figure S2). In addition, variable gene structures of BrCDPK-SnRK were observed. As shown in **Figures 3A,C**, the intron numbers of the BrCPK genes ranged from 5 to 9 with a median of 6, while the BrSnRK genes ranged from 0 to 15 with a median of 7. Interestingly, we found that 22 BrSnRK3 genes have no introns. The theoretical pI of the BrCPK gene family ranged from 4 to 9 with a median of 6, but BrSnRK proteins showed a pI range from 2 to 11 and with a median of 8 (**Figure 3B**). Other classes had complex theoretical pI ranging in value from 4 to 9 (**Figure 3B**).

Different Retention of CDPK-SnRK Genes Following WGT in *Brassica rapa*

To investigate different retention in CPK, CRK, PEPRK, PPCK, and SnRK during *B. rapa* WGT events, 44/49, 14/14, 3/3, 5/5, and 50/56 were located in the syntenic regions, respectively (**Figures 4A,B** and Supplementary Table 3). The results demonstrated that 43% (44/102) of the CPK genes were retained in the syntenic regions, relative to 44% (50/114) of the SnRK genes. The retention rates of CRK, PEPRK and PPCK are 58% (14/24), 50% (3/6), and 83% (5/6), respectively (**Figure 4F**). Additionally, we counted gene copies and analyzed the distribution of the three subgenomes by comparing the retention of CPK, CRK, PEPRK, PPCK, and SnRK (**Figure 4E**). The result showed that all PEPRK genes had more than two copies retained, which is more than the retention of the other subfamilies (42%). However, 3% of the CPK and SnRK genes were completely lost. Next, the proportions of CPK and SnRK genes retained were higher in the least fractionated (LF) subgenome than in the medium fractionated (MF1) and most fractionated (MF2) subgenome, consistent with a previous report showing that the degree of retained genes in these three subgenomes (LF, MF1, and MF2) was decreased (Wang et al., 2011). In contrast, the PEPRK and PPCK families were retained more in the MF1 subgenome

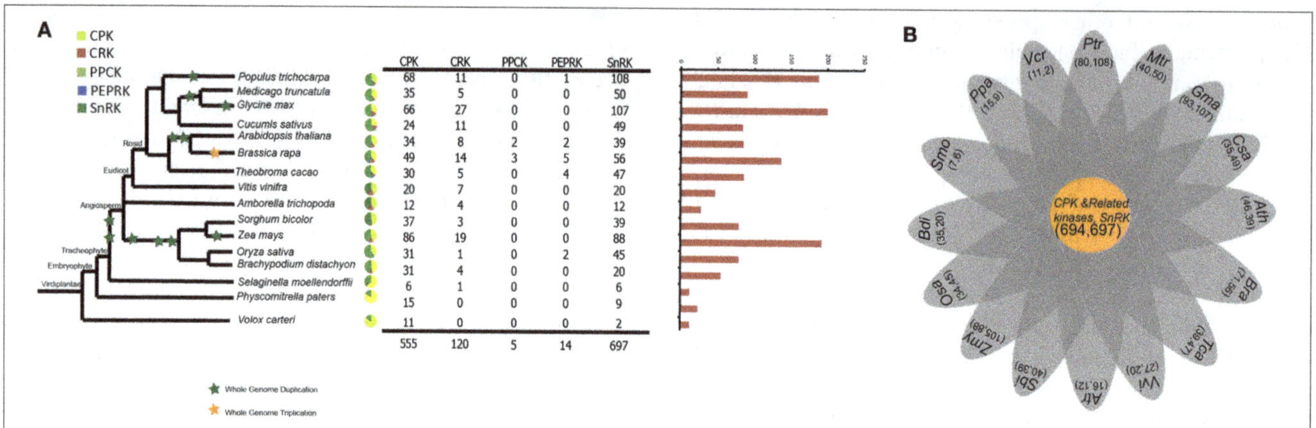

FIGURE 1 | A comparative analysis of CDPK-SnRK genes in plants. (A) The evolutionary relationships and the numerical details of the CDPK-SnRK superfamily of each species. **(B)** The Venn diagram shows the number of common gene families and genes in 16 plants.

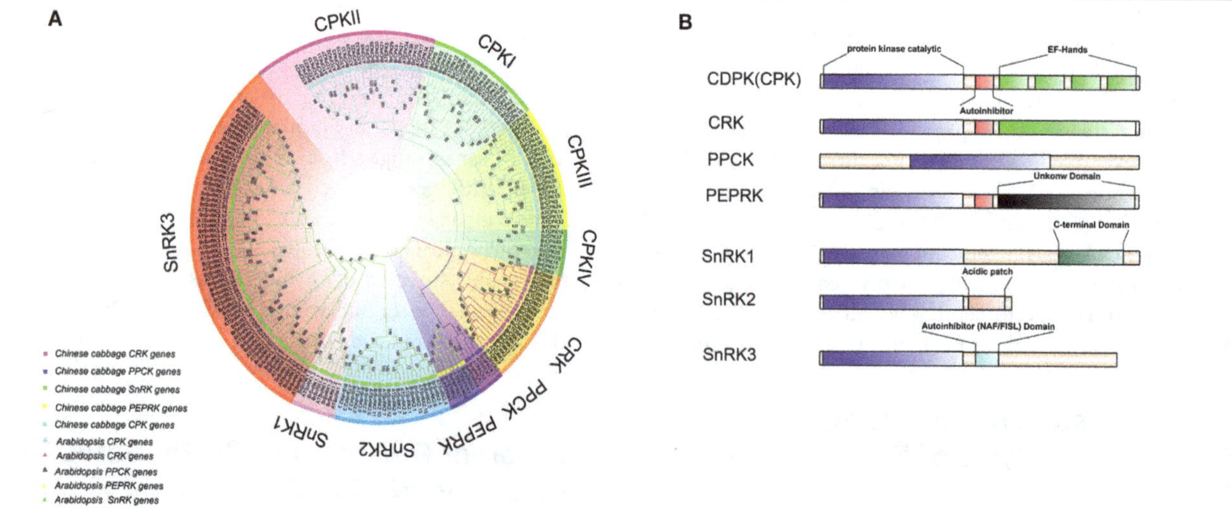

FIGURE 2 | Phylogenetic tree and major domains of CDPK-SnRK proteins. (A) Phylogenetic analysis of 212 CDPK-SnRK proteins from Chinese cabbage (127) and *Arabidopsis* (85) showing similar groups in all of the plant species. In total, 10 clades with different colors that were formed by CDPK-SnRK proteins are also marked. **(B)** The kinases are drawn with the N-terminus to the left and C-terminus to the right. Catalytic domains (blue boxes) have been aligned to facilitate comparison. Auto-inhibitory domains are shown as red boxes. EF-hands are shown as green boxes within the regulatory domains. The SnRK1 C-terminus domains are shown as dark green boxes. Acidic patch domains are shown as pink boxes. The SnRK3 NAF/FISL domains are shown as dark and light blue boxes.

than in the LF subgenome (**Figure 4F**). In summary, the results confirmed that PEPRK genes were more preferentially retained than other subfamilies and that CPK genes were retained similarly to SnRK genes during diploidization following WGT in *B. rapa*.

Furthermore, the retention rates of four CPK (*CPKI, II, III, IV*) and three SnRK (*SnRK1, 2, 3*) groups were observed. As show in **Figure 4H**, 67% of CPKIIIs and SnRK1s had more than two copies retained, which is greater than the retention of the other groups (**Figures 4H,D**). In addition, the proportion of CPKIIIs and SnRK1s retained was higher in the LF subgenome than in the other subgenomes, which once again confirmed that CPKIII and SnRK1 genes were more preferentially retained than other groups (**Figures 4C,G**).

Chromosome Distribution, *Ks* and Duplication Analysis of the CDPK-SnRK Genes in *B. rapa*

All BrCDPK-SnRK genes could be mapped onto 10 chromosomes of Chinese cabbage with a non-random distribution, except *BrPPCK3*, which is located in Scaffold000191 (**Figure 5**). On every chromosome, the proportion of BrCPK genes was similar to that of BrSnRK genes. However, Chromosome 09 contained more BrCPK genes (11 genes) than BrSnRK genes, whereas chromosomes 01 had the opposite. *B. rapa* shares two WGDs (WGD: α and β) and one whole-genome triplication event (WGT: γ) in its evolutionary history with *Arabidopsis* but has undergone an additional WGT event. Therefore, the *B. rapa* genome was further divided into three

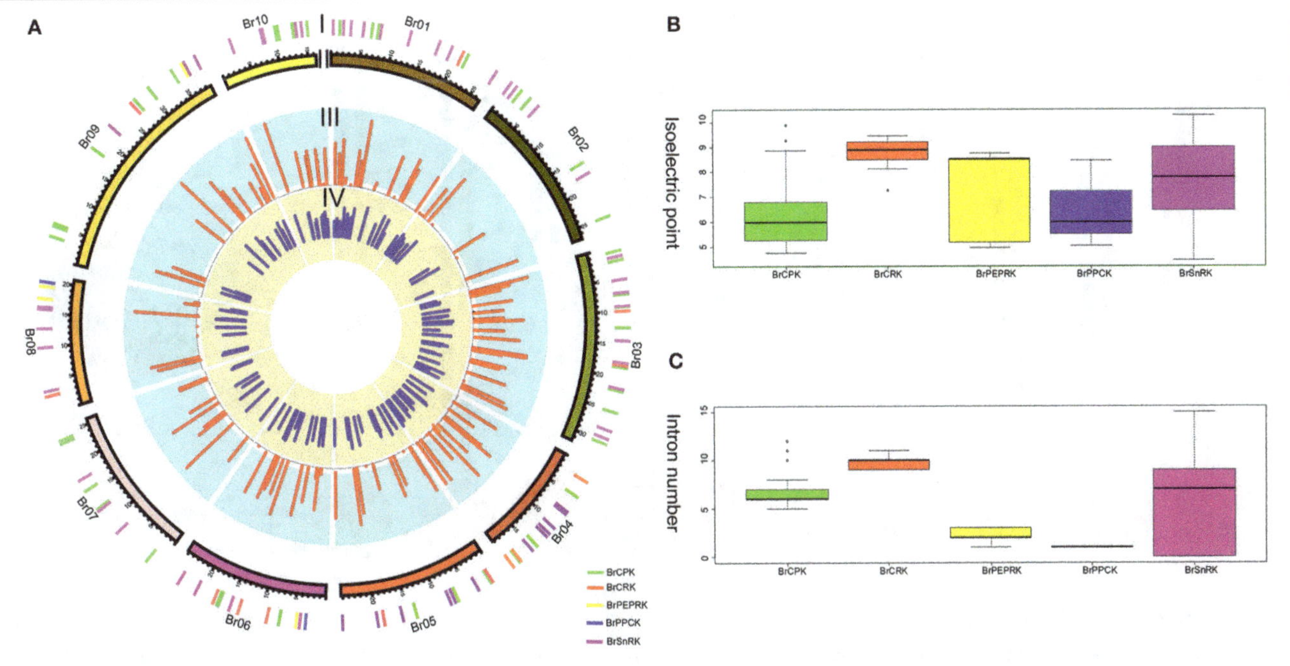

FIGURE 3 | Characteristics of isoelectric points and introns among BrCDPK-SnRK. (A) I indicates CDPK-SnRK genes within the *B. rapa* Genome. II indicates the *B. rapa* chromosome karyotype. III represents the number of introns (red) of CDPK-SnRK genes. IV indicates the Pi value (blue) of BrCDPK-SnRK genes. **(B)** The Pi value among gene classes BrCPK, BrCRK, BrPEPRK, BrPPCK, and BrSnRK. **(C)** The number of introns among gene classes BrCPK, BrCRK, BrPEPRK, BrPPCK, and BrSnRK.

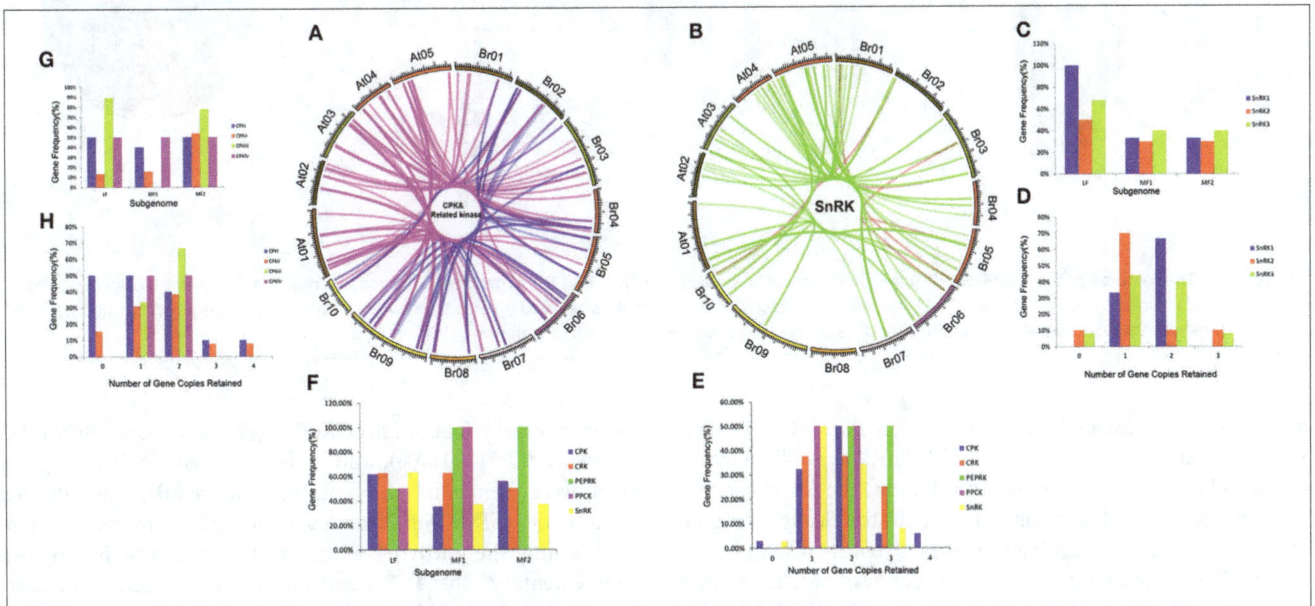

FIGURE 4 | CDPK-SnRK homologous genes in the segmental syntenic regions of the genomes of *Brassica rapa* and *Arabidopsis thaliana* and their different retention. (A) CPK, CRK, PEPRK, and PPCK syntenic gene lines are shown between the 10 *B. rapa* chromosomes (Br01-Br10) and the five *Ar. Thaliana* chromosomes (At01-At05). The pink lines represent the syntenic gene pairs between Chinese cabbage and *Arabidopsis*; the blue lines represent the syntenic genes in Chinese cabbage. **(B)** SnRK syntenic gene lines are shown between the 10 *B. rapa* chromosomes and the five *A. thaliana* chromosomes. The green lines represent the syntenic genes pairs between Chinese cabbage and *Arabidopsis*; the red lines represent the syntenic genes in Chinese cabbage. **(C)** Retention of homoeologs of SnRK genes in the three subgenomes (LF, MF1, and MF2) in *B. rapa*. LF: least fractionized subgenome; MF1: moderately fractionized subgenome; MF2: most fractionized subgenome. **(D)** Copy numbers of SnRK genes after genome triplication and fractionation in *B. rapa*. **(E)** Copy numbers of CDPK-SnRK genes after genome triplication and fractionation in *B. rapa*. **(F)** Retention of homoeologs of CDPK-SnRK genes in the three subgenomes (LF, MF1, and MF2) in *B. rapa*. **(G)** Retention of homoeologs of CPK genes in the three subgenomes (LF, MF1, and MF2) in *B. rapa*. **(H)** Copy numbers of CPK genes after genome triplication and fractionation in *B. rapa*.

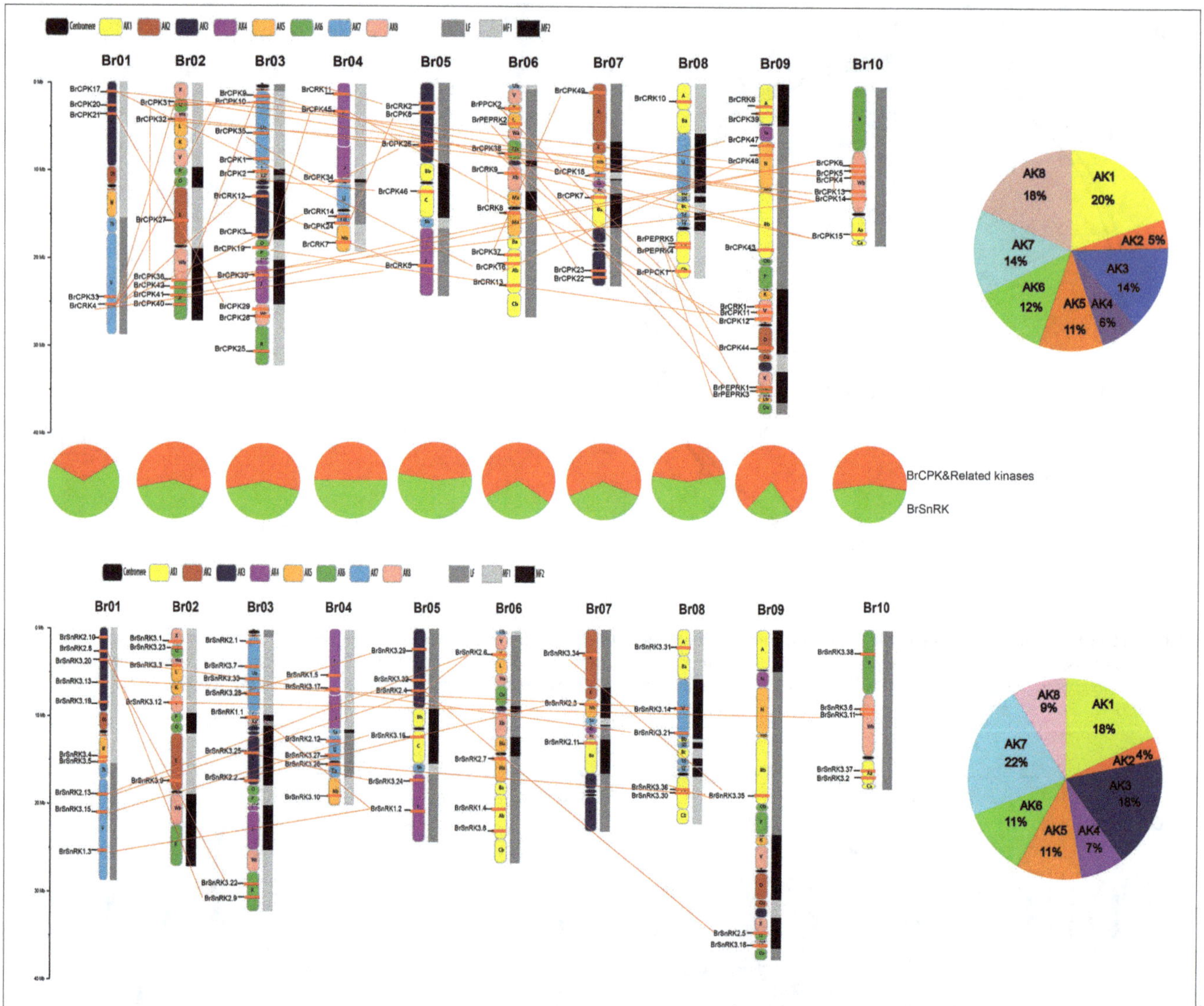

FIGURE 5 | The features of the CDPK-SnRK genes in *B. rapa*. The distribution of BrCDPK-SnRK genes on 10 chromosomes. The 8 ancestral blocks and three subgenomes, including the least fractionated (LF), medium fractionated (MF1), and most fractionated (MF2) subgenomes, were plotted as described by Cheng et al. (2013). AK represents the ancestral karyotype. The red lines connect the duplicated BrCDPK-SnRK genes.

differentially fractionated subgenomes (LF, MF1, MF2), of which LF contained more BrCPK genes and BrSnRK genes than either of the other two subgenomes. In addition, the 24 conserved ancestral genomic blocks (labeled A–X) in the *B. rapa* genome were reconstructed according to previous reports (Cheng et al., 2013). The color coding of these blocks was based on their positions in a proposed ancestral karyotype (AK1-8). We also found that most of the BrCPK genes cluster together in a region of AK1 (20%), whereas BrSnRK genes belonged to AK1 (18%) and AK3 (18%).

Furthermore, the types were identified by the MCScanX program, and the divergence times of the duplicated genes were estimated by calculating the synonymous substitution rates (*Ks*) and non-synonymous substitution rates (*Ka*). In total, 70 *BrCDPK-SnRK* duplicated gene pairs were analyzed (Supplementary Table 4). *BrCPK* (74%), *BrCRK* (86%), *BrPEPRK* (100%), *BrPPCK* (100%), and *BrSnRK* (70%) duplicated gene pairs were segmental duplications (**Figure 6B**), and all the duplicated *BrCDPK-SnRK* gene pairs had *Ka/Ks* ratios less than 1, indicating the purifying selection of these genes (**Figure 6A**, Supplementary Table 4). To understand the divergence time, the *Ks* values of the BrCPK genes ranged from 0.3 to 0.5 and had a mean of ∼0.34 (∼11 *Myr*), while the BrSnRK genes ranged from 0.2 to 0.55 and focused on ∼0.25 (∼8.5 *Myr*; **Figures 2A,B, 6A**). The divergence time of BrSnRK duplicated gene pairs was 8.49 MYA, which indicates that their divergence occurred during the *Brassica* triplication events (5 ∼ 9 *MYA*). The divergence times obtained for the BrCPK duplicated gene pairs ranged from 10 to 16.6 *MYA*, indicating that these duplications occurred during the divergence of Chinese cabbage and *Arabidopsis* (9.6–16.1 *MYA*).

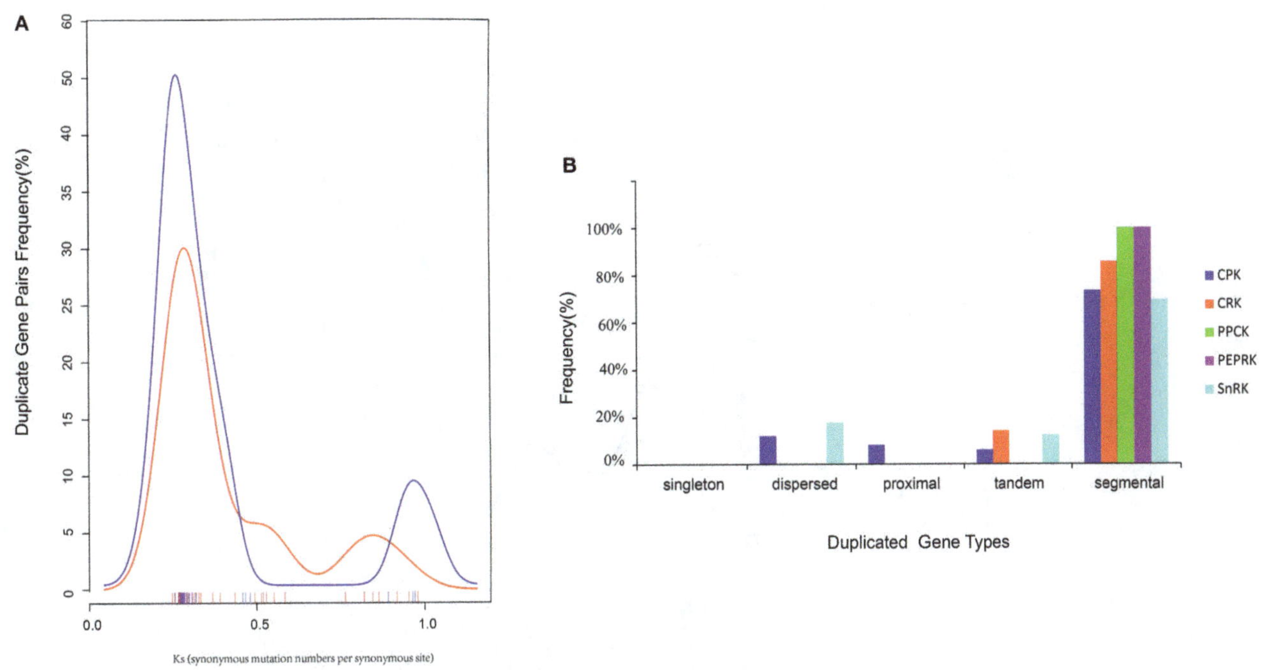

FIGURE 6 | Pairwise comparison of the *Ks* values and types for the duplicated *CDPK-SnRK* gene pairs in *Brassica rapa*. (A) The distribution of the *Ks* values for CDPK-SnRK genes between *A. thaliana* and *B. rapa*. **(B)** The different duplicated types (singleton, dispersed, proximal, tandem, and segmental) of CDPK-SnRK were counted in *B. rapa*.

Evolution Pattern of CDPK-SnRK Genes in Plants

To investigate the evolution of the CDPK-SnRK family in the plant kingdom, we selected 13 Angiospermae (8 eudicots, 4 monocots and one basal angiosperm), 3 Gymnospermae, 1 Pteridophyta, 1 Bryophyta, and 1 Chlorophyta species for comparative analysis (Figure S3). We constructed a phylogenetic tree of the *CDPK-SnRK* genes to analyze the evolutionary relationships of these species. The phylogenetic tree showed that the CDPK-SnRK genes formed five gene classes (CPK, CRK, PEPRK, PPCK, and SnRK), which is consistent with the result for *B. rapa* and *A. thaliana*. Meanwhile, we found that no CRK, PEPRK, or PPCK genes were detected in *Volvox carteri*. Therefore, the CRK, PEPRK, and PPCK gene families may only exist in land plants. To further determine the relationship among the five groups, the genetic distance was analyzed (Figures S4A,B). The box plot shows the genetic distance of CPK vs. CRK, PEPRK, and PPCK, which was smaller than SnRK vs. these groups (Figure S4B). Notably, the genetic distance between CPK and CRK was smaller than that between CPK and PEPRK or CPK and PPCK. These results indicate that CPK has a closer relationship with CRK, PEPRK, and PPCK, especially the CRK closest to CPK, which is consistent with previous reports that plant CPK and CRK may share a common evolutionary origin (Hrabak et al., 2003).

To further investigated the footprint of the CPK CRK, PPCK, PEPRK, and SnRK families, we selected four angiosperms (*C. papaya, P. trichocarpa, A. trichopoda,* and *V. vinifera*). The reason is that *Vitis vinifera, P. trichocarpa,* and *C. papaya* did not undergo α and β duplications and *A. trichopoda*, a basal angiosperm, did not undergo the γ duplication event (Jiao et al., 2011b; Albert et al., 2013; Lee et al., 2013). Phylogenetic trees in each species were constructed (Figures S5, S6). In each species, the CPK family was divided into four clades (*CPKI, CPKII, CPKIII,* and *CPKIV*), and SnRK was divided into three clades (SnRK1, SnRK2, and SnRK3). CPK, CRK, and SnRK were found to exist in *Amborella trichopoda*, which indicates that these three groups originated from duplication events prior to the γ event. However, PEPRK appeared between the salicoid duplication and the γ event. The PPCK family exists in *P. trichocarpa*, which indicates that it originated from duplication events prior to the salicoid duplication. Furthermore, due to the salicoid duplication and *Brassica*-specific WGT events, there were more CDPK-SnRK gene family members in *P. trichocarpa* and *B. rapa* than in other species (Tuskan et al., 2006; Wang et al., 2011). In general, during the course of evolution, CPK appeared most recently and expanded most rapidly. Above all, we inferred a possible evolutionary footprint of the CPK family (Figure S7).

The family size and the percentage of CPKs in five plants suggested that CPKs expanded rapidly during evolution and further expanded in the Brassicaceae (**Figure 7**). WGD is known to have important impacts on the expansion and evolution of gene families in plant genomes. However, along with the gradual increase in the CPK percentage, the genes of group III were completely lost in *V. vinifera* (**Figure 7C**). Compared with other groups, the expansion of group III was more unstable. To further investigate the relationship among the four groups,

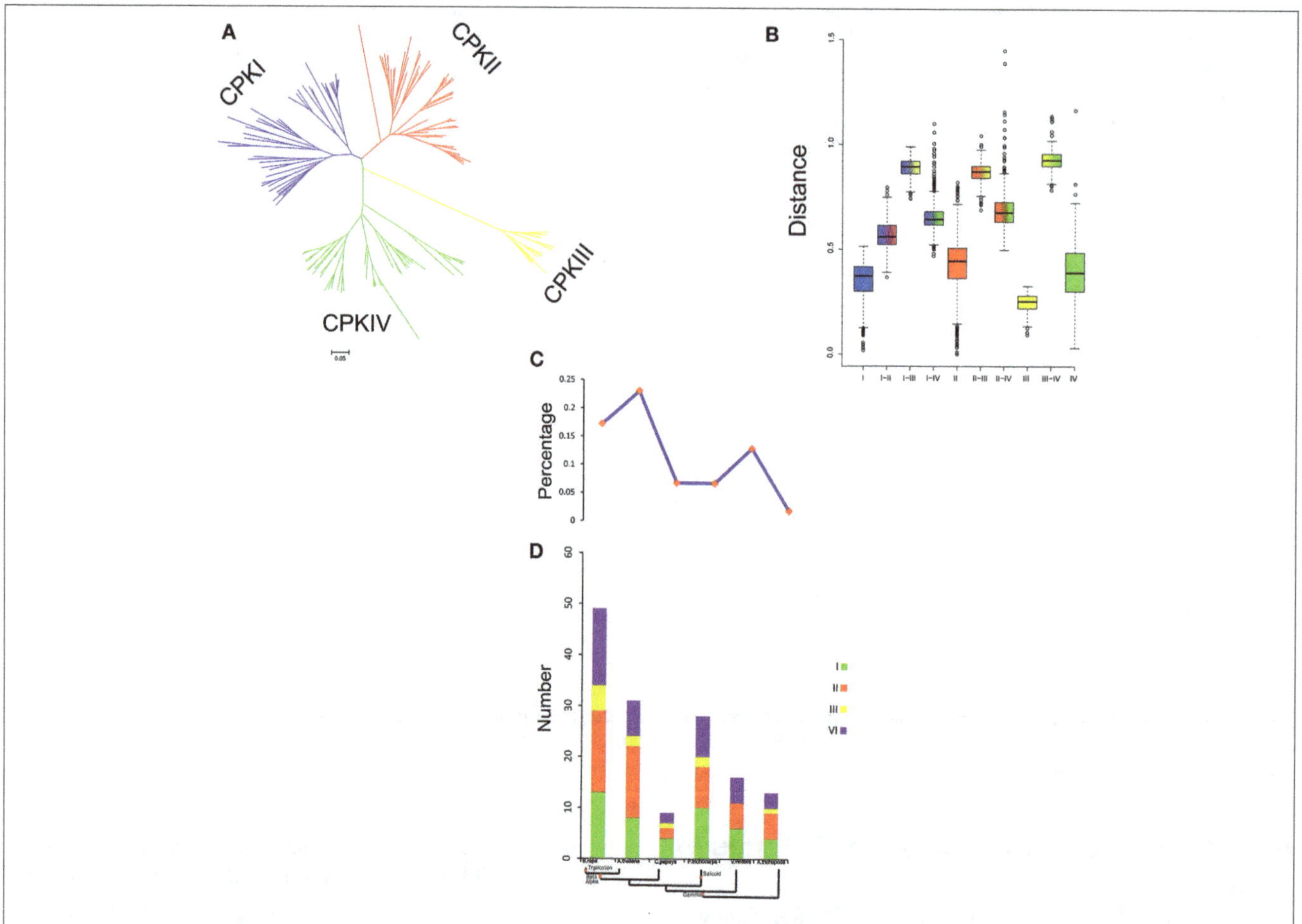

FIGURE 7 | Analysis of CPK gene evolution. (A) Phylogenetic relationships among CPK genes; **(B)** genetic distance among the different groups of CPK genes; **(C,D)** comparison of the percentage of CPK genes and copy numbers of CPK genes; CPK genes in representative species.

the genetic distance was analyzed. The results indicated that the genetic distance between CPKI and CPKIII was shorter than that between CPKI and CPKIV or that between CPKI and CPKII, and the genetic distance between CPKII and CPKIV was shorter than that between CPKII and CPKIII (**Figure 7B**). In summary, we inferred a possible evolutionary footprint of the CPK family. Before the γ event, all groups in the family (CPKI, CPKII, CPKII, CPKIV) had already appeared. The gene family further expanded within Brassicaceae. Thus, the CPK family almost doubled in size in the *B. rapa* genome compared with that of *A. trichopoda* through three duplications, one triplication, and fractionation. The expansion of groups I, II, and IV played a major role in the expansion of the CPK gene family.

Comparative Expression Pattern Analysis of the CDPK-SnRK Genes in Different Tissues of *Brassica rapa* and *Arabidopsis thaliana*

To investigate the divergence expression profiles of CDPK-SnRK genes in different organs between *A. thaliana* and *B. rapa*, including roots, stems, leaves, flowers and siliques, the expression patterns of all genes were investigated (Figure S8, Supplementary Tables 5, 6). BrCDPK-SnRK genes were found to be expressed in roots (99 BrCDPK-SnRKs; 77.95%), stems (104; 81.89%), siliques (106; 83.46%), leaves (101; 79.53%), and flowers (119; 93.7%) (**Figures 8A–D**). A total of 75 (88%) AtCDPK-SnRKs showed high expression (mean-normalized value > 1) in at least one of the five tissues (Figure S8), including roots (34 AtCDPK-SnRKs; 40.00%), stems (36; 42.35%), siliques (22; 25.88%), leaves (19; 22.35%), and flowers (16; 18.23%). Among the 116 CPKs (including 46 AtCPKs and 70 BrCDPKs), 4 (*CPK28*, *CPK29*, *CPK30*, and *CPK33*) were not expressed and 2 (*BrCPK4* and *BrCPK25*) were only expressed in flower tissue (**Figures 8A,B**). In addition, a total of 34 BrCPKs, 8 BrCRKs, 4 BrPEPRKs, and 2 BrPPCKs had high expression levels (FPKM value > 10) in at least one tissue; 13 *BrCPK*s, 2 *BrCRK*s, and 1 *BrPEPRK* were highly expressed in all 5 tissues. However, only 2 SnRK genes were expressed in one tissue; the remains 93 SnRK genes were expressed in five tissues.

Subsequently, we selected the expression patterns of genes in five tissues on the phylogenetic tree of all CDPK-SnRK genes to investigate whether the functions of homologous genes were divergent (Figure S8). All CPKI, PEPRK,

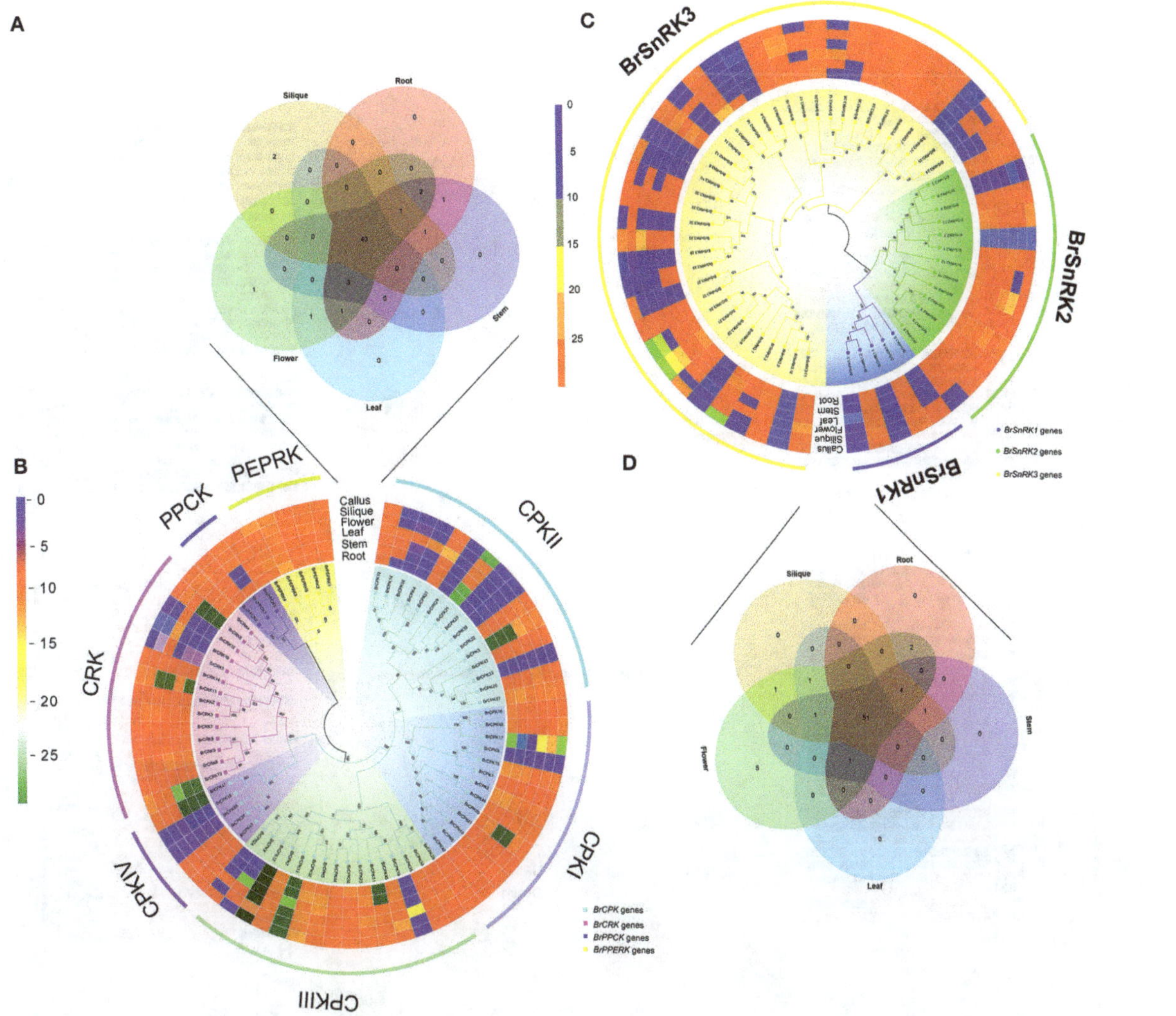

FIGURE 8 | Analysis of the CDPK-SnRK Genes in different tissues of *Brassica rapa*. (A,B) Heat map representation and hierarchical clustering of BrCPK, BrCRK, BrPEPRK, and BrPPCK genes in root, stem, leaf, flower, silique, and callus; Venn diagram depicting the distribution of shared expression of the BrCPK, BrCRK, BrPEPRK, and BrPPCK genes among five *Brassica rapa* tissues. **(C,D)** Heat map representation and hierarchical clustering of BrSnRK genes in root, stem, leaf, flower, silique, and callus; Venn diagram depicting the distribution of shared expression of the BrSnRK genes among five *Brassica rapa* tissues.

and PPCK and most CPKIII and CRK genes had high expression levels, suggesting significant roles of these genes in plant development. Most BrSnRK3s exhibited little or no expression. However, the AtSnRK3s were all expressed in five tissues, indicating that BrSnRK3 genes may have lost some functions after the duplication events (Figure S8).

Expression Divergence and Coregulatory Networks of CDPK-SnRK Genes under Multiple Treatments in *Brassica rapa*

CDPK-SnRKs are suggested to play an important role in the regulation of gene expression in response to abiotic stresses (**Figure 9**, Supplementary Table 7). To investigate the divergence information of the BrCDPK-SnRK gene family, the expression patterns following different treatments, including ABA, GA, NaCl, heat, cold, and PEG treatments, were analyzed (**Figures 9, 10**). The qRT-PCR results demonstrate that those BrCDPK-SnRK genes respond differentially to particular abiotic stresses. Sixteen percent of investigated BrCPK genes show increased expression levels upon GA at 6 h, while the other 84% of genes display downregulation or no significant changes. Meanwhile, 20% of genes were upregulated at 1 h and 6 h under GA treatment (**Figures 9A,C**). Two genes (*BrCPK4* and *BrCPK10*) were induced under both ABA and GA treatment (**Figure 9B**). During four abiotic stress treatments, excluding PEG, 80% of BrCPK genes were highly responsive to cold, NaCl, and heat. We found that, with the exceptions of *BrCPK2*, *29*, *23* and *26*, all BrCPK genes were significantly induced in response to NaCl treatment (**Figures 9D–F,H**). Under heat treatment, we found that only two genes (*BrCPK4* and *BrCPK10*) had

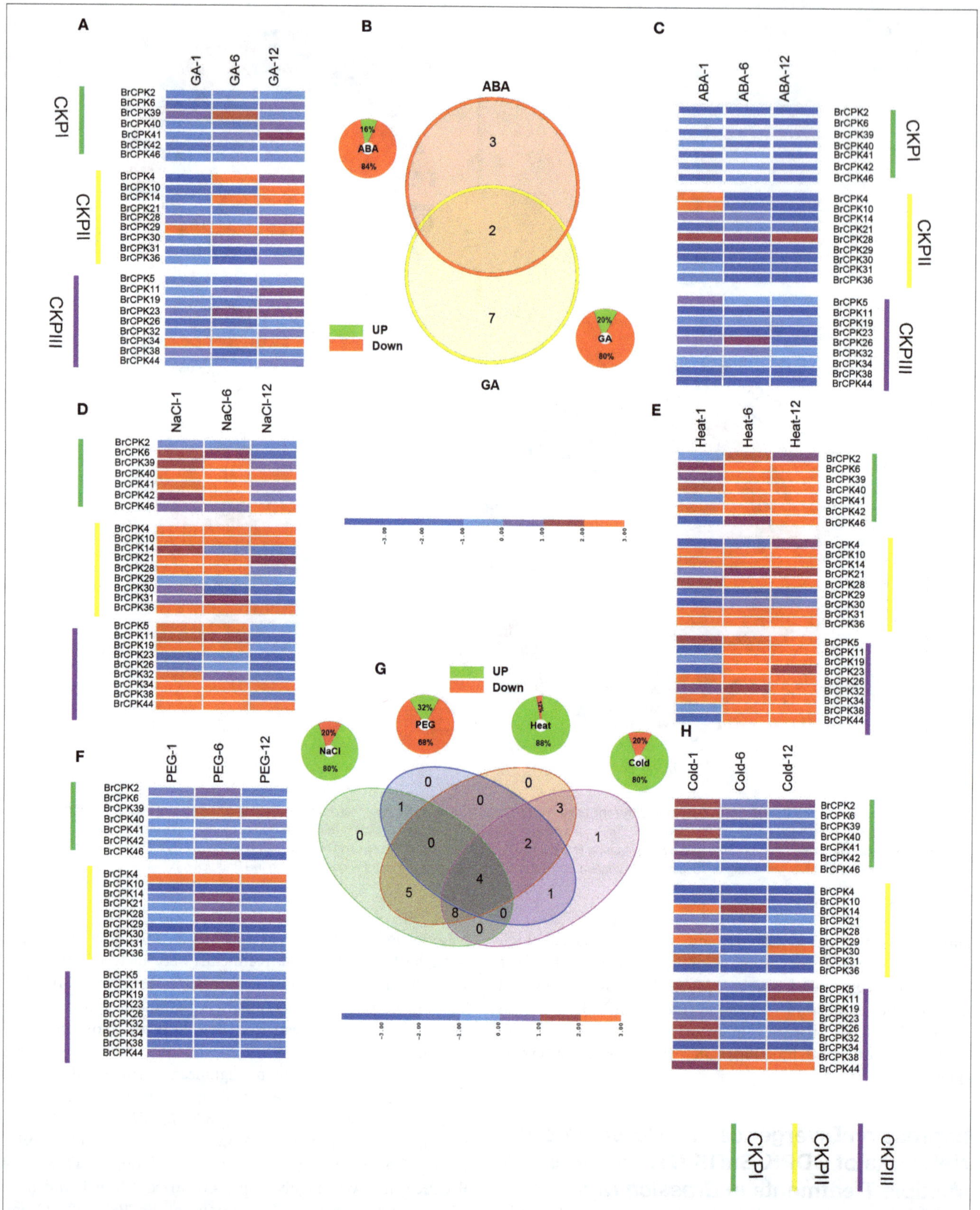

FIGURE 9 | Expression analysis of the BrCPK genes under six stress treatments in *Brassica rapa*. (A,C–F,H) Heat map representation the BrCPK genes under six stress treatments, namely, GA, ABA, NaCl, heat, PEG, and cold. **(B,G)** Venn diagram depicting the distribution of shared expression of the BrCPK genes among hormone and abiotic treatments.

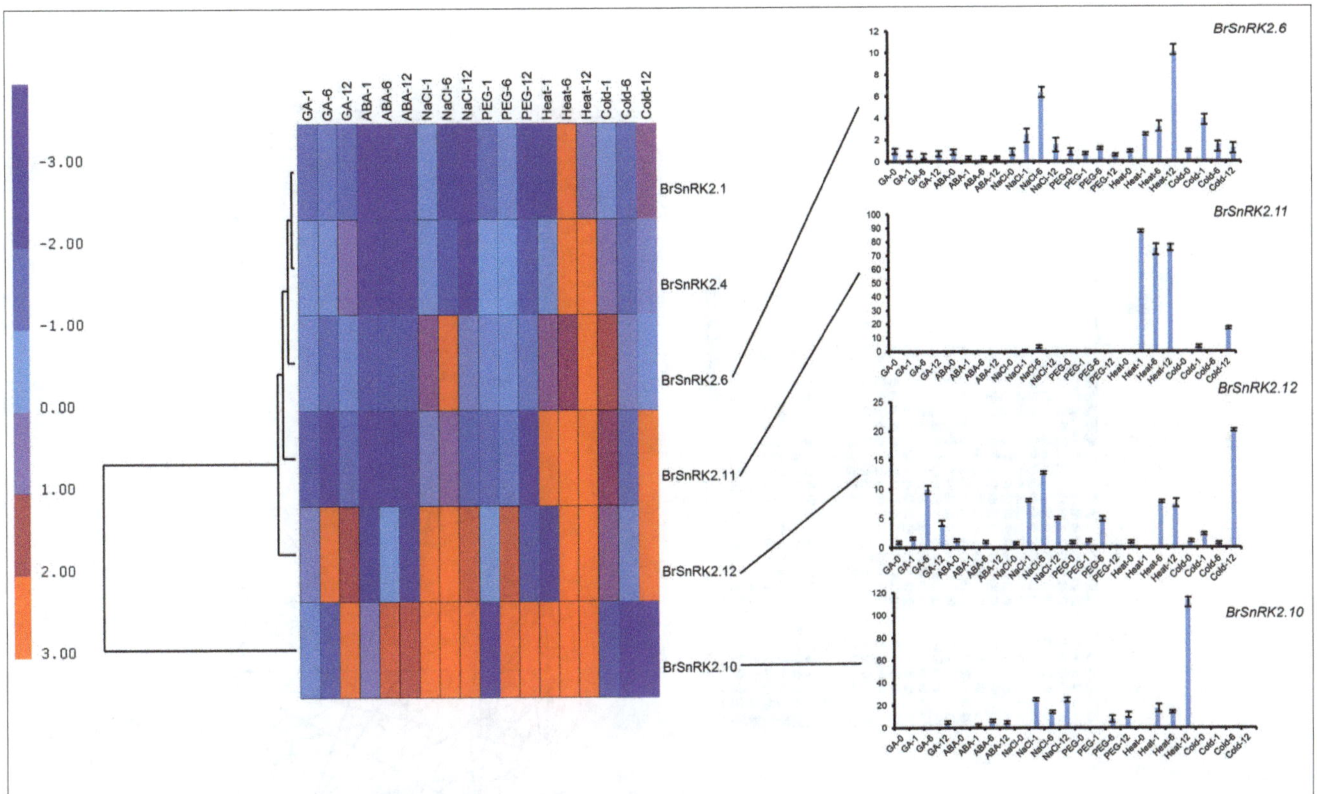

FIGURE 10 | Expression analysis of the BrSnRK2 genes under six stress treatments in *Brassica rapa*. Heat map representation the BrSnRK genes under six stress treatments, namely, GA, ABA, NaCl, heat, PEG, and cold.

no expression; other BrCPK genes were highly expressed in NaCl treatment. SnRK2 genes recognized as coding for enzymes involved in abiotic stress signal transduction in plants (Kulik et al., 2011). Therefore, we selected six BrSnRK2 genes; all of these genes had the highest expression under heat treatment (**Figure 10**). Furthermore, *BrSnRK2.12* was induced under all stress treatments (**Figure 10**).

To investigate the connections between these genes, coregulatory networks were established based on the PCCs of stress-inducible BrCDPK-SnRK gene pairs (Supplementary Table 8, **Figure 11**). All the BrCDPK-SnRKs appeared to have different degrees of positive correlation. Next, 25 BrCPKs with PCC values that were significant at the 0.05 significance level and were greater than 0.8 were collected and visualized to construct hormone and abiotic stress coregulatory networks (**Figure 11A**). Six BrSnRKs and seven BrCPKs had positive significant correlations (**Figure 11C**). Most correlations occurred among members belonging to the same group, suggesting that the gene duplication not only led to functional divergence but also enhanced the cooperative interaction of homologs to help plants to adapt to their complex environment. In addition, based on predicted Chinese cabbage BrCDPK-SnRK superfamily protein interactions, their interactions form a complex network in both *Arabidopsis thaliana* and *B. rapa*. Moreover, to study the protein interactions of stress response genes between *B. rapa* and *Arabidopsis*, STRING 9.1 was utilized (Figure S9). Figure S9 shows four complex interaction networks of CDPK-SnRKs, providing an overall view of the relationship between *B. rapa* and *Arabidopsis*.

DISCUSSION

In eukaryotes, protein kinases are involved in regulating key aspects of cellular function, including cell division, metabolism, and responses to external signals. CDPK-SnRK plays an important role in stress signal transduction in plants, such as wounding, salt or drought stress (Botella et al., 1996; Patharkar and Cushman, 2000; Saijo et al., 2000), cold (Monroy and Dhindsa, 1995; Saijo et al., 2000), hormone treatment (Abo-El-Saad and Wu, 1995; Botella et al., 1996; Davletova et al., 2001), light (Frattini et al., 1999), and pathogens (Romeis et al., 2001; Murillo et al., 2002).

The gene balance hypothesis predicts that genes whose products participate in signaling networks or macromolecular complexes or are transcription factors are more likely to be retained (Birchler and Veitia, 2007; Aad et al., 2012). In this study, we identified 49 *BrCPKs*, 14 *BrCRKs*, 3 *BrPPCKs*, 5 *BrPEPRKs*, and 56 *BrSnRKs* in the *B. rapa* genome, and they contained a higher rate of copies than the *B. rapa* whole-genome level. This result suggests that these genes had a high degree of retention following WGD. By comparing the number of different duplicated types, gene copies, and distribution of the three subgenomes, we found that all the AtCRK, AtPEPRK, and

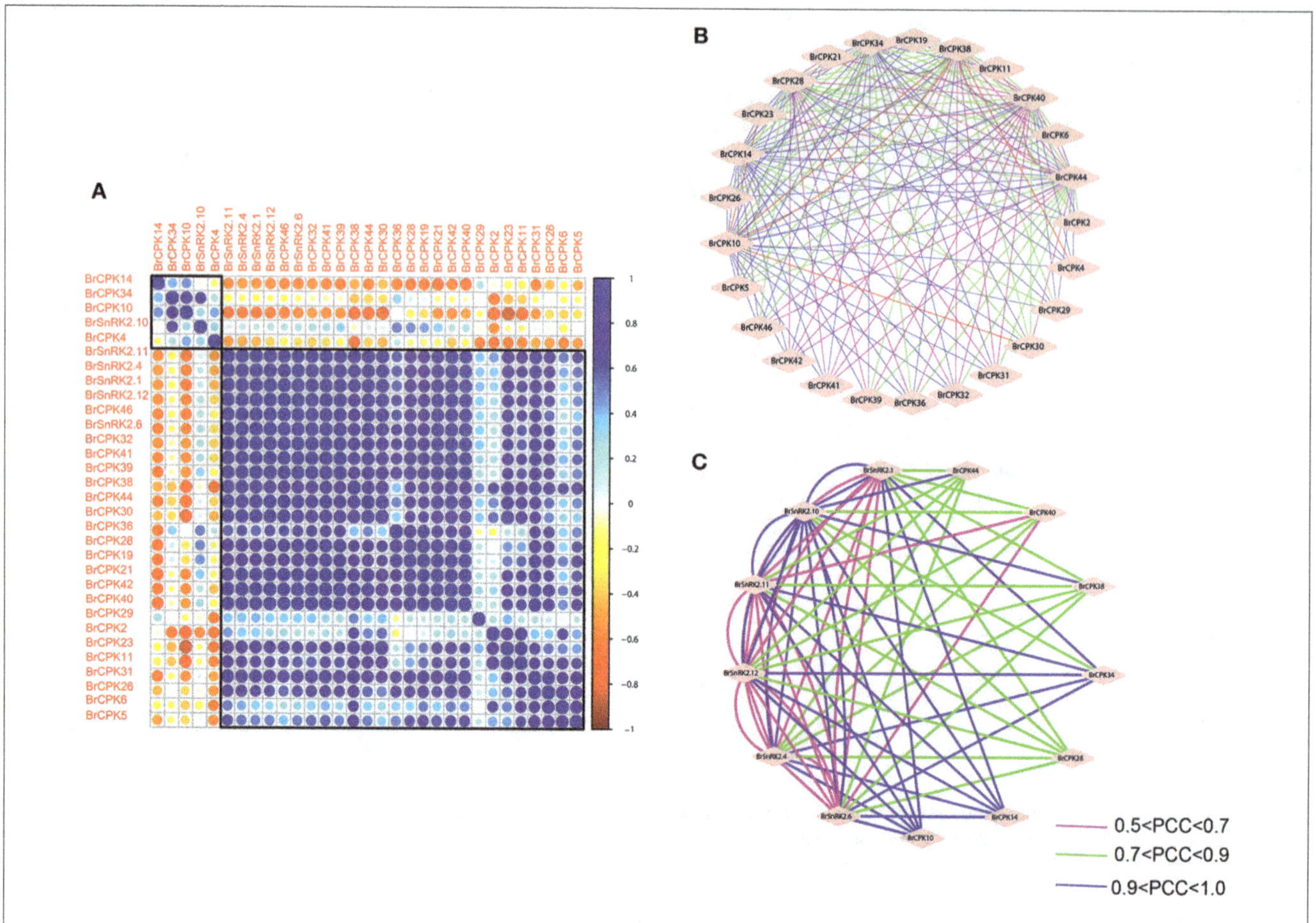

FIGURE 11 | Correlations and co-regulatory networks of 31 BrCDPK-SnRK genes under stress treatments. (A) Correlation analysis using the R package program. Each correlation is shown by the shades of blue and orange and the size of the circle shape. Blue and orange indicate a positive correlation and negative correlation, respectively. **(B,C)** Co-regulatory networks. The co-regulatory networks of 31 BrCDPK-SnRK genes under stress treatments were established based on the Pearson correlation coefficients (PCCs) of these gene pairs using transformed qPCR data. The PCC of co-regulatory gene pairs was considered significant at the 0.05 significance level (*p*-value), and different line colors and styles indicate the different significance levels of the co-regulated gene pairs.

AtPPCK orthologs were retained in *B. rapa*. In contrast, four AtCPK and two AtSnRK orthologs were completely lost. CRK, PEPRK, and PPCK were more preferentially retained than *CPK* and *SnRK*. CPK (66%) and SnRK (43%) had more than two copies retained in *B. rapa*, and more BrCPK (74%) and BrSnRK (70%) genes experienced segmental duplication. These preferentially retained CDPK-SnRK genes may have more important functions. At the same time, the important functions have been proved in previously researches.

For example, CDPK-SnRKs have been shown to have important roles in various physiological processes, including plant growth and development and abiotic and biotic stress responses in plants. The large number and variety of protein kinases that either have an EF-hand calcium-binding domain in their structure (CPK and CCaMK) or interact with EF-hand proteins (CCaMK, CaMK, CRK, and SnRK3) provides plants with a huge potential for interpreting specific calcium signals and for eliciting specific physiological responses. The SnRK1 kinase plays an important role in carbon-nitrogen interactions (Halford and Paul, 2004; Li et al., 2010) and in the transcription regulation of gene expression. In the developing tuber of potato, the expression level of *SnRK1* was higher, lower in stem and lowest in leaf. The experiment on potato provided evidence for SnRK1 to regulate the transcription (Man et al., 1997). SnRK1 influence starch biosynthesis via regulating the expression of sucrose synthase and the activity of AGPase. SnRK1 activity can respond to sucrose appropriately (Halford and Paul, 2004). Antisense gene expressions in different plant indicate that SnRK1 has very important roles in plant growth and development processes (Halford et al., 2003). The SnRK2s are about 140–160 amino acids shorter than the SnRK1s, averaging about 40 kD in size, and have a characteristic patch of acidic amino acids in their C-terminal domains (Halford et al., 2000). SnRK2 genes, which are a significant part of the ABA signal pathway, are involved in many processes that help plants resistant environmental pressures (Wang et al., 2015).The variety is further amplified in the *SnRK3* subfamily, since these kinases may interact with more than one *CBL/SCaBP* (Guo et al., 2001; Manns et al., 2001; Shi and Eberhart, 2001). Gene expression patterns can provide important clues for gene function. Therefore, the expression patterns of CDPK-SnRK genes under stresses were identified based on qRT-PCR analysis. Overall, expression patterns of CDPK-SnRK genes are dynamic in different tissues during different developmental stages and in response to abiotic stresses, demonstrating their regulatory roles in various cellular processes in *B. rapa*. This finding is consistent with the gene dosage hypothesis.

During evolutionary history, all extant angiosperms genome have undergone at least one and often multiple polyploidization events (Edger and Pires, 2009; Jiao et al., 2011a; Ohno, 2013). *Brassica rapa* experienced a complex WGD history, including γ, α, and β events, and an additional WGT event, providing an excellent chance to study the relationship between gene family fractionation and changes in plant morphotypes (Wang et al., 2011; Cheng et al., 2013). An extensive phylogenetic analysis revealed the evolutionary history of those CDPK-SnRK genes in *B. rapa* and in other plants. In total, 127 CDPK-SnRK genes were grouped in seven classes in *B. rapa*. In addition, we identified CDPK-SnRK genes in 15 other plant species representing the major clades of terrestrial plants. These phylogenetic studies suggest that those identified CPK and SnRK genes were highly conserved in each class across a wide spectrum of plants, indicating their essential regulatory roles in the plant kingdom. For the CPK gene family, the number of genes grouped into each class, gene number and genome size are summarized in **Figure 1**. Furthermore, there is a difference among the five species in class I, II, III and IV in terms of gene numbers. It suggests that those four classes underwent extensive expansion in angiosperms. Class III was entirely lost during evolution in *V. vinifera*. For the SnRK gene family, two SnRK1 genes were found in *V. carteri*, suggesting that expansion of the SnRK gene family occurred after the divergence of green algae. Meanwhile, SnRK1, SnRK2, and SnRK3 were detected in *Physcomitrella patens*, indicating that the SnRK2 and SnRK3 gene subfamilies appear to be unique to plants, which consist with previously research (Hrabak et al., 2003). Overall, for the six species investigated, the number of CPK and SnRK genes remains different, indicating different evolutionary histories accompanied by rounds of WGD and subsequent gene losses/gains by natural selection constraints.

Most land plants have undergone polyploidization that led to WGD and provided opportunities for duplicated genes to diverge in different evolutionary ways. Each of these genes subsequently experienced one of three fates: subfunctionalization, neofunctionalization, or non-functionalization (deletion or pseudogenization). These fates provided opportunities for duplicated genes to gain functional diversification, resulting in more complex organisms. The *Ka/Ks* substitution rate ratio is an indicator of the selection history on genes or gene regions. Commonly, if the value of *Ka/Ks* is lower than 1, the duplicated gene pairs may have evolved from purifying selection (also called as negative selection); $Ka/Ks = 1$ means neutral selection, while $Ka/Ks > 1$ means positive selection. In this study, the *Ka/Ks* values for the duplicated gene pairs were small (<0.01); thus, it is likely that these pairs have been under purifying selection. Neofunctionalization is an adaptive process during which one copy of a duplicated gene mutates to adopt a novel function that cannot be performed by the ancestral sequence. This mechanism can lead to the retention of both copies over long periods of time. In addition, by comparing the tissue expression pattern among the CDPK-SnRK genes in *B. rapa* and *At. thaliana*, we found that most of the duplicated genes maintained similar expression patterns, but some of the duplicated CDPK-SnRK genes demonstrated higher tissue specialization and diversification, also suggesting that CDPK-SnRKs underwent more neofunctionalization and subfunctionalization. In previous reports, even the most recently diverged paralogs differed in their catalytic efficiency, expression, and/or substrate spectrum, suggesting that duplicates have a relatively high rate of divergence in function.

In this study, we analyzed the evolutionary pattern, gene duplication, and expression divergence of the CDPK-SnRK genes in plants. We conducted a comparative evolutionary analysis with the genome information of selected plants, and our research provides new insight into the evolutionary history of the CDPK-SnRK gene family in plants. For example, CPK had a closer relationship with CRK, indicating that these two may share a common evolutionary origin; and PEPRK appeared between the salicoid duplication and the γ event. By compared expression patterns among tissues, we found that the expansion of CDPK-SnRK genes seems to be associated with increasingly complex organs in the plants' evolution. Under different stress treatments, their coregulatory networks demonstrated that the CDPK-SnRK genes enhanced the cooperative interaction of homologs to adapt to the environment. In conclusion, this study provides useful resources for functional divergence and conservation in the CDPK-SnRK gene superfamily and facilitates our understanding of the effect of polyploidy during the evolution of CDPK-SnRK genes.

METHODS

Retrieval of Sequences

The *B. rapa* sequences were downloaded from BRAD (http://brassicadb.org/brad/) (Wang et al., 2011). The *Arabidopsis* sequences were retrieved from TAIR (http://www.*arabidopsis*.org/), and the sequences of rice were retrieved from RGAP (http://rice.plantbiology.msu.edu/; Kawahara et al., 2013). The gene information on *A. trichopoda* genes was retrieved from the Amborella Genome Database (http://www.amborella.org/). The others 14 species' gene information was downloaded from Phytozome v9.1 (http://www.phytozome.net/; Goodstein et al., 2012). The homologous CDPK-SnRK genes in other species were identified through comparison with *Arabidopsis*. First, BLASTP searches were performed against the rice protein sequences using an *E*-value threshold of 1×10^{-10}. The top-ranked rice hit was used for BLASTP searches of the *Arabidopsis* proteins to confirm homologies. Starting with both *Arabidopsis* and rice homologs, BLASTP searches were performed against the proteins of other species ($e < 1 \times 10^{-10}$, identity $>70\%$). These potential sequences were analyzed using the tool SMART (http://smart.

embl-heidelberg.de/) and the NCBI database (http://www.ncbi.nlm.nih.gov/).

Identification of Gene Synteny and Duplicated CDPK-SnRK Genes

BLAST and the Multiple Collinearity Scan toolkit (MCScanX) were used for gene synteny analysis according to previous reports (Wang et al., 2012). An all-against-all BLASTP comparison provided the pairwise gene information and the *P*-value for a primary clustering. Paired segments were extended by identifying the clustered genes using dynamic programming. The potentially duplicated genes were also identified using MCScanX. The positions of *B. rapa CDPK-SnRK* genes on the blocks were verified by searching for homologous genes between *A. thaliana* and three subgenomes of *B. rapa* (LF, MF1, and MF2) in BRAD (http://brassicadb.org/brad/searchSynteny.php; Wang et al., 2011). The syntenic diagram was drawn using Circos software (Krzywinski et al., 2009).

Phylogenetic Analysis and Characterization of the CDPK-SnRK Gene Family

Phylogenetic analyses were conducted using MEGA v5.073; maximum-likelihood (ML) trees were constructed with a bootstrap value of 1,000 replications to assess the reliability of the resulting trees. The genetic distance used in this study was also calculated with MEGA v5.0. To identify conserved motifs in the complete amino acid sequences of CDPK-SnRK proteins, we used MEME software (http://meme.sdsc.edu/meme/; Bailey et al., 2009). We researched the gene structure using Gene Structure Display (GSDS, http://gsds.cbi.pku.edu.cn/). The interaction network of CDPK-SnRK proteins in Chinese cabbage was constructed with STRING software (Search Tool for the Retrieval of Interacting Genes/Proteins, http://string-db.org/; Yamada et al., 2011).

Calculation of the *Ka*/*Ks* and Dating of the Duplication Events

The duplicated CDPK-SnRK genes from *B. rapa* were aligned using MUSCLE (Edgar, 2004). The protein alignments were translated into coding sequence alignments using an in-house Perl script. *Ks* (synonymous substitution rate) and *Ka* (non-synonymous substitution rate) values were calculated based on the coding sequence alignments using the method of Nei and Gojobori as implemented in KaKs_calculator (Zhang et al., 2006). The *Ks* values of all duplicated *BrCDPK-SnRK* genes were then plotted as the density and a boxplot using the R program (Ihaka and Gentleman, 1996). The divergence time was calculated with the formula $T = Ks/2r$, with *Ks* being the synonymous substitutions per site and r being the rate of divergence for nuclear genes from plants. The value of r was taken to be 1.5×10^{-8} synonymous substitutions per site per year for dicotyledonous plants (Koch et al., 2000).

Expression Pattern Analysis for *CDPK-SnRK* Genes

For expression profiling of the CDPK-SnRK genes in *B. rapa*, we utilized the Illumina RNA-seq data that were previously generated and analyzed by Tong et al. (2013). Six tissues of *B. rapa* accession Chiifu-401-42, including callus, root, stem, leaf, flower, and silique, were analyzed. The transcript abundance was expressed as fragments per kilobase of exon model per million mapped reads (FPKM). The gene expression patterns of each tissue were analyzed using Cluster 3.0, and the expression values were log2 transformed. Finally, heat maps of hierarchical clustering were visualized using Tree View (http://jtreeview.sourceforge.net/). *A. thaliana* development expression profiling was analyzed using the AtGenExpress Visualization Tool (AVT; http://jsp.weigelworld.org/expviz/expviz.jsp) with mean-normalized values (Schmid et al., 2005). Venn diagrams were drawn using the R program.

Plant Material and Treatments

Chinese cabbage (Chiifu-401–42) was used for the experiments. The germinated seeds were grown in plastic pots in a 3:1 soil–vermiculite mixture, and the artificial growth conditions were set at 24/16°C, with a photoperiod of 16/8 h for day/night and a relative humidity of 65–70%. Four weeks later, seedlings were subjected to various treatments. For heat and cold treatment, the pots were exposed to 38°C or 4°C; the other growth conditions were same as described earlier. Meanwhile, plants were cultured with the following four treatments: (1) control; (2) 100 μM ABA; (3) 100 μM GA; (4) (w/v) polyethylene glycol (PEG) 6,000, and (5) 250 mM NaCl. All these treatments were performed over a continuous time course (0, 1, 6, 12 h). Each treatment consisted of three biological replicates. All of the samples were frozen in liquid nitrogen and stored at −70°C for RNA preparation.

RNA Isolation and qRT-PCR Analyses

Total RNA was isolated from treated frozen leaves using Trizol (Invitrogen, San Diego, CA, USA) according to the manufacturer's instructions. Specific primers used for qRT-PCR were designed by Beacon Designer 7 and are shown in Supplementary Table 9. To check the specificity of the primers, we used BLAST against the *Brassica* genome. The qRT-PCR assays were performed with three biological and technical replicates. The reactions were performed using a StepOnePlus Real-Time PCR System (Applied Biosystems, Carlsbad, CA, USA). qRT-PCR was performed according to a previous report (Song et al., 2013). Relative fold expression changes were calculated using the comparative Ct-value method.

Pearson Correlation Analyses

Pearson correlation coefficients (PCCs) of stress-inducible CDPK-SnRK gene pairs were calculated using an in-house Perl script based on log2-transformed quantitative real-time (qRT)-PCR data (Tang et al., 2013). All gene pairs whose PCC was more than 0.8 and was significant at the 0.05 significance level (*P*-value) were collected for a gene coregulatory network analysis. The coexpression networks were graphically visualized using Cytoscape based on the PCCs of these gene pairs (Shannon et al., 2003).

AUTHOR CONTRIBUTIONS

PW was responsible for the experimental design, data analysis, and manuscript writing. WD, WW, YL contributed to the experimental work. XH contributed to the interpretation of results and coordinated the study. All authors read and approved the final manuscript.

ACKNOWLEDGMENTS

This work was supported by the National Natural Science Foundation of China (No. 31330067, 31301782), the Science and Technology Pillar Program of Jiangsu Province (No. BE2013429), and the Agricultural Science and Technology Independent Innovation Funds of Jiangsu Province [CX(13)2006].

SUPPLEMENTARY MATERIAL

The Supplementary Material for this article can be found online at: http://journal.frontiersin.org/article/10.3389/fpls.2017.00162/full#supplementary-material

REFERENCES

Aad, G., Abajyan, T., Abbott, B., Abdallah, J., Khalek, S.A., Abdelalim, A., et al. (2012). Observation of a new particle in the search for the Standard Model Higgs boson with the ATLAS detector at the LHC. *Phys. Lett. B* 716, 1–29. doi: 10.1016/j.physletb.2012.08.020

Abo-El-Saad, M., and Wu, R. (1995). A rice membrane calcium-dependent protein kinase is induced by gibberellin. *Plant Physiol.* 108, 787–793. doi: 10.1104/pp.108.2.787

Albert, V. A., Barbazuk, W. B., Der, J. P., Leebens-Mack, J., Ma, H., Palmer, J. D., et al. (2013). The Amborella genome and the evolution of flowering plants. *Science* 342:1241089. doi: 10.1126/science.1241089

Alderson, A., Sabelli, P. A., Dickinson, J. R., Cole, D., Richardson, M., Kreis, M., et al. (1991). Complementation of snf1, a mutation affecting global regulation of carbon metabolism in yeast, by a plant protein kinase cDNA. *Proc. Natl. Acad. Sci. U.S.A.* 88, 8602–8605. doi: 10.1073/pnas.88.19.8602

Allen, G. J., Chu, S. P., Schumacher, K., Shimazaki, C. T., Vafeados, D., Kemper, A., et al. (2000). Alteration of stimulus-specific guard cell calcium oscillations and stomatal closing in *Arabidopsis* det3 mutant. *Science* 289, 2338–2342. doi: 10.1126/science.289.5488.2338

Anderberg, R. J., and Walker-Simmons, M. (1992). Isolation of a wheat cDNA clone for an abscisic acid-inducible transcript with homology to protein kinases. *Proc. Natl. Acad. Sci. U.S.A.* 89, 10183–10187. doi: 10.1073/pnas.89.21.10183

Asano, T., Kunieda, N., Omura, Y., Ibe, H., Kawasaki, T., Takano, M., et al. (2002). Rice SPK, a calmodulin-like domain protein kinase, is required for storage product accumulation during seed development phosphorylation of sucrose synthase is a possible factor. *Plant Cell* 14, 619–628. doi: 10.1105/tpc.010454

Asano, T., Tanaka, N., Yang, G., Hayashi, N., and Komatsu, S. (2005). Genome-wide identification of the rice calcium-dependent protein kinase and its closely related kinase gene families: comprehensive analysis of the CDPKs gene family in rice. *Plant Cell Physiol.* 46, 356–366. doi: 10.1093/pcp/pci035

Assmann, S. M., and Wang, X.-Q. (2001). From milliseconds to millions of years: guard cells and environmental responses. *Curr. Opin. Plant Biol.* 4, 421–428. doi: 10.1016/S1369-5266(00)00195-3

Bailey, T. L., Boden, M., Buske, F. A., Frith, M., Grant, C. E., Clementi, L., et al. (2009). MEME SUITE: tools for motif discovery and searching. *Nucleic Acids Res.* 37, W202–W208. doi: 10.1093/nar/gkp335

Berridge, M. J., Lipp, P., and Bootman, M. D. (2000). The versatility and universality of calcium signalling. *Nat. Rev. Mol. Biol.* 1, 11–21. doi: 10.1038/35036035

Birchler, J. A., and Veitia, R. A. (2007). The gene balance hypothesis: from classical genetics to modern genomics. *Plant Cell* 19, 395–402. doi: 10.1105/tpc.106.049338

Botella, J. R., Arteca, J. M., Somodevilla, M., and Arteca, R. N. (1996). Calcium-dependent protein kinase gene expression in response to physical and chemical stimuli in mungbean (*Vigna radiata*). *Plant Mol. Biol.* 30, 1129–1137. doi: 10.1007/BF00019547

Bowers, J. E., Chapman, B. A., Rong, J., and Paterson, A. H. (2003). Unravelling angiosperm genome evolution by phylogenetic analysis of chromosomal duplication events. *Nature* 422, 433–438. doi: 10.1038/nature01521

Bush, D. S. (1995). Calcium regulation in plant cells and its role in signaling. *Annu. Rev. Plant Biol.* 46, 95–122. doi: 10.1146/annurev.pp.46.060195.000523

Cheng, F., Mandáková, T., Wu, J., Xie, Q., Lysak, M. A., and Wang, X. (2013). Deciphering the diploid ancestral genome of the mesohexaploid *Brassica rapa*. *Plant Cell* 25, 1541–1554. doi: 10.1105/tpc.113.110486

Cheng, S.-H., Willmann, M. R., Chen, H.-C., and Sheen, J. (2002). Calcium signaling through protein kinases. The Arabidopsis calcium-dependent protein kinase gene family. *Plant Physiol.* 129, 469–485. doi: 10.1104/pp.005645

Chico, J. M., Raıces, M., Téllez-Iñón, M. A. T., and Ulloa, R. M. A. (2002). A calcium-dependent protein kinase is systemically induced upon wounding in tomato plants. *Plant Physiol.* 128, 256–270. doi: 10.1104/pp.010649

Cho, K., Agrawal, G. K., Jwa, N.-S., Kubo, A., and Rakwal, R. (2009). Rice OsSIPK and its orthologs: a "central master switch" for stress responses. *Biochem. Biophys. Res. Commun.* 379, 649–653. doi: 10.1016/j.bbrc.2008.12.107

Choi, H.-I., Park, H.-J., Park, J. H., Kim, S., Im, M.-Y., Seo, H.-H., et al. (2005). *Arabidopsis* calcium-dependent protein kinase AtCPK32 interacts with ABF4, a transcriptional regulator of abscisic acid-responsive gene expression, and modulates its activity. *Plant Physiol.* 139, 1750–1761. doi: 10.1104/pp.105.069757

Chothia, C., Gough, J., Vogel, C., and Teichmann, S. A. (2003). Evolution of the protein repertoire. *Science* 300, 1701–1703. doi: 10.1126/science.1085371

D'Angelo, C., Weinl, S., Batistic, O., Pandey, G. K., Cheong, Y. H., Schültke, S., et al. (2006). Alternative complex formation of the Ca-regulated protein kinase CIPK1 controls abscisic acid-dependent and independent stress responses in *Arabidopsis*. *Plant J. Cell Mol. Biol.* 48, 857–872. doi: 10.1111/j.1365-313X.2006.02921.x

Davletova, S., Mészáros, T., Miskolczi, P., Oberschall, A., Török, K., Magyar, Z., et al. (2001). Auxin and heat shock activation of a novel member of the calmodulin like domain protein kinase gene family in cultured *alfalfa* cells. *J. Exp. Bot.* 52, 215–221. doi: 10.1093/jexbot/52.355.215

Edgar, R. C. (2004). MUSCLE: multiple sequence alignment with high accuracy and high throughput. *Nucleic Acids Res.* 32, 1792–1797. doi: 10.1093/nar/gkh340

Edger, P. P., and Pires, J. C. (2009). Gene and genome duplications: the impact of dosage-sensitivity on the fate of nuclear genes. *Chromosome Res.* 17, 699–717. doi: 10.1007/s10577-009-9055-9

Ehrhardt, D. W., Wais, R., and Long, S. R. (1996). Calcium spiking in plant root hairs responding to Rhizobium nodulation signals. *Cell* 85, 673–681. doi: 10.1016/S0092-8674(00)81234-9

Evans, N. H., McAinsh, M. R., and Hetherington, A. M. (2001). Calcium oscillations in higher plants. *Curr. Opin. Plant Biol.* 4, 415–420. doi: 10.1016/S1369-5266(00)00194-1

Fasano, J. M., Massa, G. D., and Gilroy, S. (2002). Ionic signaling in plant responses to gravity and touch. *J. Plant Growth Regul.* 21, 71–88. doi: 10.1007/s003440010049

Frattini, M., Morello, L., and Breviario, D. (1999). Rice calcium-dependent protein kinase isoforms OsCDPK2 and OsCDPK11 show different responses to light and different expression patterns during seed development. *Plant Mol. Biol.* 41, 753–764. doi: 10.1023/A:1006316422400

Fujii, H., Chinnusamy, V., Rodrigues, A., Rubio, S., Antoni, R., Park, S.-Y., et al. (2009). *In vitro* reconstitution of an abscisic acid signalling pathway. *Nature* 462, 660–664. doi: 10.1038/nature08599

Fujii, H., and Zhu, J.-K. (2009). *Arabidopsis* mutant deficient in 3 abscisic acid-activated protein kinases reveals critical roles in growth, reproduction, and stress. *Proc. Natl. Acad. Sci. U.S.A.* 106, 8380–8385. doi: 10.1073/pnas.0903144106

Gargantini, P. R., Gonzalez-Rizzo, S., Chinchilla, D., Raices, M., Giammaria, V., Ulloa, R. M., et al. (2006). A CDPK isoform participates in the regulation of nodule number in Medicago truncatula. *Plant J.* 48, 843–856. doi: 10.1111/j.1365-313X.2006.02910.x

Goodstein, D. M., Shu, S., Howson, R., Neupane, R., Hayes, R. D., Fazo, J., et al. (2012). Phytozome: a comparative platform for green plant genomics. *Nucleic Acids Res.* 40, D1178–D1186. doi: 10.1093/nar/gkr944

Guo, D., Chen, F., Wheeler, J., Winder, J., Selman, S., Peterson, M., et al. (2001). Improvement of in-rumen digestibility of alfalfa forage by genetic manipulation of lignin O-methyltransferases. *Transgenic Res.* 10, 457–464. doi: 10.1023/A:1012278106147

Halford, N. G., Boulyz, J. P., and Thomas, M. (2000). SNF1-related protein kinases (SnRKs)—regulators at the heart of the control of carbon metabolism and partitioning. *Adv. Bot. Res.* 32, 405–434. doi: 10.1016/S0065-2296(00)32031-6

Halford, N. G., and Hardie, D. G. (1998). SNF1-related protein kinases: global regulators of carbon metabolism in plants? *Plant Mol. Biol.* 37, 735–748. doi: 10.1023/A:1006024231305

Halford, N. G., Hey, S., Jhurreea, D., Laurie, S., McKibbin, R. S., Paul, M., et al. (2003). Metabolic signalling and carbon partitioning: role of Snf1-related (SnRK1) protein kinase. *J. Exp. Bot.* 54, 467–475. doi: 10.1093/jxb/erg038

Halford, N. G., and Paul, M. J. (2004). Highly conserved protein kinases involved in the regulation of carbon and amino acid metabolism. *J. Exp. Bot.* 55, 35–42. doi: 10.1093/jxb/erh019

Hammond-Kosack, K. E., and Jones, J. (1996). Resistance gene-dependent plant defense responses. *Plant Cell* 8, 1773.

Harmon, A. C. (2003). Calcium-regulated protein kinases of plants. *Gravit. Space Biol. Bull.* 16, 83–90.

Harmon, A. C., Gribskov, M., Gubrium, E., and Harper, J. F. (2001). The CDPK superfamily of protein kinases. *New Phytol.* 151, 175–183. doi: 10.1046/j.1469-8137.2001.00171.x

Harmon, A. C., Gribskov, M., and Harper, J. F. (2000). CDPKs-a kinase for every Ca^{2+} signal? *Trends Plant Sci.* 5, 154–159. doi: 10.1016/S1360-1385(00)01577-6

Harper, J. F. (2001). Dissecting calcium oscillators in plant cells. *Trends Plant Sci.* 6, 395–397. doi: 10.1016/S1360-1385(01)02023-4

Holappa, L. D., and Walker-Simmons, M. (1995). The wheat abscisic acid-responsive protein kinase mRNA, PKABA1, is up-regulated by dehydration, cold temperature, and osmotic stress. *Plant Physiol.* 108, 1203–1210. doi: 10.1104/pp.108.3.1203

Hrabak, E. M., Chan, C. W., Gribskov, M., Harper, J. F., Choi, J. H., Halford, N., et al. (2003). The *Arabidopsis* CDPK-SnRK superfamily of protein kinases. *Plant Physiol.* 132, 666–680. doi: 10.1104/pp.102.011999

Huang, Q.-S., Wang, H.-Y., Gao, P., Wang, G.-Y., and Xia, G.-X. (2008). Cloning and characterization of a calcium dependent protein kinase gene associated with cotton fiber development. *Plant Cell Rep.* 27, 1869–1875. doi: 10.1007/s00299-008-0603-0

Hwang, I., Sze, H., and Harper, J. F. (2000). A calcium-dependent protein kinase can inhibit a calmodulin-stimulated Ca^{2+} pump (ACA2) located in the endoplasmic reticulum of *Arabidopsis*. *Proc. Natl. Acad. Sci. U.S.A.* 97, 6224–6229. doi: 10.1073/pnas.97.11.6224

Ihaka, R., and Gentleman, R. (1996). R: a language for data analysis and graphics. *J. Comput. Graphic. Stat.* 5, 299–314. doi: 10.1080/10618600.1996.10474713

Imamura, M., Takahashi, Y. T., Nakamura, N., Htwe, N. M. P. S., Zheng, S. H., Shimazaki, K. I., et al. (2008). Isolation and characterization of a cDNA coding cowpea (*Vigna unguiculata* (L) Walp.) calcineurin B-like protein-interacting protein kinase, VuCIPKJ. *Plant Biotechnol.* 25, 437–445. doi: 10.5511/plantbiotechnology.25.437

Innan, H., and Kondrashov, F. (2010). The evolution of gene duplications: classifying and distinguishing between models. *Nat. Rev. Genet.* 11, 97–108. doi: 10.1038/nrg2689

Jiao, Y., Shi, C., Edil, B. H., de Wilde, R. F., Klimstra, D. S., Maitra, A., et al. (2011a). DAXX/ATRX, MEN1, and mTOR pathway genes are frequently altered in pancreatic neuroendocrine tumors. *Science* 331, 1199–1203. doi: 10.1126/science.1200609

Jiao, Y., Wickett, N. J., Ayyampalayam, S., Chanderbali, A. S., Landherr, L., Ralph, P. E., et al. (2011b). Ancestral polyploidy in seed plants and angiosperms. *Nature* 473, 97–100. doi: 10.1038/nature09916

Jossier, M., Bouly, J. P., Meimoun, P., Arjmand, A., Lessard, P., Hawley, S., et al. (2009). SnRK1 (SNF1-related kinase 1) has a central role in sugar and ABA signalling in *Arabidopsis thaliana*. *Plant J. Cell Mol. Biol.* 59, 316–328. doi: 10.1111/j.1365-313X.2009.03871.x

Kawahara, Y., de la Bastide, M., Hamilton, J. P., Kanamori, H., McCombie, W. R., Ouyang, S., et al. (2013). Improvement of the *Oryza sativa* Nipponbare reference genome using next generation sequence and optical map data. *Rice* 6:4. doi: 10.1186/1939-8433-6-4

Kim, K. N., Cheong, Y. H., Gupta, R., and Luan, S. (2000). Interaction specificity of Arabidopsis calcineurin B-like calcium sensors and their target kinases. *Plant Physiol.* 124, 1844–1853. doi: 10.1104/pp.124.4.1844

Kim, K. N., Lee, J. S., Han, H., Choi, S. A., Go, S. J., and Yoon, I. S. (2003). Isolation and characterization of a novel rice Ca^{2+}-regulated protein kinase gene involved in responses to diverse signals including cold, light, cytokinins, sugars and salts. *Plant Mol. Biol.* 52, 1191–1202. doi: 10.1023/B:PLAN.0000004330.62660.a2

Kleinow, T., Bhalerao, R., Breuer, F., Umeda, M., Salchert, K., and Koncz, C. (2000). Functional identification of an Arabidopsis snf4 ortholog by screening for heterologous multicopy suppressors of snf4 deficiency in yeast. *Plant J.* 23, 115–122. doi: 10.1046/j.1365-313x.2000.00809.x

Knight, H., and Knight, M. R. (2001). Abiotic stress signalling pathways: specificity and cross-talk. *Trends Plant Sci.* 6, 262–267. doi: 10.1016/S1360-1385(01)01946-X

Knight, H., Trewavas, A. J., and Knight, M. R. (1996). Cold calcium signaling in *Arabidopsis* involves two cellular pools and a change in calcium signature after acclimation. *Plant Cell* 8, 489–503. doi: 10.1105/tpc.8.3.489

Koch, M. A., Haubold, B., and Mitchell-Olds, T. (2000). Comparative evolutionary analysis of chalcone synthase and alcohol dehydrogenase loci in *Arabidopsis*, *Arabis*, and related genera (Brassicaceae). *Mol. Biol. Evol.* 17, 1483–1498. doi: 10.1093/oxfordjournals.molbev.a026248

Komatsu, S., Yang, G., Khan, M., Onodera, H., Toki, S., and Yamaguchi, M. (2007). Over-expression of calcium-dependent protein kinase 13 and calreticulin interacting protein 1 confers cold tolerance on rice plants. *Mol. Genet. Genomics* 277, 713–723. doi: 10.1007/s00438-007-0220-6

Kong, X., Lv, W., Jiang, S., Zhang, D., Cai, G., Pan, J., et al. (2013). Genome-wide identification and expression analysis of calcium-dependent protein kinase in maize. *BMC Genomics* 14:433. doi: 10.1186/1471-2164-14-433

Krzywinski, M., Schein, J., Birol, I., Connors, J., Gascoyne, R., Horsman, D., et al. (2009). Circos: an information aesthetic for comparative genomics. *Genome Res.* 19, 1639–1645. doi: 10.1101/gr.092759.109

Kudla, J., Batistič, O., and Hashimoto, K. (2010). Calcium signals: the lead currency of plant information processing. *Plant Cell* 22, 541–563. doi: 10.1105/tpc.109.072686

Kulik, A., Wawer, I., Krzywińska, E., Bucholc, M., and Dobrowolska, G. (2011). SnRK2 protein kinases—key regulators of plant response to abiotic stresses. *Omics* 15, 859–872. doi: 10.1089/omi.2011.0091

Lee, T.-H., Tang, H., Wang, X., and Paterson, A. H. (2013). PGDD: a database of gene and genome duplication in plants. *Nucleic Acids Res.* 41, D1152–D1158. doi: 10.1093/nar/gks1104

Li, G., Peng, F., Zhang, L., Shi, X., and Wang, Z. (2010). Cloning and characterization of a SnRK1-encoding gene from Malus hupehensis Rehd. and heterologous expression in tomato. *Mol. Biol. Rep.* 37, 947–954. doi: 10.1007/s11033-009-9734-9

Li, A.-L., Zhu, Y.-F., Tan, X.-M., Wang, X., Wei, B., Guo, H.-Z., et al. (2008). Evolutionary and functional study of the CDPK gene family in wheat (*Triticum aestivum* L.). *Plant Mol. Biol.* 66, 429–443. doi: 10.1007/s11103-007-9281-5

Li, J., Wang, X.-Q., Watson, M. B., and Assmann, S. M. (2000). Regulation of abscisic acid-induced stomatal closure and anion channels by guard cell AAPK kinase. *Science* 287, 300–303. doi: 10.1126/science.287.5451.300

Llop-Tous, I., Domínguez-Puigjaner, E., and Vendrell, M. (2002). Characterization of a strawberry cDNA clone homologous to calcium-dependent protein kinases that is expressed during fruit ripening and affected by low temperature. *J. Exp. Bot.* 53, 2283–2285. doi: 10.1093/jxb/erf103

Ma, S.-Y., and Wu, W.-H. (2007). AtCPK23 functions in *Arabidopsis* responses to drought and salt stresses. *Plant Mol. Biol.* 65, 511–518. doi: 10.1007/s11103-007-9187-2

Man, A. L., Purcell, P. C., Hannappel, U., and Halford, N. G. (1997). Potato SNF1-related protein kinase: molecular cloning, expression analysis and peptide kinase activity measurements. *Plant Mol. Biol.* 34, 31–43. doi: 10.1023/A:1005765719873

Mathieu, J., Yant, L. J., Mürdter, F., Küttner, F., and Schmid, M. (2009). Repression of flowering by the miR172 target SMZ. *Plos Biol.* 7:e1000148. doi: 10.1371/journal.pbio.1000148

McAinsh, M. R., Evans, N. H., Montgomery, L. T., and North, K. A. (2002). Calcium signalling in stomatal responses to pollutants. *New Phytol.* 153, 441–447. doi: 10.1046/j.0028-646X.2001.00336.x

Manns, M. P., McHutchison, J. G., Gordon, S. C., Rustgi, V. K., Shiffman, M., Reindollar, R., et al. (2001). Peginterferon alfa-2b plus ribavirin compared with interferon alfa-2b plus ribavirin for initial treatment of chronic hepatitis C: a randomised trial. *Lancet* 358, 958–965. doi: 10.1016/S0140-6736(01)06102-5

Monroy, A. F., and Dhindsa, R. S. (1995). Low-temperature signal transduction: induction of cold acclimation-specific genes of *alfalfa* by calcium at 25 degrees C. *Plant Cell* 7, 321–331. doi: 10.2307/3869854

Munemasa, S., Hossain, M. A., Nakamura, Y., Mori, I. C., and Murata, Y. (2011). The *Arabidopsis* calcium-dependent protein kinase, CPK6, functions as a positive regulator of methyl jasmonate signaling in guard cells. *Plant Physiol.* 155, 553–561. doi: 10.1104/pp.110.162750

Murata, Y., Pei, Z.-M., Mori, I. C., and Schroeder, J. (2001). Abscisic acid activation of plasma membrane Ca^{2+} channels in guard cells requires cytosolic NAD (P) H and is differentially disrupted upstream and downstream of reactive oxygen species production in abi1-1 and abi2-1 protein phosphatase 2C mutants. *Plant Cell* 13, 2513–2523. doi: 10.1105/tpc.13.11.2513

Murillo, M., Gamazo, C., Goñi, M., Irache, J. M., and Blanco-Prιeto, M. (2002). Development of microparticles prepared by spray-drying as a vaccine delivery system against brucellosis. *Int. J. Pharm.* 242, 341–344. doi: 10.1016/S0378-5173(02)00212-0

Nakashima, K., Fujita, Y., Kanamori, N., Katagiri, T., Umezawa, T., Kidokoro, S., et al. (2009). Three *Arabidopsis* SnRK2 protein kinases, SRK2D/SnRK2. 2, SRK2E/SnRK2. 6/OST1 and SRK2I/SnRK2. 3, involved in ABA signaling are essential for the control of seed development and dormancy. *Plant Cell Physiol.* 50, 1345–1363. doi: 10.1093/pcp/pcp083

Ohno, S. (2013). *Evolution by Gene Duplication.* London; Berlin; Heidelberg; New York, NY: George Alien & Unwin Ltd.; Springer-Verlag.

Ohno, S., Wolf, U., and Atkin, N. B. (1968). Evolution from fish to mammals by gene duplication. *Hereditas* 59, 169–187. doi: 10.1111/j.1601-5223.1968.tb02169.x

Pandey, G. K., Grant, J. J., Cheong, Y. H., Kim, B. G., Li, L. G., and Luan, S. (2008). Calcineurin-B-like protein CBL9 interacts with target kinase CIPK3 in the regulation of ABA response in seed germination. *Mol. Plant* 1, 238–248. doi: 10.1093/mp/ssn003

Patharkar, O. R., and Cushman, J. C. (2000). A stress-induced calcium-dependent protein kinase from Mesembryanthemum crystallinum phosphorylates a two-component pseudo-response regulator. *Plant J.* 24, 679–691. doi: 10.1046/j.1365-313x.2000.00912.x

Pei, Z.-M., Murata, Y., Benning, G., Thomine, S., Klüsener, B., Allen, G. J., et al. (2000). Calcium channels activated by hydrogen peroxide mediate abscisic acid signalling in guard cells. *Nature* 406, 731–734. doi: 10.1038/35021067

Plieth, C., and Trewavas, A. J. (2002). Reorientation of seedlings in the earth's gravitational field induces cytosolic calcium transients. *Plant Physiol.* 129, 786–796. doi: 10.1104/pp.011007

Raíces, M., Ulloa, R. M., MacIntosh, G. C., Crespi, M., and Téllez-Iñón, M. T. (2003). StCDPK1 is expressed in potato stolon tips and is induced by high sucrose concentration. *J. Exp. Bot.* 54, 2589–2591. doi: 10.1093/jxb/erg282

Ray, S., Agarwal, P., Arora, R., Kapoor, S., and Tyagi, A. K. (2007). Expression analysis of calcium-dependent protein kinase gene family during reproductive development and abiotic stress conditions in rice (*Oryza sativa* L. ssp. indica). *Mol. Genet. Genomics* 278, 493–505. doi: 10.1007/s00438-007-0267-4

Rensing, S. A. (2014). Gene duplication as a driver of plant morphogenetic evolution. *Curr. Opin. Plant Biol.* 17, 43–48. doi: 10.1016/j.pbi.2013.11.002

Ritchie, S. M., Swanson, S. J., and Gilroy, S. (2002). From common signalling components to cell specific responses: insights from the cereal aleurone. *Physiol. Plant.* 115, 342–351. doi: 10.1034/j.1399-3054.2002.1150303.x

Romeis, T., Ludwig, A. A., Martin, R., and Jones, J. D. (2001). Calcium-dependent protein kinases play an essential role in a plant defence response. *EMBO J.* 20, 5556–5567. doi: 10.1093/emboj/20.20.5556

Romeis, T., Piedras, P., and Jones, J. D. (2000). Resistance gene-dependent activation of a calcium-dependent protein kinase in the plant defense response. *Plant Cell* 12, 803–815. doi: 10.1105/tpc.12.5.803

Rudd, J. J., and Franklin-Tong, V. E. (2001). Unravelling response-specificity in Ca^{2+} signalling pathways in plant cells. *New Phytol.* 151, 7–33. doi: 10.1046/j.1469-8137.2001.00173.x

Saijo, Y., Hata, S., Kyozuka, J., Shimamoto, K., and Izui, K. (2000). Over-expression of a single Ca+-dependent protein kinase confers both cold and salt/drought tolerance on rice plants. *Plant J.* 23, 319–327. doi: 10.1046/j.1365-313x.2000.00787.x

Sanders, D., Brownlee, C., and Harper, J. F. (1999). Communicating with calcium. *Plant Cell* 11, 691–706. doi: 10.1105/tpc.11.4.691

Sanders, D., Pelloux, J., Brownlee, C., and Harper, J. F. (2002). Calcium at the crossroads of signaling. *Plant Cell* 14, S401–S417. doi: 10.1105/tpc.002899

Schmid, M., Davison, T. S., Henz, S. R., Pape, U. J., Demar, M., Vingron, M., et al. (2005). A gene expression map of *Arabidopsis thaliana* development. *Nat. Genet.* 37, 501–506. doi: 10.1038/ng1543

Shannon, P., Markiel, A., Ozier, O., Baliga, N. S., Wang, J. T., Ramage, D., et al. (2003). Cytoscape: a software environment for integrated models of biomolecular interaction networks. *Genome Res.* 13, 2498–2504. doi: 10.1101/gr.1239303

Shi, Y., and Eberhart, R. C. (2001). "Fuzzy adaptive particle swarm optimization," in *Evolutionary Computation, Proceedings of the 2001 Congress on* (Boston: IEEE), 101–106.

Song, X., Li, Y., and Hou, X. (2013). Genome-wide analysis of the AP2/ERF transcription factor superfamily in Chinese cabbage (*Brassica rapa* ssp. pekinensis). *BMC Genomics* 14:573. doi: 10.1186/1471-2164-14-573

Stebbins, C. Jr. (1950). *Variation and Evolution in Plants.* New York, NY: Columbia University Press.

Szczegielniak, J., Borkiewicz, L., Szurmak, B., Lewandowska-Gnatowska, E., Statkiewicz, M., Klimecka, M., et al. (2012). Maize calcium-dependent protein kinase (ZmCPK11): local and systemic response to wounding, regulation by touch and components of jasmonate signaling. *Physiol. Plant.* 146, 1–14. doi: 10.1111/j.1399-3054.2012.01587.x

Tang, J., Wang, F., Wang, Z., Huang, Z., Xiong, A., and Hou, X. (2013). Characterization and co-expression analysis of WRKY orthologs involved in responses to multiple abiotic stresses in *Pak-choi* (*Brassica campestris* ssp. chinensis). *BMC Plant Biol.* 13:188. doi: 10.1186/1471-2229-13-188

Taylor, L. P., and Hepler, P. K. (1997). Pollen germination and tube growth. *Annu. Rev. Plant Biol.* 48, 461–491. doi: 10.1146/annurev.arplant.48.1.461

Thuleau, P., Schroeder, J. I., and Ranjeva, R. (1998). Recent advances in the regulation of plant calcium channels: evidence for regulation by G-proteins, the cytoskeleton and second messengers. *Curr. Opin. Plant Biol.* 1, 424–427. doi: 10.1016/S1369-5266(98)80267-7

Tong, C., Wang, X., Yu, J., Wu, J., Li, W., Huang, J., et al. (2013). Comprehensive analysis of RNA-seq data reveals the complexity of the transcriptome in *Brassica rapa*. *BMC Genomics* 14:689. doi: 10.1186/1471-2164-14-689

Trewavas, A. J., and Malhó, R. (1998). Ca^{2+} signalling in plant cells: the big network! *Curr. Opin. Plant Biol.* 1, 428–433. doi: 10.1016/S1369-5266(98)80268-9

Tuskan, G. A., Difazio, S., Jansson, S., Bohlmann, J., Grigoriev, I., Hellsten, U., et al. (2006). The genome of black cottonwood, *Populus trichocarpa* (Torr. & Gray). *Science* 313, 1596–1604. doi: 10.1126/science.1128691

Wang, C., Liu, Y., Li, S. S., and Han, G.-Z. (2015). Insights into the originand evolution of the plant hormone signaling machinery. *Plant Physiol.* 167, 872–886. doi: 10.1104/pp.114.247403

Wang, X., Wang, H., Wang, J., Sun, R., Wu, J., Liu, S., et al. (2011). The genome of the mesopolyploid crop species *Brassica rapa*. *Nat. Genet.* 43, 1035–1039. doi: 10.1038/ng.919

Wang, Y., Tang, H., DeBarry, J. D., Tan, X., Li, J., Wang, X., et al. (2012). MCScanX: a toolkit for detection and evolutionary analysis of gene synteny and collinearity. *Nucleic Acids Res.* 40, e49–e49. doi: 10.1093/nar/gkr1293

Yamada, T., Letunic, I., Okuda, S., Kanehisa, M., and Bork, P. (2011). iPath2. 0: interactive pathway explorer. *Nucleic Acids Res.* 39(Suppl. 2), W412–W415. doi: 10.1093/nar/gkr313

Zhang, Z., Li, J., Zhao, X.-Q., Wang, J., Wong, G. K.-S., and Yu, J. (2006). KaKs_Calculator: calculating Ka and Ks through model selection and model averaging. *Genomics Proteomics Bioinformatics* 4, 259–263. doi: 10.1016/S1672-0229(07)60007-2

Zuo, R., Hu, R., Chai, G., Xu, M., Qi, G., Kong, Y., et al. (2013). Genome-wide identification, classification, and expression analysis of CDPK and its closely related gene families in poplar (*Populus trichocarpa*). *Mol. Biol. Rep.* 40, 2645–2662. doi: 10.1007/s11033-012-2351-z

Conflict of Interest Statement: The authors declare that the research was conducted in the absence of any commercial or financial relationships that could be construed as a potential conflict of interest.

11

Expression of the *KNOTTED HOMEOBOX* Genes in the Cactaceae Cambial Zone Suggests their Involvement in Wood Development

Jorge Reyes-Rivera[1†], Gustavo Rodríguez-Alonso[2†], Emilio Petrone[1], Alejandra Vasco[1], Francisco Vergara-Silva[3], Svetlana Shishkova[2*] and Teresa Terrazas[1*]

[1] Departamento de Botánica, Instituto de Biología, Universidad Nacional Autónoma de México, Mexico City, Mexico, [2] Departamento de Biología Molecular de Plantas, Instituto de Biotecnología, Universidad Nacional Autónoma de México, Cuernavaca, Mexico, [3] Jardín Botánico, Instituto de Biología, Universidad Nacional Autónoma de México, Mexico City, Mexico

Edited by:
Federico Valverde, Spanish National Research Council, Spain

Reviewed by:
David Smyth, Monash University, Australia
Ykä Helariutta, University of Helsinki, Finland

****Correspondence:***
Teresa Terrazas
tterrazas@ib.unam.mx
Svetlana Shishkova
sveta@ibt.unam.mx

† These authors have contributed equally to this work.

The vascular cambium is a lateral meristem that produces secondary xylem (i.e., wood) and phloem. Different Cactaceae species develop different types of secondary xylem; however, little is known about the mechanisms underlying wood formation in the Cactaceae. The *KNOTTED HOMEOBOX (KNOX)* gene family encodes transcription factors that regulate plant development. The role of class I *KNOX* genes in the regulation of the shoot apical meristem, inflorescence architecture, and secondary growth is established in a few model species, while the functions of class II *KNOX* genes are less well understood, although the *Arabidopsis thaliana* class II KNOX protein KNAT7 is known to regulate secondary cell wall biosynthesis. To explore the involvement of the *KNOX* genes in the enormous variability of wood in Cactaceae, we identified orthologous genes expressed in species with fibrous (*Pereskia lychnidiflora* and *Pilosocereus alensis*), non-fibrous (*Ariocarpus retusus*), and dimorphic (*Ferocactus pilosus*) wood. Both class I and class II *KNOX* genes were expressed in the cactus cambial zone, including one or two class I paralogs of *KNAT1*, as well as one or two class II paralogs of *KNAT3-KNAT4-KNAT5*. While the *KNOX* gene *SHOOTMERISTEMLESS* (*STM)* and its ortholog *ARK1* are expressed during secondary growth in the *Arabidopsis* and *Populus* stem, respectively, we did not find *STM* orthologs in the Cactaceae cambial zone, which suggests possible differences in the vascular cambium genetic regulatory network in these species. Importantly, while two class II *KNOX* paralogs from the *KNAT7* clade were expressed in the cambial zone of *A. retusus* and *F. pilosus*, we did not detect *KNAT7* ortholog expression in the cambial zone of *P. lychnidiflora*. Differences in the transcriptional repressor activity of secondary cell wall biosynthesis by the *KNAT7* orthologs could therefore explain the differences in wood development in the cactus species.

Keywords: Cactaceae, dimorphic wood, fibrous wood, KNAT7 ortholog, KNOX transcription factors, non-fibrous wood, vascular cambium, wood lignification

INTRODUCTION

As reservoirs of pluripotent cells, meristems have played a leading role in the diversification of angiosperm growth forms. The shoot and root apical meristems maintain primary growth in plants, while the lateral meristems, comprised of the vascular cambium and cork cambium in Eudicotyledons and gymnosperms, are involved in secondary growth (Evert, 2006). The vascular cambium maintains a population of initial (stem) cells, which divide asymmetrically to generate two daughter cells; one maintains the cambial initial identity, while the other divides again and the daughters differentiate to generate secondary phloem or xylem (Groover et al., 2010; Miyashima et al., 2013). Vascular cambium derivatives are thought to have influenced speciation and diversification events (Spicer and Groover, 2010; Lucas et al., 2013). In the Cactaceae, the traits of the secondary xylem (i.e., wood) suggest that it has evolved by heterochronic processes where a change in the timing of developmental processes leads to morphological differences between species (Altesor et al., 1994). The larger species (≥1.5 m in height) in this family of succulent plants have fibrous wood with vessel elements in the xylem similar to those typically derived from vascular cambium, with a similar wood chemical composition to that of other woody dicot species (Reyes-Rivera et al., 2015). On the other hand, the wood found in smaller species is generally scarce and non-fibrous, with abundant wide-band tracheids and vessel elements similar to those typical of proto- and metaxylem. In these species, the level of wood lignification is insignificant and the lignin has a heterogeneous chemical composition (Reyes-Rivera et al., 2015). Many species of Cactaceae have dimorphic wood, where one type of wood is produced in the juvenile stages and a different structure is formed in the adult stages of development. To the best of our knowledge, this phenomenon is unique among Eudicotyledons and is related to the globose and globose-depressed growth forms of some Cactaceae species (Mauseth and Plemons, 1995). In species with dimorphic wood that produce wide-band tracheids, these always develop first, before the fibrous or parenchymatous wood is produced (Mauseth and Plemons, 1995; Loza-Cornejo and Terrazas, 2011). The mechanisms shaping the development of the vascular cambium and its derivatives in the Cactaceae are mostly unknown; nevertheless, it was suggested that the wide variation in wood anatomy of different cacti species might be attributed to a variation of gene expression patterns and gene expression level (Mauseth and Plemons, 1995; Mauseth, 2006; Landrum, 2008; Vázquez-Sánchez and Terrazas, 2011).

The interplay of diverse factors in the regulation of vascular cambium activity has been reported previously (Liu et al., 2014; Nieminen et al., 2015), and the roles of growth regulators, including auxin, cytokinin, and ethylene, in this process are well established. Later in development, genes regulating cell expansion, secondary xylem differentiation, lignification, and secondary wall deposition contribute to wood formation (reviewed in Růžička et al., 2015; Ye and Zhong, 2015; Zhong and Ye, 2015). Transcription profiling of the vascular cambium in aspen (*Populus tremula*) uncovered similarities between the gene regulatory networks operating in the shoot apical meristem and the vascular cambium. In particular, it was reported that four members of the *KNOTTED1-LIKE HOMEOBOX (KNOX)* gene family, namely *PttKNOX1*, *PttSHOOT MERISTEMLESS (PttSTM)*, *PttKNOX2*, and *PttKNOX6*, are highly expressed in both tissues (Schrader et al., 2004). Functional analysis suggests that the *Populus* orthologs of *Arabidopsis thaliana (Arabidopsis)* class I *KNOX* genes *STM* and *KNOTTED-LIKE HOMEOBOX OF ARABIDOPSIS THALIANA 1 (KNAT1)/BREVIPEDICELLUS (BP)*, *PttSTM/ARBORKNOX 1 (ARK1)/ARK1a* and *PttKNOX1/ARK2*, respectively, regulate secondary growth (Groover et al., 2006; Du et al., 2009; Liu et al., 2015). While the roles of class I *KNOX* genes in the regulation of shoot apical meristem, inflorescence architecture, and compound leaf development are well established, the functions of class II *KNOX* genes are less well understood. In general, class I transcripts are less abundant than class II *KNOX* transcripts, and are expressed in specific regions of meristems, particularly in the shoot apical meristem and leaf meristems. By contrast, class II transcripts are found in differentiating cells and all mature plant organs (Serikawa et al., 1997; reviewed by Hay and Tsiantis, 2010; Arnaud and Pautot, 2014), but not in the shoot apical meristem (Furumizu et al., 2015). Moreover, the dark-green serrated leaf phenotype of the class II *knat3 knat4 knat5* triple loss-of-function mutant in *Arabidopsis* is similar to that of the class I gain-of-function mutants (Furumizu et al., 2015), suggesting opposing functions for genes of the two classes. Here, we report that class I and class II *KNOX* genes are expressed in the cambial zone, consisting of vascular cambium and derived cells, of Cactaceae species with fibrous (*Pereskia lychnidiflora* and *Pilosocereus alensis*), non-fibrous (*Ariocarpus retusus*), and dimorphic (*Ferocactus pilosus*) wood. We also present the phylogeny of class I and class II KNOX proteins encoded in the sequenced plant genomes retrieved from the Phytozome database, confirming monophyly of class I and class II KNOX proteins, assigning the Cactaceae KNOX proteins into clades, and exploring the number of paralogs of the plant species in every KNOX clade.

MATERIALS AND METHODS

Plant Material and Tissue Sampling

Four Cactaceae species with different wood types were used in this study: *Pereskia lychnidiflora* DC. (subfamily: Pereskoideae); *Pilosocereus alensis* (F. A. C. Weber) Byles & G. D. Rowley (subfamily: Cactoideae, tribe: Cereeae); *Ariocarpus retusus* Scheidw (Cactoideae, tribe: Cacteae); and *Ferocactus pilosus* (Galeotti ex Salm-Dyck) Werderm. (Cactoideae, Cacteae). Samples were collected from the cambial zone of adult individuals growing in Mexico. For the tall species (*F. pilosus*, *P. lychnidiflora*), the cambial zone was harvested from one individual of both species in the field (*F. pilosus* at the Mexican Plateau, San Luis Potosí state, location 23°19′31″N; 100°33′30″W, voucher TT890 MEXU; and *P. lychnidiflora* at the Isthmus of Tehuantepec, Oaxaca state, location 16°22′52″N; 95°19′03″W, voucher TT966 MEXU), with samples collected during the rainy season to ensure that the vascular cambium was active. All surrounding tissues including the cortex were removed as described by Liu et al. (2014). The cambial zone was peeled off with a disposable microtome knife and immediately frozen in liquid nitrogen. One whole individual of both *A. retusus* and *P. alensis* were placed in pots containing their native soil and transported from their natural habitat to the laboratory (*A. retusus* from the Mexican Plateau, Nuevo León state, location 23°23′39″N; 100°21′57″W, voucher SA1976 MEXU; and *P. alensis* from the Pacific coast, Michoacán state, location 18°46′41″N; 103°08′05″W, voucher JRR3 MEXU). Species identification was confirmed by S. Arias [Instituto de Biología, Universidad Nacional Autónoma de México (IB, UNAM)], the leading taxonomy specialist of the Cactaceae family in Mexico. The cambial zone was harvested as described above. The cambial zone samples were stored at −80°C until RNA extraction. For expression analysis by RT-PCR, one adult plant of *A. retusus, F. pilosus*, and *P. lychnidiflora*, one 2-year-old juvenile plant of the first two species, and two 5-year-old young individuals of *A. retusus*, previously collected from the same locations, were donated by the Botanical Garden IB, UNAM. For the root tip transcriptome of *Pachycereus pringlei* (S. Watson) Britton & Rose, plants were germinated from seeds collected near Bahia Kino, Sonora state, and donated by F. Molina-Freaner and J. F. Martínez-Rodríguez (Instituto de Ecología, campus Hermosillo, UNAM).

For the wood maceration, fine wood chips (2 mm thickness) were obtained every 5 mm, from the region closest to the pith (young wood) to the section closest to the vascular cambium (mature wood), and each region was processed separately. The non-fibrous *A. retusus* samples were placed in 2-mL microcentrifuge tubes filled with Franklin solution (5:1:4 acetic acid:hydrogen peroxide:water; Ruzin, 1999), while the dimorphous *F. pilosus* samples and the fibrous *P. alensis* and *P. lychnidiflora* were placed into tubes containing Jeffrey solution (equal volumes of 10% chromic acid and 10% nitric acid; Berlyn and Miksche, 1976). For each species, 0.2 g of the tissue was used. The samples were then incubated at 56°C for 4 h (non-fibrous wood) and 24 h (dimorphic and fibrous wood). Additionally, all samples were sonicated in a Branson 200 ultrasonic cleaner (Branson Ultrasonics, Danbury, CT, USA) until completely macerated, washed with water, and dehydrated using a series of ethanol washes at 50, 70, 90%, and absolute concentrations. The *A. retusus* macerations were stained with a 0.1% aqueous solution of toluidine blue for 12 h (Ruzin, 1999) and mounted onto slides. The macerations of the wood samples from another three species were stained with Safranin for 2 h and mounted onto slides using synthetic resin (Entellan, Merck Millipore, Darmstadt, Germany). All wood elements were photographed using a BX51 optical microscope (Olympus Corporation, Tokyo, Japan) and the images were analyzed using Image-Pro Plus v. 6.1 software (Media Cybernetics, Inc., Bethesda, MD, USA).

RNA Extraction, cDNA Synthesis, and PCR Amplification

Total RNA was extracted using TRIzol Reagent (Invitrogen, Carlsbad, CA, USA), according to the manufacturer's protocol, including the optional step of centrifugation before the separation of phases, or using the Spectrum Plant Total RNA Isolation Kit (Sigma–Aldrich, St. Louis, MO, USA). The cDNA was synthesized using SuperScript II Reverse Transcriptase (Invitrogen), according to the manufacturer's instructions. Degenerate PCR primers were used to amplify the *KNOX* genes. Primers for amplifying putative cacti orthologs of *KNAT3* were designed (KNAT3_Cact_F: 5′-GAGAGRAATAATGGCWTATCATC-3′ and KNAT3_Cact-R: 5′-CCTTCTGGTTCTACTTCCCTC-3′) based on the alignments of nucleotide sequences encoding the class II KNOX proteins *KNAT3* (*Arabidopsis*) and its closest homologs from *Beta vulgaris* and *Pachycereus pringlei.* For the Cactaceae orthologs of *KNAT1*, primers designed by Du et al. (2009) were used. PCR products were purified with Sephadex Centri-sep columns (Thermo Fisher Scientific, Waltham, MA, USA) as instructed by the manufacturer. The amplified and purified products were sequenced in a 3500xL Genetic Analyzer sequencer (Applied Biosystems, Foster City, CA, USA) using the PCR primers. Platinum Taq polymerase (Thermo Fischer Scientific, Waltham, MA, USA) was used for PCR reactions. Primers used for RT-PCR are listed in **Supplementary Table S1**. RNA-seq was performed at the Beijing Genome Institute, Hong Kong; the vascular cambium and root tip transcriptomes were *de novo* assembled using Trinity v. 2.2[1] and CLC Genomic Workbench v. 7.5 (Qiagen[2]), respectively.

Sequence Alignment and Phylogenetic Analysis

KNOX-like protein sequences were retrieved from the Phytozome database[3] (v. 11; last accessed on May 9, 2016) using the PhytoMine tools. All proteins with KNOX (IPR005540 [InterProscan definition]/PF03790 [Pfam definition] and IPR005541/PF03791), ELK (IPR005539/PF03789), and HD domains (IPR009057/PF05920) were retrieved. *B. vulgaris* sequences were retrieved using tBLASTn from the RefBeet v. 1.2 genome assembly (The *Beta vulgaris* resource[4], Dohm et al., 2013) using a BLOSUM80 substitution matrix (Henikoff and Henikoff, 1992; at the time, *B. vulgaris* was the only species from the order Caryophyllales with a sequenced genome). Chimeric sequences and those from the early release genomes were discarded. After that, KNOX protein sequences from the tree species *Betula luminifera* and *Juglans nigra* were added. For Cactaceae species, the KNOX protein sequences used were deduced from the amplified and sequenced PCR fragments (see previous section), from our RNA-seq data and the *de novo* assembly of the cambial zone transcriptome of the four species reported in this work (the same RNA samples were used as starting material for the cambial zone transcriptome assembly), from the root tip transcriptome of the Cactoideae subfamily species *Pachycereus pringlei* (the analysis of the transcriptomes will be reported elsewhere), and from the recently published *Lophophora williamsii* transcriptome (Ibarra-Laclette et al., 2015). The resulting sequences were

[1]https://github.com/trinityrnaseq/trinityrnaseq/releases

[2]https://www.qiagenbioinformatics.com/products/clc-genomics-workbench/

[3]http://phytozome.jgi.doe.gov

[4]http://bvseq.molgen.mpg.de/index.shtml

aligned with Clustal Omega[5] and the alignment file was manually edited. Alignment positions with more than 30% gaps were not included in the analysis. The identity and similarity values (%) were obtained from a pairwise alignment (Needle-EMBOSS[6]), with a complete gap deletion for each pair.

The phylogeny was reconstructed with MEGA7 (Kumar et al., 2016). A maximum likelihood algorithm based on the JTT substitution model (Felsenstein, 1981) was used to resolve the phylogenetic relationship of the KNOX proteins derived from the Cactaceae species and the plant species with sequenced genomes. The resulting topology was statistically tested with a 1000 replicate bootstrap for both the complete and the selected datasets. The NWK files were visualized and edited on FigTree[7] (v. 1.4.2). The BELL 1 protein of *Arabidopsis* (AT5G41410.1) was used as the outgroup. Nucleotide sequences reported in this work were deposited in GenBank under the accession numbers KX891335-KX891338 for *F. pilosus*; KX891339-KX891343 and KX891349 for *A. retusus*; KX891344-KX891346 for *P. lychnidiflora*; KX891347 and KX891348 for *P. alensis*; and KX870027-KX870031 for *P. pringlei*.

RESULTS

Morphological Analysis of Xylem Cells

To examine the detailed cellular features of the secondary xylem in the four species studied, we performed a morphological analysis of xylem cells using wood macerations. The young xylem of *A. retusus* with non-fibrous wood and *F. pilosus* with dimorphic wood forms wide-band tracheids and vessel elements with annular and helical secondary wall patterns formed during early development. The mature fibers and vessel elements with pseudoscalariform and alternate pitting were formed in adult plants during late development. No differences were observed between young and mature wood of *P. alensis* and *P. lychnidiflora*, with both having fibrous wood. The tracheary elements of *A. retusus* and young *F. pilosus* had a low proportion of lignified cell wall, similar to those of proto- and metaxylem (i.e., primary xylem), whereas vessel elements and fibers of mature *F. pilosus*, *P. alensis*, and *P. lychnidiflora*, had a higher proportion of lignified cell walls. This analysis allowed the unequivocal identification of each secondary xylem cell type and its cell wall features (**Figure 1**) and thus confirmed the wood types we previously identified using an anatomical analysis of wood sections (Reyes-Rivera et al., 2015). The variability in size and pitting of vessel elements associated with the presence of fibers was found in species with dimorphic and fibrous wood.

KNOX Family Genes Are Expressed in the Cactaceae Cambial Zone

To explore the involvement of the *KNOX* family in the enormous variability of wood morphologies in the Cactaceae, we looked for orthologs of these genes expressed in the Cactaceae cambial zone, which comprises the vascular cambium and recently derived cells. RT-PCR using degenerate primers for the class I *KNOX* genes resulted in a major amplification product of the predicted molecular weight in the four Cactaceae species studied. The amino acid sequences inferred from the amplified and sequenced fragments of these genes (*ArKNOX1a* from *Ariocarpus retusus*, *FpKNOX1a* from *Ferocactus pilosus*, *PaKNOX1a* from *Pilosocereus alensis*, and *PlKNOX1a* from *Pereskia lychnidiflora*) correspond to nearly 70% of the ARK2 and KNAT1 protein sequences of *Populus* and *Arabidopsis*, respectively, including part of the KNOX1 domain and the entirety of the KNOX2, ELK, and HD domains [**Figure 2** (The last letters in the gene names refer to: "a," amplification as a method of gene isolation; "e," the inferred coding sequence extended after alignment with sequences resulting from the RNA-seq; and "r," RNA-seq as a method of gene identification)].

Fragments of class II *KNOX* genes were successfully amplified by RT-PCR with degenerate primers from the three species from the Cactoideae subfamily, but were not isolated from *P. lychnidiflora*. The amino acid sequences inferred from the amplified and sequenced fragments of these three genes, *ArKNOX3a*, *FpKNOX3a*, and *PaKNOX3a*, cover nearly 95% of the KNAT3 protein sequence, including the KNOX1, KNOX2, and ELK domains, as well as most of the HD domain (**Figure 3**).

In addition to the seven transcripts mentioned above, eight more *KNOX* transcripts were *de novo* assembled from our RNA-seq data of the cambial zone samples of three Cactaceae species, *A. retusus*, *F. pilosus*, and *P. lychnidiflora*. The amino acid sequences deduced from the amplified transcripts of *F. pilosus* were later extended by alignment with the assembled transcripts, and were therefore renamed *FpKNOX1ae* and *FpKNOX3ae* (**Figures 2–5**). Expression of all *KNOX* transcripts in the cambial zone of *A. retusus*, *F. pilosus*, and *P. lychnidiflora* was confirmed by RT-PCR (**Supplementary Figure S1**). Remarkably, in the cambial zone of 2-year-old juvenile plants of both *A. retusus* and *F. pilosus*, one of the two orthologs of *KNOX3*, namely, *ArKNOX3a* and *FpKNOX3ae*, and of *KNOX7*, namely, *ArKNOX7r1* and *FpKNOX7r2*, were expressed. No expression of *KNOX1* orthologs was detected in young plants. At this stage, only very scarce accumulation of wood, represented by wide-band tracheids and vessel elements, was detected in these species. From the six *KNOX* genes expressed in the vascular cambium of adult *A. retusus* plant, expression of only *ArKNOX7r1* was detected in the tubercle of 5-year-old plant (data not shown).

Phylogenetic Analysis of the *KNOX* Family

Initially, 524 KNOX-like protein sequences were retrieved from the Phytozome database. Chimeric sequences and those from draft genome releases were filtered out, and the inferred protein sequences of Cactaceae, the trees *B. luminifera* and *J. nigra*, as well as those retrieved from the *B. vulgaris* genome, were included. The resulting matrix contained 478 aligned protein sequences belonging to 45 species with sequenced genomes, six Cactaceae species, and two species of tree. The identifiers of the sequences retrieved from the Phytozome database are listed in **Supplementary Table S2**. The maximum likelihood phylogeny

[5]http://www.ebi.ac.uk/Tools/msa/clustalo/

[6]http://www.ebi.ac.uk/Tools/psa/emboss_needle/

[7]http://tree.bio.ed.ac.uk/

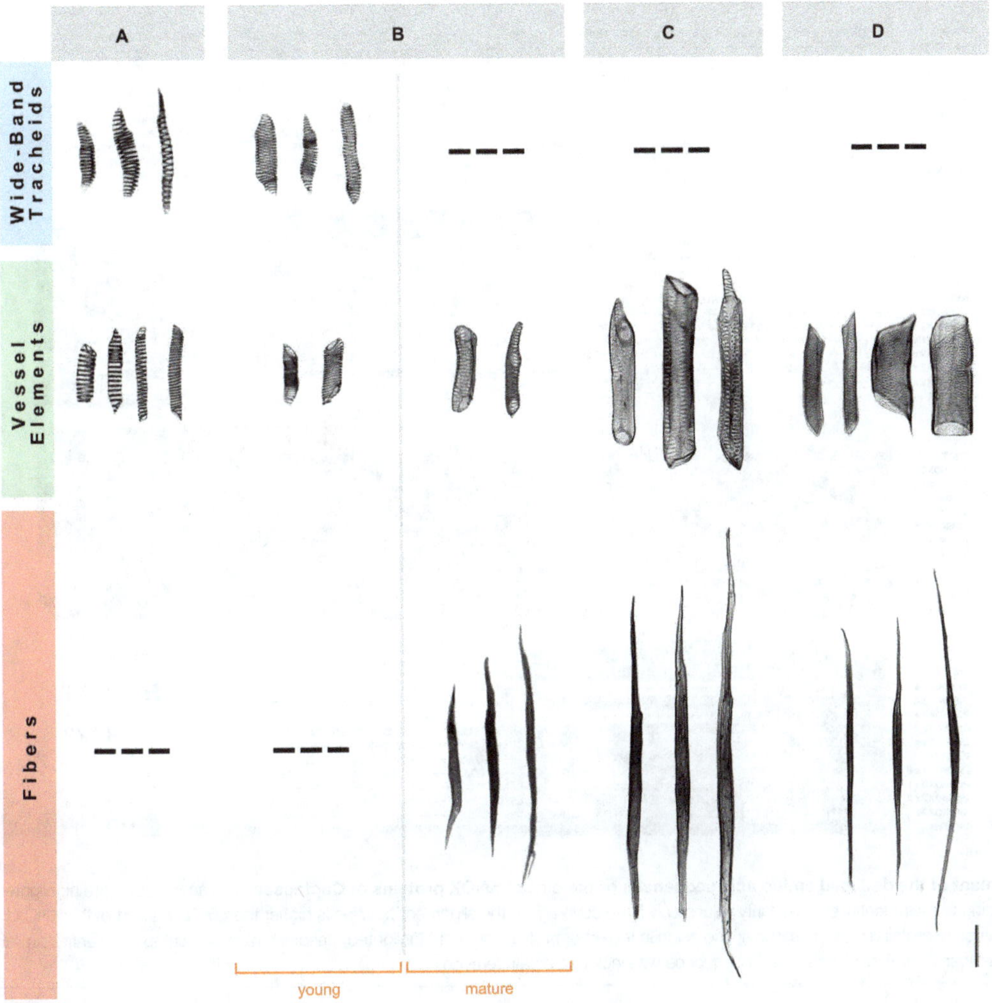

FIGURE 1 | Morphology of the secondary xylem cells in four species of Cactaceae. (A) Non-fibrous wood of *Ariocarpus retusus*, from a 15-cm tall specimen. **(B)** Dimorphic wood of *Ferocactus pilosus*, from a 90-cm tall specimen. **(C)** Fibrous wood of *Pilosocereus alensis*, from a 1.5-m tall specimen. **(D)** Fibrous wood of *Pereskia lychnidiflora*, from a 4-m tall specimen. Only mature wood is illustrated in **(A,C,D)**, as no differences were observed between young and mature wood of these species. Scale bar: 100 μm.

grouped the KNOX proteins into five main clades (**Figures 4, 5**), which were named according to their putative *Arabidopsis* ortholog: i.e., the KNAT1 clade (gray), KNAT2-KNAT6 clade (cyan), STM clade (pink), KNAT3-KNAT4-KNAT5 clade (blue), and KNAT7 clade (green). The first three clades belong to the class I KNOX and the last two belong to the class II KNOX proteins. **Figure 5** depicts the phylogenetic relationships of the subset of KNOX proteins from *Arabidopsis*, the Cactaceae, and *B. vulgaris*. *Arabidopsis* was used as the reference species, with *B. vulgaris* from order Caryophyllales, to which the Cactaceae family belongs, included as the closest sister taxon with a sequenced genome.

The molecular phylogenetic analysis confirmed that the four Cactaceae class 1 *KNOX* genes amplified from the cambial zone by RT-PCR with degenerate primers (*ArKNOX1a*, *FpKNOX1ae*, *PaKNOX1a*, and *PlKNOX1a*) are part of the KNAT1 clade (**Figures 4, 5**); moreover, two of the *de novo* assembled transcripts, *ArKNOX1r* and *PlKNOX1r*, also belonged to the KNAT1 clade. The class II *ArKNOX3a*, *FpKNOX3ae*, and *PaKNOX3a* genes expressed in the vascular zone of the three species fall into the KNAT3-KNAT4-KNAT5 clade, along with two other class II *KNOX* genes, *ArKNOX3r* and *PlKNOX3r* (the only *Pereskia* class II gene identified in this study). Four *de novo* assembled transcripts, *ArKNOX7r1*, *ArKNOX7r2*, *FpKNOX7r1*, and *FpKNOX7r2*, belong to the KNAT7 clade (**Figures 4, 5** and **Table 1**). The identity and similarity percentages for the proteins encoded by the identified Cactaceae genes are shown in **Supplementary Figure S2**.

Of the five *KNOX* transcripts detected in the *P. pringlei* root tip, both of the class I *KNOX* genes were found in the KNAT2-KNAT6 clade, one class II sequence was attributed to the KNAT3-KNAT4-KNAT5 clade, and two more belonged to the KNAT7 clade. Among the five *KNOX* genes identified in the recently published *L. williamsii* transcriptome

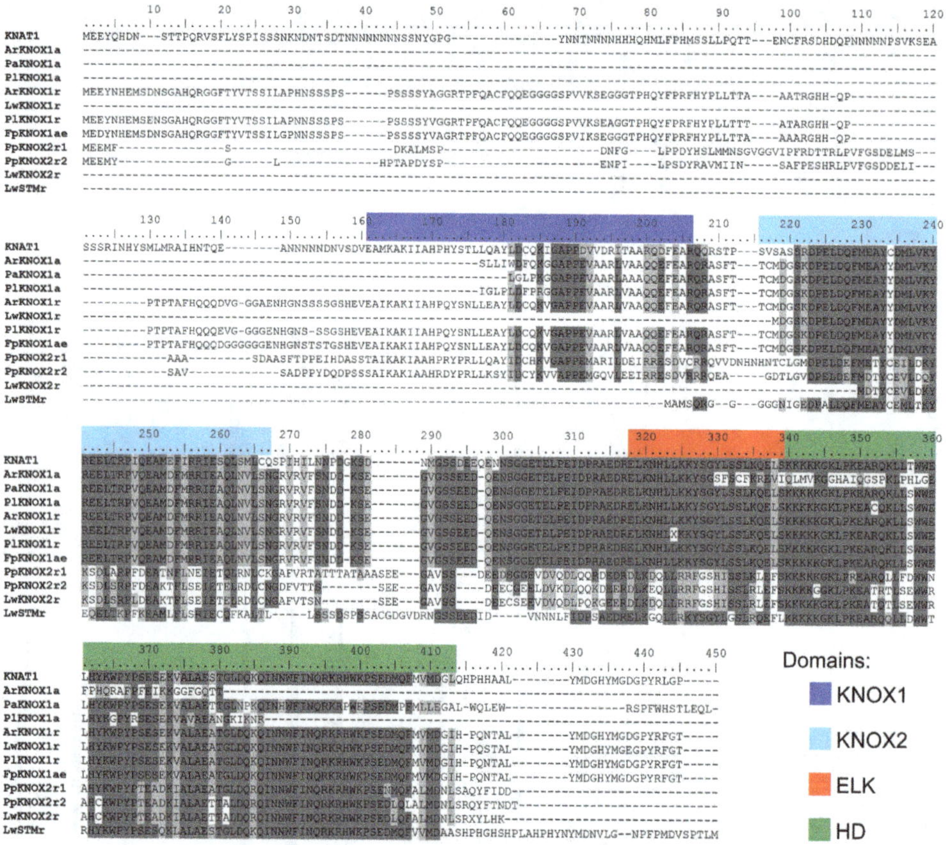

FIGURE 2 | Alignment of the deduced amino acid sequences of the class I KNOX proteins of Cactaceae. The amino acids are highlighted with a grayscale background according to their identity and similarity values only if the coverage in the alignment position is higher than 55%. As part of the KNOX1 domain is not covered in the incomplete protein sequences, the amino acids in the other proteins are not highlighted, despite a high percentage of identity/similarity. Domains were defined with InterProscan. The *Arabidopsis* KNAT1 sequence was included as a reference.

(Ibarra-Laclette et al., 2015), one was found in each of the five clades (**Figures 4, 5** and **Table 1**). In the global KNOX phylogeny (**Figure 4**), *B. vulgaris* sequences were always resolved as sister to the Cactaceae sequences. In the phylogeny for the subset of KNOX proteins (**Figure 5**), *B. vulgaris* class I KNOX proteins were resolved as sister to the Cactaceae sequences in all three clades, while this was not the case for two clades of class II KNOX. The only *B. vulgaris* paralog of KNAT3, KNAT4, and KNAT5 fell in the subclade of the *A. thaliana* sequences. Within the KNAT7 clade, the *B. vulgaris* sequence represented the sister sequence for the Cactaceae KNOX7r2 subclade, whereas the KNOX7r1 sequences grouped as a separate subclade. The possible reasons of this small inconsistency could be the incomplete sequences of many Cactaceae proteins, as well as the very restricted number of sequences in this phylogeny (**Figure 5**).

DISCUSSION

Although, it is well established that some KNOX transcription factors are important for the maintenance of the shoot apical meristem, their role in vascular cambium activity and secondary growth is less well understood. Four class I *KNOX* genes were reported to be expressed in the vascular cambium of *P. tremula* (Schrader et al., 2004; Du et al., 2009), while the expression of a class II *KNOX* gene was detected in the vascular cambium of *J. nigra* (Huang et al., 2009). It was therefore of particular interest to determine whether genes of both classes are expressed in the vascular zones of cacti, and whether different paralogs are expressed in species producing different types of wood. By performing RT-PCR with degenerate primers in the four Cactaceae species (*A. retusus*, *F. pilosus*, *P. lychnidiflora*, and *P. alensis*) and RNA-seq with subsequent *de novo* transcriptome assembly for three of these species, we have found, and consequently confirmed by RT-PCR (**Supplementary Figure S1**), that the transcripts of both class I and class II *KNOX* genes are present in the cambial zone of the adult individuals in all four species (**Table 1**). The possibility exists that other *KNOX* genes could be expressed at very low levels in the Cactaceae vascular zone, impeding successful *de novo* assembly. Genome sequencing of species from the Cactaceae family will facilitate a more detailed analysis of the *KNOX* expression patterns. To annotate the Cactaceae KNOX sequences, a phylogenetic tree was constructed

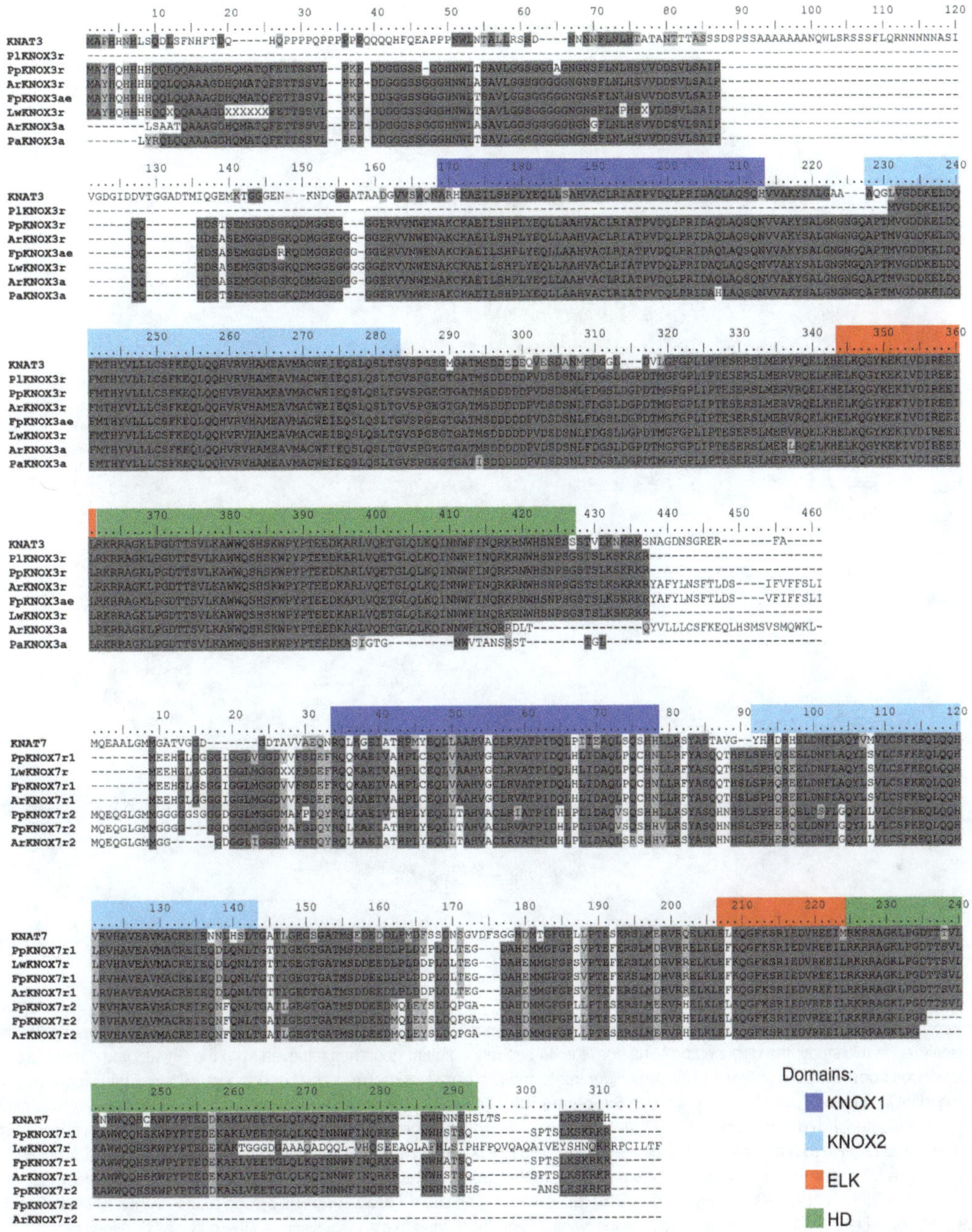

FIGURE 3 | Alignment of the deduced amino acid sequences of the class II KNOX proteins of Cactaceae and some class II KNOX proteins of *Arabidopsis*. The amino acid scores are highlighted with a grayscale background according to their identity and similarity only if the coverage in the alignment position is higher than 55%. As part of the KNOX1 domain is not covered in the incomplete protein sequences, the amino acids in the other proteins are not highlighted, despite a high percentage of identity/similarity. Domains were defined with InterProscan.

that included KNOX sequences from sequenced angiosperm genomes from the Phytozome database and from the *B. vulgaris* genome as a species from the order Caryophyllales, to which the Cactaceae family belongs. Expression of all *KNOX* genes in the cambial zone of adult *A. retusus*, *F. pilosus*, and *P. lychnidiflora* plants was confirmed by RT-PCR (**Supplementary Figure S1**). Meanwhile, in the tubercle of 5-year-old *A. retusus* plant, we detected expression of only one of the six *ArKNOX* genes reported in this work, *ArKNOX7r1*. Expression of two *KNOX* genes, the orthologous *ArKNOX7r1/FpKNOX7r1* and *ArKNOX3a/FpKNOX3ae*, was detected in the vascular cambium of juvenile *A. retusus* and *F. pilosus* plants, when they were just starting to accumulate wood. This finding further suggests that the *KNOX* genes, reported in this work, are involved in the formation of mature wood, and that the *KNOX7r1* paralog has broader functions in Cactaceae development.

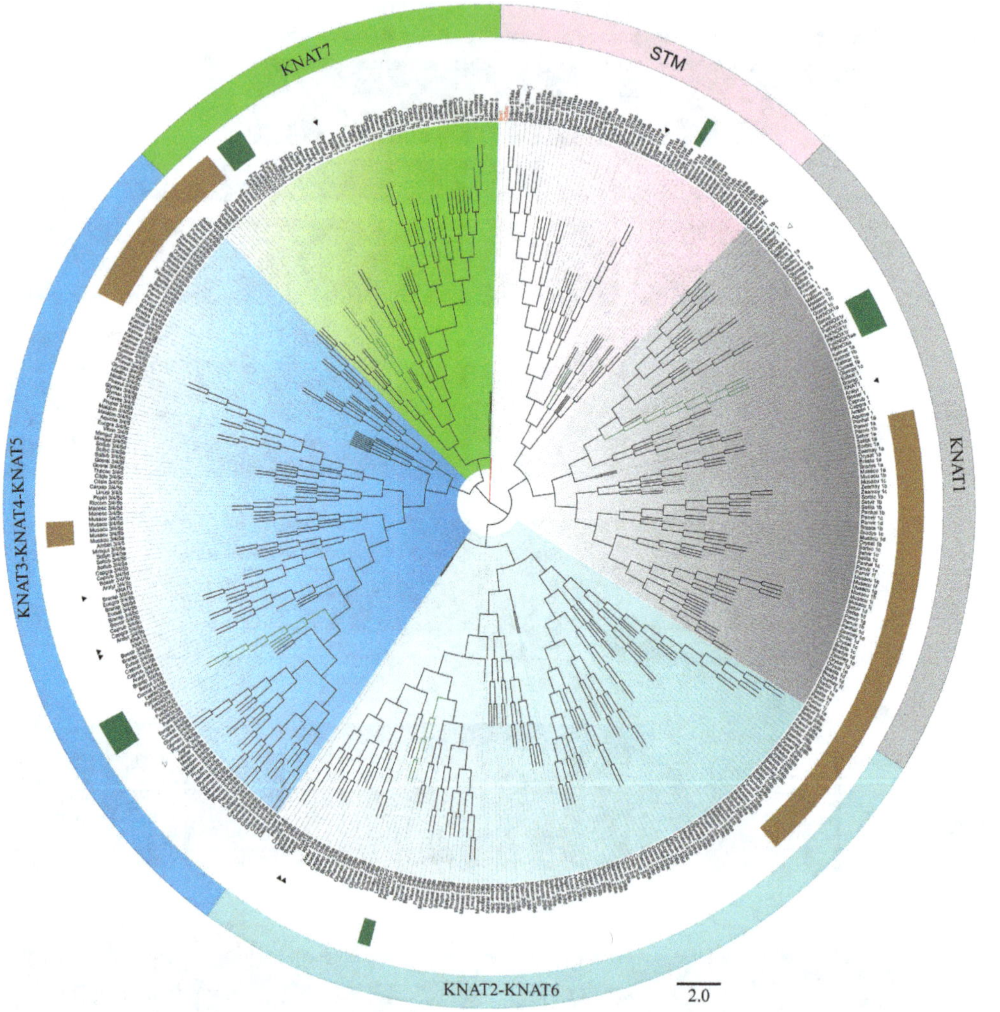

FIGURE 4 | Maximum likelihood phylogenetic tree of KNOX proteins. Phylogeny of the complete dataset, incorporating the deduced proteins of the five Cactaceae species analyzed in this study; the deduced proteins from the 44 sequenced plant genomes retrieved from the Phytozome database and the *Beta vulgaris* genome; sequences from *Betula luminifera* and *Juglans nigra* trees; and proteins deduced from the *Lophophora williamsii* transcriptome (Ibarra-Laclette et al., 2015). The gene identifiers for each sequence are listed in **Supplementary Table S2**. The branches for the Cactaceae and *Beta vulgaris* sequences are represented in green and indicated with green ribbons. The sequences from monocots (Poaceae and Musaceae) are indicated by brown ribbons. The BEL1 protein from *Arabidopsis thaliana* was used to root the tree and is indicated in red. The KNOX protein from the unicellular green alga *Ostreococcus lucimarinus* at the base of the tree is also indicated in red. The five main clades are named according to their putative *Arabidopsis* ortholog, indicated by a black arrowhead. The class I KNOX proteins belong to the KNAT1 (gray), KNAT2-KNAT6 (cyan), and STM (pink) clades. The class II KNOX proteins belong to the KNAT3-KNAT4-KNAT5 (blue) and KNAT7 (green) clades. *Populus* class I *KNOX* genes and the *Juglans* class II gene mentioned in the discussion are the only *KNOX* genes reported to be involved in wood formation in these tree species and are indicated by open arrowheads. The scale bar depicts the number of substitutions per sequence.

The six class I *KNOX* genes found to be expressed in the cambial zone of the four cactus species are putative orthologs of the *Arabidopsis BP/KNAT1* gene, as they belong to the KNAT1 clade. The *Populus* gene *ARK2/PttKNOX1,* which is expressed in the vascular cambium and developing xylem (Schrader et al., 2004; Du et al., 2009), also belongs to this clade (white arrowhead in **Figure 4**). Previous work has suggested that *ARK2* is involved in vascular cambium activity and secondary growth; *Populus* plants constitutively overexpressing *ARK2* had a wider cambial zone and decreased differentiation of the lignified secondary xylem, while the downregulation of *ARK2* resulted in the early differentiation of lignified secondary xylem cells (Du et al., 2009).

Importantly, we did not find putative orthologs of the *Arabidopsis STM* and *KNAT2-KNAT6* genes expressed in the Cactaceae cambial zone. By contrast, both the *ARK1a* and *ARK1b* paralogs in *Populus* (orthologous to *STM*) are shown to be expressed in the vascular cambium (Popgenie database[8]). *ARK1* overexpression in a hybrid *Populus* resulted in pleiotropic phenotypes, including the slower differentiation of cambium-derived cells (Groover et al., 2006). While the *KNAT1* ortholog *ARK2* is expressed in both the vascular cambium and the developing xylem of *Populus*, *ARK1a* was shown to be

[8]http://popgenie.org/

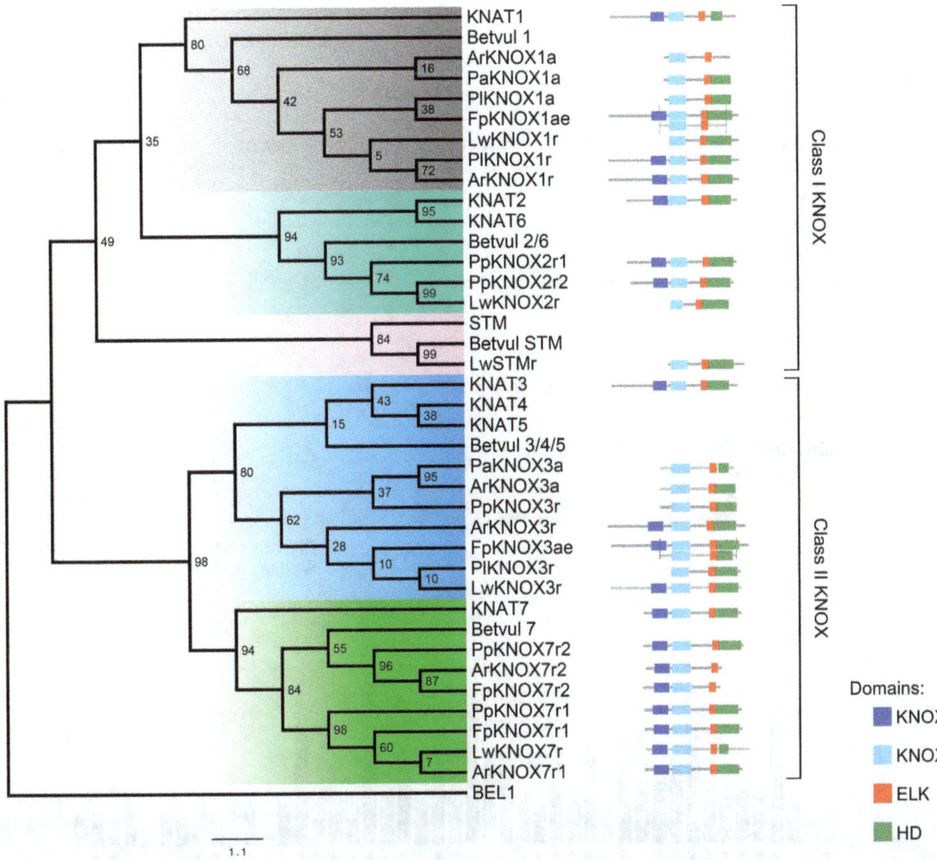

FIGURE 5 | Phylogenetic tree of the selected dataset of KNOX proteins from *Arabidopsis*, the Cactaceae, and *Beta vulgaris*. The class I KNOX proteins belong to the KNAT1 (gray), KNAT2-KNAT6 (cyan), and STM (pink) clades. The class II KNOX proteins belong to the KNAT3-KNAT4-KNAT5 (blue) and KNAT7 (green) clades. The domain positions for Cactaceae sequences are displayed on the right alongside the KNAT ortholog. The ELK domain in KNAT7 proteins is recognized by InterProscan but not by PFAM tools; therefore, the InterProscan database was used for the domain definition. For *FpKNOX1ae* and *FpKNOX3ae,* the portion of the original amplified sequence is shown below the assembled sequence. Note that it was not possible to obtain the full KNOX1 domain sequence for the Cactaceae sequences ending in "a"; therefore, their phylogenetic position might be slightly changed when the complete protein sequence is available. Bootstrap values from 1000 trees are displayed at the nodes. The scale depicts the number of substitutions per sequence.

downregulated in non-meristematic secondary vascular tissues (Groover et al., 2006), thus showing more similarity to the *STM* expression pattern in the shoot apex. *STM* expression was previously also detected in the *Arabidopsis* inflorescence stems during induced secondary growth (Ko and Han, 2004). As expected, we did not find a *STM* ortholog in the *P. pringlei* root tip transcriptome, while two *P. pringlei* paralogs orthologous to KNAT2-KNAT6 were expressed in the root tip (**Figures 4, 5** and **Table 1**). The *STM* putative ortholog was present in the *L. williamsii* shoot and root transcriptome, suggesting that *STM* might be expressed in some Cactaceae tissues, most probably in the shoot apical meristem. The absence of putative *STM* orthologs in the cambial zone transcriptome suggests possible differences in the vascular cambium genetic regulatory network in the Cactaceae versus *Arabidopsis* and *Populus*. Interestingly, monocots do not have orthologs of *Arabidopsis STM*; the grass genes involved in shoot apical meristem maintenance (maize: *KNOTTED1* and *ROUGH SHEATH1*; rice: *ORYZA SATIVA HOMEOBOX1* (*OSH1*) and *OSH15*; Tsuda et al., 2011; Bolduc et al., 2014) belong to the KNAT1 clade (**Figure 4**). Moreover, monocot *KNOX* genes clustered as a single subclade within the KNAT1, KNAT2-KNAT6, and KNAT7 clades, while within the KNAT3-KNAT4-KNAT5 clade two subclades, one from Poaceae and another from Musaceae, were present (brown ribbons on **Figure 4**).

For the three species used for RNA-seq and the *de novo* transcriptome assembly, two class I paralogs were identified in the vascular zone of *A. retusus* and *P. lychnidiflora*, while only one was identified in *F. pilosus* (**Figures 2, 4, 5** and **Table 1**). The missing class I *KNOX* paralog could have been lost from *F. pilosus* as an evolutionary developmental process, enabling speciation after a gene duplication in the Cactaceae, a phenomenon that has been well documented in other families. Using data from sequenced genomes (**Supplementary Table S2**), we showed that the number of class I and class II *KNOX* genes in a moss, a lycophyte, and 43 angiosperm species varies significantly, particularly between Eudicotyledonous species (**Figure 6**). As there is still no Cactaceae species with a sequenced

TABLE 1 | Number of *KNOX* genes from this study expressed in the cambial zone, root tip, and shoot/root of Cactaceae species.

		Number of *KNOX* orthologs				
		Class I clades			Class II clades	
Species	**Genes expressed in**	***STM***	***KNAT1***	***KNAT2-KNAT6***	***KNAT3-KNAT4-KNAT5***	***KNAT7***
Pereskia lychnidiflora	Cambial zone	0	2	0	1	0
Ariocarpus retusus	Cambial zone	0	2	0	2	2
Ferocactus pilosus	Cambial zone	0	1	0	1	2
Pilosocereus alensis[a]	Cambial zone	0	1	0	1	0
Pachycereus pringlei	Root tip	0	0	2	1	2
Lophophora williamsii[b]	Shoot and root	1	1	1	1	1

[a] *No RNA-seq data available.* [b] *For comparison, the data from Ibarra-Laclette et al. (2015) were used.*

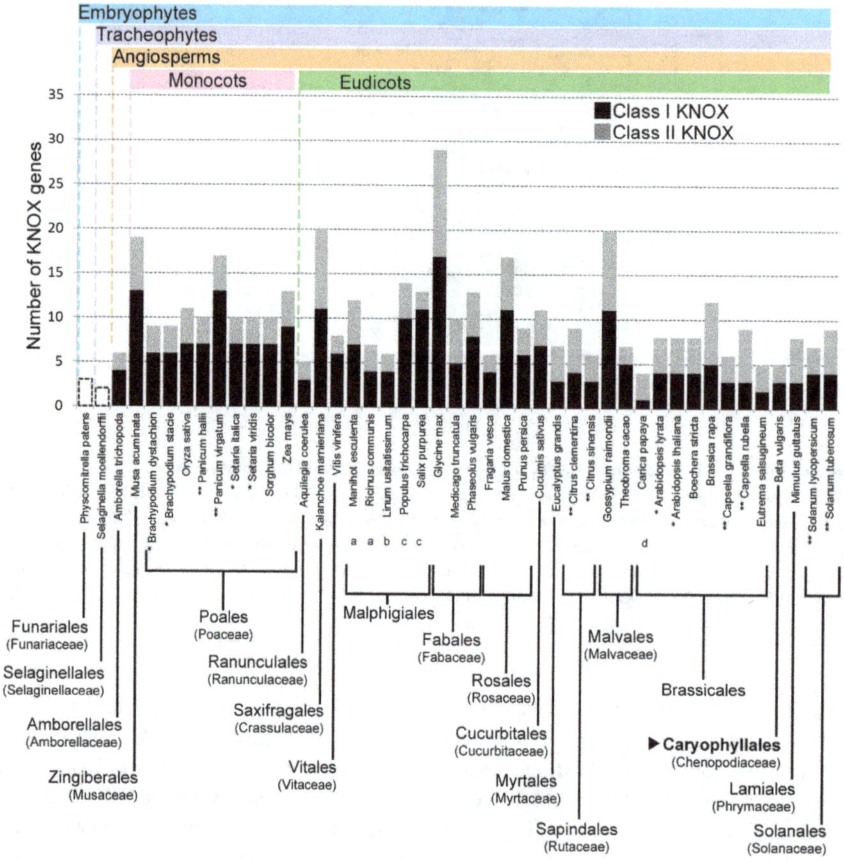

FIGURE 6 | Number of *KNOX* genes in species from different plant taxa. The number of both class I and class II *KNOX* genes are shown for the angiosperms, while for *Physcomitrella patens* and *Selaginella moellendorffii*, only the total numbers of *KNOX* genes are indicated. When all the species from an order belonged to the same family, the family is indicated in parentheses below the order. Where different families of an order are depicted, they are indicated with lowercase letters below the species name. For the order Malpighiales, the included families are: (a) Euphorbiaceae, (b) Linaceae, and (c) Salicaceae. For the order Brassicales, one species belongs to the (d) Caricaceae family, while the rest belong to the Brassicaceae family. The order Caryophyllales, to which the Cactaceae family belongs, is depicted in bold and indicated by a black triangle. Cactaceae species are not included, because there are no Cactaceae species with sequenced genomes.
* Species from the same genus with the same number of *KNOX* genes. ** Species from the same genus with different numbers of *KNOX* genes. Note that the number of *KNOX* genes belonging to class I, class II, or both, can be different.

genome, Cactaceae were not included in this analysis. All of the analyzed angiosperm species have several class II genes, and all but *Carica papaya* have more than one class I gene. There are 29 *KNOX* genes in *Glycine max* (17 class I and 12 class II genes) but only four in *Carica* (one class I and three class II genes). Furthermore, different species can have a different number of paralogs in every clade (**Supplementary Table S2** and **Figure 5**). The number of *KNOX* genes can also vary within species of the same family; for instance, within diploid species of the Brassicaceae family (which includes the largest number of species with sequenced genomes), the number of *KNOX* genes varies from five in *Eutrema salsugineum* to 12 in *Brassica rapa*. Even species within a single genus can have variable numbers of *KNOX* genes; while the *Arabidopsis* species *A. thaliana* and *A. lyrata* both possess eight *KNOX* genes, there are six and nine genes in the *Capsella grandiflora* and *C. rubella* genomes, respectively (**Figure 6** and **Supplementary Table S2**).

We found that two putative orthologs of the *Arabidopsis* class II *KNOX* paralogs *KNAT3*, *KNAT4*, and *KNAT5* were expressed in the cambial zone of *A. retusus*, but just one was expressed in the cambial zone of *F. pilosus* and *P. lychnidiflora* (**Figures 3–5** and **Table 1**). Only one putative paralog expressed in the root tip of *P. pringlei* belonged to this class. One paralog from the published shoot and root transcriptome of *L. williamsii* (Ibarra-Laclette et al., 2015) also belongs to this class. Remarkably, among the three species used for cambial zone transcriptome assembly, two paralogs in *A. retusus* and *F. pilosus* belonged to the KNAT7 clade, whereas no putative *KNAT7* ortholog was found in the cambium of *P. lychnidiflora* (*P. alensis* was not considered as the transcriptome assembly was not performed for this species). KNAT7 is a transcriptional repressor of secondary cell wall biosynthesis (Liu et al., 2014). *P. lychnidiflora* has fibrous wood with significant cell wall lignification, *A. retusus* has non-fibrous wood with abundant wide-band tracheids and less lignification, while in the dimorphic wood of *F. pilosus*, the wide-band tracheids develop before more lignified cells. Our findings therefore suggest that the repressor activity of the *KNAT7* orthologs during secondary cell wall biosynthesis could explain the differences in wood type among these species, including the lower lignification level observed in *A. retusus* and *F. pilosus*.

We found that the transcripts of both *KNOX* gene classes are expressed in the cambial zone of the four Cactaceae species studied, and therefore could be expressed either in different cell types or in the same cell type within the vascular zone. Several KNOX proteins have been shown to selectively move from one cell layer to another (reviewed by Han et al., 2014), suggesting that even if the class I and II genes are transcribed in different cell types, the encoded proteins could coexist in the same cell as a result of protein trafficking. In any case, as no evidence of mutual repression was found when class I and class II *KNOX* loss-of-function mutations of *Arabidopsis* were combined (Furumizu et al., 2015), KNOX proteins belonging to different classes could exert their functions within the same cell. KNOX proteins are known to interact with a sister group of proteins from the BEL1-LIKE-HOMEODOMAIN (BHL, or BELL from the founding *BELL 1* gene) family; both KNOX and BELL proteins belong to the TALE (three amino acid loop extension) superclass of homeobox proteins. The KNOX-BELL heterodimer formation affects its cellular localization and the KNOX target selection (reviewed by Di Giacomo et al., 2013; Arnaud and Pautot, 2014). From heterologous expression experiments it was proposed that particular KNOX proteins could interact with different BELL partners leading to numerous combinations with distinct activities, and thus regulating different sets of targets, including transcription factors and hormonal pathways, and ultimately influence multiple plant developmental processes. Recently, however, a more specific selectivity was suggested for the *in vivo* KNOX-BELL interactions (Furumizu et al., 2015), which could also enhance the properties of each heterodimerization partner; thus, the interaction between the transcriptional repressor KNAT7 and its partner BHL6 could enhance their activity in repressing genes involved in secondary cell wall biosynthesis (Liu et al., 2014).

Knowledge of the processes of duplication and subfunctionalization of the regulatory genes helps us understand the evolution of the different aspects of plant development (Pires and Dolan, 2012), including the evolution of vascular development. Our work contributes to the elucidation of the mechanisms by which different wood types are formed in the Cactaceae, and provides insight into the evolutionary history of *KNOX* genes in different angiosperm species and their role in the speciation of land plants.

AUTHOR CONTRIBUTIONS

TT, JR-R, AV, FV-S, GR-A, and SS designed the work; TT, JR-R, EP, AV, GR-A, and SS performed the experiments and analyzed the data; GR-A and JR-R prepared the figures; and TT and SS wrote the manuscript. All the authors have read and approved the manuscript.

ACKNOWLEDGMENTS

Funding was provided by the DGAPA-PAPIIT, UNAM (grants IN209012 and 210115 to TT and IN207115 to SS) and by Consejo Nacional de Ciencia y Tecnología (CONACyT) [CB2014-240055 grant to SS; Ph.D. scholarships 220343 to JR-R; 367139 to GR-A and 700638 to EP, and a postdoc grant (220343) to JR-R]. We thank Jorge Nieto Sotelo for the valuable comments, Paul Gaytan, Eugenio Lopez and personnel of the DNA synthesis and sequencing unit of the IBt-UNAM for oligonucleotide synthesis, and Laura Márquez Valdelamar for DNA sequencing.

SUPPLEMENTARY MATERIAL

The Supplementary Material for this article can be found online at: http://journal.frontiersin.org/article/10.3389/fpls.2017.00218/full#supplementary-material

FIGURE S1 | Class I and II KNOX transcript expression detected in cambial zone samples by end point RT-PCR.

FIGURE S2 | The identity and similarity matrix of the deduced KNOX proteins identified in this study. (A) Class I proteins. **(B)** Concatenated KNOX1, KNOX2, ELK, and HD domains of class I proteins. **(C)** Class II proteins. **(D)** Concatenated KNOX1, KNOX2, ELK, and HD domains of class II proteins.

TABLE S1 | PCR primers used in this study.

TABLE S2 | The identifiers of the KNOX sequences used for molecular phylogenetic analysis.

REFERENCES

Altesor, A., Silva, C., and Ezcurra, E. (1994). Allometric neoteny and the evolution of succulence in cacti. *Bot. J. Linn. Soc.* 114, 283–292. doi: 10.1111/j.1095-8339.1994.tb01937.x

Arnaud, N., and Pautot, V. (2014). Ring the BELL and tie the KNOX: roles for TALEs in gynoecium development. *Front. Plant Sci.* 5:93. doi: 10.3389/fpls.2014.00093

Berlyn, G. P., and Miksche, J. P. (1976). *Botanical Microtechnique and Cytochemistry*. Ames, IA: Iowa State University Press.

Bolduc, N., Tyers, R. G., Freeling, M., and Hake, S. (2014). Unequal redundancy in maize *knotted1* homeobox genes. *Plant Physiol.* 164, 229–238. doi: 10.1104/pp.113.228791

Di Giacomo, E., Iannelli, M. D., and Frugis, G. (2013). TALE and shape: how to make a leaf different. *Plants* 2, 317–342. doi: 10.3390/plantas2020317

Dohm, J. C., Minoche, A. E., Holtgräwe, D., Capella-Gutiérrez, S., Zakrzewski, F., Tafer, H., et al. (2013). The genome of the recently domesticated crop plant sugar beet (*Beta vulgaris*). *Nature* 505, 546–549. doi: 10.1038/nature12817

Du, J., Mansfield, S. D., and Groover, A. T. (2009). The *Populus* homeobox gene *ARBORKNOX2* regulates cell differentiation during secondary growth. *Plant J.* 60, 1000–1014. doi: 10.1111/j.1365-313X.2009.04017.x

Evert, R. F. (2006). *Esau's Plant Anatomy: Meristems, Cells, and Tissues of the Plant Body: Their Structure, Function, and Development*, 3rd Edn. Hoboken, NJ: John Wiley & Sons.

Felsenstein, J. (1981). Evolutionary trees from DNA sequences: a maximum likelihood approach. *J. Mol. Evol.* 17, 368–376. doi: 10.1007/BF01734359

Furumizu, C., Alvarez, J. P., Sakakibara, K., and Bowman, J. L. (2015). Antagonistic roles of *KNOX1* and *KNOX2* genes in patterning the land plant body plan following an ancient gene duplication. *PLoS Genet.* 11:e1004980. doi: 10.1371/journal.pgen.1004980

Groover, A. T., Mansfield, S. D., DiFazio, S. P., Dupper, G., Fontana, J. R., Millar, R., et al. (2006). The *Populus* homeobox gene *ARBORKNOX1* reveals overlapping mechanisms regulating the shoot apical meristem and the vascular cambium. *Plant Mol. Biol.* 61, 917–932. doi: 10.1007/s11103-006-0059-y

Groover, A. T., Nieminen, K., Helariutta, Y., and Mansfield, S. D. (2010). "Wood formation in *Populus*," in *Genetics and Genomics of Populus, Plant Genetics and Genomics: Crops and Models 8*, eds S. Jansson, R. Bhalerao, and A. Groover (Berlin: Springer Science+Business Media), doi: 10.1007/978-1-4419-1541-2_10

Han, X., Kumar, D., Chen, H., Wu, S., and Kim, J. Y. (2014). Transcription factor-mediated cell-to-cell signalling in plants. *J. Exp. Bot.* 65, 1737–1749. doi: 10.1093/jxb/ert422

Hay, A., and Tsiantis, M. (2010). *KNOX* genes: versatile regulators of plant development and diversity. *Development* 137, 3153–3165. doi: 10.1242/dev.030049

Henikoff, S., and Henikoff, J. G. (1992). Amino acid substitution matrices from protein blocks. *Proc. Natl. Acad. Sci. U.S.A.* 89, 10915–10919.

Huang, Z., Meilan, R., and Woeste, K. (2009). A *KNAT3*-like homeobox gene from *Juglans nigra* L., *JnKNAT3*-like, highly expressed during heartwood formation. *Plant Cell Rep.* 28, 1717–1724. doi: 10.1007/s00299-009-0771-6

Ibarra-Laclette, E., Zamudio-Hernández, F., Pérez-Torres, C. A., Albert, V. A., Ramírez-Chávez, E., Molina-Torres, J., et al. (2015). *De novo* sequencing and analysis of *Lophophora williamsii* transcriptome, and searching for putative genes involved in mescaline biosynthesis. *BMC Genomics* 16:657. doi: 10.1186/s12864-015-1821-9

Ko, J., and Han, K. (2004). *Arabidopsis* whole-transcriptome profiling defines the features of coordinated regulations that occur during secondary growth. *Plant Mol. Biol.* 55, 433–453. doi: 10.1007/s11103-004-1051-z

Kumar, S., Stecher, G., and Tamura, K. (2016). MEGA7: molecular evolutionary genetics analysis version 7.0 for bigger datasets. *Mol. Biol. Evol.* 33, 1870–1874. doi: 10.1093/molbev/msw054

Landrum, J. V. (2008). Wide-band tracheids from a southern African succulent and their responses to varying light intensities: a preadaptation for future water stress? *Int. Bot.* 4, 99–103. doi: 10.3923/ijb.2008.99.103

Liu, L., Filkov, V., and Groover, A. (2014). Modeling transcriptional networks regulating secondary growth and wood formation in forest trees. *Physiol. Plant.* 151, 156–163. doi: 10.1111/ppl.12113

Liu, L., Zinkgraf, M., Petzold, H. E., Beers, E. P., Filkov, V., and Groover, A. (2015). The *Populus* ARBORKNOX1 homeodomain transcription factor regulates woody growth through binding to evolutionarily conserved target genes of diverse function. *New Phytol.* 205, 682–694. doi: 10.1111/nph.13151

Loza-Cornejo, S., and Terrazas, T. (2011). Morfo-anatomía de plántulas en especies de Pachycereeae: ¿hasta cuándo son plántulas? *Bol. Soc. Bot. Méx.* 88, 1–13.

Lucas, W. J., Groover, A., Lichtenberger, R., Furuta, K., Yadav, S. R., Helariutta, Y., et al. (2013). The plant vascular system: evolution, development and functions. *J. Integr. Plant Biol.* 55, 294–388. doi: 10.1111/jipb.12041

Mauseth, J. D. (2006). Structure-function relationships in highly modified shoots of Cactaceae. *Ann. Bot.* 98, 901–926.

Mauseth, J. D., and Plemons, B. J. (1995). Developmentally variable, polymorphic woods in cacti. *Am. J. Bot.* 82, 1199–1205.

Miyashima, S., Sebastian, J., Lee, J.-Y., and Helariutta, Y. (2013). Stem cell function during plant vascular development. *EMBO J.* 32, 178–193. doi: 10.1038/emboj.2012.301

Nieminen, K., Blomster, T., Helariutta, Y., and Mähönen, A. P. (2015). Vascular cambium development. *Arabidopsis Book* 13:e0177. doi: 10.1199/tab.0177

Pires, N. D., and Dolan, L. (2012). Morphological evolution in land plants: new designs with old genes. *Philos. Trans. R. Soc. B.* 367, 508–518. doi: 10.1098/rstb.2011.0252

Reyes-Rivera, J., Canché-Escamilla, G., Soto-Hernández, M., and Terrazas, T. (2015). Wood chemical composition in species of Cactaceae: the relationship between lignification and stem morphology. *PLoS ONE* 10:e0123919. doi: 10.1371/journal.pone.0123919

Růžička, K., Ursache, R., Hejátko, J., and Helariutta, Y. (2015). Xylem development–from the cradle to the grave. *New Phytol.* 207, 519–535. doi: 10.1111/nph.13383

Ruzin, E. S. (1999). *Plant Microtechnique and Microscopy*. Oxford: Oxford University Press.

Schrader, J., Nilsson, J., Mellerowicz, E., Berglund, A., Nilsson, P., Hertzberg, M., et al. (2004). A high-resolution transcript profile across the wood-forming meristem of poplar identifies potential regulators of cambial stem cell identity. *Plant Cell* 16, 2278–2292. doi: 10.1105/tpc.104.024190

Serikawa, K. A., Martinez-Laborda, A., Kim, H.-S., and Zambryski, P. C. (1997). Localization of expression of *KNAT3*, a class 2 knotted1-like gene. *Plant J.* 11, 853–861.

Spicer, R., and Groover, A. (2010). Evolution of development of vascular cambia and secondary growth. *New Phytol.* 186, 577–592. doi: 10.1111/j.1469-8137.2010.03236.x

Tsuda, K., Ito, Y., Sato, Y., and Kurata, N. (2011). Positive autoregulation of a *KNOX* gene is essential for shoot apical meristem maintenance in rice. *Plant Cell* 23, 4368–4381. doi: 10.1105/tpc.111.090050

Vázquez-Sánchez, M., and Terrazas, T. (2011). Stem and wood allometric relationships in Cacteae (Cactaceae). *Trees* 25, 755–767. doi: 10.1007/s00468-011-0553-y

Ye, Z.-H., and Zhong, R. (2015). Molecular control of wood formation in trees. *J. Exp. Bot.* 66, 4119–4131. doi: 10.1093/jxb/erv081

Zhong, R., and Ye, Z.-H. (2015). Secondary cell walls: biosynthesis, patterned deposition and transcriptional regulation. *Plant Cell Physiol.* 56, 195–214. doi: 10.1093/pcp/pcu140

Conflict of Interest Statement: The authors declare that the research was conducted in the absence of any commercial or financial relationships that could be construed as a potential conflict of interest.

12

Genome-Wide Analysis of Soybean JmjC Domain-Containing Proteins Suggests Evolutionary Conservation Following Whole-Genome Duplication

Yapeng Han [1,2†], *Xiangyong Li* [1,2†], *Lin Cheng* [1,2], *Yanchun Liu* [1,2], *Hui Wang* [1,2], *Danxia Ke* [1,2], *Hongyu Yuan* [1,2], *Liangsheng Zhang* [3,4*] and *Lei Wang* [1,2*]

[1] *College of Life Sciences, Xinyang Normal University, Xinyang, China,* [2] *Institute for Conservation and Utilization of Agro-bioresources in Dabie Mountains, Xinyang Normal University, Xinyang, China,* [3] *Center for Genomics and Biotechnology, Haixia Institute of Science and Technology, Fujian Agriculture and Forestry University, Fuzhou, China,* [4] *Key Laboratory of Genetics, Breeding and Multiple Utilization of Corps (Fujian Agriculture and Forestry University), Ministry of Education, Fujian Provincial Key Laboratory of Haixia Applied Plant Systems Biology, Fujian Agriculture and Forestry University, Fuzhou, China*

Edited by:
José M. Romero,
University of Seville, Spain

Reviewed by:
David Smyth,
Monash University, Australia
Ning Zhang,
Food and Drug Administration, USA

***Correspondence:**
Lei Wang
wangleibio@xynu.edu.cn;
wangleibio@126.com
Liangsheng Zhang
zls@tongji.edu.cn

[†] *These authors have contributed equally to this work.*

Histone modifications, such as methylation and demethylation, play an important role in regulating chromatin structure and gene expression. The JmjC domain-containing proteins, an important family of histone lysine demethylases (KDMs), play a key role in maintaining homeostasis of histone methylation *in vivo*. In this study, we performed a comprehensive analysis of the *jumonji C* (*JmjC*) gene family in the soybean genome and identified 48 *JmjC* genes (*GmJMJs*) distributed unevenly across 18 chromosomes. Phylogenetic analysis showed that these JmjC domain-containing genes can be divided into eight groups. *GmJMJs* within the same phylogenetic group share similar exon/intron organization and domain composition. In addition, 16 duplicated gene pairs were formed by a *Glycine*-specific whole-genome duplication (WGD) event approximately 13 million years ago (Mya). By investigating the expression profiles of these gene pairs in various tissues, we showed that the expression pattern is conserved in the polyploidy-derived JmjC duplicates, demonstrating that the majority of *GmJMJs* were preferentially retained after the most recent WGD event and suggesting important roles for demethylase duplications in soybean evolution. These results shed light on the evolutionary history of this family in soybean and provide insights into the *JmjCs* which will be helpful to reveal their functions in controlling soybean development.

Keywords: soybean (*Glycine max* L.), *JmjC* gene family, genome-wide analysis, phylogeny, gene structure, expression pattern

INTRODUCTION

Histone methylation and demethylation have important roles in regulating transcription, genome integrity, and epigenetic inheritance (Klose et al., 2006; Klose and Zhang, 2007; Liu et al., 2010). Histone methylation can occur at various lysine and arginine residues, including K4, K9, K27, K36, and K79 in histone H3 and K20 in histone H4 (Allis et al., 2007). Histone methylation, which is mainly catalyzed by protein families that contain PRMT and SET domains, can have both activating and repressive effects on chromatin function (Ahmad and Cao, 2012; Zhang and Ma, 2012). Two kinds of demethylase are involved in the homeostasis of methylation in organisms. Lysine Specific Demethylase 1 (LSD1) was the first histone demethylase identified and is a member of the flavin-dependent amine oxidase family (Lee et al., 2005; Metzger et al., 2005; Chen et al., 2011). Genes in the second class of histone demethylases have a JmjC domain with which they catalyze histone lysine demethylation through oxidative reactions dependent on ferrous ion (Fe(II)) and α-ketoglutarate (α-KG) (Elkins et al., 2003; Trewick et al., 2005).

Plant JmjC proteins are known to play important roles in regulating epigenetic processes and in growth and development (Klose et al., 2006; Kouzarides, 2007). Many members of the *JmjC* gene family from different plant species have been characterized. In *Arabidopsis*, AtJMJ11/ELF6 (EARLY FLOWERING 6) is a repressor in the photoperiodic flowering pathway, and its loss-of-function mutation causes early flowering (Noh et al., 2004; Yu et al., 2008). Its relative, AtJMJ12/REF6 (RELATIVE OF EARLY FLOWERING 6) has an opposite effect in the regulation of flowering time (Noh et al., 2004; Yu et al., 2008; Lu et al., 2011a). Loss-of-function mutation of REF6 leads to increased expression of the flowering repressor FLC (FLOWERING LOCUS C) and hence late flowering (Lu et al., 2011a). In addition, AtJMJ14, an active histone H3K4 demethylase (Lu et al., 2010; Yang et al., 2010), was also implicated in preventing early flowering by repressing the expression of *FLOWERING LOCUS T* (*FT*) and its homologs. Recently, Ning reported that AtJMJ14 associates with two NAC transcription factors, NAC050 and NAC052, and co-occupies hundreds of common target genes, resulting in H3K4 demethylation and transcriptional repression (Ning et al., 2015). Apart from controlling flowering time, there is also evidence that AtJMJ14 functions in RNA silencing and cell-to-cell movement of an RNA silencing signal (Lu et al., 2010). The histone H3K9 demethylase AtJMJ25/IBM1 (INCREASE IN BONSAI METHYLATION 1) (Wang et al., 2013; Shen et al., 2014a) protects genes from CHG (H represents A, T, or G) hypermethylation by CMT3 (CHROMOMETHYLASE 3). Gain-of-function mutants of *AtJMJ15* showed enhanced salt tolerance, in contrast with increased salt sensitivity in the loss-of-function mutant (Shen et al., 2014a). In addition, *AtJMJ14* and *AtJMJ15* have also been shown to be involved in the control of flowering time (Yang et al., 2012). AtJMJ30/JMJD5, an evening-expressed gene, is the sole AtJMJ protein to show a robust circadian rhythm of expression (Mockler et al., 2007; Michael et al., 2008; Jones and Harmer, 2011; Lu et al., 2011b). The role of *AtJMJ30* as a genetic regulator of period length in the *Arabidopsis* circadian clock was confirmed by analysis of loss- and gain-of-function mutants (Lu et al., 2011b). In tomato, a similar role is played by *JMJ524*, which standardizes the circadian clock and also alters GA response to regulate stem elongation (Li et al., 2015). In *Medicago truncatula*, *MtJMJC5* (Medtr4g066020), an ortholog of *AtJMJ30/JMJD5*, may play a role in epigenetic regulation of the link between the circadian clock and cold signaling (Shen et al., 2016). In rice, OsJMJ705 is a biotic stress-responsive H3K27me2/3 demethylase that may remove H3K27me3 from marked defense-related genes and increase their basal and induced expression during pathogen infection (Li et al., 2013). *OsJMJ706*, encodes a heterochromatin-associated H3K9 demethylase, is reported to involve in the regulation of flower development in rice (Sun and Zhou, 2008).

Soybean [*Glycine max* (L.) Merr.] is one of the most economically important crop species in the world. Its genome has undergone two rounds of whole-genome duplication (WGD; Schlueter et al., 2004; Schmutz et al., 2010; Vanneste et al., 2014; Liu et al., 2015); one occurred approximately 59 Mya and was shared by other legumes such as *Medicago* and *Lotus*, while the other was specific to *Glycine* and occurred around 13 Mya. Thus, about 75% of the genes in the soybean genome have multiple paralogs (Schmutz et al., 2010; Severin et al., 2011; Singh and Jain, 2015), making it an excellent model for studying the evolution of duplicate genes following polyploidy. Here, we systemically identify the *JmjC* gene family members in soybean (*G. max*), *Medicago* (*M. truncatula*), and *Lotus* (*Lotus japonicus*) and subsequently analyze the evolutionary relationships between these genes among the three legumes, *Oryza sativa*, and *Arabidopsis*. In addition, we study the *GmJMJs* in further detail, including subfamily classification, gene structures, chromosomal distribution, duplication patterns, conserved residues, and expression profiling. We propose that demethylases exhibit conservative functions through duplication events. Our data will facilitate future studies to elucidate the exact biological functions of the *GmJMJs*.

MATERIALS AND METHODS

Identification of JmjC Domain-Containing Proteins in Soybean and Other Legumes

The *G. max* 2.0 genome database (https://phytozome.jgi.doe.gov/pz/#) was searched to identify JmjC domain-containing proteins using Basic Local Alignment Search Tool algorithms (BLASTP) with a threshold of *e*-value < 1e-10, using the published *Arabidopsis* (21) and *O. sativa* (20) JmjC domain-containing protein sequences (**Table S1**) as queries (Huang et al., 2016). All obtained protein sequences were examined for the presence of the JmjC (PF02373, SM00558) domain using the Hidden Markov Model (HMM) of Pfam (Finn et al., 2016) (http://pfam.sanger.ac.uk/search), and SMART (Letunic et al., 2015) (http://smart.embl-heidelberg.de/). Sequences with obvious errors and/or JmjC domain length of <90 amino acids were removed manually. Following the same approach, putative *M. truncatula* and *L. japonicus* JmjC domain-containing proteins were identified from Phytozome v10 (https://phytozome.jgi.doe.gov/pz/portal.html) and the *L. japonicus* genome assembly build 3.0 (http://www.kazusa.or.jp/lotus/), respectively.

Phylogenetic Analysis

Multi-species phylogenetic tree was constructed using MEGA 6.0 (Tamura et al., 2013) with the Neighbor-Joining (NJ) method, and bootstrap analysis was conducted using 1000 replicates with the p-distance model. The JmjC domain alone was used to set up the phylogenetic tree and define the groups (**Figure S1**). And then with the aim to obtain a better phylogeny within each group, we added additional conserved domains to JmjC domain (**Table S2**) to construct the phylogenetic tree (**Figures 1A, 4A**). Multiple sequences alignments were performed using ClustalW with default parameters in MEGA 6.0.

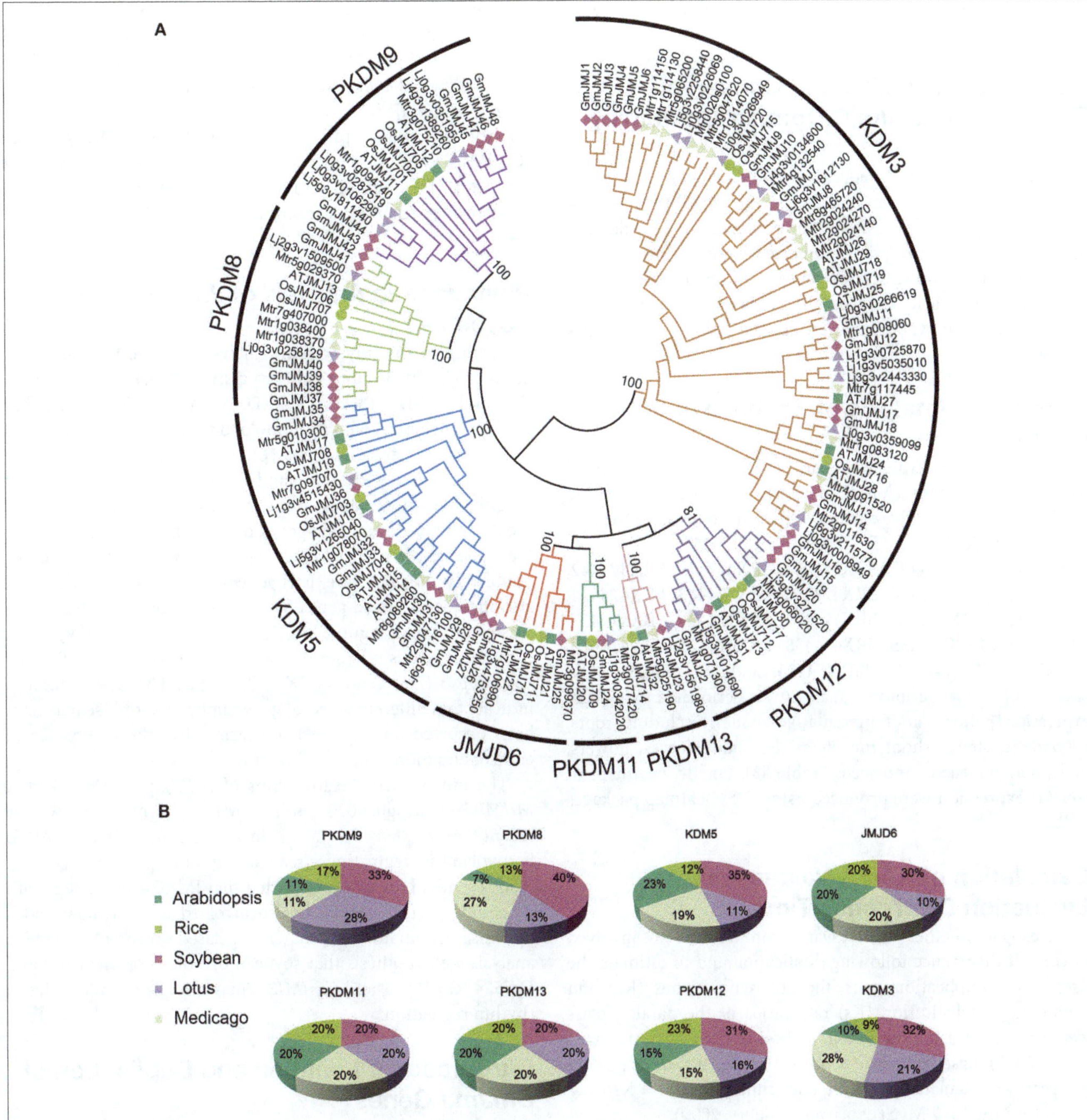

FIGURE 1 | Phylogenetic relationship and distribution of JmjC-domain containing proteins from five plant species. (A) Phylogenetic relationship of JmjC-domain containing proteins among soybean, *Medicago*, *Lotus*, rice, and *Arabidopsis*. Both multiple sequences alignment and phylogenetic tree were performed by MEGA6.0. The value at the nodes represents bootstrap values from 1000 replicates. Different groups were shown by different color. **(B)** Percentage representation of JmjC-domain containing proteins across the five plant species within each group. Colors correspond to the plant taxa as listed in the left.

Chromosomal Locations and Gene Structure of *JmjC* Genes

The locations of the JmjC domain-containing genes on the soybean chromosomes were plotted using the MapChart software. The location information of each JmjC domain-containing gene on each chromosome was determined from the soybean genome annotation file (Gmax_275_Wm82.a2.v1.gene.gff3). The blocks regarded as recent duplications were obtained from SoyBase (Grant et al., 2010) (http://www.soybase.org/). The presence of introns and exons was also annotated according to the soybean genome annotation file. Schematic diagrams were pictured by using GSDS2.0 (Gene Structure Display Server http://gsds.cbi.pku.edu.cn/).

Conserved Domains and Conserved Residues in the JmjC Domain-Containing Proteins

To explore the full-length sequences of JmjC domain-containing proteins, NCBI CDD (Marchler-Bauer et al., 2015) (http://www.ncbi.nlm.nih.gov/cdd/), SMART (http://smart.embl-heidelberg.de/), and Pfam (http://pfam.xfam.org/) were performed with default parameters to search for conserved domains. To identify conserved amino acid residues for interaction with co-factors, the sequences of JmjC domain were aligned using the DNAMAN software.

Expression Analysis of Soybean *JmjC* Genes

To determine the expression patterns of the *JmjC* genes in soybean tissues, transcriptome data was downloaded from the NCBI Short Read Archive database under the following accession numbers: SRX474427, SRX474441, SRX474445, SRX474430, SRX474431, SRX474433, SRX474432, SRX474439, SRX474442, SRX474419, SRX474428, SRX474440, SRX474443, SRX474424, SRX474423, SRX474422, SRX474434, SRX474436, SRX474437, SRX474416, SRX474435, SRX474438, SRX474421, SRX474420, SRX474446, SRX474444, SRX474426, and SRX474429 (Shen et al., 2014b). Transcriptome analysis was performed to identify expression patterns in representative tissues, including roots, cotyledons, stems, shoot meristems, leaf buds, leaves, flowers, pods, pod and seeds, and seeds (**Table S3**). Finally, heatmaps of *GmJMJ* expression were produced using the pheatmap packages in R.

Calculation of Ka/Ks-Values and Evaluation Divergence Time

To investigate whether positive Darwinian selection was involved in *GmJMJ* divergence following duplication and to estimate the date of the duplication pairs, the non-synonymous (Ka) and synonymous substitution (Ks) rate ratios of the paralog pairs were calculated using the YN00 method of the PAML program (Yang, 2007). Based on a rate of 6.1×10^{-9} substitutions per site per year, we calculated the divergence time (T) as $T = \text{Ks}/(2 \times 6.1 \times 10^{-9}) \times 10^{-6}$ Mya (Lynch and Conery, 2000).

RESULTS

Identification of *JmjC* Gene Family in Soybean

Using the combined methods, we identified a total of 48 *GmJMJs*, which is more than twice the number found in *Arabidopsis* (21) or rice (20) (Lu et al., 2008). To better understand the expansion and evolutionary history of *GmJMJs*, the same methods were used to search for *JmjC* genes in two other legumes, *Medicago* and *Lotus*. We identified 33 *Medicago* and 27 *Lotus JmjC* genes, which is still less than the number found in soybean.

A variety of information about *GmJMJs*, such as different version of gene codes, gene length, isoelectric point (pI), and molecular weight (Mw) and so on, were listed in **Table S4**. For example, the identified *GmJMJs* encode proteins ranging from 284 (GmJMJ2) to 1831 (GmJMJ35) amino acids, with the isoelectric point (pI) varying from 4.91 (GmJMJ23) to 9.25 (GmJMJ37) and the molecular weight (Mw) varying from 32.2 kD (GmJMJ2) to 209.2 kD (GmJMJ35). *GmJMJ1* was excluded from further analyses, as there is no annotation data available for it.

Phylogenetic Analysis of *JmjC* Genes in Soybean

According to the phylogenetic analysis (**Figure 1A**), the *JmjC* genes can be divided into eight groups: PKDM9, PKDM8, KDM5, JMJD6, PKDM11, PKDM13, PKDM12, and KDM3. KDM3 has the most members, with 57 homologous *JmjC* genes, and KDM5 is the second largest group, containing 26 *JmjC* genes. The smallest clades are PKDM11 and PKDM13, which both consist of only five *JmjC* genes, one from each species. In general, most of the clades include genes from all five species, although the clades are also enriched in particular species. For example, doubled *GmJMJ* pairs are sister genes to *AtJMJ28* in a clade which also includes doubled genes from the other legumes, forming a cluster of several legume *JMJ* genes with a single *AtJMJ* gene. Likewise, there is a larger of percentage of soybean (40%) than *Arabidopsis* (7%) genes in PKDM8 (**Figure 1B**). These findings indicate that different levels of gene duplication or lose may have been occurred among the five species after the divergence of eudicot and monocot.

The phylogenetic relationships of *AtJMJ30*, *OsJMJ717*, and *MtJMJC5* (Mtr4g066020) in PKDM12 are consistent with a recent report (Shen et al., 2016). Shen et al. showed that *MtJMJC5* is involved in regulating circadian rhythm (Shen et al., 2016). And recently, Li et al also reported that JMJ524, consistent with its counterparts AtJMJ30, is also involved in a circadian clock response in tomato (Li et al., 2015). Based on the phylogenetic analysis we hypothesis that soybean orthologs of *MtJMJC5* and *JMJ524 (GmJMJ19* and *GmJMJ20)* may also play similar roles in rhythm regulation.

Chromosome Location and Duplication of *GmJmJ* Genes

Compared to other species, soybean has an extensively expanded *GmJMJ* family with more than twice as many *JmjCs* as rice and *Arabidopsis* (**Table 1**). We carried out a comprehensive analysis of the *GmJMJs* with the aim of understanding their duplication status and identifying duplicated gene pairs. First, *GmJMJ* pairs located in a pair of paralogous blocks formed by *Glycine*-specific WGD were considered as candidate duplicate gene pairs. As shown in **Figure 2**, all 47 *GmJMJs* (except *GmJMJ1*) were randomly located on 18 of the 20 soybean chromosomes. For example, chromosome 10 possesses six *GmJMJs*, chromosomes 4, 6, 7, 8, and 15 each contain three *GmJMJs*, and chromosome 2, 3, 5, 12, 13, 14, and 17 each have only one *GmJMJ*. In total, we found that a large proportion (41 of 47) of the *GmJMJs* (linked by purple

TABLE 1 | JmjC gene distribution among species that were used in this study.

Species name	Abbr.	PKDM9	PKDM8	KDM5	JMJD6	PKDM11	PKDM13	PKDM12	KDM3	All	Genome size (mb)	Chromosomes
Glycine max	Gma	6	6	9	3	1	1	4	18	48	1100	40
Lotus japonicus	Lj	5	2	3	1	1	1	2	12	27	472	12
Medicago truncatula	Mtr	2	4	5	2	1	1	2	16	33	500	16
Arabidopsis thaliana	At	2	1	6	2	1	1	2	6	21	125	10
Oryza sativa	Os	3	2	3	2	1	1	3	5	20	466	24

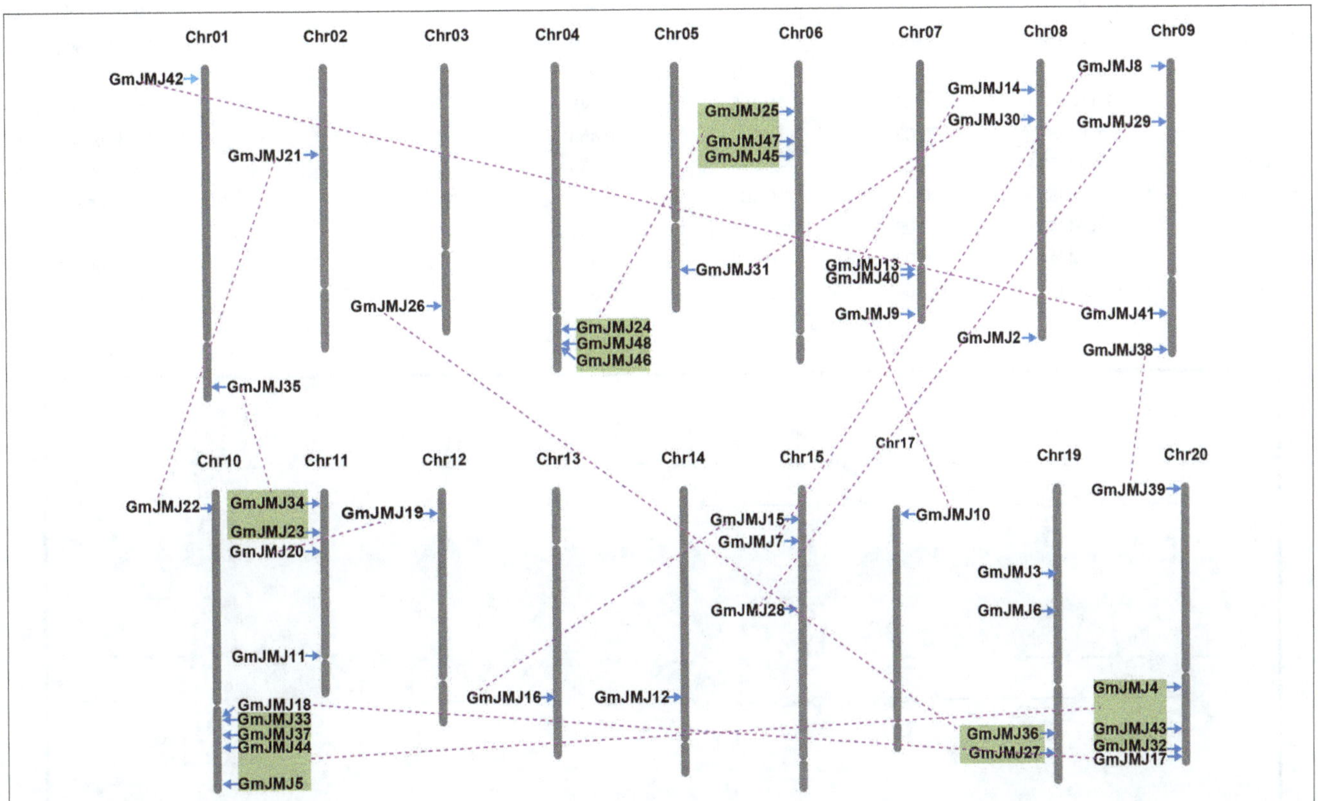

FIGURE 2 | Chromosomal locations, region duplication, and predicted cluster for soybean JmjC genes. The chromosome number is indicated above each chromosome. Throat on the chromosomes indicate the position of centromere. Only those chromosomes bearing JmjC genes (18) were represented. Segmental duplicated genes are indicated by purple lines. Colored boxes indicate the gene blocks generated by the most recently WGD event. The figure was produced and modified using the MapChart and Adobe Illustrator, respectively.

lines in **Figure 2**) was distributed preferentially in duplicated blocks. These 41 genes be considered as candidates of the most recent *Glycine*-specific WGD and used in the next analysis. Second, close phylogenetic relationships have been shown among all candidate *GmJMJs* (**Figure S2**). Third, duplication types of all *GmJMJs* have been obtained by MCscanX programs (Wang et al., 2012) (**Table S5**). Fourth, collinearity analysis (**Figure 3**) was carried out among each candidate duplicated gene pairs. The candidate pairs were considered as created by the recent Glycine-specific WGD duplications with at least three paralogous gene pairs along the flanking regions. In conclusion, we identified that 16 *GmJMJ* pairs were formed by the most recent *Glycine*-specific WGD (**Table 2**).

Based on the divergence rate of 6.1×10^{-9} synonymous mutations per synonymous site per year which has been proposed for soybean (Lynch and Conery, 2000), among the 48 *JmjCs* in soybean, 73% (35 of 48) represented WGD/segmental duplication genes. Ks-value was calculated for estimating the separation time of each paralogous gene pair. All Ks-values ranged from 0.068 to 0.18, which was consisted with whole genome duplication events at round 13 Mya. In addition, our divergence time analyses showed that duplications among 16 paralogous pairs occurred between 5.6 and 15.5 Mya, with an average of 9.7 Mya (**Table 2**).

The history of selection acting on coding sequences can also be measured based on the ratio of non-synonymous to synonymous substitutions (Ka/Ks) (Li et al., 1981). Ka and Ks can be estimated using a number of substitution models and methods, and the estimates are sensitive to these choices and other complications such as the GC content of the sequences and their genomic

TABLE 2 | Divergence between JmjC gene pairs in soybean.

Gene1	Gene2	Ka	Ks	Ka/Ks	Duplication date (Mya)	Duplication type
GmJMJ4	GmJMJ5	0.101	0.186	0.541	15.3	WGD or segmental
GmJMJ7	GmJMJ8	0.072	0.189	0.381	15.5	WGD or segmental
GmJMJ9	GmJMJ10	0.046	0.076	0.069	6.2	WGD or segmental
GmJMJ13	GmJMJ14	0.034	0.090	0.380	7.4	WGD or segmental
GmJMJ15	GmJMJ16	0.026	0.068	0.381	5.6	WGD or segmental
GmJMJ17	GmJMJ18	0.018	0.169	0.104	13.8	WGD or segmental
GmJMJ19	GmJMJ20	0.028	0.147	0.191	12.1	WGD or segmental
GmJMJ21	GmJMJ22	0.072	0.126	0.574	10.3	WGD or segmental
GmJMJ26	GmJMJ27	0.043	0.171	0.250	14.0	WGD or segmental
GmJMJ28	GmJMJ29	0.036	0.103	0.346	8.4	WGD or segmental
GmJMJ30	GmJMJ31	0.035	0.077	0.458	6.3	WGD or segmental
GmJMJ32	GmJMJ33	0.026	0.074	0.359	6.0	WGD or segmental
GmJMJ34	GmJMJ35	0.024	0.101	0.241	8.3	WGD or segmental
GmJMJ41	GmJMJ42	0.026	0.104	0.250	8.5	WGD or segmental
GmJMJ43	GmJMJ44	0.030	0.092	0.333	7.5	WGD or segmental
GmJMJ47	GmJMJ48	0.045	0.130	0.349	10.7	WGD or segmental

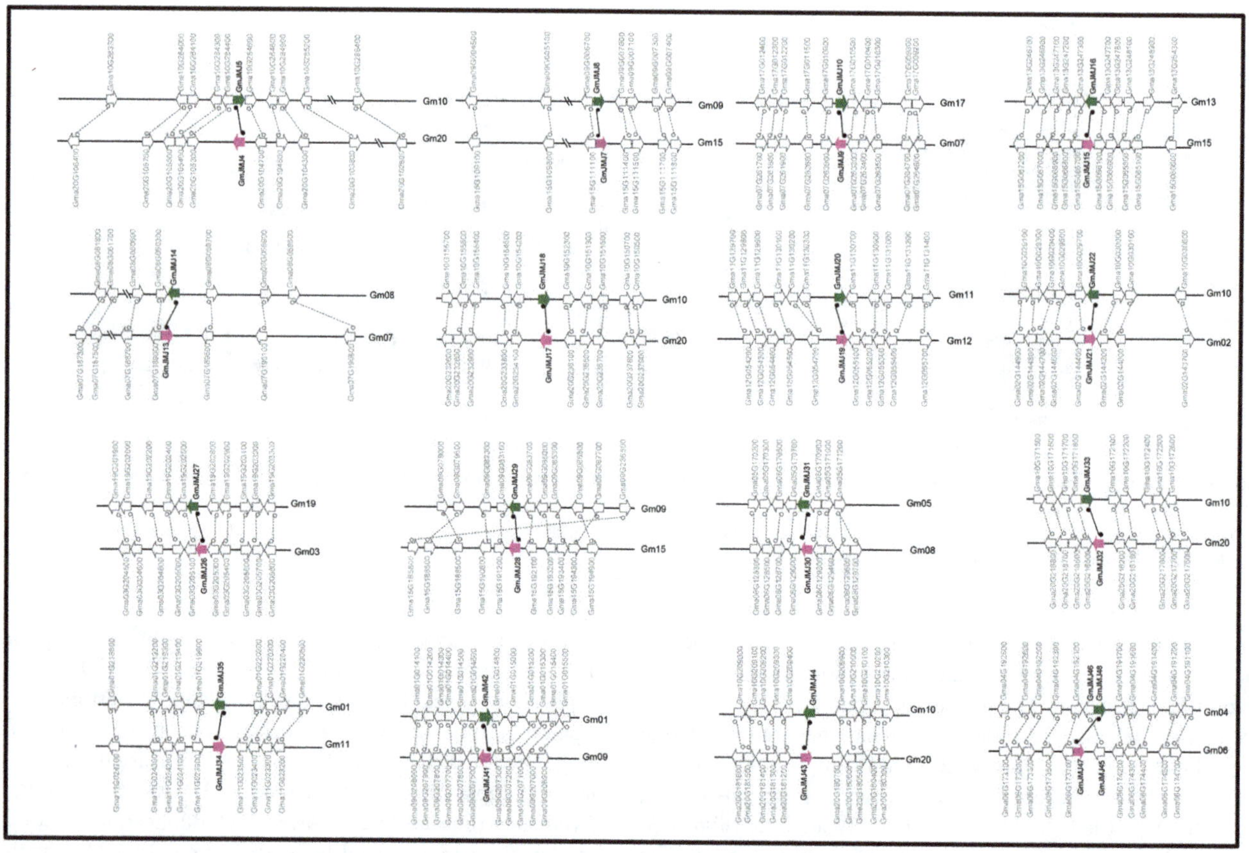

FIGURE 3 | Collinearity analysis of duplicated gene pairs formed by Glycine-specific WGD event. Green and purple arrows represented duplicated JmjC gene pairs. The direction of the arrow indicates the position of the gene in the positive (to left) and negative (to right) chain of DNA.

context (Bustamante et al., 2002). Ka is usually much smaller than Ks, so a pair of sequences will have Ka/Ks << 1 if both sequences have been under purifying selection, Ka/Ks < 1 if one sequence has been under purifying selection but the other drifting naturally, and in rare, Ka/Ks > 1 when both sites are under positive seletion (Juretic et al., 2005). As shown in **Table 2**, the average Ka/Ks-values of the *GmJMJ* gene pairs were 0.36. Five paralog pairs have small Ka/Ks ratios (< 0.3), most Ka/Ks ratios in the range from 0.3 to 0.7, and none of them > 1.

Exon–Intron Structure and Domain Architecture of *GmJmJ* Genes

Structural divergence has been very prevalent in duplicate genes and, in many cases, has led to the generation of functionally distinct paralogs (Lynch and Conery, 2000). To better understand the structural diversity of the *GmJMJs* following duplication events, the exon/intron structures (**Figure 4B**) were compared using Gene Structure Display Server 2.0 (http://gsds.cbi.pku.edu.cn/). Our analysis clearly revealed that most of the paralogs share a similar gene structure. For example, 12 gene pairs (*GmJMJ9/-10, GmJMJ13/-14, GmJMJ15/-16, GmJMJ19/-20, GmJMJ21/-22, GmJMJ28/-29, GmJMJ30/-31, GmJMJ32/-33, GmJMJ34/-35, GmJMJ41/-42, GmJMJ43/-44,* and *GmJMJ47/-48*) were found to have highly consistent gene structures, including the numbers of exons/introns and the length of exons. However, there were some differences in intron lengths and in the 5′ UTR region, which is related to the regulation of expression. For example, *GmJMJ9/-10* and *GmJMJ30/-31* both present a large divergence in the length of their 5′ UTR, implying that a subtle distinction in function in the development and growth of soybean may have appeared between the two paralogs. In addition, four gene pairs (*GmJMJ4/-5, GmJMJ7/-8, GmJMJ17/-18,* and *GmJMJ26/-27*) had greater changes in their structural organization, especially in the numbers of exons.

We also studied the proteins encoded by the *GmJMJs*, using the full-length protein sequences of JmjCs as queries in CDD (Marchler-Bauer et al., 2015), SMART (Letunic et al., 2015), and Pfam (Finn et al., 2016) with default parameters in order to gain more insights into the diversity of the domain architecture, as shown in **Figure 4C**. These proteins all share a JmjC domain. The JmjN domain was the second most widespread domain, appearing in the majority of members of three groups, KDM5, PKDM8, and PKDM9. This domain, which is not adjacent with JmjC, was identified in the jumonji family (Balciunas and Ronne, 2000), and its interaction with the JmjC catalytic domain was found to be important for Jhd27 (also known as KDM5), a H3K4-specific demethylase in budding yeast (Huang et al., 2010; Quan et al., 2011). In PKDM9, the zf-C2H2 domain, which contains two cysteines and two histidines that coordinate a zinc atom to create a compact nucleic acid-binding domain (Chrispeels et al., 2000), was found in four tandem repeats. Furthermore, another zinc-finger domain, zf-C5HC2, was identified in PKDM8 and KDM5. Three groups, PKDM11, PKDM12 and PKDM13, all have only one domain (JmjC) in their full-length sequence, and can be grouped together as "JmjC domain-only proteins," in keeping with previous studies (Klose et al., 2006; Lu et al., 2008; Huang et al., 2016). Two-thirds of the members of KDM5 have FYRN and FYRC domains, which may harbor chromatin-binding activity (Lu et al., 2008) or contribute to JmjC function by interacting with other proteins. For example, it has been reported that the functional specificity of AtJMJ14 in flowering time control is based on the specificity of its interaction with transcription factors through the FYRC domain (Ning et al., 2015). Two GmJMJ proteins, GmJMJ34 and GmJMJ35, have the ARID domain (AT-rich interaction domain), which has been implicated in sequence-specific DNA binding (Gregory et al., 1996).

Strangely, we found that not only the gene pairs which share a similar gene structure but also the four pairs which had greater differences in gene structural organization all had consistent domain architectures. For example, *GmJMJ26/-27* share a relatively consistent functional domain in the full-length protein sequences. This implies that although the gene structure of JmjC family may change through evolution, their protein structures and functions were conserved.

Conserved Amino Acid Residues in Active Sites of GmJmJ

Fe(II) iron and α-KG are needed as cofactors by JmjC-domain proteins to carry out their demethylase activity (Chen et al., 2006; Huang et al., 2016). A total of five amino acid residues are needed to bind these cofactors; three residues (His188, Glu/Asp190, and His276) bind to the Fe (II) cofactor and two other residues (Thr/Phe185 and Lys206) bind to α-KG. With the aim of clarifying whether the conserved residues interacting with cofactors had diverged among *GmJMJs*, we aligned the domain sequences of JmjC proteins from soybean and *Arabidopsis*.

Based on the alignments, we grouped these proteins into two groups according to amino acids at the conserved sites. The first group, which includes PKDM8, PKDM9, and KDM5, has the conserved amino acids His (H), Glu (E), and His (H) for Fe(II) binding, and Phe (F) and Lys (K) for α-KG binding (**Figure 5A**), while the second group, which includes JMJD6, PKDM13, PKDM11, PKDM12, and KDM3, has conserved the residues His (H), Asp (D), and His (H) for Fe(II) and Thr (T) and Lys (K) for α-KG (**Figure 5B**). Both forms are compatible with histone demethylation activity (Lu et al., 2008). In general, most members have conserved residues for interacting with cofactors, although there are some exceptions. For example, a substitution can be seen in the first sites in PKDM12, where Thr (T) was changed into Ser (S) in GmJMJ21/22 and AtJMJ31. However, Thr and Ser have similar physical and chemical properties, so the ability to bind to cofactors may not have changed despite the presence of a different amino acid. Furthermore, the first site to interact with α-KG is absent in JMJD6, which is consistent with findings in rice (Lu et al., 2008). The detection of this change in all three plants, soybean, rice, and *Arabidopsis*, suggests that it may have occurred in the ancestor of these plants and is necessary for their common function. Overall, the high conservation in the interaction sites implies a significant role for these sites in the demethylase activity of the *JmjC* gene family.

Expression Profiles of *JmjC* Genes in Soybean

To investigate the tissue-specific expression profiles of *GmJMJs*, transcriptome data (Shen et al., 2014b) were studied in 10 tissues at different developmental stages including roots, cotyledons, stems, shoot meristems, leaf buds, leaves, flowers, pods, pod

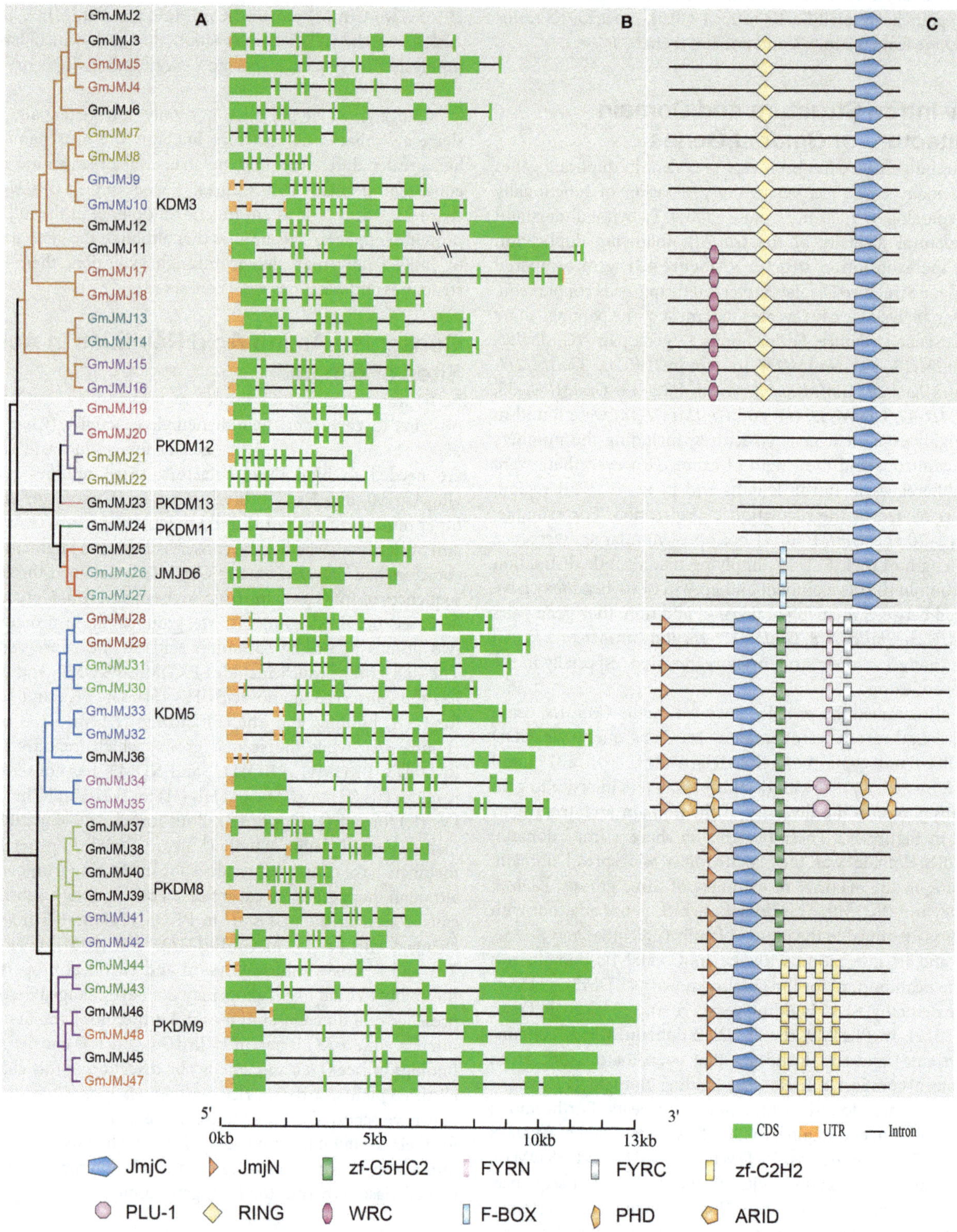

FIGURE 4 | Phylogenetic analysis, gene structure, and domain architecture of GmJMJs. (A) Phylogenetic tree construction of GmJMJs based on the JmjC domain amino acid sequences. Name of genes marked in same color are a pair of paralogs. **(B)** Exon/intron structures of *GmJMJs* genes. The black line refers introns, the green box represents exons, and the orange box refers UTR. Over-longed introns were represented with slash–slash. The sizes of exons and introns can be estimated using the scale at the bottom. **(C)** The domain architecture of the full-length JmjC-domain containing proteins. JmjC, Jumonji C domain; JmjN, Jumonji N domain; PHD, plant homeobox domain; ARID, AT-rich interaction domain; zf-C2H2, Zinc finger of C2H2-type; zf-C5HC2, Zinc finger of C5HC2-type; FYRN, "FY-rich" domain N-terminal; FYRC, "FY-rich" domain C-terminal; WRC, Trp, Arg, and Cys domain; RING, (Really interesting new gene) finger domain.

FIGURE 5 | Two alignment groups (A,B) of JmjC domain sequences. (Only the representative sequences were shown, other sequences can be seen in **Figure S3**). The conserved residues compatible with the demethylation activity within the Fe(II) binding site are highlighted in red and those in the αKG binding site are indicated in yellow. The sequences with black, gray, light gray background indicated identical 100%, conservative (75–99%), and block (50–74%) similarity of amino acid residues, respectively.

and seeds, and seeds. As indicated in **Figure 6**, the expression patterns of the *GmJMJs* can be divided into three clusters, C1–C3. C1 genes show hardly any expression in almost all of the tissues except certain expression in flower, implying that these genes may possess certain specific function in flower after WGD events. C2 can be divided into two subgroups, C2-Sub1 and C2-Sub2, according to their expression levels, with lower expression levels in C2-Sub1 than C2-Sub2. Genes in C3 show a low expression level in all tissues investigated. We found that genes, which belong to the same clade in the phylogenetic tree (**Figure 1A**), were sometimes dispersed in different clusters based on expression (**Figure 6**). For example, the four genes in C1 (*GmJMJ37*, *GmJMJ38*, *GmJMJ39*, and *GmJMJ40*) grouped together with *GmJMJ41* and *GmJMJ42* in a clade, indicating that they may have been produced through evolution after WGD events, but clustered into two clusters and show different expression patterns, implying they have acquired different functions after the duplication event.

In addition, we determined the expression profiles of the recently duplicated *JmjC* gene pairs in 10 tissues. Most of the paralogs generally have the same expression pattern. For further analysis, we divided the duplicated genes into three types based on their detailed expression patterns, shown with a blue, green, and red box in **Figure 7**. The 12 paralogs in the blue box all have a relatively low expression level in four tissues (pods, podseeds, roots, and seeds) compared with other tissues. These paralogs also show a complex expression pattern in another six tissues examined, but almost all have high expression in the flower, leaf, and shoot meristem. For example, the two copies *GmJMJ4* and *GmJMJ5* both have high expression in flowers, leaf buds and shoot meristems. The expression pattern of these genes can therefore be pictured as similar to the shape of the letter "M." The three paralogs in the green box only show high expression in one organ, such as the flower, seed, and root. For instance, *GmJMJ7/-8 and GmJMJ41-42* both have a high expression level in flower and roots, respectively. The third box contains only one paralog pair (*GmJMJ26* and *GmJMJ27*), and the expression level of one copy (*GmJMJ27*) is higher than the other copy (*GmJMJ26*) in all tissues. Overall, the paralogs show conserved expression profiles, demonstrating that the *JmjC* gene family has conserved its functions through duplication events.

DISCUSSION

As histone demethylases, JmjC domain-containing proteins play essential roles in histone modification, which is a significant part of epigenetics (Klose et al., 2006; Chen et al., 2011). To date, many efforts on the *JmjC* gene family have been undertaken to elucidate their evolutionary history in a wide variety of plant species, such as *Arabidopsis* (Lu et al., 2008; Zhao et al., 2015), rice (Lu et al., 2008; Zong et al., 2013), and *Fragaria vesca* (Gu et al., 2016). However, little is known about the *JmjC* gene family in soybean. In this study, we performed a comprehensive analysis of *GmJMJs*, including their phylogenetic relationships, gene structure, domain architecture, chromosome location, duplication patterns, and expression profiles.

Phylogeny and Domain Architectures of JmjCs in Soybean

In total, 48 *JmjC* genes were identified in the soybean genome, which is larger than other model plants or the other two legumes examined. The number of JmjCs from each species in each group is summarized in **Table 1**. In most groups, there is still a larger number of JmjCs from soybean than any other species, indicating that these groups may have different evolutionary history among the five species. PKDM11 and PKDM13, two exception groups, both of which including a single gene from each species, may have no duplication or loss after divergence from *Arabidopsis*. The phylogenetic analysis of *JmjC* genes among five plant species showed that each group contains *JmjCs* from all species investigated, four eudicots and one monocot, revealing that the JmjC family may have already existed before the divergence of these two lineages. And combining the phylogeny with the time

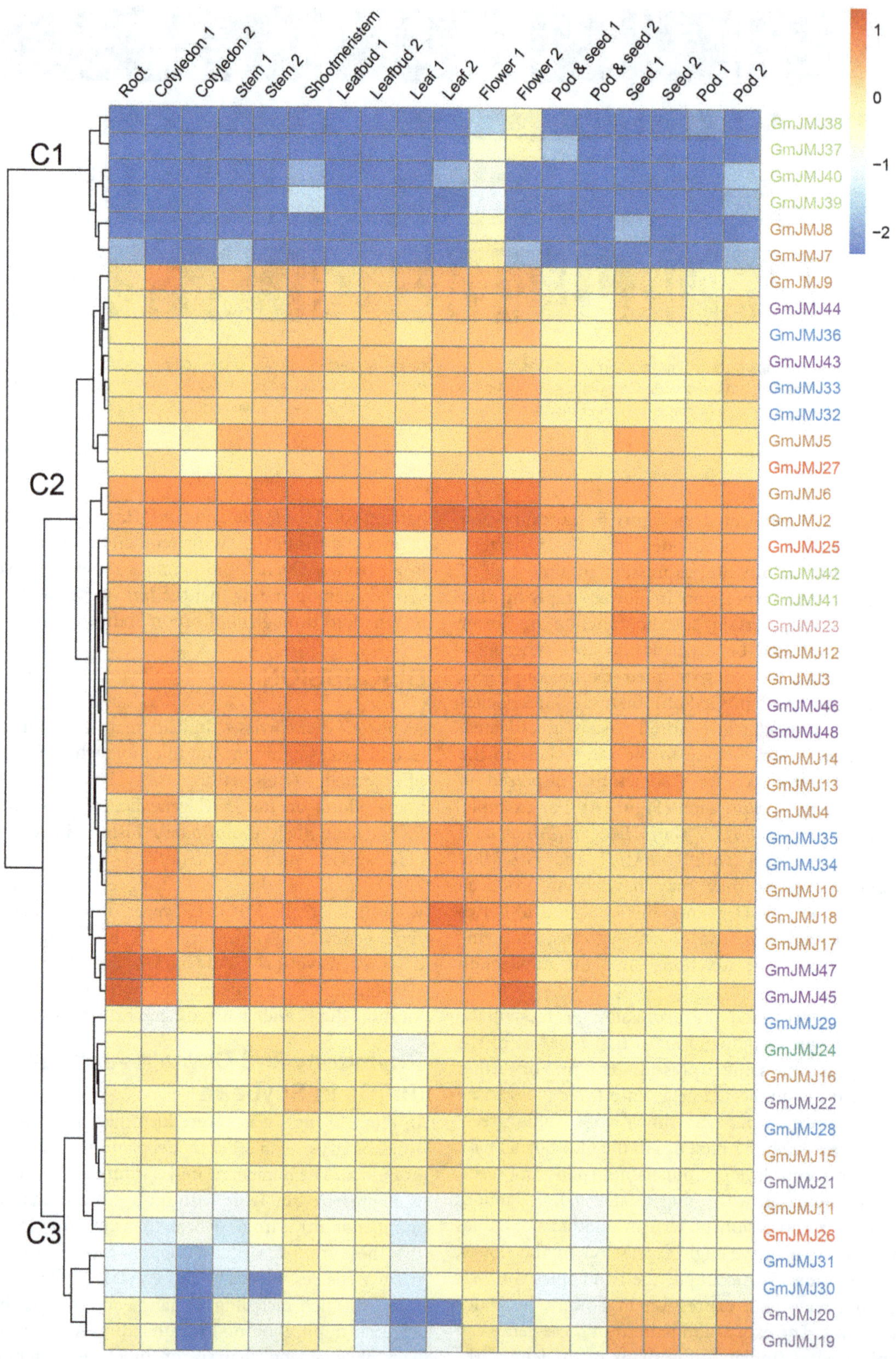

FIGURE 6 | Heatmaps representing the expression profiles of *GmJMJs* in several tissues. The numbers following the tissues indicated different developmental stages. The color scale on the right indicates expression values, blue indicating low transcript abundance, and red indicating high levels of transcript abundance.

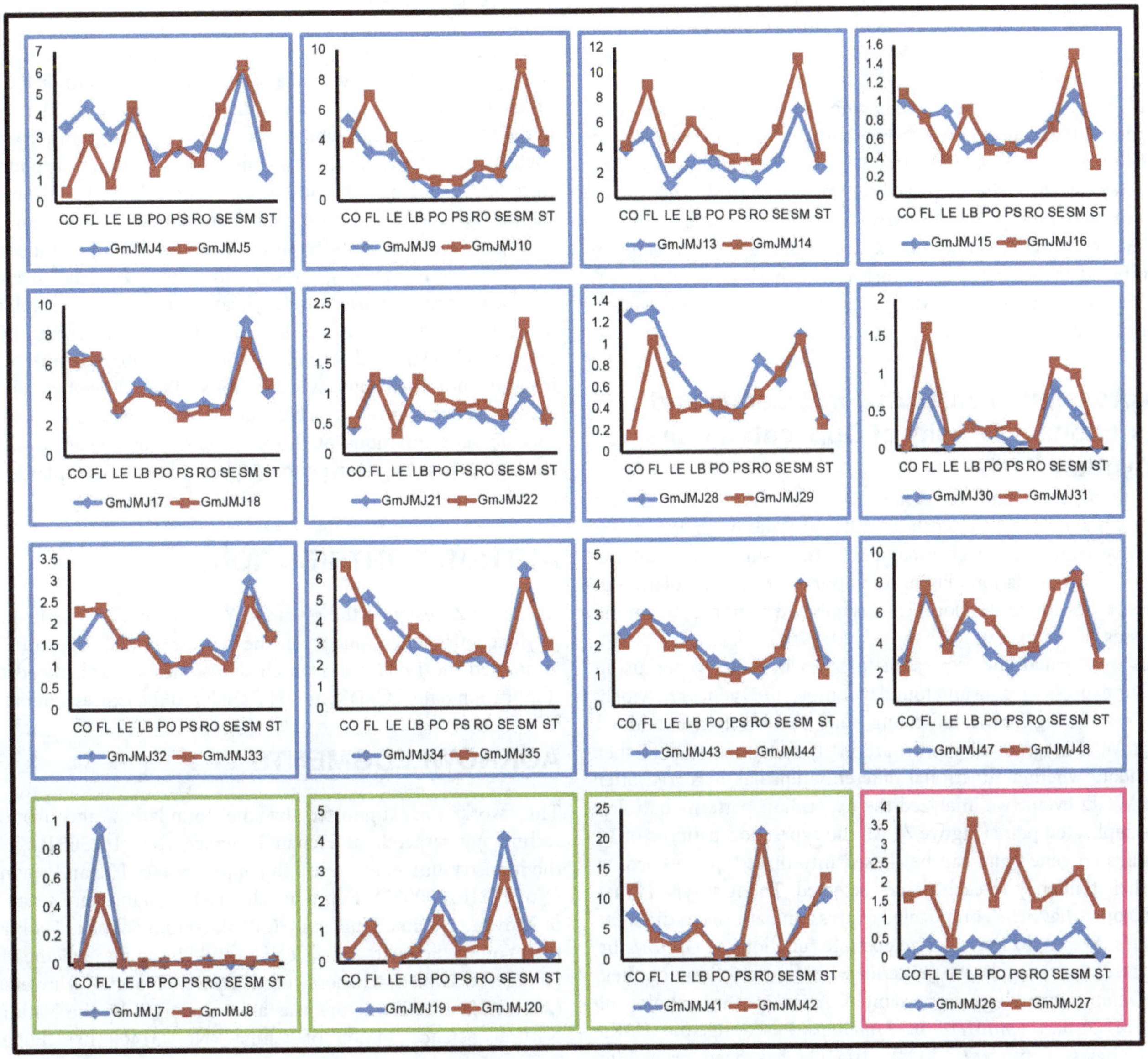

FIGURE 7 | Three trends of expression patterns of duplicated *GmJMJ* gene pairs. X-axis indicates representative tissues and Y-axis represents scale. The sample used are Cotyledon1(CO), Flower1 (FL), Leaf1(LE), LeafBud1(LB), Pod1(PO), Pod and Seed1(PS), Root1(RO), Seed1(SE), ShootMeristem(SM), Stem1(ST).

estimation, 16 *GmJMJs* pairs, formed by the most recent WGD of soybean, were identified.

Although many *GmJMJ* genes were produced by WGD events, there has been little differentiation in their gene structure, domain architecture and the conserved residues acting with cofactors. Another interesting phenomenon is that even outside the conserved coactive-sites we found GmJMJs to have higher amino acid similarity with AtJMJs from the same sub-cluster rather than with GmJMJs from other sub-clusters even though these genes were all grouped in one clade. For example, in PKDM8, GmJMJ41 and GmJMJ42 have the same residues as AtJMJ13 around the Lys (K) site, a Cys (C) and Ser (S), whereas GmJMJ37/-38/-39/-40 have different residues. The unconserved sites around the coactive-sites indicate that the JmjCs may use different ways to bind to the co-factors. However, further evidence is needed to demonstrate and understand this mechanism.

WGDs Contributed to *JmjC* Gene Expansion in *Glycine Max*

WGD/segmental duplication and tandem duplication might lead to duplicated gene pairs on the DNA level. Previous studies have shown that the soybean genome has undergone two rounds of WGD (Schmutz et al., 2010). In this study, we have demonstrated that 16 soybean paralogous pairs derived from the second WGD, which suggests that the WGD duplication might be the main mechanism of *JmjC* gene family expansion and functional diversity during the evolution of soybean. This result is consistent with some other gene families not only *AAT* gene family (Cheng et al., 2016), GST supergene family (Liu et al., 2015), and receptor-like kinase genes (Zhou et al., 2016) in soybean, but also SET domain family in *Populus trichocar* (Lei et al., 2012; Zhang and Ma, 2012), and 14-3-3 family genes and AP2/ERF superfamily in *M. truncatula* (Qin et al., 2016; Shu et al., 2016). And in *Arabidopsis*, previous studies have proposed that more than 90% regulatory genes increased due to WGD (Maere et al., 2005). However, the dispersed duplications and retro-transpositions played the most important role in the evolution of *JmjC* genes in *F. vesca* (Gu et al., 2016). Furthermore, KDM3 group is preferentially expanded in the soybean genome compared to other groups, consistent with *F. vesca* (Gu et al., 2016), indicating that KDM3 group genes may have evolved to meet some unique regulatory needs.

Expression Profiles of *JmjC* Genes and Functional Diversity of Duplicated Pairs in Soybean

In angiosperms and vertebrates, both of the *JmjC* and *SET* genes, maintaining homeostasis of the histones methylation, are the key regulators of chromatin structure, suggesting that the epigenetic modulation playing an important role in regulation of gene expression in developmental stages and responses to abiotic stresses (Lei et al., 2012; Zhang and Ma, 2012; Qian et al., 2015). We investigated the expression profiles of *JmjC* genes using public expression data and found that most *JmjC* genes are widely expressed (**Figure 6**), indicating that these genes, remained after WGD/segmental events, are likely functional. To further elucidate whether functional differentiation has occurred after the WGD event, we analyzed the expression patterns between the duplicated pairs (**Figure 7**). All the expression patterns of 16 duplicated gene pairs can be classed into three types according to their tendency in each tissues detected. The first type is that two copies have the complicated expression patterns in different tissues. We could hypothesize possible functions of *GmJMJs* by coupled their expression patterns with the functions of their *Arabidopsis* orthologs. For example, *AtJMJ24*, the ortholog of *GmJMJ17* and *GmJMJ18* in *Arabidopsis,* is a histone H3K9 demethylase (Lu et al., 2008). *AtJMJ24* has been proved to promote basal level transcription of endogenous silenced loci by counteracting H3K9me (Deng et al., 2015). Therefore, *GmJMJ17* and *GmJMJ18* might have the functions similar to *AtJMJ24* in reinforcing the silence. The second type showed that both copies have high expression in one organ, such as the flower, seed, and root. The third type is that one duplicate was expressed at higher levels than the other one nearly in all tissues, implying that the former one has stronger function than the latter, and implying that it may play important roles in regulating broad developmental or reproductive stages. Above all, the expression patterns among the duplicated pairs are relatively conserved, suggesting little functional differentiation has occurred following the WGD event. However, some *JmjC* genes were specific to soybean, for example, *GmJMJ4* and *GmJMJ5*. All of these genes have abundant transcripts in soybean and are expressed at different levels in different tissues. These results indicate that in *Arabidopsis* their counterparts may be lost and the functions might have been performed by other genes.

CONCLUSIONS

Here, we performed comprehensive and evolutionary analyses of *JmjC* gene family in soybean, and provided detailed information on its members. A total of 48 putative *JmjC* genes were identified in the soybean genome, which represented non-random across all soybean genome chromosomes and majority of them expanded from WGD/segmental duplication rather than the dispersed duplications. The exon/intron compositions and domain arrangements were considerably conserved among members in the same groups or subgroups. Many duplicated genes present similar expression patterns in soybean tissues detected implying functional conservation. The close phylogenetic relationship between *GmJMJs* and *AtJMJs* in the same subgroup provided insights into their putative functions. Taken together, all of these results provided valuable clues in future efforts to identify specific gene functions of this gene family and gene diversity among different genotype of soybean and other plants in *Leguminosae.*

AUTHOR CONTRIBUTIONS

LW and LZ designed the research; YH, XL, and LW performed phylogenetic analysis and wrote the manuscript; LC, YL, and HW annotated the *JmjC* genes on chromosomes and calculated the duplication date; LC, DK, and HY analyzed the expression data.

ACKNOWLEDGMENTS

This work was supported by the foundation and frontier technology research of Henan Province (No. 162300410257), the industry-university-research cooperation of Henan Province (No. 162107000032), Funding scheme for young core teachers of Xinyang Normal University (2015), Nanhu Scholars Program for Young Scholars of XYNU, and Funding scheme for young backbone teachers of Xinyang Normal University (2016GGJS-13). This work was also supported by the National Natural Science Funds of China (No. 31400213). Fujian-Taiwan Joint Innovative Center for Germplasm Resources and Cultivation of Crop [Fujian 2011 Program, (2015)75] to LZ.

SUPPLEMENTARY MATERIAL

The Supplementary Material for this article can be found online at: http://journal.frontiersin.org/article/10.3389/fpls.2016.01800/full#supplementary-material

Figure S1 | Phylogenetic relationship of JmjC-domain containing proteins from five plant species by using the JmjC domain alone.

Figure S2 | Phylogenetic relationships of *GmJMJs*.

Figure S3 | Alignment of JmjC domain sequences. The conserved residues compatible with the demethylation activity within the Fe(II) binding site are highlighted in red and those in the αKG binding site are indicated in yellow. The sequences with black, gray, light gray background indicated identical 100%, conservative (75–99%), and block (50–74%) similarity of amino acid residues, respectively.

Table S1 | Protein sequences of Arabidopsis, *Oryza sativa* JmjC domain-containing protein used as blast queries.

Table S2 | The domain sequences used in the phylogenetic trees.

Table S3 | The expression data of *GmJmjC* genes in various organs.

Table S4 | Detailed information of soybean *JmjC* family genes.

Table S5 | The duplication types of *GmJmjC* genes calculating by MCScanX program.

REFERENCES

Ahmad, A., and Cao, X. (2012). Plant PRMTs broaden the scope of arginine methylation. *J. Genet. Genomics* 39, 195–208. doi: 10.1016/j.jgg.2012.04.001

Allis, C. D., Berger, S. L., Cote, J., Dent, S., Jenuwien, T., Kouzarides, T., et al. (2007). New nomenclature for chromatin-modifying enzymes. *Cell* 131, 633–636. doi: 10.1016/j.cell.2007.10.039

Balciunas, D., and Ronne, H. (2000). Evidence of domain swapping within the jumonji family of transcription factors. *Trends Biochem. Sci.* 25, 274–276. doi: 10.1016/S0968-0004(00)01593-0

Bustamante, C. D., Nielsen, R., and Hartl, D. L. (2002). A maximum likelihood method for analyzing pseudogene evolution: Implications for silent site evolution in humans and rodents. *Mol. Biol. Evol.* 19, 110–117.

Chen, X., Hu, Y., and Zhou, D. X. (2011). Epigenetic gene regulation by plant Jumonji group of histone demethylase. *Biochim. Biophys. Acta* 1809, 421–426. doi: 10.1016/j.bbagrm.2011.03.004

Chen, Z., Zang, J., Whetstine, J., Hong, X., Davrazou, F., Kutateladze, T. G., et al. (2006). Structural insights into histone demethylation by JMJD2 family members. *Cell* 125, 691–702. doi: 10.1016/j.cell.2006.04.024

Cheng, L., Yuan, H. Y., Ren, R., Zhao, S. Q., Han, Y. P., Zhou, Q. Y., et al. (2016). Genome-wide identification, classification, and expression analysis of amino acid transporter gene family in *Glycine max*. *Front. Plant Sci.* 7:515. doi: 10.3389/fpls.2016.00515

Chrispeels, H. E., Oettinger, H., Janvier, N., and Tague, B. W. (2000). AtZFP1, encoding *Arabidopsis thaliana* C2H2 zinc-finger protein 1, is expressed downstream of photomorphogenic activation. *Plant Mol. Biol.* 42, 279–290. doi: 10.1023/A:1006352809700

Deng, S., Xu, J., Liu, J., Kim, S. H., Shi, S., and Chua, N. H. (2015). JMJ24 binds to RDR2 and is required for the basal level transcription of silenced loci in Arabidopsis. *Plant J.* 83, 770–782. doi: 10.1111/tpj.12924

Elkins, J. M., Hewitson, K. S., McNeill, L. A., Seibel, J. F., Schlemminger, I., Pugh, C. W., et al. (2003). Structure of factor-inhibiting hypoxia-inducible factor (HIF) reveals mechanism of oxidative modification of HIF-1 alpha. *J. Biol. Chem.* 278, 1802–1806. doi: 10.1074/jbc.C200644200

Finn, R. D., Coggill, P., Eberhardt, R. Y., Eddy, S. R., Mistry, J., Mitchell, A. L., et al. (2016). The Pfam protein families database: towards a more sustainable future. *Nucleic Acids Res.* 44, D279–D285. doi: 10.1093/nar/gkv1344

Grant, D., Nelson, R. T., Cannon, S. B., and Shoemaker, R. C. (2010). SoyBase, the USDA-ARS soybean genetics and genomics database. *Nucleic Acids Res.* 38, D843–D846. doi: 10.1093/nar/gkp798

Gregory, S. L., Kortschak, R. D., Kalionis, B., and Saint, R. (1996). Characterization of the dead ringer gene identifies a novel, highly conserved family of sequence-specific DNA-binding proteins. *Mol. Cell. Biol.* 16, 792–799. doi: 10.1128/MCB.16.3.792

Gu, T., Han, Y., Huang, R., McAvoy, R. J., and Li, Y. (2016). Identification and characterization of histone lysine methylation modifiers in *Fragaria vesca*. *Sci. Rep.* 6:23581. doi: 10.1038/srep23581

Huang, F., Chandrasekharan, M. B., Chen, Y.-C., Bhaskara, S., Hiebert, S. W., and Sun, Z.-W. (2010). The JmjN domain of Jhd2 is important for its protein stability, and the plant homeodomain (PHD) finger mediates its chromatin association independent of H3K4 methylation. *J. Biol. Chem.* 285, 24548–24561. doi: 10.1074/jbc.M110.117333

Huang, Y., Chen, D., Liu, C., Shen, W., and Ruan, Y. (2016). Evolution and conservation of JmjC domain proteins in the green lineage. *Mol. Genet. Genomics* 291, 33–49. doi: 10.1007/s00438-015-1089-4

Jones, M. A., and Harmer, S. (2011). JMJD5 functions in concert with TOC1 in the arabidopsis circadian system. *Plant Signal. Behav.* 6, 445–448. doi: 10.4161/psb.6.3.14654

Juretic, N., Hoen, D. R., Huynh, M. L., Harrison, P. M., and Bureau, T. E. (2005). The evolutionary fate of MULE-mediated duplications of host gene fragments in rice. *Genome Res.* 15, 1292–1297. doi: 10.1101/gr.4064205

Klose, R. J., Kallin, E. M., and Zhang, Y. (2006). JmjC-domain-containing proteins and histone demethylation. *Nat. Rev. Genet.* 7, 715–727. doi: 10.1038/nrg1945

Klose, R. J., and Zhang, Y. (2007). Regulation of histone methylation by demethylimination and demethylation. *Nat. Rev. Mol. Cell Biol.* 8, 307–318. doi: 10.1038/nrm2143

Kouzarides, T. (2007). Chromatin modifications and their function. *Cell* 128, 693–705. doi: 10.1016/j.cell.2007.02.005

Lee, M. G., Wynder, C., Cooch, N., and Shiekhattar, R. (2005). An essential role for CoREST in nucleosomal histone 3 lysine 4 demethylation. *Nature* 437, 432–435. doi: 10.1038/nature04021

Lei, L., Zhou, S. L., Ma, H., and Zhang, L. S. (2012). Expansion and diversification of the SET domain gene family following whole-genome duplications in *Populus trichocarpa*. *BMC Evol. Biol.* 12:51. doi: 10.1186/1471-2148-12-51

Letunic, I., Doerks, T., and Bork, P. (2015). SMART: recent updates, new developments and status in 2015. *Nucleic Acids Res.* 43, D257–D260. doi: 10.1093/nar/gku949

Li, J., Yu, C., Wu, H., Luo, Z., Ouyang, B., Cui, L., et al. (2015). Knockdown of a JmjC domain-containing gene JMJ524 confers altered gibberellin responses by transcriptional regulation of GRAS protein lacking the DELLA domain genes in tomato. *J. Exp. Bot.* 66, 1413–1426. doi: 10.1093/jxb/eru493

Li, T., Chen, X., Zhong, X., Zhao, Y., Liu, X., Zhou, S., et al. (2013). Jumonji C domain protein JMJ705-mediated removal of histone H3 lysine 27 trimethylation is involved in defense-related gene activation in rice. *Plant Cell* 25, 4725–4736. doi: 10.1105/tpc.113.118802

Li, W.-H., Gojobori, T., and Nei, M. (1981). Pseudogenes as a paradigm of neutral evolution. *Nature* 292, 237–239. doi: 10.1038/292237a0

Liu, C., Lu, F., Cui, X., and Cao, X. (2010). Histone methylation in higher plants. *Annu. Rev. Plant Biol.* 61, 395–420. doi: 10.1146/annurev.arplant.043008.091939

Liu, H.-J., Tang, Z.-X., Han, X.-M., Yang, Z.-L., Zhang, F.-M., Yang, H.-L., et al. (2015). Divergence in enzymatic activities in the soybean gst supergene family provides new insight into the evolutionary dynamics of whole-genome duplicates. *Mol. Biol. Evol.* 32, 2844–2859. doi: 10.1093/molbev/msv156

Lu, F., Cui, X., Zhang, S., Jenuwein, T., and Cao, X. (2011a). Arabidopsis REF6 is a histone H3 lysine 27 demethylase. *Nat. Genet.* 43, 715–719. doi: 10.1038/ng.854

Lu, F., Cui, X., Zhang, S., Liu, C., and Cao, X. (2010). JMJ14 is an H3K4 demethylase regulating flowering time in Arabidopsis. *Cell Res.* 20, 387–390. doi: 10.1038/cr.2010.27

Lu, F., Li, G., Cui, X., Liu, C., Wang, X. J., and Cao, X. (2008). Comparative analysis of JmjC domain-containing proteins reveals the potential histone demethylases in arabidopsis and rice. *J. Integr. Plant Biol.* 50, 886–896. doi: 10.1111/j.1744-7909.2008.00692.x

Lu, S. X., Knowles, S. M., Webb, C. J., Celaya, R. B., Cha, C., Siu, J. P., et al. (2011b). The Jumonji C domain-containing protein JMJ30 regulates period length in the Arabidopsis circadian clock. *Plant Physiol.* 155, 906–915. doi: 10.1104/pp.110.167015

Lynch, M., and Conery, J. S. (2000). The evolutionary fate and consequences of duplicate genes. *Science* 290, 1151–1155. doi: 10.1126/science.290.5494.1151

Maere, S., De, Bodt, S., Raes, J., Casneuf, T., Van, Montagu,. M., Kuiper, M., et al. (2005). Modeling gene and genome duplications in Eukaryotes. *Proc. Natl. Acad. Sci. U.S.A.* 102, 5454–5459. doi: 10.1073/pnas.0501102102

Marchler-Bauer, A., Derbyshire, M. K., Gonzales, N. R., Lu, S., Chitsaz, F., Geer, L. Y., et al. (2015). CDD: NCBI's conserved domain database. *Nucleic Acids Res.* 43, D222–D226. doi: 10.1093/nar/gku1221

Metzger, E., Wissmann, M., Yin, N., Müller, J. M., Schneider, R., Peters, A. H., et al. (2005). LSD1 demethylates repressive histone marks to promote androgen-receptor-dependent transcription. *Nature* 437, 436–439. doi: 10.1038/nature04020

Michael, T. P., Mockler, T. C., Breton, G., McEntee, C., Byer, A., Trout, J. D., et al. (2008). Network discovery pipeline elucidates conserved time-of-day-specific cis-regulatory modules. *PLoS Genet.* 4:e14. doi: 10.1371/journal.pgen.0040014

Mockler, T. C., Michael, T. P., Priest, H. D., Shen, R., Sullivan, C. M., Givan, S. A., et al. (2007). The DIURNAL project: DIURNAL and circadian expression profiling, model-based pattern matching, and promoter analysis. *Cold Spring Harb. Symp. Quant. Biol.* 72, 353–363. doi: 10.1101/sqb.2007.72.006

Ning, Y.-Q., Ma, Z.-Y., Huang, H.-W., Mo, H., Zhao, T.-T., Li, L., et al. (2015). Two novel NAC transcription factors regulate gene expression and flowering time by associating with the histone demethylase JMJ14. *Nucleic Acids Res.* 43, 1469–1484. doi: 10.1093/nar/gku1382

Noh, B., Lee, S.-H., Kim, H.-J., Yi, G., Shin, E.-A., Lee, M., et al. (2004). Divergent roles of a pair of homologous jumonji/zinc-finger-class transcription factor proteins in the regulation of Arabidopsis flowering time. *Plant Cell* 16, 2601–2613. doi: 10.1105/tpc.104.025353

Qian, S., Wang, Y., Ma, H., and Zhang, L. (2015). Expansion and functional divergence of Jumonji C-containing histone demethylases: significance of duplications in ancestral angiosperms and vertebrates. *Plant Physiol.* 168, 1321–1337. doi: 10.1104/pp.15.00520

Qin, C., Cheng, L. M., Shen, J., Zhang, Y., Cao, H., Lu, D., et al. (2016). Genome-wide identification and expression analysis of the 14-3-3 family genes in *Medicago truncatula. Front. Plant Sci.* 7:320. doi: 10.3389/fpls.2016.00320

Quan, Z., Oliver, S. G., and Zhang, N. (2011). JmjN interacts with JmjC to ensure selective proteolysis of Gis1 by the proteasome. *Microbiology* 157, 2694–2701. doi: 10.1099/mic.0.048199-0

Schlueter, J. A., Dixon, P., Granger, C., Grant, D., Clark, L., Doyle, J. J., et al. (2004). Mining EST databases to resolve evolutionary events in major crop species. *Genome* 47, 868–876. doi: 10.1139/g04-047

Schmutz, J., Cannon, S. B., Schlueter, J., Ma, J., Mitros, T., Nelson, W., et al. (2010). Genome sequence of the palaeopolyploid soybean. *Nature* 463, 178–183. doi: 10.1038/nature08670

Severin, A. J., Cannon, S. B., Graham, M. M., Grant, D., and Shoemaker, R. C. (2011). Changes in twelve homoeologous genomic regions in soybean following three rounds of polyploidy. *Plant Cell* 23, 3129–3136. doi: 10.1105/tpc.111.089573

Shen, Y., Silva, N., Audonnet, L., Servet, C., Wei, W., and Zhou, D.-X. (2014a). Over-expression of histone H3K4 demethylase gene JMJ15 enhances salt tolerance in Arabidopsis. *Front. Plant Sci.* 5:290. doi: 10.3389/fpls.2014.00290

Shen, Y., Wu, X., Liu, D., Song, S., Liu, D., and Wang, H. (2016). Cold-dependent alternative splicing of a Jumonji C domain-containing gene MtJMJC5 in Medicago truncatula. *Biochem. Biophys. Res. Commun.* 474, 271–276. doi: 10.1016/j.bbrc.2016.04.062

Shen, Y., Zhou, Z., Wang, Z., Li, W., Fang, C., Wu, M., et al. (2014b). Global dissection of alternative splicing in paleopolyploid soybean. *Plant Cell* 26, 996–1008. doi: 10.1105/tpc.114.122739

Shu, Y., Liu, Y., Zhang, J., Song, L., and Guo, C. (2016). Genome-wide analysis of the AP2/ERF superfamily genes and their responses to abiotic stress in *Medicago truncatula. Front. Plant Sci.* 6:1247. doi: 10.3389/fpls.2015.01247

Singh, V. K., and Jain, M. (2015). Genome-wide survey and comprehensive expression profiling of Aux/IAA gene family in chickpea and soybean. *Front. Plant Sci.* 6:918. doi: 10.3389/fpls.2015.00918

Sun, Q., and Zhou, D.-X. (2008). Rice jmjC domain-containing gene JMJ706 encodes H3K9 demethylase required for floral organ development. *Proc. Natl. Acad. Sci. U.S.A.* 105, 13679–13684. doi: 10.1073/pnas.0805901105

Tamura, K., Stecher, G., Peterson, D., Filipski, A., and Kumar, S. (2013). MEGA6: molecular evolutionary genetics analysis version 6.0. *Mol. Biol. Evol.* 30, 2725–2729. doi: 10.1093/molbev/mst197

Trewick, S. C., McLaughlin, P. J., and Allshire, R. C. (2005). Methylation: lost in hydroxylation? *EMBO Rep.* 6, 315–320. doi: 10.1038/sj.embor.7400379

Vanneste, K., Baele, G., Maere, S., and Van de Peer, Y. (2014). Analysis of 41 plant genomes supports a wave of successful genome duplications in association with the Cretaceous-Paleogene boundary. *Genome Res.* 24, 1334–1347. doi: 10.1101/gr.168997.113

Wang, X., Duan, C.-G., Tang, K., Wang, B., Zhang, H., Lei, M., et al. (2013). RNA-binding protein regulates plant DNA methylation by controlling mRNA processing at the intronic heterochromatin-containing gene IBM1. *Proc. Natl. Acad. Sci. U.S.A.* 110, 15467–15472. doi: 10.1073/pnas.1315399110

Wang, Y., Tang, H., Debarry, J. D., Tan, X., Li, J., Wang, X., et al. (2012). MCScanX: a toolkit for detection and evolutionary analysis of gene synteny and collinearity. *Nucleic Acids Res.* 40, e49. doi: 10.1093/nar/gkr1293

Yang, H., Mo, H., Fan, D., Cao, Y., Cui, S., and Ma, L. (2012). Overexpression of a histone H3K4 demethylase, JMJ15, accelerates flowering time in Arabidopsis. *Plant Cell Rep.* 31, 1297–1308. doi: 10.1007/s00299-012-1249-5

Yang, W., Jiang, D., Jiang, J., and He, Y. (2010). A plant-specific histone H3 lysine 4 demethylase represses the floral transition in Arabidopsis. *Plant J.* 62, 663–673. doi: 10.1111/j.1365-313X.2010.04182.x

Yang, Z. (2007). PAML 4: phylogenetic analysis by maximum likelihood. *Mol. Biol. Evol.* 24, 1586–1591. doi: 10.1093/molbev/msm088

Yu, X., Li, L., Li, L., Guo, M., Chory, J., and Yin, Y. (2008). Modulation of brassinosteroid-regulated gene expression by Jumonji domain-containing proteins ELF6 and REF6 in Arabidopsis. *Proc. Natl. Acad. Sci. U.S.A.* 105, 7618–7623. doi: 10.1073/pnas.0802254105

Zhang, L., and Ma, H. (2012). Complex evolutionary history and diverse domain organization of SET proteins suggest divergent regulatory interactions. *New Phytol.* 195, 248–263. doi: 10.1111/j.1469-8137.2012.04143.x

Zhao, W., Shafiq, S., Berr, A., and Shen, W. H. (2015). Genome-wide gene expression profiling to investigate molecular phenotypes of Arabidopsis mutants deprived in distinct histone methyltransferases and demethylases. *Genom. Data* 4, 143–145. doi: 10.1016/j.gdata.2015.04.006

Zhou, F., Guo, Y., and Qiu, L. J. (2016). Genome-wide identification and evolutionary analysis of leucine-rich repeat receptor-like protein kinase genes in soybean. *BMC Plant Biol.* 16:58. doi: 10.1186/s12870-016-0744-1

Zong, W., Zhong, X., You, J., and Xiong, L. (2013). Genome-wide profiling of histone H3K4-tri-methylation and gene expression in rice under drought stress. *Plant Mol. Biol.* 81, 175–188. doi: 10.1007/s11103-012-9990-2

Conflict of Interest Statement: The authors declare that the research was conducted in the absence of any commercial or financial relationships that could be construed as a potential conflict of interest.

13

Evolution of Daily Gene Co-expression Patterns from Algae to Plants

Pedro de los Reyes[1]*, Francisco J. Romero-Campero*[1,2]*, M. Teresa Ruiz*[1]*, José M. Romero*[1] *and Federico Valverde*[1*]

[1] *Plant Development Unit, Institute for Plant Biochemistry and Photosynthesis, Consejo Superior de Investigaciones Científicas, Universidad de Sevilla, Seville, Spain,* [2] *Department of Computer Science and Artificial Intelligence, Universidad de Sevilla, Seville, Spain*

Edited by:
Verónica S. Di Stilio, University of Washington, United States

Reviewed by:
Sven B. Gould, Heinrich Heine University, Germany
Andrew Millar, University of Edinburgh, United Kingdom

***Correspondence:**
Federico Valverde
federico.valverde@ibvf.csic.es

Daily rhythms play a key role in transcriptome regulation in plants and microalgae orchestrating responses that, among other processes, anticipate light transitions that are essential for their metabolism and development. The recent accumulation of genome-wide transcriptomic data generated under alternating light:dark periods from plants and microalgae has made possible integrative and comparative analysis that could contribute to shed light on the evolution of daily rhythms in the green lineage. In this work, RNA-seq and microarray data generated over 24 h periods in different light regimes from the eudicot *Arabidopsis thaliana* and the microalgae *Chlamydomonas reinhardtii* and *Ostreococcus tauri* have been integrated and analyzed using gene co-expression networks. This analysis revealed a reduction in the size of the daily rhythmic transcriptome from around 90% in *Ostreococcus*, being heavily influenced by light transitions, to around 40% in *Arabidopsis*, where a certain independence from light transitions can be observed. A novel Multiple Bidirectional Best Hit (MBBH) algorithm was applied to associate single genes with a family of potential orthologues from evolutionary distant species. Gene duplication, amplification and divergence of rhythmic expression profiles seems to have played a central role in the evolution of gene families in the green lineage such as *Pseudo Response Regulators* (*PRRs*), *CONSTANS-Likes* (*COLs*), and *DNA-binding with One Finger* (*DOFs*). Gene clustering and functional enrichment have been used to identify groups of genes with similar rhythmic gene expression patterns. The comparison of gene clusters between species based on potential orthologous relationships has unveiled a low to moderate level of conservation of daily rhythmic expression patterns. However, a strikingly high conservation was found for the gene clusters exhibiting their highest and/or lowest expression value during the light transitions.

Keywords: daily rhythmic genes, evolution, co-expression networks, systems biology, circadian, *Arabidopsis*, *Chlamydomonas*, *Ostreococcus*

INTRODUCTION

The evolution of the green lineage of photosynthetic organisms from unicellular green algae to land plants is subject to intense research. Recently, genomic analysis have been applied to study key events in the evolution of the green lineage such as terrestrialization (Delwiche and Cooper, 2015; de Vries et al., 2016). Nevertheless, the integration of genomics and transcriptomics to study the evolution of gene expression patterns in the green lineage has only recently been explored (Romero-Campero et al., 2013; Ruprecht et al., 2017). Photosynthetic organisms are particularly influenced by daily light/dark transitions as their main energy income is a very demanding light-dependent process. Therefore, those plant ancestors (whose extant representative species are unicellular green algae) that were able to anticipate daily fluctuations and schedule their physiological processes accordingly, evolved into the present plant species (Mora-García et al., 2017). Transcriptomics have become a powerful tool to study the global influence of daily light/dark cycles in photosynthetic organisms. Thus, in higher plants, it has been described that between a third and a half of their gene expression is regulated by the circadian clock (Covington et al., 2008; Michael et al., 2008), whereas in algae between 80 and 90% of the transcriptome follows light-dependent daily rhythmic patterns (Monnier et al., 2010; Zones et al., 2015). It seems then that daily rhythmic regulation of gene expression reached a maximum in algae and has substantially decreased in Spermatophytes. Key biological processes such as starch and sugar metabolism exhibit daily rhythmic patterns in both plants and algae (Bläsing et al., 2005; Sorokina et al., 2011), but these analysis remain fragmented and focused on individual species. In order to better understand the evolution of daily rhythmic gene expression between algae and plants, confirm the tendency to decrease daily regulation and determine the evolution of daily regulated biological processes from algae to plants, a series of comparative Systems Biology analysis integrating genomics and transcriptomics has been used in this study.

The availability of massive amounts of transcriptomic data obtained from different species under equivalent environmental conditions has enabled the use of comparative transcriptomics methodologies to study the evolution of key biological processes (Trachana et al., 2010). In this work, database available RNA-seq and microarray data generated over 24 h periods in neutral days (ND: 12 h of light/12 h of dark) and long days (LD: 16 h of light/8 h of dark) conditions from three different model photosynthetic species (*Arabidopsis thaliana, Chlamydomonas reinhardtii,* and *Ostreococcus tauri*) have been integrated and analyzed. These photosynthetic eukaryotes represent distant phylogenetic groups. Among the *Chlorophyta* algae division, the marine Prasinophyceae microalga *Ostreococcus* is considered a representative of ancient microalgae, one of the smallest eukaryote (picoeukaryote) and constitutes an important part of sea phytoplankton (Derelle et al., 2006; Palenik et al., 2007). Due to its small genome (13 Mb), whose sequencing has recently been improved (Blanc-Mathieu et al., 2014), its non-flagellar small body (0.8 μm), its single copy organella (Henderson et al., 2007) and its planktonic life style, it has been considered as the smallest possible photosynthetic eukaryote (Raven et al., 2013) and promising organism for Systems Biology (Thommen et al., 2015). *Chlamydomonas* is a Chlorophyceae microalga and has been used as a model for photosynthetic organisms' studies for many years (Harris, 2001). *Chlamydomonas* has two polar flagella (10 μm body), a much bigger genome than *Ostreococcus* (120 Mb; Merchant et al., 2007) and lives in sweet water environments. Omics are currently intensely used to explore *Chlamydomonas* potential in biotechnological applications (Aucoin et al., 2016). As a representative Spermatophyte or seed plant, *Arabidopsis* has several characteristics that make it especially useful for Systems Biology studies in general and circadian experiments in particular, as well as a bigger genome and a complex physiology in comparison to microalgae (Van Norman and Benfey, 2009; Koornneef and Meinke, 2010).

Arabidopsis has been extensively used as a model to describe the basic aspects of plant circadian clock regulation over daily rhythms (Millar, 2016; Nohales and Kay, 2016). In a few words, it is formed by three interlocked positive/negative feedback loops. CIRCADIAN CLOCK-ASSOCIATED 1 (CCA1) and LATE ELONGATED HYPOCOTYL (LHY) myb transcription factors constitute the positive/negative "morning loop" together with PRR9, PRR7 and PRR5 (Nakamichi et al., 2012; Liu et al., 2013, 2016; Kamioka et al., 2016). The negative "central loop" is constituted by CCA1/LHY repression over the gene *TIME OF CAB EXPRESSION 1* (*TOC1*) that, in time, represses the "morning loop" genes and activates evening-regulated genes in a 24 h-long feed-back loop (Huang et al., 2012). Over this basic core, the circadian genes: *GIGANTEA* (*GI*), *ZEITLUPE* (*ZTL*), *EARLY LIGHT FLOWERING 3* (*ELF3*), *ELF4*, *LUX ARRHYTMO* (*LUX*) among others, constitute the "evening loop" that refines the clock and allows for the complex response to the changing daily conditions, including light and temperature inputs (Miyazaki et al., 2015). Outputs of this clock are cell wall synthesis, photosynthetic and starch metabolism genes, among many others (Adams and Carré, 2011). Taking *Arabidopsis* as model, a much simpler clock has been described in *Chlamydomonas* and *Ostreococcus*, where only some genes of the higher plant circadian clock have been identified so far (Mittag et al., 2005; Corellou et al., 2009). In this way, most evolutionary studies have focused on phylogenetic analyses of the key genes regulating the circadian clock (Serrano-Bueno et al., 2017) whereas the analysis of global rhythmic patterns conservation and evolution among different photosynthetic species still remains to be explored.

In this work, phylogenomic and transcriptomic data integration and analysis have been performed by gene co-expression networks construction (Romero-Campero et al., 2013; Gehan et al., 2015; Ruprecht et al., 2017) and a novel algorithm for the identification of potential orthologues called Multiple Best Bi-directional Hit (MBBH). Using clustering techniques, specific gene clusters or modules that showed a rhythmic daily regulation have been identified. Interestingly, these clusters consist of groups of highly co-expressed genes involved in particular biological processes such as cell cycle progression, photosynthesis and ribogenesis, revealing a significant temporal organization in their specific gene co-expression patterns. By comparing the gene modules identified in the different gene co-expression networks obtained for each species, it was possible to determine which biological processes have conserved a daily rhythmic co-expression pattern over the green lineage and which ones have evolved into different patterns. Additionally, a web based software tool, CircadiaNET (http://viridiplantae.ibvf.csic.es/circadiaNet/) has been developed that will allow researchers to independently analyze their circadian genes of interest studying the biological processes they are potentially involved in, the conservation or evolution of the gene co-expression patterns they follow, as well as the transcription factor (TF) binding sites that are significantly present in their promoters.

MATERIALS AND METHODS

Identification of Putative Orthologous Proteins

The protein sequences of the three photosynthetic species analyzed in this study were downloaded from publicly available databases. *Chlamydomonas* v5.5 and *Arabidopsis* TAIR10 proteins were downloaded from Phytozome (http://www.phytozome.net/) (Goodstein et al., 2012), while *Ostreococcus* proteins v2 were downloaded from ORCAE (http://bioinformatics.psb.ugent.be/orcae/) (Sterck et al., 2012; Blanc-Mathieu et al., 2014). Using the tools available from the Pfam database (http://pfam.xfam.org/) (Punta et al., 2012), with default parameters (*E*-value 1.0), the protein domains in all protein sequences were identified. Potential homologous proteins between species were identified based on sequence similarity. We developed a variant of the Bidirectional Best Hit (BBH) that takes into account several candidates. We called this variant Multiple Bidirectional Best Hit (MBBH) and happen to be specially suited for duplication-enriched species such as *Arabidopsis*. The algorithm receives as input two fasta files containing the protein sequences of the two species to compare and the number *N* of multiple bidirectional best hits to consider (**Figure 1**). In this study *N* was fixed to 20. For each protein $Prot_i^1$ (encoded by $Gene_i^1$, deep green) from the first species, the *N* proteins (encoded by $Gene_{i,k}^2$, red) from the second species, $Prot_{i,1}^2$, $Prot_{i,2}^2$,...,$Prot_{i,N}^2$, exhibiting the highest sequence similarity are selected using the Needleman-Wunsch algorithm implemented in the R Bioconductor package Biostrings (Pagès et al., 2016). These proteins will be called "initial best hits." The same process is carried out for species 2: For each protein $Prot_i^2$ (encoded by $Gene_{i,k}^2$, red) from the second species, the *N* proteins from the first species, $Prot_h^1$, $Prot_i^1$,...,$Prot_N^1$, exhibiting the highest sequence similarity are selected. Next, for each $Prot_i^1$ from the first species its initial best hits, $Prot_{i,1}^2$, $Prot_{i,2}^2$,...,$Prot_{i,N}^2$, are filtered. Only those $Prot_{i,k}^2$ that present $Prot_i^1$ in their *N* initial best hits are kept. The same process is carried out for each protein $Prot_j^2$ from species 2 keeping only the initial best hits that present $Prot_j^2$ among their *N* initial best hits. An additional filtering process is performed by removing from the best bidirectional hits those that do not share at least one domain with the corresponding protein. Finally, we also assign putative orthology to proteins showing no MBBH target, but sharing exactly identical number and order of one or more, non-overlapping pfam domains. This means that due to the specificity of pfam domains, the presence of a single domain does not immediately imply the assignation of a putative orthologue as would be the case with specific higher plant domains such as NAM (**Figure 3C**) or specific microalgal domains. In this way, our approach combines global sequence similarity with

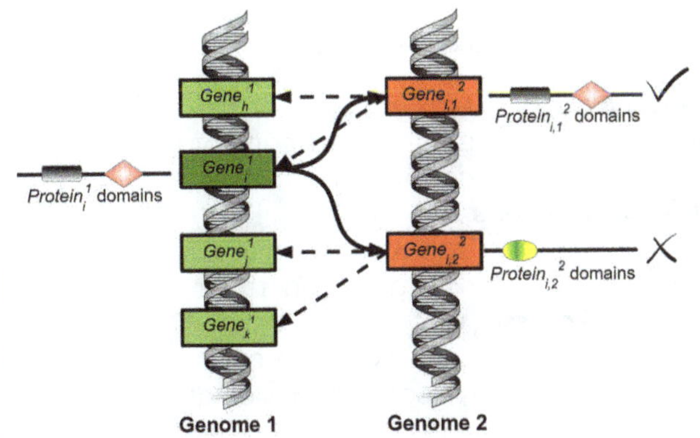

FIGURE 1 | Explicative diagram of the Multiple Bidirectional Best Hits tool. For the *Protein*$_i^1$ encoded by the *Gene*$_i^1$ from the first species (deep green), the algorithm selects the genes (red) encoding the *N* proteins that exhibit the highest sequence similarity (*protein*$^2_{i,1}$, *protein*$^2_{i,2}$,...*protein*$^2_{i,n}$). In this study, *N* was set to 20. These 20 proteins are called initial best hits. The same process is carried out for species 2. For each initial hit from species 2, the *N* genes (red) coding for proteins showing the highest sequence similarity are selected. Then, the initial best hits of the genes (green) coding for *protein*$_i^1$ from species 1 are filtered. Only those genes coding for proteins that exhibit *protein*$_i^1$ in their initial best hits are kept (marked by a √ symbol). Thus, bidirectionality is required. Additionally, the best bidirectional hits that do not share at least one domain with the corresponding protein are removed (marked by an X symbol).

protein domain structure and, additionally, allows for multiple best hits in order to capture the evolution of a single gene into a multi-gene family. In general, MBBH could be a useful tool for those researchers trying to identify orthologues from any given gene in different species, even when they belong to distantly related ones. The tools developed in this article will be operative in the web site: (http://viridiplantae.ibvf.csic.es/circadiaNet/).

Transcriptomic Data Acquisition and Processing

For the three photosynthetic species analyzed in this study, transcriptomic data comprising time series of 48 or 72 h were collected from the GEO database (https://www.ncbi.nlm.nih.gov/geo/) (Barrett et al., 2013). In the case of *Arabidopsis*, two different data sets were used. The first one identified with the accession number GSE3416 (Bläsing et al., 2005) consists of microarray data taken in ND conditions (12:12). The second one identified with the accession number GSE43865 (Rugnone et al., 2013) consists of RNA-seq data taken in LD conditions (16:8). For *Chlamydomonas*, a single RNA-seq data set was collected identified with the accession number GSE71469 (Zones et al., 2015) taken in ND conditions. For *Ostreococcus*, a single microarray data set was used, identified with the accession number GSE16422 (Monnier et al., 2010) taken in ND conditions. The microarray data were processed using the Robust Multi-array Average (RMA) implemented in the Bioconductor R package affy (Gautier et al., 2004). The RNA-seq data was processed following the Tuxedo protocol (Trapnell et al., 2012). Reads were mapped to the reference genomes downloaded from Phytozome using Tophat. Transcripts were

TABLE 1 | Primers employed in QPCR experiments.

	Forward sequence	Reverse sequence	Amplified fragment size (bp)
***Arabidopsis* GENES**			
UBQ10	5′-GAAGTTCAATGTTTCGTTTCATGT-3′	5′-GGATTATACAAGGCCCCAAAA-3′	145
HY5	5′-GCTGAAGAGGTTGTTGAGGA-3′	5′-TCTCCAAGTCTTTCACTCTGTTT-3′	101
***Chlamydomonas* GENES**			
CrTUB	5′-GTTGCATCGTTAGCGTGGACG-3′	5′-GCAGCAGCCAATGTTCAGACT-3′	170
CrHY5	5′-GGCATGAACCCTAGCTTCC-3′	5′-CATCATGTCCTCGCTGATGT-3′	144
***Ostreococcus* GENES**			
OtEF1α	5′-GACGCGACGGTGGATCAA-3′	5′-CGACTGCCATCGTTTTACC-3′	202
OtHY5	5′-GAGGGAAGATTGGGAAGAAGAA-3′	5′-CCTTCTCCAACGCCTGAAT-3′	81

assembled using Cufflinks. Default parameters were used for the different software tools. Gene expression levels were measured as FPKM (Fragments Per Kilobase of exon per Million fragments mapped). Since RNA-seq and microarray gene expression levels have different measurement units these data were normalized by subtracting the mean and dividing by the standard deviation using the R function scale from the base package. This allows representing both types of data on a common scale. Samples were collected every 3 or 4 h in the data analyzed in this study. In order to produce a smooth representation of the gene expression profiles in heatmap graphs these data were linearly interpolated to produce time series consisting of 24 h using the R function approx from the stats package. In heatmaps, genes were sorted according to their expression profile similarity using hierarchical clustering.

Identification and Clustering of Rhythmic Daily Gene Expression Patterns

The detection of significant periodic patterns in the transcriptomic data analyzed was performed using RAIN (Rhythmicity Analysis Incorporating Nonparametric methods). This Bioconductor R package consists of functions that implement robust nonparametric methods for the detection of rhythms with arbitrary wave forms and pre-specified periods (Thaben and Westermark, 2014). The main function *rain* was used with the following parameters: A numeric matrix comprising the gene expression levels from the different biological replicates of the 24 h time series; a sampling interval of 4 h for *Arabidopsis* and 3 h for *Chlamydomonas* and *Ostreococcus*; a period of 24 h; and a number of three replicates for *Arabidopsis* and *Ostreococcus* and two for *Chlamydomonas*. A significance level $\alpha = 1\%$ was chosen to assume that a certain gene exhibits a rhythmic daily expression pattern. The wave form of a gene exhibiting rhythmic daily expression was characterized using its peak (time point where the highest expression value is reached) and its trough (time point where the lowest expression value is reached). Daily rhythmic genes were thus classified into 16 different clusters according to their peaks and troughs at a particular time of the day, that is, all genes in the same cluster present their peaks and troughs at the same time point.

Gene Ontology Term and Pathway Enrichment Analysis

Gene Ontology (GO) terms associated to each *Arabidopsis* and *Chlamydomonas* gene were downloaded from Phytozome. For *Ostreococcus*, GO terms were downloaded from ORCAE. The Bioconductor R package topGO (Alexa and Rahnenfuhrer, 2016) was used to determine GO terms significantly enriched in different gene sets. The entire genome of the corresponding species was used as gene background. The statistical significance test selected was Fisher's exact test with a significance level $\alpha = 5\%$. The web based tool REVIGO (Supek et al., 2011) was used to remove redundancy from the enriched GO terms and produced a summary. The identification of enriched pathways in *Arabidopsis* gene sets was performed with the Bioconductor R package clusterProfiler (Yu et al., 2012). The enrichKEGG function was used with Bejamin-Hochberg as *p*-value correction method for the multiple testing and a *q*-value cutoff of 0.05.

Gene Co-expression Network Construction, Visualization, and Analysis

Gene co-expression was measured using the correlation between gene expression profiles over the 24 h time series. Two daily rhythmic genes were assumed to be co-expressed when the Pearson correlation coefficient between their expression profiles was greater than 0.95. A gene co-expression network was constructed for each species where the nodes represent daily rhythmic genes and an edge is drawn between nodes when the corresponding genes are co-expressed according to the previous criterion. Cytoscape, a software tool for the representation and analysis of complex networks (Shannon et al., 2003), was used to visualize the gene co-expression networks applying the Prefuse Force-Directed Layout. The analysis of the networks was performed using the R package igraph (Csárdi and Nepusz, 2006). The scale-free property was tested using linear regression over the logarithmic transform of the degree distribution. The small-world property was tested by generating 10^4 random scale-free networks with the same number of nodes and edges as the corresponding network, using the *barabasi.game* function from the igraph package.

Module Conservation Analysis

Two daily rhythmic genes from two different species determined as potential orthologues using MBBH were defined as exhibiting a conserved daily pattern when both belonged to the same cluster (they presented their peak and trough in the same time interval) or when the Pearson correlation coefficient between their expression profiles were higher than 0.98. The conservation among the co-expression patterns of two different sets of genes from two different species were computed according to the *summary composite conservation statistic* "Zsummary" as defined in Langfelder et al. (2011). A Zsummary value lower than 2 indicates no conservation, a Zsummary value 2–10 implies moderate conservation, while a Zsummary greater than 10 constitutes evidence of a great level of conservation. The R package WGCNA (Langfelder and Horvath, 2012) and the function implemented therein were used.

Plant, Algal Material, and Growth Conditions

Three independent biological replicates for plants and algae were grown in a model SG-1400 phytotron (Radiber SA, Spain) under LD conditions with temperatures ranging from 22°C (day) to 18°C (night) and 75 $\mu Em^{-2}s^{-1}$ light intensity. *Arabidopsis thaliana* Col-0 wild type seeds were incubated 4 days at 4°C in the dark before sowing in MS plates. 12-day-old seedlings were collected every 4 h during a 24 h period. Time points were denoted as Zeitgeber Time *N* (ZTN) indicating the time point *N* hours after the lights are switched on in the phytotron (ZT0), mimicking dawn. *Chlamydomonas reinhardtii* wild type CW15 was grown in flasks with minimal Sueoka medium for 12 days. Similarly, *Ostreococcus tauri* wild type RCC 745 was grown in flasks with Keller medium (Keller et al., 1987) for 12 days. Algae were harvested every 4 h during a 24 h period.

RNA Isolation and QPCR

RNA was isolated from *Arabidopsis* seedlings (0.1 g leaf tissue), *Chlamydomonas* and *Ostreococcus* (pellet of 25 ml culture at exponential phase) using a modified TRIZOL (Invitrogen) protocol as described by the manufacturer. Briefly, the sample was mixed with 1 ml of TRIZOL and 0.2 ml of chloroform and the mixture was centrifuged at 16,000 g for 10 min at 4°C. The supernatant was treated with 1 volume of 100% (v/v) 2-propanol, incubated 15 min at room temperature and centrifuged at 16,000 g for 10 min at 4°C. The pellet was dissolved in 0.75 ml 3 M LiCl, incubated for $t > 10$ min at room temperature and centrifuged at 16,000 g for 10 min at 4°C. This pellet was washed with 80% (v/v) ethanol and centrifuged at 16,000 g for 10 min at 4°C. The final RNA sample was suspended in 21 μl of DEPC treated water and 1 μl quantified employing a ND-1000 Spectrophotometer (Nanodrop).

One micro gram of TRIZOL-isolated RNA was used to synthesize cDNA employing the Quantitec® Reverse kit (Qiagen) following the instructions recommended by the manufacturer. cDNA samples were diluted to a final concentration of 10 ng/μL and stored at −20°C until QPCR was performed. Primers to amplify the 3′ translated region of *AtHY5*, *CrHY5*, *OtHY5*, including *AtUBQ10*, *CrTUB*, and *OtEF1α* as housekeeping genes (**Table 1**) were designed using the Oligo analyzer program from Integrated DNA Technologies (http://eu.idtdna.com/analyzer/Applications/OligoAnalyzer/). QPCR was performed in a Multicolor Real-Time PCR Detection System iQTM5 (Bio-Rad) in a 10 μL reaction: primer concentration 0.2 μM, 10 ng cDNA and 5 μL SensiFAST TM SYBR & Fluorescein Kit (Bioline). The QPCR program consisted in (i) 1 cycle (95°C, 2 min); (ii) 40 cycle (95°C, 5 s; 60°C, 10 s and 72°C, 6 s) (iii) 1 cycle (72°C, 6 s). Fluorescence was measured at the end of each extension step and the melting curve was performed between 55 and 95°C. Three biological replicates with three technical replicates from each species were used for every time point. The QPCR results were estimated using the ddCt R Bioconductor package (Zhang et al., 2015).

RESULTS AND DISCUSSIONS

Most *Ostreococcus* Proteins Present Potential Orthologues in *Arabidopsis* and *Chlamydomonas*, but This Is Not the Case with the Other Two Species

The identification of potential orthologues between two species constitutes one of the most important bottlenecks in comparative genomics and transcriptomics (Dessimoz et al., 2012). The Bidirectional Best Hit (BBH) algorithm is the most commonly used method for the automated identification of putative orthologues. In spite of its simplicity, BBH is highly accurate when dealing with bacterial and archaeal genomes (Wolf and Koonin, 2012). Nevertheless, BBH does not perform optimally, missing as much as 60% of orthologues, in gene-duplication-enriched genomes such as those of plants and animals (Dalquen and Dessimoz, 2013). In this work, in order to identify orthology among distantly related species, we have developed a variant of the BBH algorithm termed Multiple Bidirectional Best Hit (MBBH). MBBH assigns to each gene g_i^1 from the first species $k \leq N$ potential orthologues from the second species, $g_1^2,...,g_k^2$, based on sequence similarity of the proteins they encode if, and only if, the query gene g_i^1 is among the *N* most similar genes from the first species (Section Materials and Methods, **Figure 1**).

Using MBBH tool, more than 97% of *Ostreococcus* proteins could be ascribed to a potential orthologue either in *Arabidopsis* or *Chlamydomonas* (Supplementary Table 1). This is in agreement with the reduced and compact *Ostreococcus* genome (Palenik et al., 2007) and suggests, as has been proposed (Raven et al., 2013), that this marine picoeukaryote contains the minimal genome for a functional photosynthetic eukaryote and that it is almost entirely regulated by the circadian clock (see below). Confirming this idea, a GO term enrichment analysis over the set of genes without potential *Arabidopsis* or *Chlamydomonas* orthologues did not produce any significant result, indicating that these genes are not involved in any specific biological process. In fact, the highest number of genes without potential orthologues was located on chromosomes 2 and 19. Some of these genes codified for a number of unnamed products (ostta02g03690, ostta02g05500, ostta02g00800) and a group of Tc1-like/mariner transposases (ostta02g01245, ostta02g01247, ostta02g02355).

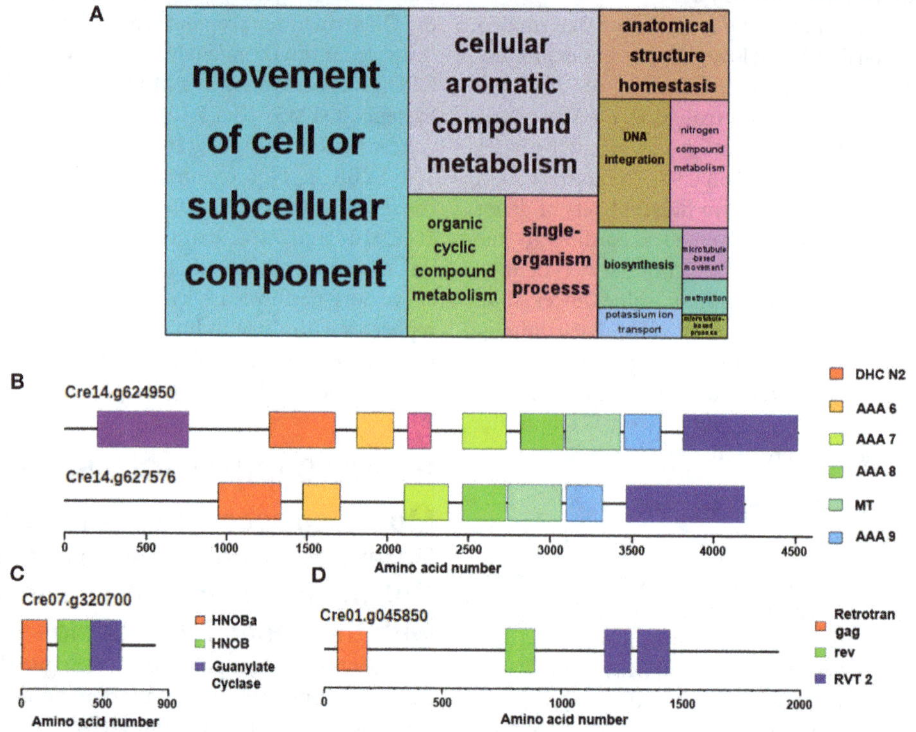

FIGURE 2 | Functional annotation of genes exclusively identified in *Chlamydomonas*. No potential *Ostreococcus* or *Arabidopsis* orthologues were identified for 15% of *Chlamydomonas* proteins. **(A)** Non-redundant GO term enrichment analysis over the set of *Chlamydomonas* genes with no potential orthologue in the other species, suggesting that they are mainly involved in "*movement of cell or subcellular component*," "*organic cyclic compound metabolism*," and "*DNA integration*". Each rectangle area in the treemap represents the −log10 (*p*-value) for the corresponding GO-term. **(B)** Domain structure of the proteins encoded by *Flagellar inner arm dynein 1* (Cre14.g624950) and *axonemal dynein heavy chain 6* (Cre14.g627576), two *Chlamydomonas*-specific flagellar proteins absent in *Ostreococcus* and *Arabidopsis* involved in "*movement of cell*" or "*subcellular component.*" **(C)** Domain structure of class III guanylyl and adenylyl cyclase encoded by *CYG11* (Cre07.g320700), a *Chlamydomonas*-specific protein involved in "*organic cyclic compound metabolism*." This protein shows a higher homology to animal cyclases than to plant ones. **(D)** Domain structure of the protein encoded by *gag-pol-related retrotransposon* (Cre01.g045850), an example of *Chlamydomonas* specific proteins associated to the "*DNA integration*" GO term. Color boxes represent domains identified in pfam database including their identification codes.

This supports previous analysis on the heterogeneity of *Ostreococcus* genome that identified chromosomes 2 and 19 as different from the other chromosomes in terms of GC content, codon usage and number of transposable elements (Derelle et al., 2006).

Following the same comparative analysis, approximately 85% of *Chlamydomonas* proteins present potential orthologues in *Ostreococcus* and *Arabidopsis* (Supplementary Table 2). Therefore, a set of approximately 2,600 *Chlamydomonas* genes that codify for specific proteins could not be traced to any orthologue in the other two species. An ontology enrichment analysis performed over this set of genes using the R Bioconductor package topGO and summarized with REVIGO revealed that the top three most significant and non-redundant GO terms were "*Movement of cell or subcellular component*," "*Organic cyclic compound metabolism*," and "*DNA integration*" (**Figure 2A**). The same set of genes were associated with the GO terms "*Movement of cell or subcellular component*," "*Anatomical structure homeostasis*" and "*single-organism process.*" Therefore, the last two GO terms were considered redundant and only "*Movement of cell or subcellular component*" is discussed. Similarly, the GO terms "*Organic cyclic compound metabolism*" and "*cellular aromatic compound metabolism*" included the same list of associated genes than "*Organic cyclic compound metabolism*," so only the latter is discussed. Genes codifying flagellar proteins such as *flagellar inner arm dynein 1* (Cre14.g624950) and *axonemal dynein heavy chain 6* (Cre14g.627576) (**Figure 2B**) are representative of the group "Movement of cell or subcellular component." *Chlamydomonas* cells exhibit two polar flagella whereas no flagellar-like structures are present in *Ostreococcus* or *Arabidopsis* cells, explaining the lack of orthologues in the non-flagellated organisms. Besides, the *Chlamydomonas* specific genes annotated with the GO term "*Organic cyclic compound metabolism*" include the class III guanylyl and adenylyl cyclase family including genes such as *adelynate/guanylate cyclase CYG11* (Cre07.g320700) (**Figure 2C**). These enzymes that catalyze the synthesis of cGMP and cAMP in animals are absent in plants, hence the lack of potential orthologues in *Ostreococcus* and *Arabidopsis*. In fact, this concurs with the idea that some *Chlamydomonas* genes are closer to animal than to plant ones (Merchant et al., 2007). On the other hand, the unique *gag-pol-related* *retrotransposon* (Cre01g.045850) (**Figure 2D**) is an example of the *Chlamydomonas* specific genes associated with *"DNA integration"* GO term, suggesting that this set of viral pathogens infecting *Chlamydomonas* is specific for the alga.

The analysis also identified potential *Ostreococcus* and *Chlamydomonas* orthologues for approximately 75% of *Arabidopsis* proteins (Supplementary Table 3), suggesting that one fourth of the *Arabidopsis* proteins are not present in microalgae and have potentially been acquired during the course of higher plant evolution. In accordance with this idea, a GO term enrichment analysis showed that these *Arabidopsis*-specific proteins are involved in the following top three most significant non-redundant terms: *"Cell communication," "Cell wall organization,"* and *"Multiorganism reproductive process"* (**Figure 3A**). Similar to the previous result, the GO term "Signaling" was found to share the same gene list with "Cell communication"; "Cellular glucan metabolism" presented the same associated genes as "Cell wall organization" and the set of genes assigned to "Biological regulation" and "Multiorganism reproductive process" were identical. Therefore, the former GO terms were considered redundant and only the latter GO terms are discussed. These GO terms correspond to complex multicellular plants features that are absent in microalgae. This way, key proteins involved in cell wall biogenesis and cellular glucan metabolism in *Arabidopsis* such as *Pectin Methylesterase 1* (*PME1*, At1g53840) lack *Ostreococcus* and *Chlamydomonas* potential orthologues (**Figure 3B**). Indeed, the evolutionary history of *PME* genes has been previously studied, establishing their appearance in multicellular Charophyte algae and supporting their absence in more primitive algae (Wang et al., 2013). A case of proteins annotated as involved in *"Multi-cellular processes"* without potential orthologues in *Ostreococcus* and *Chlamydomonas* is *Arabidopsis NAC domain containing protein 98* (*ANAC098*, At5g53950) (**Figure 3C**). The family of NAC TFs is one of the largest in plants and is involved in multiple key developmental processes such as floral development. Again, this TF family first originated in Charophytes (Zhu et al., 2012) and is absent in *Chlamydomonas* and *Ostreococcus*. Finally, the *Recognition of Peronospora Parasitica 1* (*RPP1*, At3g44480) is an example of a specific *Arabidopsis* protein involved in *"Cell communication/signaling"* and is part of the set of specific *Arabidopsis* genes related to pathogen resistance and programmed cell death (Schreiber et al., 2016; **Figure 3D**).

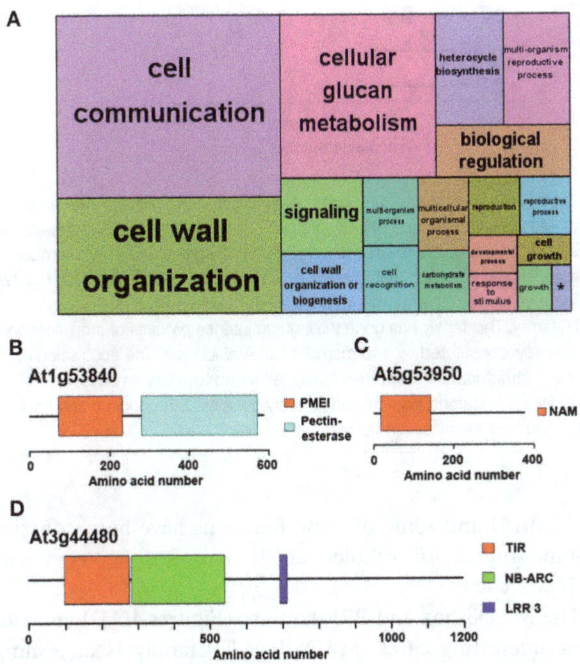

FIGURE 3 | Functional annotation of genes exclusively identified in *Arabidopsis*. No potential *Chlamydomonas* or *Ostreococcus* orthologues were identified for a quarter of *Arabidopsis* proteins. **(A)** Non-redundant GO term enrichment analysis over the set of *Arabidopsis*-specific genes suggests that they are mainly involved in *"cell communication/signaling," "cell wall organization," "cellular glucan metabolism"* and *"multiorganism reproductive process."* (*) *"single-organism developmental process."* Each rectangle area in the treemap represents the −log10 (*p*-value) for the corresponding GO-term. **(B)** Domain structure of the protein encoded by *Pectin Methylesterase 1*, (*PME1*, At1g53840), an example of an *Arabidopsis* specific protein involved in *"cell wall organization and glucan metabolism."* **(C)** An instance of an *Arabidopsis* specific protein annotated with *"multi-organism reproductive process"* is the protein encoded by *NAC domain containing protein 98* (*ANAC098*, At5g53950) involved in floral development. **(D)** Domain structure of the protein encoded by *Recognition of Peronospora Parasitica 1* (*RPP1*, At3g44480), one of the specific *Arabidopsis* proteins involved in *"cell communication/signaling."* Color boxes represent domains identified in pfam database including their identification codes.

Large TF Families in *Arabidopsis* Have Single Orthologues in Microalgae Suggesting Gene Amplification and Functional Diversification Processes

The approach to define potential orthologues by MBBH can detect several candidates for any given gene, identifying multiple orthologous genes that could have appeared from processes of gene duplication. In *Arabidopsis*, on average, 5.19 and 7.83 genes could be ascribed to a homolog in *Chlamydomonas* and *Ostreococcus*, indicating that *Arabidopsis* gene families are, on average, five to eight times larger than in those organisms. This concurs with the idea that whole genome duplication and gene duplication events were crucial in the evolution of plant gene families (Romero-Campero et al., 2013; Rensing, 2014). Interestingly, this process is particularly frequent in TF families and MBBH tool has efficiently detected functional domains in the protein sequences of the three species under study and identified the TF family they belong to. In general, their sizes coincided with the data available in the Plant Transcription Factor Database, PlantTFDB (Jin et al., 2017), confirming the accuracy of this approach. On average, less than 4 and 7 protein members formed each *Ostreococcus* and *Chlamydomonas* TF family respectively, whereas a media of 30 members constituted *Arabidopsis* TF families, further supporting the idea of multiple duplication events.

Two clear examples of amplification in TFs are the DOF and COL protein families (**Figures 4A,B**) that present one single member in *Chlamydomonas* and two members in *Ostreococcus*,

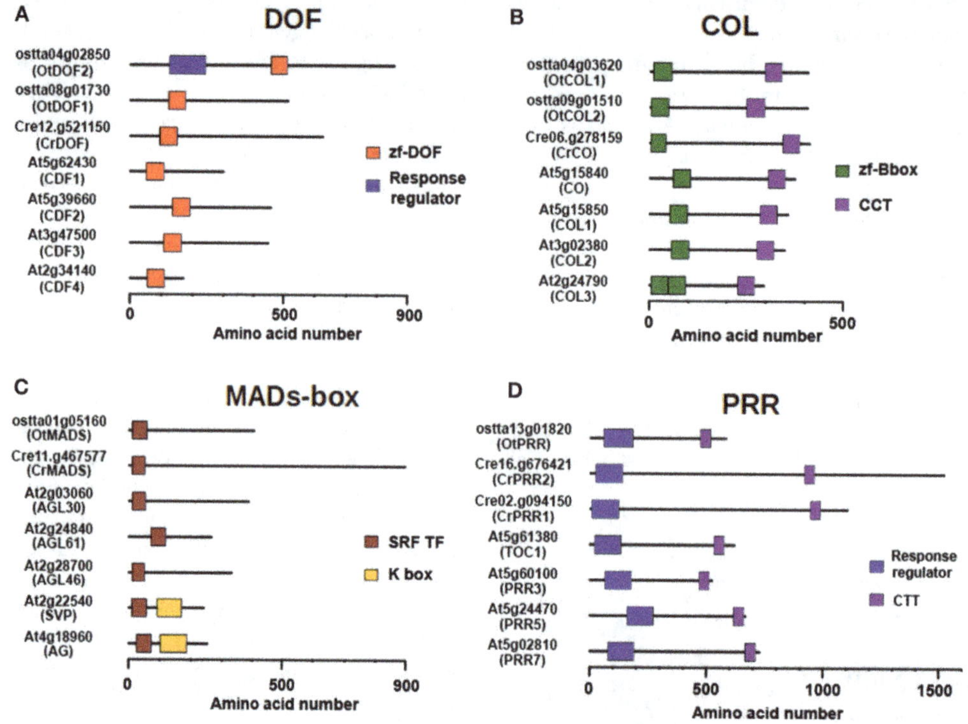

FIGURE 4 | Amplification of TF families through evolution. *Arabidopsis* TF families are respectively on average five and eight times larger than *Chlamydomonas* or *Ostreococcus* ones, indicating an evolutionary history with multiple gene duplication events. **(A)** Domain structure of DOF proteins. *Ostreococcus* has two members; one of them exhibiting a Response Regulator domain in the N-terminal (OtDOF2), while it is constituted by a single member in *Chlamydomonas*. Of the 47 members of the family in *Arabidopsis*, the closest homologs to algal DOFs are the CDFs, involved in photoperiodic flowering. **(B)** Domain structure of the COL TF family. It has two members in *Ostreococcus*, a single member in *Chlamydomonas* and 22 in *Arabidopsis*. Some variation in the number of B-boxes in the N-terminal domain is observed. CO and closest homologs are shown. **(C)** Domain structure of the MADS-box TF family. This family has gone through an intense process of amplification from the single copy gene present in *Ostreococcus* and *Chlamydomonas* to the multi-gene family, constituted by 146 members, in *Arabidopsis*. The acquisition of novel protein domains in *Arabidopsis* would have given rise to different subfamilies. **(D)** Domain structure of PRR protein family. Although relatively small in *Arabidopsis*, with around 10 members, it has also been submitted to a process of gene amplification starting from the single copy gene in *Ostreococcus* and two genes in *Chlamydomonas*. Color boxes represent domains identified in pfam database including their identification codes.

respectively, whereas in *Arabidopsis* 47 DOF and 22 COL proteins are present. A subset of plant *DOF* genes are regulated by the clock (*CYCLYNG DOF FACTORS, CDFs*) and, in time, control the daily expression of *CONSTANS* (*CO*) in *Arabidopsis* and potato during photoperiodic flowering (Imaizumi et al., 2005; Fornara et al., 2009; Kloosterman et al., 2013). This constitutes a conserved DOF-CO module in Spermatophytes that is also conserved in *Chlamydomonas* (Lucas-Reina et al., 2015). In *Ostreococcus* it might be different. While the *Chlamydomonas* and *Arabidopsis* orthologues presented a single DOF domain, one of the two *Ostreococcus* DOF proteins (OtDOF2, ostta04g02850) presented an additional N-terminal Response Regulator domain with a potential phosphoaceptor aspartic acid-aspartic acid-lysine (DDK) motif (**Figure 4A**). This suggests that OtDOF2 could be part of a phosphorelay system (Djouani-Tahri et al., 2011). The *COL* gene family (**Figure 4B**) seemed to have appeared in microalgae, with a single representative in *Chlamydomonas* (*CrCO*, Cre06.g278159) (Valverde, 2011) and two putative orthologues in *Ostreococcus* (*OtCOL1*, ostta04g03620 and *OtCOL2*, ostta09g01510). In *Chlamydomonas CrCO* is involved in the control of the cell cycle, starch synthesis and oil content (Serrano et al., 2009; Deng et al., 2015) and some of these functions have been conserved in some *Arabidopsis* orthologues (Romero-Campero et al., 2013; Ortiz-Marchena et al., 2014).

The MADS-box and PRR families (**Figures 4C,D**) constitute other interesting cases. MADS-box TF family is an example of frequent amplification over the course of plant evolution, since only a single copy is present in *Chlamydomonas* and *Ostreococcus*, whereas in *Arabidopsis* this family contains 146 members. MADS-boxes have been classified into different subfamilies depending on the recruitment of new protein domains besides the conserved MADS DNA-binding domain, such as the K-boxes (**Figure 4C**), and that may explain the ample functional diversity of these TF family in higher plants. On the other hand, modern plant PRRs (**Figure 4D**) contain an N-terminal Pseudo Response Regulator domain and a C-terminal CCT domain, but algal proteins are different. Spermatophytic PRRs are similar in size to COLs and, like them, seem to have experienced a similar amplification process from a single gene copy in *Ostreococcus* (*OtPRR*, ostta13g01820) and two genes in *Chlamydomonas* (*CrPPR1*, Cre02g.094150; *CrPRR2*, Cre16.g676421) to a multi-gene family in *Arabidopsis*. Nevertheless, while in *Ostreococcus* and *Chlamydomonas* proteins, a potential true phosphoaceptor DDK motif is present, suggesting that the algal proteins still retain part of the ancestral phosphorelay signaling mechanism, this motif is missing in the higher plant domain, constituting a "Pseudo" Response Regulator (Mizuno and Nakamichi, 2005; Satbhai et al., 2011).

To study to what extent the conservation in protein sequence and domain structure is accompanied by conservation in the expression profiles, transcriptomic data consisting of 24 h time series for the three species under study was analyzed. Not surprisingly, differences in the expression profiles of *Arabidopsis* genes were observed, some genes retaining the same one as in *Ostreococcus* and *Chlamydomonas*, while others acquired completely new expression profiles (**Figure 5**). This way, in the *DOF* family (**Figure 5A**) the expression profile of the *Arabidopsis* genes *CDF1* and *CDF2* have retained similar expression profiles as *CrDOF* and *OtDOF1* with a peak around ZT0 and a trough around ZT12. However, *CDF4* and *OtDOF2* peak around ZT21, while *CDF3* exhibits an expression profile with a peak at ZT3 and trough at ZT18, not observed in *Chlamydomonas* or *Ostreococcus* genes. A similar situation is observed for the COL (**Figure 5B**), MADS-box (**Figure 5C**) and PRR (**Figure 5D**) TF families. The *Arabidopsis COL* genes *COL1*, *COL2* and *COL3* exhibit very similar expression profiles to *OtCOL1* and *CrCO*, showing a trough at ZT12 (**Figure 5B**). On the other hand, *OtCO2* and *CO* present distinctly different expression profiles with their minimum expression levels at ZT0 and ZT6, respectively. The MADS-box family presents a great diversification in the expression profile of their members, with a substantial difference between the expression patterns of *Arabidopsis*, *Chlamydomonas* and *Ostreococcus* genes (**Figure 5C**). In the *PRR* family, *OtPRR* presents the same expression pattern as *Arabidopsis PRR1* (*TOC1*) and *PRR3*, peaking at ZT12, whereas *PRR5* and *PRR7* peak earlier, at around ZT9. However, *CrPRR1* and *CrPRR2* present completely different profiles, peaking at ZT21 and ZT0, respectively (**Figure 5D**). This could imply that, in general, gene amplification is accompanied by expression profile diversification and hence, functional diversification (Romero-Campero et al., 2013).

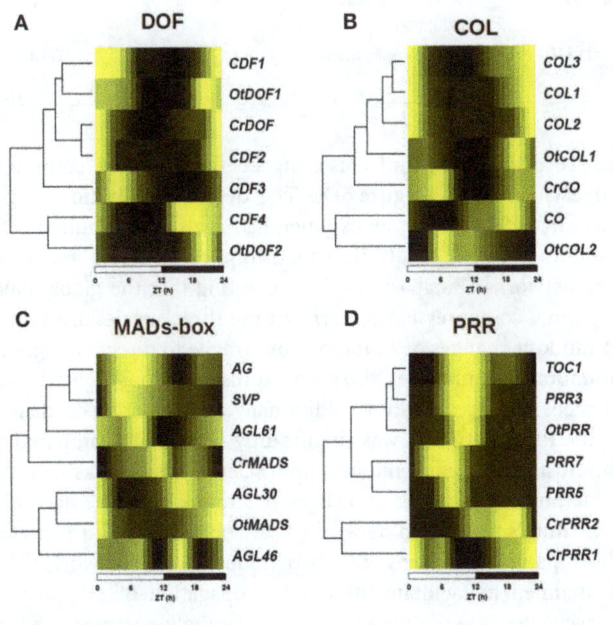

FIGURE 5 | Diversification of expression profiles after gene amplification. Heat plots showing the diversification in the expression profiles of the different members of TF gene families from figure 4 in a ND photoperiod. A color scale from deep yellow (maximum expression) to black (minimal expression) is used. Hierarchical clustering is used to sort genes according to the similarity between their expression profiles. **(A)** Expression profile of *DOF* gene family. **(B)** Expression profile of *COL* gene family. **(C)** Expression profile of *MADS*-box gene family. **(D)** Expression profile of *PRR* gene family. Notice that while some *Arabidopsis* genes have retained similar expression patterns as *Ostreococcus* and *Chlamydomonas* genes, others have acquired completely new expression profiles. Below each graphic, a time scale in Zeitgeber Time (ZT) in hours (h) is shown.

Most Genes in *Chlamydomonas* and *Ostreococcus* Exhibit Light-Dependent Daily Rhythmic Expression Patterns, but Only 40% of the *Arabidopsis* Transcriptome Shows a Periodic Expression

Recently, massive amounts of transcriptomic data for photosynthetic organisms under diverse environmental and physiological conditions have been produced. One of the most studied environmental signals is the alternation of light and dark cycles. Nevertheless, the analysis of the diurnal changes in the transcriptome has been independently performed in different species, making necessary the application of Molecular Systems Biology techniques to integrate and compare them. In this study we have analyzed microarray data collected over 24 h periods in ND conditions for *Ostreococcus* (Monnier et al., 2010) and *Arabidopsis* (Bläsing et al., 2005), as well as RNA-seq data for *Chlamydomonas* (Zones et al., 2015) in ND conditions and *Arabidopsis* in LD conditions (Rugnone et al., 2013). Details on transcriptomic data processing are described in the Materials and Methods section. In order to determine the effect of the alternation of light and dark periods over the transcriptome we identified genes exhibiting rhythmic or oscillating expression patterns with a period of approximately 24 h in the three species under study. Nonparametric methods implemented in the Bioconductor R package RAIN (Rhythmicity Analysis Incorporating Nonparametric methods) were used to detect significant gene expression patterns with arbitrary wave forms and a pre-specified period of 24 h (Thaben and Westermark, 2014), see Materials and Methods section for details.

The analysis revealed that more than 90% of the *Ostreococcus* genes targeted in the microarray exhibited daily rhythmic patterns (Supplementary Table 1). No GO term significantly enriched in the non-periodic genes was identified. This suggests that practically the entire *Ostreococcus* transcriptome is periodic and that most *Ostreococcus* biological processes are strongly affected by the alternation of light and dark cycles. In *Chlamydomonas* approximately 70% of the genes were found to follow significant daily rhythmic expression patterns (Supplementary Table 2). The GO term enrichment analysis over the non-periodic *Chlamydomonas* genes identified only a few significant biological processes. The two most significant processes were "*DNA integration*" and "*Defense response to virus*" including genes such as the reverse transcriptase Cre05.g235102, the DNA ligase Cre06.g277801 and the 2′-5′ oligoadenylate synthetase Cre15.g641050. These have been identified as biological processes induced by biotic stimuli, and are thus independent from the abiotic inputs from light/dark cycles. Finally, only 43.18% of the *Arabidopsis* genes showed significant circadian patterns according to our analysis (Supplementary Table 3), suggesting that during the evolution of higher plants many biological processes were uncoupled from the external influence of alternating light and dark cycles. These percentages of rhythmic genes are largely in agreement with previous results (Bläsing et al., 2005; Monnier et al., 2010; Zones et al., 2015). The pathway enrichment analysis performed over the *Arabidopsis* daily rhythmic genes revealed several key significant pathways influenced by rhythmic changes (**Table 2**). As expected, the most significant pathway represented in the enrichment was the one termed "*Circadian rhythms in plants.*" Also expected were the pathways "*Porphyrin and chlorophyll metabolism*" and "*Pentose phosphate pathway*". These pathways are involved in photosynthesis and hence are expected to be highly regulated by light/dark cycles exhibiting rhythmic patterns. Key metabolic pathways such as "*Fatty acid degradation*" and "*Starch and sucrose metabolism*" were also detected as significantly enriched in daily rhythmic genes indicating that *Arabidopsis* metabolism is still highly affected by alternating periods of light and dark following rhythmic patterns. Somehow surprising was the identification of the "*Alpha-Linolenic acid metabolism*" and "*Plant hormone signal transduction*" among the enriched GO groups. This result suggests either that these pathways are ancient or that during the course of plant speciation new or exclusive plant pathways involved in hormone synthesis and sensing were recruited and acquired a daily rhythmic regulatory pattern. Curiously, although the course of evolution seems to have made the clock more complex in *Arabidopsis*, a lower percentage of the transcriptome seems to be regulated in a light-dependent periodic manner. It could be that a precise, fine tuning of the clock was recruited to make new processes independent from light/dark transitions in higher plants. More circadian experiments in different tissues or developmental stages may be needed to explain this apparent paradox.

With the aim of representing the daily rhythmic genes and their complex co-expression relationships, a gene co-expression network for each photosynthetic species analyzed in this study was constructed (**Figures 6A–C**). Nodes represent daily rhythmic genes and edges co-expression relationships, so that two circadian genes are assumed to be co-expressed when the Pearson correlation index between their 24 h expression profiles is greater than 0.95 and an edge is drawn between them. The three gene co-expression networks were visualized using the Prefuse Force Directed layout implemented in the software tool Cytoscape (Shannon et al., 2003). Interestingly, the three different gene co-expression networks acquired the same ring-like structure, capturing the chronological relationship between the co-expression patterns of the periodic genes. The basic topological parameters in network analysis, namely node

TABLE 2 | Significantly enriched pathways among the *Arabidopsis* circadian genes.

Pathway ID	Description	*q*-value	Ratio circadian/total
ath04712	Circadian rhythm-plant	$4.20 \cdot 10^{-6}$	29/36
ath00030	Pentose phosphate pathway	$5.25 \cdot 10^{-3}$	35/58
ath00052	Galactose metabolism	$1.54 \cdot 10^{-2}$	32/55
ath00590	Arachidonic acid metabolism	$1.54 \cdot 10^{-2}$	13/17
ath00966	Glucosinolate biosynthesis	$1.54 \cdot 10^{-2}$	14/19
ath00860	Porphyrin and chlorophyll metabolism	$1.76 \cdot 10^{-2}$	28/48
ath00360	Phenylalanine metabolism	$1.84 \cdot 10^{-2}$	25/42
ath00071	Fatty acid degradation	$2.66 \cdot 10^{-2}$	24/41
ath00500	Starch and sucrose metabolism	$2.66 \cdot 10^{-2}$	66/139
ath00130	Ubiquinone and other terpenoid-quinone biosynthesis	$2.66 \cdot 10^{-2}$	21/35
ath00260	Glycine, serine and threonine metabolism	$3.54 \cdot 10^{-2}$	37/72
ath00592	alpha-Linolenic acid metabolism	$3.54 \cdot 10^{-2}$	21/36
ath01200	Carbon metabolism	$3.62 \cdot 10^{-2}$	114/262
ath01230	Biosynthesis of amino acids	$3.62 \cdot 10^{-2}$	111/255
ath04075	Plant hormone signal transduction	$3.62 \cdot 10^{-2}$	118/273
ath01210	2-Oxocarboxylic acid metabolism	$4.66 \cdot 10^{-2}$	37/74

degree distribution and clustering coefficient, were computed for each network (**Figure 6D**). The degree distribution of the three networks follows an exponential negative distribution with p-values below $2.2 \cdot 10^{-16}$, which implies that they are scale-free networks (Barabási, 2015), suggesting that the global daily rhythmic co-expression patterns in the three species are robust to random changes or mutations but fragile to direct changes or mutations in hub genes, those co-expressed with a high number of genes (Aoki et al., 2007). Additionally, the clustering coefficient of the three networks was significantly high when compared to the clustering coefficient of random scale-free networks with the same number of nodes and edges. Therefore, all three networks constitute small-world networks (Barabási, 2015) with expected short paths connecting any two nodes of the network. This is assumed to facilitate the quick propagation of information between genes with different periodic expression patterns.

Daily Rhythmic Genes Clusters Constitute Transcriptional Programs That Confer Temporal Separation to Different Biological Processes with Different Levels of Conservation between *Ostreococcus*, *Chlamydomonas*, and *Arabidopsis*

The daily rhythmic expression pattern of individual genes can be described by using its peak, the time point when the expression

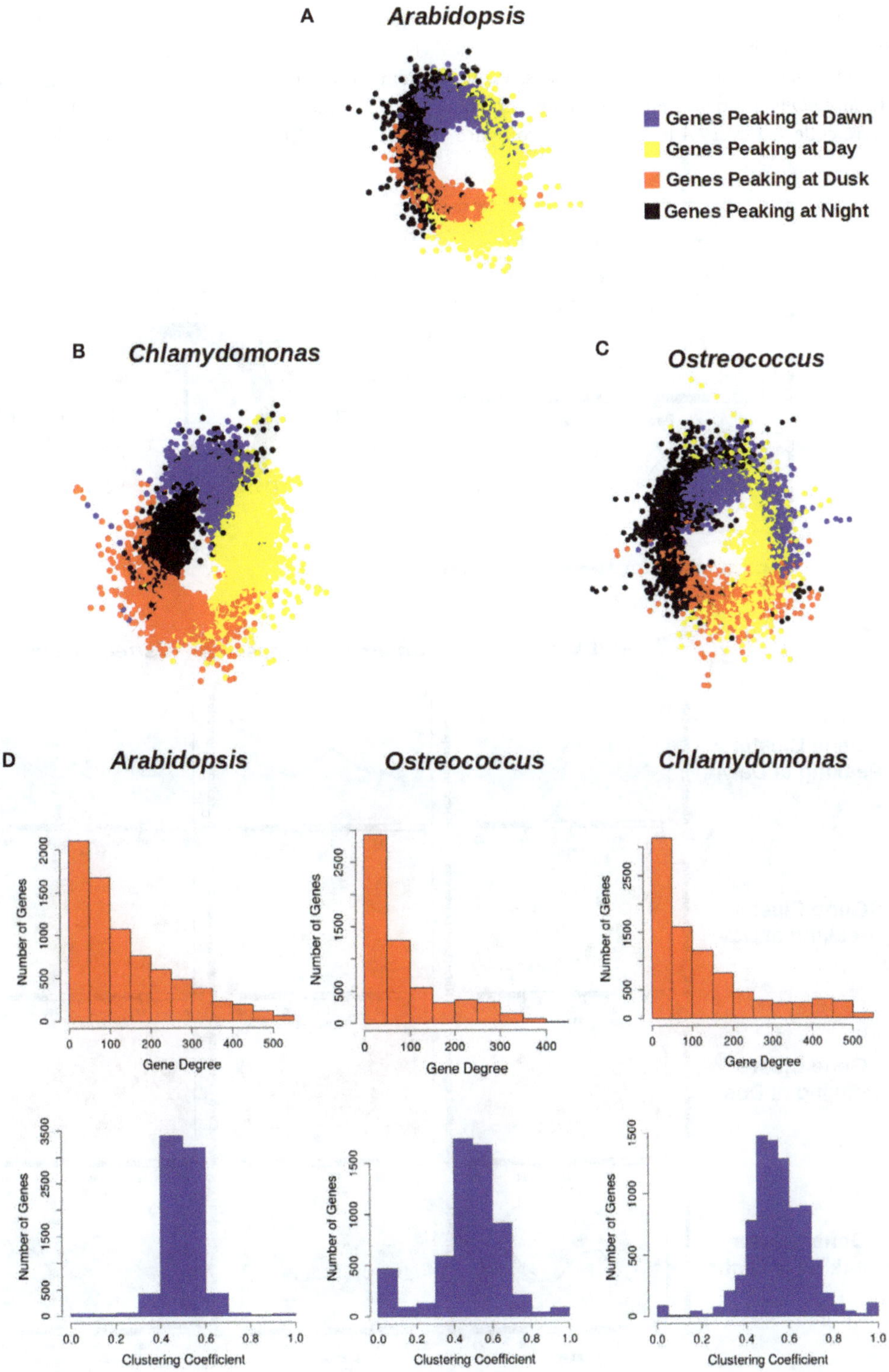

FIGURE 6 | Circadian gene co-expression networks. Genes are represented by nodes and edges represent co-expression between them (Pearson correlation index between their 24 h-expression profiles greater than 0.95). The three networks acquired a ring-like structure mirroring the periodic expression pattern of daily rhythmic genes. **(A)** *Arabidopsis* daily rhythmic gene co-expression network comprised 7639 circadian genes with 537027 co-expression relationships. **(B)** *Chlamydomonas* daily rhythmic gene co-expression network consisted of 10338 circadian genes and 573980 co-expression interactions. **(C)** *Ostreococcus* daily rhythmic gene co-expression network comprised 5782 nodes and 222493 co-expression relationships. **(D)** The topological analysis of the three gene co-expression network shows that they are scale-free and small-world networks: (1) Their degree distributions (above, red) follow an exponential negative distribution. (2) Their average clustering coefficient (below, blue) is significantly greater than random networks with equal number of nodes and edges. Gene cluster with a peak at Dawn is represented in blue, that with a peak during the Day in yellow, the one showing a peak at Dusk in red and that with a peak at Night in black.

level is highest and its trough, the time point when the expression level is lowest (**Figure 7A**, Supplementary Tables 1–3). The transcriptomic data used in the co-expression networks was collected in ND conditions, but using different time sets. While *Ostreococcus* and *Chlamydomonas* data were collected every 3 h, *Arabidopsis* data were collected every 4 h. In order to compare the expression profiles from the different species, a 24 h day was divided into four temporal intervals that contained all points from the different time series. Therefore, Dawn (blue) was defined as the time interval from ZT21 to ZT3; Day (yellow) was considered from ZT3 to ZT9; Dusk (red) consisted of the time period from ZT9 to ZT15 and Night (black) was assumed

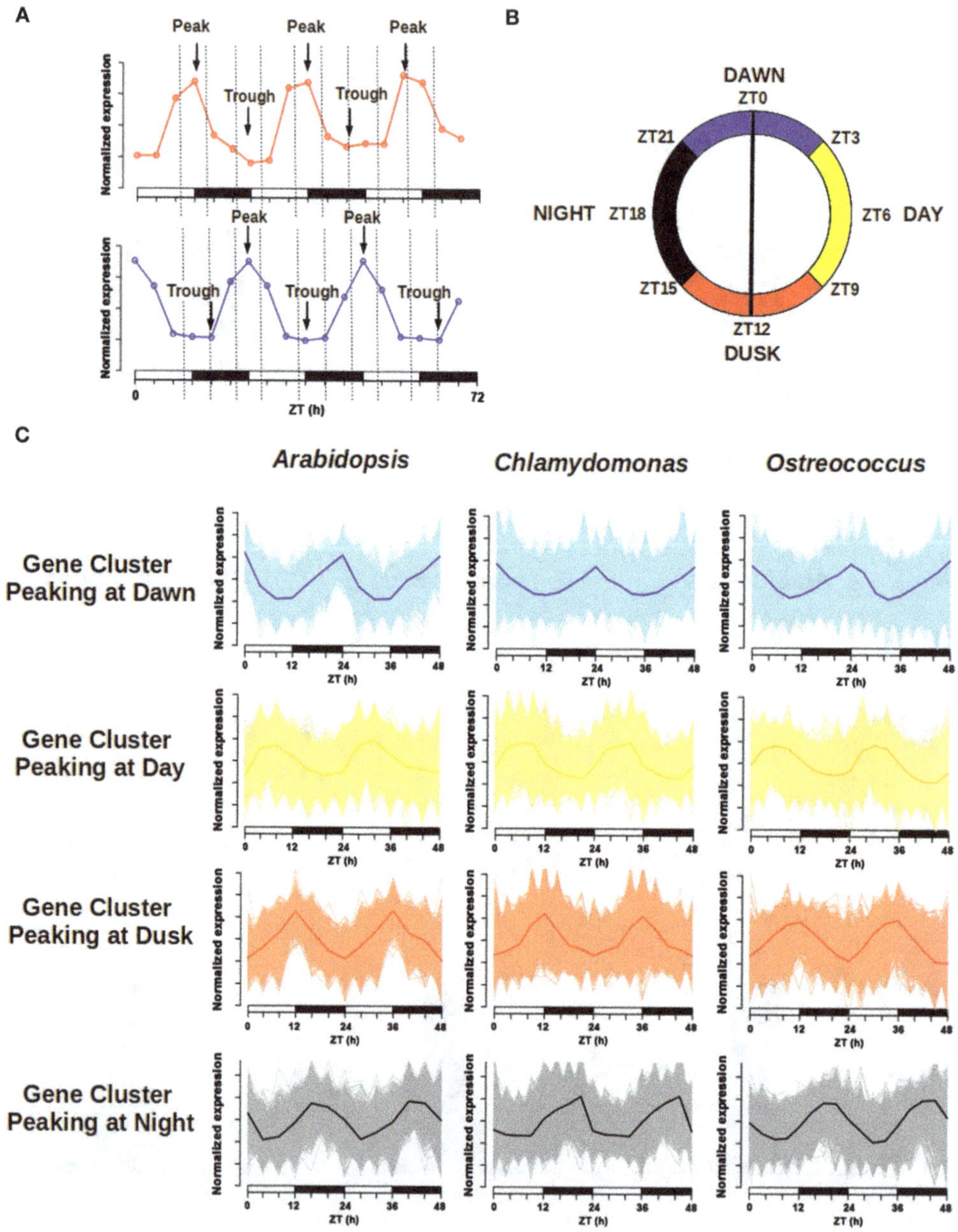

FIGURE 7 | Daily rhythmic gene clustering. The wave form of daily periodic genes expression profile can be characterized by its peak (maximum) and trough (minimum). **(A)** Example of gene expression for two daily rhythmic genes in a 72 h course. Time is shown in hours (h) as Zeitgeber Time (ZT). The first one exhibits its peak around the light/dark transition and its trough around the dark/light, while the second one presents a symmetric profile. **(B)** The 24 h day was divided into four different time intervals: "Dawn" was defined as the time interval from ZT21 to ZT3 (blue); "Day" was considered from ZT3 to ZT9 (yellow); "Dusk" consisted of the time period from ZT9 to ZT15 (red) and "Night" was assumed to be from ZT15 to ZT21 (black). **(C)** Four different clusters were defined for visualization purposes following the same color code: genes peaking at Dawn, Day, Dusk, and Night. A 48 h course normalized expression profile of every gene in each cluster was represented, as well as the average gene expression profile for each cluster. Note that the wave form of the average expression profile for each gene cluster is similar in the three different species.

to be from ZT15 to ZT21 (**Figure 7B**). Thus, daily rhythmic genes could be classified into 16 different clusters according to the time interval where their peak was located (Dawn, Day, Dusk, and Night) and the time interval containing their trough (again Dawn, Day, Dusk, and Night). Due to the amount of data engaged, the graphic representation of these 16 clusters can be confusing, so, in order to facilitate their visualization, we decided to merge all clusters peaking at the same interval into a single one. This way, for each generalized gene cluster we represented the expression profile of every gene and the average expression profile of the entire cluster (**Figure 7C**). Interestingly, no major differences were apparent between the average expression profiles from the three different species under analysis. Furthermore, when we colored the location of these clusters in the corresponding gene co-expression networks (**Figures 6A–C**), we observed that these clusters presented the same localization in the three different networks. This is a strong support for the veracity of our approach and shows, at a single view, that daily rhythmic genes associate in true temporal sets of genes performing synchronous actions and roughly reflecting the natural course of a 24 h clock. In this analysis the boundaries separating clusters were more sharply defined in *Arabidopsis* and *Chlamydomonas* than *Ostreococcus* that showed a higher degree of genes mixing between temporal clusters (**Figures 6A–C**). This supports the already established idea that the primitive *Ostreococcus* clock is not as efficient as those of the other more complex organisms.

The two largest gene clusters in *Ostreococcus* and *Chlamydomonas* contain genes peaking at Dusk with their troughs at Dawn (1620 and 3751 genes, respectively) and genes peaking at Dawn with their troughs at Dusk (1744 and 2869 genes, respectively; **Table 3**). The genes of these two clusters comprise 56.85 and 49.98% of the entire set of daily rhythmic genes in *Ostreococcus* and *Chlamydomonas*, respectively and concur with the common notion that the light/dark and dark/light transitions play central roles in daily rhythmic gene expression regulation in both algae. Surprisingly, in *Arabidopsis* these two gene clusters are not the largest ones, comprising only 22.51% of all periodic genes. This suggests a degree of uncoupling between regulation of circadian gene expression and light/dark and dark/light transitions in higher plants. Nevertheless, the largest gene cluster in *Arabidopsis* is made up of 1209 genes peaking at Dawn with their troughs in the Day, indicating that the dark/light transition still plays a key role in the regulation of daily rhythmic genes in *Arabidopsis*. The second largest cluster in *Arabidopsis* contains 1014 genes whose expression peaks take place at Night and troughs during the Day. This suggests certain independence in the regulation of daily rhythmic gene expression in *Arabidopsis* from the light/dark transitions since a large number of genes present their peaks or troughs in the periods of light and dark.

We performed GO term enrichment analysis over the different gene clusters (**Table 3**) in order to determine the biological processes that were carried out by the different rhythmic gene clusters. Different clusters were enriched in distinct biological processes indicating a temporal separation among them. Some biological processes show enrichment in the same clusters for the three different species such as "*DNA metabolic process*" (gene clusters with peak at Dusk and trough at Dawn), "*photosynthesis*" (gene clusters with peak at Day and trough at Night) and "*carbohydrate catabolic process*" (gene clusters that have the peak during the Dawn and trough at Dusk) indicating a conservation in their daily rhythmic expression patterns (**Table 2**). On the contrary, an anticipation, delay or uncoupling from periodic expression of certain biological processes was apparent when the three species were compared.

Cell cycle-related GO terms such as "*DNA metabolic process*" were enriched in the gene cluster with peak at Dusk and trough at Dawn in the three species, including genes such as *REPLICATION PROTEIN A* (Cre16.g65100, ostta18g01440 and At4g19130), *ORIGIN RECOGNITION COMPLEX 1* (Cre10.g455600, ostta04g05220, and At1g26840) and *CELL DIVISION CYCLE 45* (Cre06.g270250, ostta04g04640, and At3g25100) that showed very similar expression profiles in the three species (**Figure 8A**). Interestingly, the GO terms "*DNA replication*" and "*DNA metabolic process*" are also enriched in the gene cluster with peak at Day and trough at Dawn only in *Ostreococcus* and *Arabidopsis* indicating an apparent anticipation or broader gene expression peaks in these two species with respect to *Chlamydomonas*, where narrow peaks are observed at the light/dark transition (**Figure 8A**). For instance, *DNA POLYMERASE A4* (Cre07.g312350, ostta13g02040, and At5g41880), *MINI CHROMOSOME MAINTENANCE 2* (Cre07.g338000, ostta01g02580, and At1g44900), *CYCLIN A1* (Cre03.g207900, ostta02g00150, and At1g44110) present a narrow peak centered on ZT12 in *Chlamydomonas*, whereas in *Ostreococcus* and *Arabidopsis* they present a broader peak centered on ZT6. This could also indicate a better culture synchronization in *Chlamydomonas* than *Ostreococcus* and the lack of synchronicity between the different tissues in *Arabidopsis*. The GO term "*Photosynthesis*" appeared significantly enriched in the gene clusters with peak at the Day and trough at Night in the three different species (**Table 3**). Specifically, *PHOTOSYSTEM I SUBUNIT D* (Cre05.g238332, ostta10g03280, and At1g03130) and *PHOTOSYSTEM II SUBUNIT O* (Cre09.g396213, ostta14g00150, and At3g50820) exhibit very similar expression patterns with broader peaks in *Ostreococcus* and *Arabidopsis* than in *Chlamydomonas* (**Figure 8B**). Only small variations were detected, for instance, in the expression profiles of genes codifying for components of plastid ATP synthase and b6f complex. *ATP CHLOROPLAST SYNTHASE 1* (*ATPC*, Cre06.g259900, ostta09g01080, and At4g04640) and *PHOTOSYNTHETIC ELECTRON TRANSFER C* (*PETC*, Cre11g.467689, ostta07g02450, and At4g03280) present a peak at dawn and trough at dusk in *Chlamydomonas* and *Ostreococcus*, while in *Arabidopsis ATPC1* is delayed, peaking during the day and *PETC* anticipates dawn, peaking at night (**Figure 8B**).

Ribogenesis, the process of ribosome making, is another interesting example of conservation and evolution of the periodic pattern of a biological process. In fact, "*Ribosome biogenesis*" seems to be enriched mainly in the gene cluster with peak at Dawn and a trough during the Day in *Ostreococcus* and *Chlamydomonas*, whereas in *Arabidopsis* is

TABLE 3 | GO term enrichment in the circadian gene clusters.

	Arabidopsis thaliana			*Ostreococcus tauri*			*Chlamydomonas reinhardtii*		
	Size	**Enriched GO terms**	**Genes**	**Size**	**Enriched GO terms**	**Genes**	**Size**	**Enriched GO terms**	**Genes**
Peak dawn	0			422	Protein catabolic process	*ostta11g00020*	1,426	Carboxylic acid metabolic process	*Cre16.g687350*
Trough dawn					Transcription from RNA polymerase II promoter	*ostta14g01300*		Cellular amino acid metabolic process	*Cre01.g029250*
Peak dawn	1,209	Carbohydrate catabolic process	*AT4G17090*	1,384	RNA metabolic process	*ostta10g00100*	1,801	Ribosome biogenesis	*Cre13.g573050*
Trough day		Monocarboxylic acid metabolic process	*AT4G25700*		Ribosome biogenesis	*ostta06g01560*		RNA metabolic process	*Cre01.g012350*
Peak dawn	811	Response to auxin	*AT1G29440*	1,744	Tetrapyrrole metabolic process	*ostta17g00130*	2,869	Tetrapyrrole metabolic process	*Cre06.g306300*
Trough dusk		Carbohydrate biosynthetic process	*AT1G70290*		Carbohydrate catabolic process	*ostta01g03040*		Carbohydrate catabolic process	*Cre01.g006950*
Peak dawn	132	Regulation of transcription	*AT5G62430*	739	Protein localization to membrane	*ostta03g05390*	1,212	Protein targeting to ER	*Cre16.g683950*
Trough night		Organic cyclic compound metabolic process	*AT1G20330*		Nucleotide metabolic process	*ostta15g00150*		Cyclic nucleotide metabolic process	*Cre02.g074150*
Peak day	980	DNA replication	*AT4G02060*	997	DNA metabolic process	*ostta03g03780*	1,987	Protein catabolic process	*Cre12.g531100*
Trough dawn		Actin cytoskeleton organization	*AT1G13180*		Establishment of localization	*ostta02g01890*		Golgi vesicle transport	*Cre10.g447350*
Peak day	127			141	Protein localization to membrane	*ostta01g00410*	256	Peptide biosynthetic process	*Cre10.g421600*
Trough day					Response to stress	*ostta08g03450*		Vitamin metabolic process	*Cre02.g090150*
Peak day	222	Tricarboxylic acid metabolic process	*AT5G04950*	514	Tetrapyrrole biosynthetic process	*ostta02g02380*	962	Photosynthesis	*Cre10.g425900*
Trough dusk		Isoprenoid metabolic process	*AT4G15560*		Photosynthesis	*ostta14g00150*		Tetrapyrrole biosynthetic process	*Cre06.g306300*
Peak day	935	Photosynthesis	*AT4G10340*	1,060	Signal transduction	*ostta04g01810*	1,467	Photosynthesis	*Cre16.g673650*
Trough night		Ribosome biogenesis	*AT5G16750*		DNA replication	*ostta01g02580*		Carbohydrate derivative	*Cre12.g508700*
					Photosynthesis	*ostta14g00150*		Biosynthetic process	
Peak dusk	907	DNA metabolic process	*AT2G42120*	1,620	DNA metabolic process	*ostta02g00690*	3,751	DNA metabolic process	*Cre01.g017450*
Trough dawn		Carbohydrate catabolic process	*AT2G32290*		Carbohydrate metabolic process	*ostta05g04330*		Carbohydrate biosynthetic process	*Cre12.g508700*
Peak dusk	432	Vitamin metabolic process	*AT2G29630*	437	Microtubule-based process	*ostta10g00680*	1,573	Mitotic cell cycle	*Cre02.g086650*
Trough day		Peroxisome organization	*AT1G47750*					Microtubule-based process	*Cre10.g456350*
Peak dusk	196	Peptide biosynthetic process	*AT4G29430*	696	Vesicle-mediated transport	*ostta09g03760*	882	Vesicle-mediated transport	*Cre02.g087551*
Trough night					DNA replication	*ostta04g04640*		Chromatin organization	*Cre16.g668200*
Peak nightt	122	Hexose metabolic process	*AT1G24280*	540	RNA processing	*ostta11g03270*	1,488	Phosphorylation	*Cre10.g423200*
Trough dawn					Protein modification process	*ostta11g02830*		Nucleotide biosynthetic process	*Cre13.g606250*
Peak night	1,014	Dephosphorylation	*AT1G71860*	1,234	Ribosome biogenesis	*ostta09g00820*	1,479	Protein metabolic process	*Cre01.g035500*

(Continued)

TABLE 3 | Continued

	Arabidopsis thaliana			Ostreococcus tauri			Chlamydomonas reinhardtii		
	Size	**Enriched GO terms**	**Genes**	**Size**	**Enriched GO terms**	**Genes**	**Size**	**Enriched GO terms**	**Genes**
Trough day		Autophagy	*AT3G60640*		Vitamin metabolic process	*ostta09g00820*		Regulation of cellular process	*Cre12.g489000*
Peak night Trough dusk	177	Hexose metabolic process Response to stimulus	*AT5G35790* *AT1G10470*	689	Gene expression Peptide biosynthetic process	*ostta18g01220* *ostta02g01970*	951	Protein metabolic process Signal transduction	*Cre01.g001400* *Cre13.g571200*
Peak night Trough night	3			60	L-serine metabolic process	*ostta12g03060*	57	Cellular respiration	*Cre13.g568800*

only significantly detected in the gene cluster with peak at Day and trough at Night (**Table 3**). A closer inspection into ribogenesis in *Chlamydomonas* reveals an uncoupling between the genesis of cytosolic, plastid and mitochondrial ribosomes (Zones et al., 2015). Cytosolic ribosome components peak at Dusk (i.e., *CrRPL5*, Cre14.g621450), plastid ribosome components peak at Dawn (i.e., *CrPRPL1*, Cre02.g088900) and mitochondrial ribosome component present a broad peak at Dawn (i.e., *CrMRPL21*, Cre09.g388550) (**Figure 8C**). This uncoupling is less evident in *Ostreococcus* where genes codifying for components of cytosolic (i.e., *OtRPL5*, ostta15g01160), plastid (i.e., *OtPRPL1*, ostta12g02760) and mitochondrial (i.e., *OtMRPL21*, ostta01g04030) ribosomes exhibit a peak at Dawn and a trough at Day (**Figure 8C**). No apparent synchronization is observed between genes codifying for ribosome components in *Arabidopsis* (**Figure 8C**). Therefore, the analysis shows how the light/dark cycles in ancient algae synchronized all ribosome synthesis at the same time point. In *Chlamydomonas* ribogenesis was divided into three different stages depending on the nuclear, chloroplast or mitochondrial genome control, while *Arabidopsis* showed a ribosome biogenesis almost independent from daily rhythms.

The conservation and evolution of the central transcriptional regulators in circadian/photoperiod response in *Arabidopsis* was analyzed. A strong conservation was observed between the expression profiles of the key circadian/photoperiod regulators in *Ostreococcus* and *Arabidopsis* with some variations in *Chlamydomonas* (**Figure 8D**). For instance, the three different species showed similar expression profiles of two central photoperiodic genes; *DOF* (Cre12.g51440, ostta04g02850, At5g62430) and *COL* (Cre06.g278159, ostta04g03620, At5g15840) peaking at Dawn, with their trough at Dusk (**Figure 8D**). Remarkably, two of the central genes in the circadian clock; *CCA1* (At2g46830, ostta06g01220) and *TOC1* (At5g61380, ostta13g01820) exhibit the same symmetric expression profiles in *Ostreococcus* and *Arabidopsis*, while in *Chlamydomonas* the putative orthologues *CrCCA1* (Cre12.g514400) and *CrTOC1* (Cre02.g094150) detected in our analysis present the same expression profile, peaking at Dawn with their troughs at Dusk (**Figure 8D**). This suggests an independent and divergent evolution in *Chlamydomonas* of the core regulatory circuit of circadian rhythms as previously suggested (Mittag et al., 2005). Nevertheless, the full validation of these results would require more extensive analysis and experimental work.

Other relevant genes involved in photomorphogenesis such as the bZIP TF *ELONGATED HYPOCOTYL 5* (*HY5*) and *HY5-HOMOLOG* (*HYH*) exhibit the same expression pattern in *Chlamydomonas* (Cre07.g318050, Cre10.g438850), *Ostreococcus* (ostta03g00340, ostta01g03880), and *Arabidopsis* (At5g11260, At3g17609), peaking at Dawn with their troughs at Day (**Figures 8D, 11C,D**).

It was also interesting to observe how the clustering approach worked when the photoperiod changed from ND to LD (**Figure 9**). This way, when the expression profile by RNAseq of a 24 h Col-0 plant grown in LD (red) was compared with the profile in ND (blue), more than 41% of the total genes

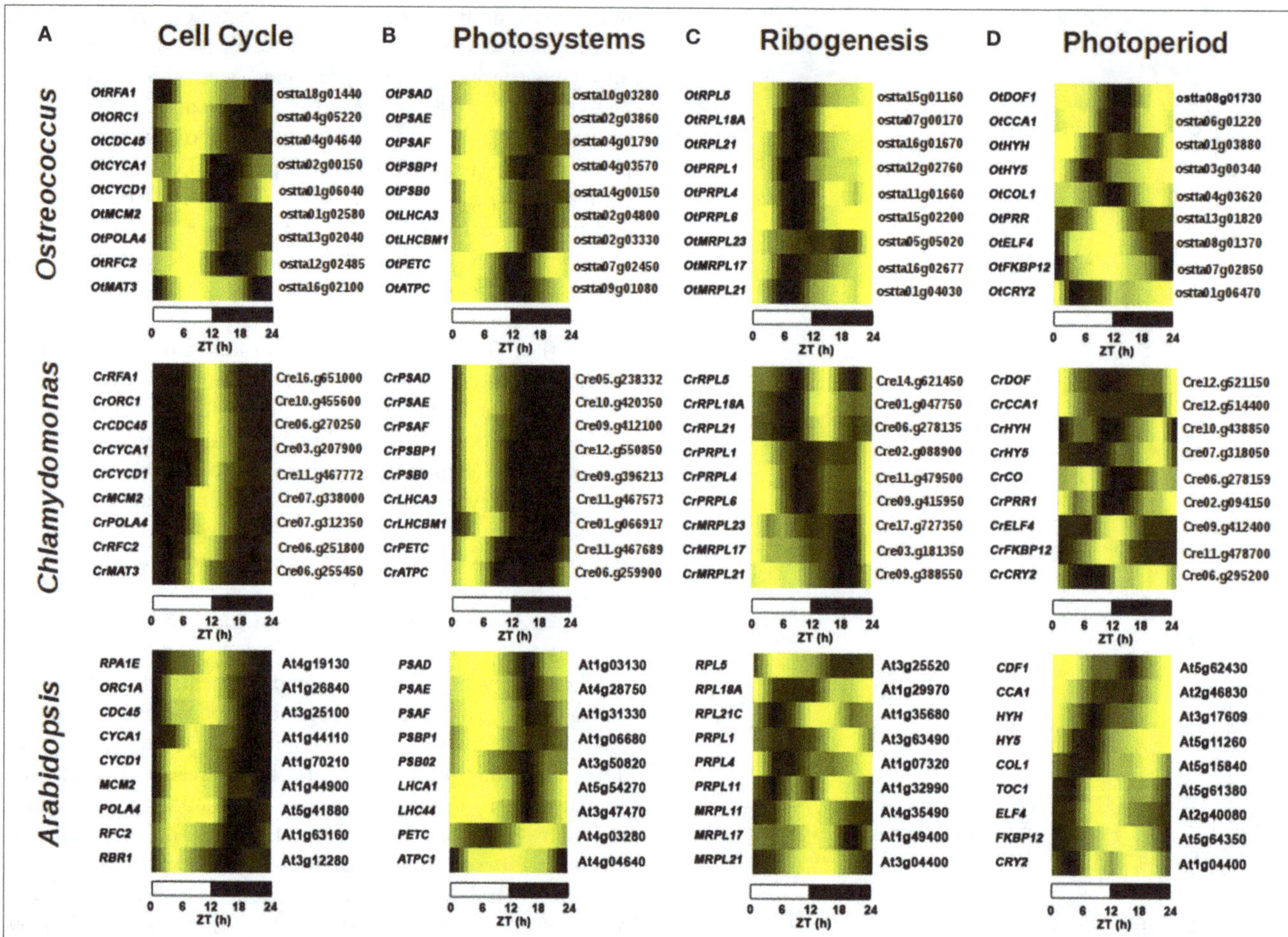

FIGURE 8 | Comparison of gene expression patterns in key biological processes in *Chlamydomonas*, *Ostreococcus* and *Arabidopsis*. Heatmaps are used to represent gene expression with genes sorted according to their expression profile similarity using hierarchical clustering as in **Figure 5**. **(A)** Cell cycle genes such as *REPLICATION PROTEIN A* (*RPA*) and *ORIGIN RECOGNITION COMPLEX 1* (*ORC1*) exhibit a peak at Dusk and trough at Dawn in the three species. In *Ostreococcus* and *Arabidopsis* genes such as *DNA POLYMERASE A4* (*POLA4*) and *MINI CHROMOSOME MAINTENANCE 2* (*MCM2*) are expressed previously to those in *Chlamydomonas*, presenting broader peaks expanding from Day to Dusk with respect to the narrow peaks at Dusk observed in *Chlamydomonas*. **(B)** Photosystem components such as *PHOTOSYSTEM I SUBUNIT D* (*PSAD*) and *PHOTOSYSTEM II SUBUNIT O* (*PSBO*) follow similar expression patterns in the three species with peaks at Day and troughs at Night. Only small variations were detected in photosynthesis-involved genes such as the expression profiles of *ATP CHLOROPLAST SYNTHASE 1* (*ATPC1*) and *PHOTOSYNTHETIC ELECTRON TRANSFER C* (*PETC*). These genes present a peak at Dawn and trough at Dusk in both algae, while in *Arabidopsis*, *ATPC1* is delayed, peaking during the day and *PETC* anticipates dawn peaking at night. **(C)** Ribogenesis-involved genes present their peaks at Dawn and troughs at Day in the algae, such as *PLASTID RIBOSOMAL PROTEIN L1* (*PRPL1*) and *MITOCHONDRIAL RIBOSOMAL PROTEIN L21* (*MRPL21*). Cytosolic ribogenesis is uncoupled from organella ribogenesis in *Chlamydomonas* with genes such as *RIBOSOMAL PROTEIN L5*, *L21* (*RPL5*, *RPL21*), peaking at Dusk. This uncoupling is not evident in *Ostreococcus* and no apparent synchronization is observed between genes codifying for ribosome components in *Arabidopsis*. **(D)** Photoperiod/circadian central transcriptional regulators exhibit highly similar expression profiles in the three species. *DOF*, *CCA1*, and *COL* present the same expression pattern; peaking at Dawn with their troughs at Dusk. Some genes, such as *OtPRR* and *TOC1*, present very similar profiles (peaking at Dusk with their troughs at Dawn) in *Ostreococcus* and *Arabidopsis*, whereas in *Chlamydomonas CrPRR* follows a symmetric pattern peaking at Dawn and a trough at Dusk.

expressed in LD presented a daily rhythmic profile (11157), while the percentage was reduced to 37% of genes in ND (7640). Out of the 7640 genes showing a periodic regulation in ND, 5143 (deep purple, 67%) could be also identified in LD, while approximately half of the 11157 genes in LD (6014, 54%) did not follow a daily rhythmic pattern in ND. This could mean that increasing the photoperiod also augments the daily rhythmic regulation of the transcriptome in higher plants, although never reaching the numbers shown in microalgae. When we organized the periodic-expressed genes in clusters, based on their peak and trough during the day as in **Figure 8**, a clear clock-wise distribution was observed (**Figure 9B**). This seems to indicate that the basic organization of the daily rhythms is independent of the photoperiod. However, when a temporal-clustered gene co-expression network was constructed with the intersection genes between daily rhythmic genes in LD and ND (5143 genes, deep purple) a diffused circadian pattern of expression was observed, suggesting a displacement or alteration in the daily rhythmic expression of a substantial number of genes due to the extended photoperiod (**Figure 9C**).

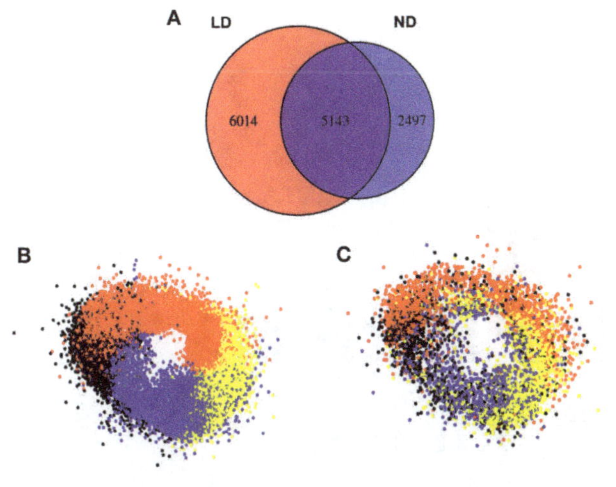

FIGURE 9 | The clustering of the clock-regulated genes does not change substantially between ND and LD conditions in *Arabidopsis*. **(A)** Venn diagrams representing genes in LD (red) and ND (blue) clusters. In LD conditions a higher percentage of genes (41%) follow a circadian expression than in ND (33%). Most genes identified as circadian in ND conditions (7640) exhibit circadian patterns also in LD (5143, 76%). Nevertheless, new genes are identified as circadian in LD (6014, 55%) indicating a higher circadian dependence in LD. **(B)** In the gene co-expression network composed of circadian genes in LD (11157), the same ring-like structure as in ND conditions is observed. This is constituted by the sequential distribution of the clusters constituted by genes peaking at Dawn (red), Day (yellow), Dusk (blue), and Night (black). **(C)** In the representation of the clusters obtained in ND conditions intersecting with the gene co-expression network in LD conditions (deep purple, 5143 genes), an apparent conservation of the circadian pattern is observed.

Module and Pathway Conservation Reveals High Level of Conservation between *Ostreococcus* and *Chlamydomonas* and a Moderate Level of Conservation between Microalgae and *Arabidopsis*

With the aim of determining highly likely functional orthologues between the three different species, information regarding co-expression patterns was integrated with the results obtained previously using the MBBH method. Similar approaches based on the integration of sequence similarity, gene expression profiles and co-expression patterns have been recently successfully applied to determine functional orthologues (Romero-Campero et al., 2013; Das et al., 2016). Thus, it was assumed that two MBBH potential orthologous periodic genes from two different species exhibited a conserved daily pattern when both presented their peak and trough in the same time interval (i.e., same circadian cluster) or when the Pearson correlation coefficient between their expression profiles was higher than 0.98. Therefore, such two genes could be named as expresologues (Das et al., 2016).

According to this criterion approximately 34% of *Arabidopsis* daily rhythmic genes presented an expresologue in *Ostreococcus* or *Chlamydomonas*. This suggests a high level of conservation in spite of the large evolutionary distance between flowering plants and microalgae and presumably, along the plant evolutionary lineage. Interestingly, only the *Arabidopsis* merged cluster of daily rhythmic genes peaking at Dusk was significantly enriched in *Ostreococcus* and *Chlamydomonas* expresologues (p-value of 4.84×10^{-5} by Fisher's exact test), suggesting that most of these conserved genes maintain a strong influence of light/dark transitions on their expression control. In *Ostreococcus*, 83.62% of the daily rhythmic genes present an expresologue in *Chlamydomonas* but just 36.16% have an expresologue in *Arabidopsis*. In *Chlamydomonas*, 52.71% of the daily rhythmic genes present an expresologue in *Ostreococcus* whereas only 19.05% have an *Arabidopsis* expresologue, which supports a more divergent evolution in this species when compared to *Ostreococcus*.

To study the conservation of daily rhythmic patterns in the different clusters among the three species, beyond the comparison between individual gene expression profiles, the Summary Composite Conservation Statistic (Zsummary) was computed as defined in Langfelder et al. (2011) (**Figure 10A**). A Zsummary value lower than 2 indicates no conservation, a Zsummary value 2–10 implies a moderate conservation, while Zsummary greater than 10 constitutes evidence of a great level of conservation. For each daily rhythmic gene cluster, the Zsummary for the six different possible comparisons (*Arabidopsis* vs. *Ostreococcus*, *Arabidopsis* vs. *Chlamydomonas*, *Ostreococcus* vs. *Arabidopsis*, *Ostreococcus* vs. *Chlamydomonas*, *Chlamydomonas* vs. *Arabidopsis* and *Chlamydomonas* vs. *Ostreococcus*) was computed and the corresponding average and standard deviation was plotted. In **Figure 10A**, it can be observed that clusters with a peak at Dawn present a moderate level of daily rhythmic pattern co-expression conservation. In fact, the highest level of conservation was obtained for the cluster with peak at Dawn and trough at Dusk and the cluster peak at Dusk and trough at Dawn (Zsummary > 10). This suggests that evolution has mostly conserved periodic genes with a strong influence from the light/dark and dark/light transitions in their expression profiles, again hinting the importance of these transitory states in plant physiology and their conservation across the entire green lineage.

To corroborate the conservation of co-expression patterns in the three different species, two central pathways identified as daily-regulated in *Arabidopsis* were chosen (**Table 2**). For these gene sub-clusters, smaller gene co-expression networks were constructed, namely key regulators in circadian/photoperiod system (**Figure 10B**) and central enzymes in starch/sucrose metabolism (**Figure 10C**). In these networks, where red edges indicate positive correlations and blue edges indicate negative correlations, a general conservation between the plots can be observed (**Figures 10B,C**). The circadian/photoperiod plot revealed a high conservation among the three sub-clusters. Nevertheless, a general higher conservation between *Arabidopsis* and *Ostreococcus* display, when compared to *Chlamydomonas* could be observed, indicating again that this microalgae has slightly differentiated from the core *Arabidopsis* model clock (**Figure 10B**). On the contrary, daily-regulated starch metabolic

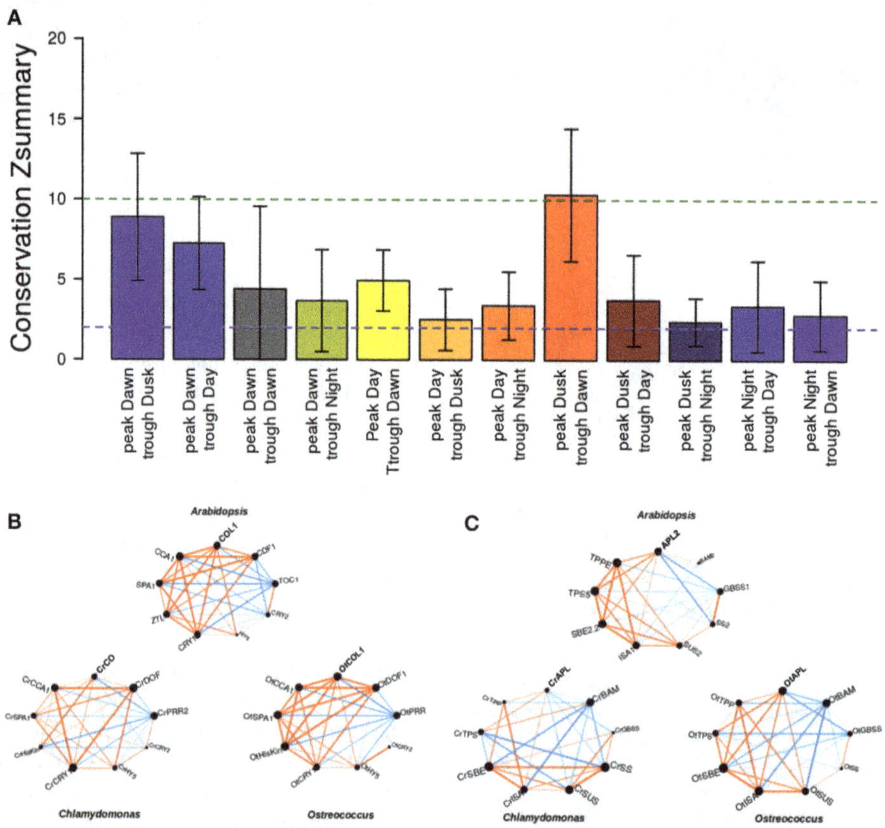

FIGURE 10 | Conservation of circadian patterns in *Arabidopsis*, *Chlamydomonas,* and *Ostreococcus*. **(A)** The average and standard deviation of the Zsummary conservation score between the three different species was computed for all clusters capturing different daily rhythmic patterns. Only clusters that exhibit a periodic pattern with peak or trough at Dusk or Dawn are significantly conserved over the three different species. **(B)** The co-expression pattern of the key circadian/photoperiod regulators exhibits a high level of conservation in the three species, but it is more evident for *Arabidopsis* and *Ostreococcus*. Red edges represent positive correlation between the corresponding gene expression profiles whereas blue edges indicate negative correlation. Edge width is proportional to the absolute value of the correlation between the corresponding expression profiles. **(C)** The co-expression patterns between key enzymes in starch/sucrose metabolism are highly conserved among the three different species with slight differences in *Chlamydomonas*.

genes showed a higher conservation degree between *Ostreococcus* and *Chlamydomonas*, maybe revealing that through the course of evolution, additional regulatory steps were incorporated into the higher plant starch synthesis control (**Figure 10C**). On the whole, this sub-cluster visualization allowed us to compare different subset of genes that had showed a correlation in our general analysis (Supplementary Tables 1–3) further indicating levels of conservation and divergence among the species and constituting efficient tools to initiate the study on the evolution of particular processes.

To further test the veracity of our approach, the promoters of genes showing a high level of conservation both in sequence similarity (**Figures 11A,B**, leftmost) and co-expression patterns (**Figures 11A,B**, middle) in the three species were analyzed. As observed in **Figures 11A,B** rightmost, the promoters of *COLs* (**Figure 11A**) and *GBSSs* (**Figure 11B**) orthologues showed also conservation in the TF binding sites identified in the corresponding gene promoters. This result suggests that through evolution, in order to maintain the regulation of these functional networks, whole set of genes have to evolve synchronously and reveal why these networks are so resilient to change even in long evolutionary time scales. Finally, in order to provide an independent validation of the conservation of these daily rhythmic patterns and the capacity to predict orthology in modern plants and algae, the normalized expression of the *Arabidopsis* bZIP gene *HY5* and the putative orthologues detected in this study in *Chlamydomonas* (*CrHY5*) and *Ostreococcus* (*OtHY5*) from the ND microarray analysis (**Figure 11C**) were plotted and compared with the expression detected in a 24 h LD course by QPCR experiments (**Figure 11D**). Both plots showed a remarkable similarity, except that the LD expression profiles of HY5 and orthologues prolonged the daily minimal expression levels compared to the ND profile until the last hours of light, when the mRNA levels started to rise again. Therefore, the expression profiles of *HY5* and orthologues showed a very clear conserved pattern, as had been shown in our previous analysis (**Figure 8**) indicating that the three genes have a high chance of expressing functional orthologues, a line of research that is now being followed in the laboratory.

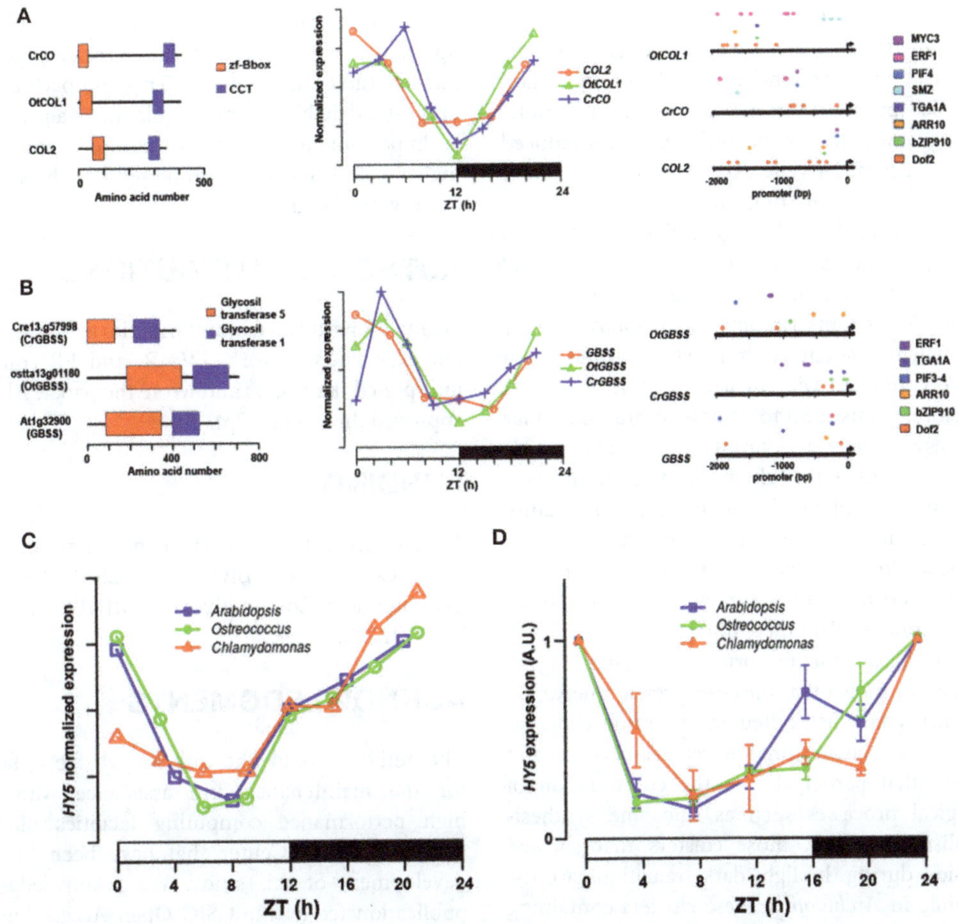

FIGURE 11 | Putative TF orthologues show a conserved protein domain structure, promoter distribution and circadian expression profile in algae and higher plants. **(A)** Gene structure (leftmost), expression in 24 h ND course (middle) and TF binding site in the promoter of *Arabidopsis COL2* gene and putative orthologues from the two microalgae. **(B)** As above, showing *Arabidopsis GBSS* gene and putative orthologues. **(C)** Normalized expression level in 24 h ND photoperiod from the microarray and RNAseq experiments of *HY5* gene from *Arabidopsis* (blue), *CrHY5* from *Chlamydomonas* (red) and *OtHY5* from *Ostreococcus* (green). Notice the high similarity in the expression profile of the three species. **(D)** Expression levels (Arbitrary Units) of *HY5* (blue) from *Arabidopsis*, *CrHY5* from *Chlamydomonas* (red), and *OtHY5* from *Ostreococcus* (green) in a 24 h course in LD by QPCR. Notice how *HY5* circadian expression pattern is also conserved in the three species, although the daily expression lengthens, strongly supporting the idea that they are true orthologues. Time is shown in hours (h) as Zeitgeber Time (ZT). For each time point of the analysis, three biological replicates and three technical replicates were analyzed. Each point shows standard error bars ± s.e.m.

CONCLUSIONS

As the amount of transcriptomics and phylogenomics data accumulates, the complexity of the evolution of the mechanisms governing the control of gene expression in eukaryotes becomes more evident. A paradigm of a specific pattern that controls the transcriptome occurs under alternating light/dark cycles (Millar, 2016). In photosynthetic organisms this control is particularly important due to their dependence on sunlight for most of their main physiological processes and thus, a complex periodic gene expression mechanism can be found as early as in microalgae (Mittag et al., 2005; Corellou et al., 2009). Here, a Systems Biology approach has been used to dissect this dependence and find out mechanisms anciently controlled by daily rhythms and those that have acquired a new regulation. To achieve this, MBBH tool was developed to help discern orthology from species evolutionarily very distant as algae and modern plants. When used in parallel with gene co-expression networks, a very strong toolset to understand how different processes have conserved a periodic regulation through long evolutionary distances has been built. Furthermore, a web-based tool has been implemented to allow the study of the daily rhythmic evolution of any set of genes from algae to modern plants in a similar way to previous web-based tools that allow researchers to compare diurnal expression profiling between monocots and dicots (Mockler et al., 2007). Together with the available pipelines and databases presented in former papers (Romero-Campero et al., 2013, 2016) they constitute strong resources for the research community.

Using these tools, we confirmed that primitive picoeukaryote microalgae, such as *Ostreococcus*, govern most of its transcriptome (a practical 100%) by a daily rhythmic mechanism under alternating light/dark cycles (Monnier et al., 2010). Other microalgae have reduced this light dependence, but the number of genes regulated by light is still significantly high. This is the case of *Chlamydomonas* (Harris, 2001), which can perform some complex physiological processes independently of light/dark cycles, now controlled by other factors such as external biotic stimuli. Following a similar tendency, *Arabidopsis* has reduced the temporal dependence of its gene expression control under light/dark cycles to just over a third of its transcriptome (Michael et al., 2008) but is still significantly higher than the 10–15% of genes showing a periodic control in mammals (Lowrey and Takahashi, 2011). Nevertheless, it has been reported that up to 89% of the *Arabidopsis* transcriptome could follow a daily rhythmic pattern when different environmental signals, such as temperature, are combined with light/dark cycles (Michael et al., 2008). Therefore, plants seem to possess several molecular mechanisms to integrate diverse environmental signals beside light/dark cycles to exert daily rhythmic regulation over their transcriptome that are missing in microalgae. Temperature seems to be a key factor among these signals. The lack of potential microalgal orthologues for key genes in the evening loop such as GI and ZTL, that exert the temperature compensation to the clock in *Arabidopsis*, could explain these differences.

Finally, the clustering analysis performed attending to gene rhythmic expression has shown that the co-expression networks in the three organisms are distributed in the same order as the clock, giving the circadian timer a real associative and temporal dimension that participates in the coordination of different physiological processes such as ribosome synthesis or starch metabolism. In algae, those clusters that showed maximum expression during the light/dark transitions are the most abundant, while in *Arabidopsis*, these clusters containing orthologous genes, seem to be expressed during the dark or light periods. It seems then that *Arabidopsis* has been able to predict dawn and dusk in a better way than algae and has advanced gene expression in order to anticipate these transitions. Moreover, when a global conservation analysis is performed, only genes with their peaks and troughs at dawn and dusk are strongly conserved between microalgae and plants, again suggesting a diversification in the temporal control of gene expression during their evolution. Finally, we have also shown the powerful predictable capacity of the approach by picking up putative orthologues that may result in good candidates for further studies in the future. This approach could help to better understand daily rhythmic regulation in algal systems, which may be important to understand the more complex mechanisms of higher plants and shade some light on how these mechanisms have evolved in the green lineage.

AUTHOR CONTRIBUTIONS

Pd and FR performed bioinformatics analysis. Pd and MR carried out experimental work. FV, JR, and FR conceived the study, interpreted the results and wrote the paper. All authors read and approved the manuscript.

FUNDING

The authors would like to thank funding from projects BIO2011-28847-C02-00 and BIO2014-52425-P (Spanish Ministry of Economy and Competitiveness, MINECO) partially supported by FEDER funding.

ACKNOWLEDGMENTS

The authors would like to acknowledge Rafael Álvarez-Romo for the maintenance and assistance with the use of the high performance computing facilities of the CIC-Cartuja Data Processing Center that have been instrumental to the development of this study. We acknowledge support of the publication fee by the CSIC Open Access Publication Support Initiative through its Unit of Information Resources for Research (URICI).

SUPPLEMENTARY MATERIAL

The Supplementary Material for this article can be found online at: http://journal.frontiersin.org/article/10.3389/fpls.2017.01217/full#supplementary-material

REFERENCES

Adams, S., and Carré, I. A. (2011). Downstream of the plant circadian clock: output pathways for the control of physiology and development. *Essays Biochem.* 49, 53–69. doi: 10.1042/bse0490053

Alexa, A., and Rahnenfuhrer, J. (2016). *TopGO: Enrichment Analysis for Gene Ontology*. R package version 2.26.20.

Aoki, K., Ogata, Y., and Shibata, D. (2007). Approaches for extracting practical information from gene coexpression networks in plant biology. *Plant Cell Physiol.* 48, 381–390. doi: 10.1093/pcp/pcm013

Aucoin, H. R., Gardner, J., and Boyle, N. R. (2016). Omics in Chlamydomonas for biofuel production. *Subcell. Biochem.* 86, 447–469. doi: 10.1007/978-3-319-25979-6_18

Barabási, A. L. (2015). *Network Science*. Cambridge, MA: Cambridge University Press.

Barrett, T., Wilhite, S. E., Ledoux, P., Evangelista, C., Kim, I. F., Tomashevsky, M., et al. (2013). NCBI GEO: archive for functional genomics data sets—update. *Nucleic Acids Res.* 41, D991–D995. doi: 10.1093/nar/gks1193

Blanc-Mathieu, R., Verhelst, B., Derelle, E., Rombauts, S., Bouget, F.-Y., Carré, I., et al. (2014). An improved genome of the model marine alga *Ostreococcus tauri* unfolds by assessing Illumina *de novo* assemblies. *BMC Genomics* 15:1103. doi: 10.1186/1471-2164-15-1103

Bläsing, O. E., Gibon, Y., Günther, M., Höhne, M., Morcuende, R., Osuna, D., et al. (2005). Sugars and circadian regulation make major contributions to the global regulation of diurnal gene expression in *Arabidopsis*. *Plant Cell* 17, 3257–3281. doi: 10.1105/tpc.105.035261

Corellou, F., Schwartz, C., Motta, J. P., Djouani-Tahri, E. B., Sanchez, F., and Bouget, F. Y. (2009). Clocks in the green lineage: comparative functional analysis of the circadian architecture of the picoeukaryote *Ostreococcus*. *Plant Cell* 21, 3436–3449. doi: 10.1105/tpc.109.068825

Covington, M. F., Maloof, J. N., Straume, M., Kay, S. A., and Harmer, S. L. (2008). Global transcriptome analysis reveals circadian regulation of key pathways in plant growth and development. *Genome Biol.* 9:130. doi: 10.1186/gb-2008-9-8-r130

Csárdi, G., and Nepusz, T. (2006). *The Igraph Software Package for Complex Network Research*. Inter Journal Complex Systems (1695).

Dalquen, D. A., and Dessimoz, C. (2013). Bidirectional best hits miss many orthologs in duplication-rich clades such as plants and animals. *Genome Biol. Evol.* 5, 1800–1806. doi: 10.1093/gbe/evt132

Das, M., Haberer, G., Panda, A., Das Laha, S., Ghosh, T. C., and Schäffner, A. R. (2016). Expression pattern similarities support the prediction of orthologs retaining common functions after gene duplication events. *Plant Physiol.* 171, 2343–2357. doi: 10.1104/pp.15.01207

de Vries, J., Stanton, A., Archibald, J. M., and Gould, S. B. (2016). Streptophyte terrestralization in light of plastid evolution. *Trends Plant Sci.* 21, 467–476. doi: 10.1016/j.tplants.2016.01.021

Delwiche, C. F., and Cooper, E. D. (2015). The evolutionary origin of a terrestrial flora. *Curr. Biol.* 25, 899–910. doi: 10.1016/j.cub.2015.08.029

Deng, X., Fan, X., Li, P., and Fei, X. (2015). A photoperiod-regulating gene CONSTANS is correlated to lipid biosynthesis in *Chlamydomonas reinhardtii*. *Biomed. Res. Int.* 2015:715020. doi: 10.1155/2015/715020

Derelle, E., Ferraz, C., Rombauts, S., Rouzé, P., Worden, A. Z., Robbens, S., et al. (2006). Genome analysis of the smallest free-living eukaryote *Ostreococcus tauri* unveils many unique features. *Proc. Natl. Acad. Sci. U.S.A.* 103, 11647–11652. doi: 10.1073/pnas.0604795103

Dessimoz, C., Gabaldon, T., Roos, D. S., Sonnhammer, E. L. L., and Herrero, J. (2012). Towards community standards in the quest for orthologs. *Bioinformatics* 28, 900–904. doi: 10.1093/bioinformatics/bts050

Djouani-Tahri, E. B., Christie, J. M., Sanchez-Ferandin, S., Sanchez, F., Bouget, F.-Y., and Corellou, F. (2011). A eukaryotic LOV-histidine kinase with circadian clock function in the picoalga *Ostreococcus*. *Plant J.* 65, 578–588. doi: 10.1111/j.1365-313X.2010.04444.x

Fornara, F., Panigrahi, K. C. S., Gissot, L., Sauerbrunn, N., Rühl, M., Jarillo, J. A., et al. (2009). Arabidopsis DOF transcription factors act redundantly to reduce CONSTANS expression and are essential for a photoperiodic flowering response. *Dev. Cell* 17, 75–86. doi: 10.1016/j.devcel.2009.06.015

Gautier, L., Cope, L., Bolstad, B. M., and Irizarry, R. A. (2004). affy—analysis of Affymetrix GeneChip data at the probe level. *Bioinformatics* 20, 307–315. doi: 10.1093/bioinformatics/btg405

Gehan, M. A., Greenham, K., Mockler, T. C., and McClung, C. R. (2015). Transcriptional networks-crops, clocks and abiotic stress. *Curr. Opin. Plant Biol.* 24, 39–46. doi: 10.1016/j.pbi.2015.01.004

Goodstein, D. M., Shu, S., Howson, R., Neupane, R., Hayes, R. D., Fazo, J., et al. (2012). Phytozome: a comparative platform for green plan genomics. *Nucleic Acids Res.* 40, D1178–D1186. doi: 10.1093/nar/gkr944

Harris, E. H. (2001). Chlamydomonas as a model organism. *Annu. Rev. Plant Physiol. Plant Mol. Biol.* 52, 363–406. doi: 10.1146/annurev.arplant.52.1.363

Henderson, G. P., Gan, L., and Jensen, G. J. (2007). 3-D ultrastructure of *O. tauri*: electron cryotomography of an entire eukaryotic cell. *PLoS ONE* 2:e749. doi: 10.1371/journal.pone.0000749

Huang, W., Pérez-García, P., Pokhilko, A., Millar, A. J., Antoshechkin, I., Riechmann, J. L., et al. (2012). Mapping the core of the *Arabidopsis* circadian clock defines the network structure of the oscillator. *Science* 336, 75–79. doi: 10.1126/science.1219075

Imaizumi, T., Schultz, T. F., Harmon, F. G., Ho, L. A., and Kay, S. A. (2005). FKF1 F-box protein mediates cyclic degradation of a repressor of CONSTANS in *Arabidopsis*. *Science* 309, 293–297. doi: 10.1126/science.1110586

Jin, J. P., Tian, F., Yang, D. C., Meng, Y. Q., Kong, L., Luo, J. C., et al. (2017). PlantTFDB 4.0: toward a central hub for transcription factors and regulatory interactions in plants. *Nucleic Acids Res.* 45, 1040–1045. doi: 10.1093/nar/gkw982

Kamioka, M., Takao, S., Suzuki, T., Taki, K., Higashiyama, T., Kinoshita, T., et al. (2016). Direct repression of evening genes by CIRCADIAN CLOCK-ASSOCIATED 1 in the *Arabidopsis* circadian clock. *Plant Cell* 28, 696–711. doi: 10.1105/tpc.15.00737

Keller, M. D., Selvin, R. C., Claus, W., and Guillard, R. R. L. (1987). Media for the culture of marine ultraphytoplankton. *J. Phycol.* 23, 633–638. doi: 10.1111/j.1529-8817.1987.tb04217.x

Kloosterman, B., Abelenda, J. A., Gomez, M. D. M. C., Oortwijn, M., de Boer, J. M., Kowitwanich, K., et al (2013). Naturally occurring allele diversity allows potato cultivation in northern latitudes. *Nature* 495, 246–250. doi: 10.1038/nature11912

Koornneef, M., and Meinke, D. (2010). The development of *Arabidopsis* as a model plant. *Plant J.* 61, 909–921. doi: 10.1111/j.1365-313X.2009.04086.x

Langfelder, P., and Horvath, S. (2012). Fast R functions for robust correlations and hierarchical clustering. *J. Stat. Softw.* 46, 1–17. doi: 10.18637/jss.v046.i11

Langfelder, P., Luo, R., Oldham, M. C., and Horvath, S. (2011). Is my network module preserved and reproducible? *PLoS Comput. Biol.* 7:e1001057. doi: 10.1371/journal.pcbi.1001057

Liu, T., Carlsson, J., Takeuchi, T., Newton, L., and Farré, E. M. (2013). Direct regulation of abiotic responses by the *Arabidopsis* circadian clock component PRR7. *Plant J.* 76, 101–114. doi: 10.1111/tpj.12276

Liu, T., Newton, L., Liu, M. J., Shiu, S. H., and Farré, E. M. (2016). A G-Box-like motif is necessary for transcriptional regulation by circadian pseudo-response regulators in *Arabidopsis*. *Plant Physiol.* 170, 528–539. doi: 10.1104/pp.15.01562

Lowrey, P. L., and Takahashi, J. S. (2011). Genetics of circadian rhythms in mammalian model organisms. *Adv. Genet.* 74, 175–230. doi: 10.1016/b978-0-12-387690-4.00006-4

Lucas-Reina, E., Romero-Campero, F. J., Romero, J. M., and Valverde, F. (2015). An evolutionarily conserved DOF-CONSTANS module controls plant photoperiodic signaling. *Plant Physiol.* 168, 561–574. doi: 10.1104/pp.15.00321

Merchant, S. S., Prochnik, S. E., Vallon, O., Harris, E. H., Karpowicz, S. J., Witman, G. B., et al. (2007). The Chlamydomonas genome reveals the evolution of key animal and plant functions. *Science* 318, 245–250. doi: 10.1126/science.1143609

Michael, T. P., Mockler, T. C., Breton, G., McEntee, C., Byer, A., Trout, J. D., et al. (2008). Network discovery pipeline elucidates conserved time-of-day-specific cis-regulatory modules. *PLoS Genet.* 4:e14. doi: 10.1371/journal.pgen.0040014

Millar, A. J. (2016). The intracellular dynamics of circadian clocks reach for the light of ecology and evolution. *Annu. Rev. Plant Biol.* 67, 595–618. doi: 10.1146/annurev-arplant-043014-115619

Mittag, M., Kiaulehn, S., and Johnson, C. H. (2005). The circadian clock in *Chlamydomonas reinhardtii*. What is it for? What is it similar to? *Plant Physiol.* 137, 399–409. doi: 10.1104/pp.104.052415

Miyazaki, Y., Takase, T., and Kiyosue, T. (2015). ZEITLUPE positively regulates hypocotyl elongation at warm temperature under light in *Arabidopsis thaliana*. *Plant Signal. Behav.* 10:e998540. doi: 10.1080/15592324.2014.998540

Mizuno, T., and Nakamichi, N. (2005). Pseudo-Response Regulators (PRRs) or True Oscillator Components (TOCs). *Plant Cell Physiol.* 46, 677–685. doi: 10.1093/pcp/pci087

Mockler, T. C., Michael, T. P., Priest, H. D., Shen, R., Sullivan, C. M., Givan, S. A., et al. (2007). The DIURNAL project: DIURNAL and circadian expression profiling, model-based pattern matching, and promoter analysis. *Cold Spring Harb. Symp. Quant. Biol.* 72, 353–363. doi: 10.1101/sqb.2007.72.006

Monnier, A., Liverani, S., Bouvet, R., Jesson, B., Smith, J. Q., Mosser, J., et al. (2010). Orchestrated transcription of biological processes in the marine picoeukaryote *Ostreococcus* exposed to light/dark cycles. *BMC Genomics* 22:192. doi: 10.1186/1471-2164-11-192

Mora-García, S., de Leone, M. J., and Yanovsky, M. (2017). Time to grow: circadian regulation of growth and metabolism in photosynthetic organisms. *Curr. Opin. Plant Biol.* 35, 84–90. doi: 10.1016/j.pbi.2016.11.009

Nakamichi, N., Kiba, T., Kamioka, M., Suzuki, T., Yamashino, T., Higashiyama, T., et al. (2012). Transcriptional repressor PRR5 directly regulates clock-output pathways. *Proc. Natl. Acad. Sci. U.S.A.* 109, 17123–17128. doi: 10.1073/pnas.1205156109

Nohales, M. A., and Kay, S. A. (2016). Molecular mechanisms at the core of the plant circadian oscillator. *Nat. Struct. Mol. Biol.* 23, 1061–1069. doi: 10.1038/nsmb.3327

Ortiz-Marchena, M. I., Albi, T., Lucas-Reina, E., Said, F. E., Romero-Campero, F. J., Cano, B., et al. (2014). Photoperiodic control of carbon distribution during the floral transition in *Arabidopsis*. *Plant Cell* 26, 565–584. doi: 10.1105/tpc.114.122721

Pagès, H., Aboyoun, P., Gentleman, R., and DebRoy, S. (2016). *Biostrings: String Objects Representing Biological Sequences, and Matching Algorithms*. R package version 2.42.41.

Palenik, B., Grimwood, J., Aerts, A., Rouzé, P., Salamov, A., Putnam, N., et al. (2007). The tiny eukaryote *Ostreococcus* provides genomic insights into the paradox of plankton speciation. *Proc. Natl. Acad. Sci. U.S.A.* 104, 7705–7710. doi: 10.1073/pnas.0611046104

Punta, M., Coggill, P., Eberhardt, R., Mistry, J., Tate, J., Boursnell, C., et al. (2012). The Pfam protein families database. *Nucleid Acids Res.* 40, 290–301. doi: 10.1093/nar/gkr1065

Raven, J. A., Beardall, J., Larkum, A. W. D., and Sánchez-Baracaldo, P. (2013). Interactions of photosynthesis with genome size and function. *Philos. Trans. R. Soc. Lond. B Biol. Sci.* 368:20120264. doi: 10.1098/rstb.2012.0264

Rensing, S. A. (2014). Gene duplication as a driver of plant morphogenetic evolution. *Curr. Opin. Plant Biol.* 17, 43–48. doi: 10.1016/j.pbi.2013.11.002

Romero-Campero, F. J., Lucas-Reina, E., Said, F. E., Romero, J. M., and Valverde, F. (2013). A contribution to the study of plant development evolution based on gene co-expression networks. *Front. Plant Sci.* 4:291. doi: 10.3389/fpls.2013.00291

Romero-Campero, F. J., Perez-Hurtado, I., Lucas-Reina, E., Romero, J. M., and Valverde, F. (2016). ChlamyNET: a Chlamydomonas gene co-expression network reveals global properties of the transcriptome and the early setup of key co-expression patterns in the green lineage. *BMC Genomics* 17:227. doi: 10.1186/s12864-016-2564-y

Rugnone, M. L., Faigón-Soverna, A., Sanchez, S. E., Schlaen, R. G., Hernando, C. E., Seymour, D. K., et al. (2013). LNK genes integrate light and clock signaling networks at the core of the *Arabidopsis* oscillator. *Proc. Natl. Acad. Sci. U.S.A.* 110, 12120–12125. doi: 10.1073/pnas.1302170110

Ruprecht, C., Proost, S., Hernandez-Coronado, M., Ortiz-Ramirez, C., Lang, D., Rensing, S. A., et al. (2017). Phylogenomic analysis of gene co-expression networks reveals the evolution of functional modules. *Plant J.* 90, 447–465. doi: 10.1111/tpj.13502

Satbhai, S. B., Yamashino, T., Okada, R., Nomoto, Y., Mizuno, T., Tezuka, Y., et al. (2011). Pseudo-response regulator (PRR) homologues of the moss *Physcomitrella patens*: insights into the evolution of the PRR family in land plants. *DNA Res.* 18, 39–52. doi: 10.1093/dnares/dsq033

Schreiber, K. J., Bentham, A., Williams, S. J., Kobe, B., and Staskawicz, B. J. (2016). Multiple domain associations within the *Arabidopsis* immune receptor RPP1 regulate the activation of programmed cell death. *PLOS Pathog.* 12:e1005769. doi: 10.1371/journal.ppat.1005769

Serrano, G., Herrera-Palau, R., Romero, J. M., Serrano, A., Coupland, G., and Valverde, F. (2009). Chlamydomonas CONSTANS and the evolution of plant photoperiodic signaling. *Curr. Biol.* 19, 359–368. doi: 10.1016/j.cub.2009.01.044

Serrano-Bueno, G., Romero-Campero, F. J., Lucas-Reina, E. I., Romero, J. M., and Valverde, F. (2017). Evolution of photoperiod sensing in plants and algae. *Curr. Opin. Plant Biol.* 37, 10–17. doi: 10.1016/j.pbi.2017.03.007

Shannon, P., Markiel, A., Ozier, O., Baliga, N. S., Wang, J. T., Ramage, D., et al. (2003). Cytoscape: a software environment for integrated models of biomolecular interaction networks. *Genome Res.* 13, 2498–2504. doi: 10.1101/gr.1239303

Sorokina, O., Corellou, F., Dauvillée, D., Sorokin, A., Goryanin, I., Ball, S., et al. (2011). Microarray data can predict diurnal changes of starch content in the picoalga *Ostreococcus*. *BMC Systems Biol.* 5:36. doi: 10.1186/1752-0509-5-36

Sterck, L., Billiau, K., Abeel, T., Rouzé, P., and Van der Peer, Y. (2012). ORCAE: online resource for community annotation of eukaryotes. *Nat. Methods* 9:1041. doi: 10.1038/nmeth.2242

Supek, F., Bošnjak, M., Škunca, N., and Šmuc, T. (2011). REVIGO summarizes and visualizes long lists of gene ontology terms. *PLoS ONE* 6:e21800. doi: 10.1371/journal.pone.0021800

Thaben, P. F., and Westermark, P. O. (2014). Detecting rhythms in time series with RAIN. *J. Biol. Rhythms* 29, 391–400. doi: 10.1177/07487304145 53029

Thommen, Q., Pfeuty, B., Schatt, P., Bijoux, A., Bouget, F. Y., and Lefranc, M. (2015). Probing entrainment of *Ostreococcus tauri* circadian clock by green and blue light through a mathematical modeling approach. *Front. Genet.* 6:65. doi: 10.3389/fgene.2015.00065

Trachana, K., Jensen, L. J., and Bork, P. (2010). Evolution and regulation of cellular periodic processes: a role for paralogues. *EMBO Rep.* 11, 233–238. doi: 10.1038/embor.2010.9

Trapnell, C., Roberts, A., Goff, L., Pertea, G., Kim, D., Kelley, D. R., et al. (2012). Differential gene and transcript expression analysis of RNA-seq experiments with TopHat and Cufflinks. *Nat. Protoc.* 7, 562–578. doi: 10.1038/nprot.2012.016

Valverde, F. (2011). CONSTANS and the evolutionary origin of photoperiodic timing of flowering. *J. Exp. Bot.* 62, 2453–2463. doi: 10.1093/jxb/erq449

Van Norman, J. M., and Benfey, P. N. (2009). *Arabidopsis thaliana* as a model organism in systems biology. *Wiley Interdiscip. Rev. Syst. Biol. Med.* 1, 372–379. doi: 10.1002/wsbm.25

Wang, M., Yuan, D., Gao, W., Li, Y., Tan, J., and Zhang, X. (2013). A comparative genome analysis of PME and PMEI families reveals the evolution of pectin metabolism in plant cell walls. *PLoS ONE* 8:e72082. doi: 10.1371/journal.pone.0072082

Wolf, Y. I., and Koonin, E. V. (2012). A tight link between orthologs and bidirectional best hits in bacterial and archaeal genomes. *Genome Biol. Evol.* 4, 1286–1294. doi: 10.1093/gbe/evs100

Yu, G., Wang, L., Han, Y., and He, Q. (2012). clusterProfiler: an R package for comparing biological themes among gene clusters. *OMICS* 16, 284–287. doi: 10.1089/omi.2011.0118

Zhang, J. D., Biczok, R., and Ruschhaupt, M. (2015). *ddCt: The ddCt Algorithm for the Analysis of Quantitative Real-Time PCR (qRT-PCR).* R package version 1.30.30.

Zhu, T., Nevo, E., Sun, D., and Peng, J. (2012). Phylogenetic analyses unravel the evolutionary history of NAC proteins in plants. *Evolution* 66, 1833–1848. doi: 10.1111/j.1558-5646.2011.01553.x

Zones, J. M., Blaby, I. K., Merchant, S. S., and Umen, J. G. (2015). High-resolution profiling of a synchronized diurnal transcriptome from *Chlamydomonas reinhardtii* reveals continuous cell and metabolic differentiation. *Plant Cell* 27, 2743–2769. doi: 10.1105/tpc.15.00498

Conflict of Interest Statement: The authors declare that the research was conducted in the absence of any commercial or financial relationships that could be construed as a potential conflict of interest.

14

Evolutionary Analysis of DELLA-Associated Transcriptional Networks

*Asier Briones-Moreno[1], Jorge Hernández-García[1], Carlos Vargas-Chávez[2], Francisco J. Romero-Campero[3,4], José M. Romero[4], Federico Valverde[4] and Miguel A. Blázquez[1]**

[1] *Instituto de Biología Molecular y Celular de Plantas, Consejo Superior de Investigaciones Científicas – Universidad Politécnica de Valencia, Valencia, Spain,* [2] *Institute for Integrative Systems Biology (I2SysBio), University of Valencia, Valencia, Spain,* [3] *Department of Computer Science and Artificial Intelligence, Universidad de Sevilla, Sevilla, Spain,* [4] *Instituto de Bioquímica Vegetal y Fotosíntesis, Consejo Superior de Investigaciones Científicas – Universidad de Sevilla, Sevilla, Spain*

Edited by:
Stefan de Folter,
Center for Advanced Research, The National Polytechnic Institute, Cinvestav-IPN, Mexico

Reviewed by:
Marie Monniaux,
Max Planck Institute for Plant Breeding Research (MPG), Germany
Frank Wellmer,
Trinity College, Dublin, Ireland
Jean-Michel Davière,
UPR2357 Institut de Biologie Moléculaire des Plantes (IBMP), France

***Correspondence:**
Miguel A. Blázquez
mblazquez@ibmcp.upv.es

DELLA proteins are transcriptional regulators present in all land plants which have been shown to modulate the activity of over 100 transcription factors in Arabidopsis, involved in multiple physiological and developmental processes. It has been proposed that DELLAs transduce environmental information to pre-wired transcriptional circuits because their stability is regulated by gibberellins (GAs), whose homeostasis largely depends on environmental signals. The ability of GAs to promote DELLA degradation coincides with the origin of vascular plants, but the presence of DELLAs in other land plants poses at least two questions: what regulatory properties have DELLAs provided to the behavior of transcriptional networks in land plants, and how has the recruitment of DELLAs by GA signaling affected this regulation. To address these issues, we have constructed gene co-expression networks of four different organisms within the green lineage with different properties regarding DELLAs: *Arabidopsis thaliana* and *Solanum lycopersicum* (both with GA-regulated DELLA proteins), *Physcomitrella patens* (with GA-independent DELLA proteins) and *Chlamydomonas reinhardtii* (a green alga without DELLA), and we have examined the relative evolution of the subnetworks containing the potential DELLA-dependent transcriptomes. Network analysis indicates a relative increase in parameters associated with the degree of interconnectivity in the DELLA-associated subnetworks of land plants, with a stronger effect in species with GA-regulated DELLA proteins. These results suggest that DELLAs may have played a role in the coordination of multiple transcriptional programs along evolution, and the function of DELLAs as regulatory 'hubs' became further consolidated after their recruitment by GA signaling in higher plants.

Keywords: gene co-expression networks, integrative molecular systems biology, evo–devo, transcriptional regulation, plant signaling

INTRODUCTION

Higher plants are characterized by a particularly flexible capacity to adapt to multiple environmental conditions. In other words, environmental signals are very efficient modulators of plant developmental decisions. This ability is generally assumed to be based on at least two mechanistic features: the presence of an extensive and sensitive repertoire of elements that perceive environmental signals (such as light photoreceptors covering a wide range of wavelengths), and the high degree of interconnectivity between the different signaling pathways to allow cellular integration of variable information (Casal et al., 2004).

Evidence has accumulated in recent years about the important role that plant hormones play in the translation of environmental signals into developmental decisions. On one hand, it has become evident that hormone pathways share common components with the pathways that transduce light and other environmental signals (Jaillais and Chory, 2010); and, on the other hand, hormones have been shown to participate in the regulation of developmental processes all throughout a plant's life cycle (Alabadi et al., 2009). In this context, gibberellins (GAs) and DELLA proteins are a paradigmatic example of the mechanisms that allow environmental signal integration. DELLA proteins constitute a small clade within the GRAS family of loosely defined plant specific nuclear proteins (Vera-Sirera et al., 2015). Their name was coined on the basis of a short stretch of amino acids (D-E-L-L-A) in their N-terminal region, which is tightly conserved among all higher plant species. They also present additional conserved motifs, such as the VHYNP domain, two leucine heptad repeats which may mediate protein–protein interactions, a putative nuclear localization signal, and a putative SH2 phosphotyrosine-binding domain, among others (Vera-Sirera et al., 2015). It has been shown in *Arabidopsis thaliana* and rice that recognition of GAs by their GID1 receptor allows physical interaction with DELLA proteins and promotes their degradation via the proteasome. In *A. thaliana*, loss of DELLA function mimics the phenotype of plants treated with an excess of GAs, both anatomically and also at the transcriptional level (Schwechheimer, 2011; Locascio et al., 2013b). Work in the past few years has established that DELLAs regulate transcription through the interaction with more than 100 transcription factors (TFs) in *A. thaliana* (de Lucas et al., 2008; Feng et al., 2008; Crocco et al., 2010; Hou et al., 2010; Gallego-Bartolomé et al., 2012; Daviere et al., 2014; Marin-de la Rosa et al., 2014, 2015; Resentini et al., 2015). In some cases, interaction with the TF inhibits its ability to bind DNA, while in other cases DELLAs seem to act as co-activators (Locascio et al., 2013b; Daviere and Achard, 2016). For all the cases examined in detail, the DELLA region responsible for the interaction with the TFs is the C-terminal region of the protein, the GRAS domain. Given that GA levels are strongly regulated by environmental signals such as light, temperature and photoperiod (Hedden and Thomas, 2012; Colebrook et al., 2014), cellular DELLA levels seem to be a proxy for the environmental status faced by plants (Claeys et al., 2014). Changes in DELLA levels could in turn differentially modulate distinct sets of TFs and their target genes in various developmental contexts. The promiscuous interaction with TFs, and the observation that *A. thaliana dellaKO* mutants display constitutive growth even under stress, and suffer from increased sensitivity to several types of stress factors such as salinity, cold, or fungal attacks (Alabadí et al., 2004; Achard et al., 2006, 2007, 2008a,b; Cheminant et al., 2011) suggests that DELLAs are potentially important 'hubs' in the transcriptional network that regulates the balance between growth and stress tolerance in higher plants.

Previous interest in the evolution of DELLA proteins is restricted to the question on how they were recruited to mediate cellular signaling by GAs. Based on phylogenetic analyses and shallow molecular analysis with fern and moss orthologs, it seems that the GA/GID1/DELLA module originated with early diverging tracheophytes (Wang and Deng, 2014). For instance, the *Selaginella* genus possesses the ability to synthesize GAs, a GID1 GA receptor, and a DELLA protein (Wang and Deng, 2014), which is sensitive to GA-induced degradation, even when introduced in an angiosperm, such as *A. thaliana* (Hirano et al., 2007; Yasumura et al., 2007). On the other hand, the DELLA proteins that existed in other land plants before the emergence of vascular plants were not involved in GA signaling. First, there are no *bona-fide DELLA* genes in algae and, second, the genomes of bryophytes like *Physcomitrella patens* encode DELLA proteins that lack the canonical 'DELLA motif' (Wang and Deng, 2014), and PpDELLAs are not sensitive to GAs when introduced in *A. thaliana* (Yasumura et al., 2007). However, the ability of DELLA proteins to modulate transcriptional programs relies on the GRAS domain through which interactions with TFs occur, and the evolution of this activity has not been addressed before.

In an attempt to identify the possible function of ancestral DELLAs and to delineate how evolution has shaped the functions of the GA/DELLA module in higher plants, we have addressed the analysis of the transcriptional networks potentially regulated by DELLAs in several species. For this reason, we have used gene co-expression networks, in which genes are represented as nodes, and if two genes exhibit a significant correlation value for co-expression, the corresponding nodes are joined by an edge. Importantly, if a node is a TF, first neighbors can be confidently taken as targets for that particular TF (Franco-Zorrilla et al., 2014). Therefore, the analysis of topological parameters of a gene co-expression network is an interesting tool that may reveal information about the function and evolutionary history of transcriptional programs (Aoki et al., 2007; Usadel et al., 2009).

Here we have investigated the properties of networks formed by DELLA-interacting TFs and their co-expressing genes in *A. thaliana*, and compared them with the orthologous networks in three other plant species: (i) *Solanum lycopersicum* (possessing a fully operative GA/DELLA module); (ii) *P. patens* (possessing GA-independent DELLA functions); and (iii) *Chlamydomonas reinhardtii* (without GA perception or DELLAs) (**Figure 1A**). All the parameters examined suggest that the functions regulated by DELLA-interacting TFs (and thus DELLAs themselves) have increased their level of coordination along evolution.

RESULTS AND DISCUSSION

Construction of Networks and Subnetworks

Gene expression data from RNA sequencing (RNA-seq) experiments in *A. thaliana*, *S. lycopersicum*, *P. patens*, and *C. reinhardtii* were obtained from the Gene Expression Omnibus, and gene co-expression networks were inferred for

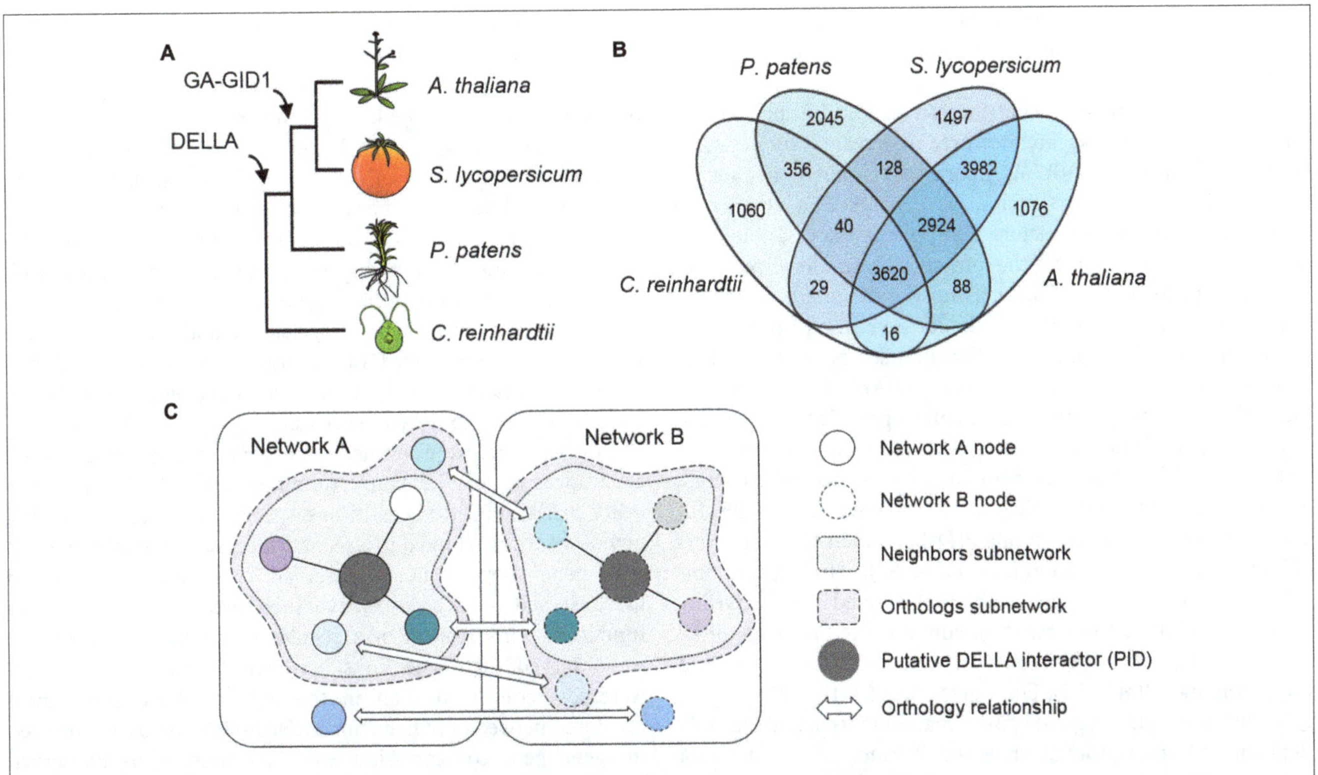

FIGURE 1 | Phylogenetic relationships between the chosen species. (A) Representation of the species tree indicating the origin of key elements related to the gibberellin signaling pathway. **(B)** Venn's diagram showing the number of OrthoMCL groups in which genes of each species are present. **(C)** Schematic representation of the basis for subnetwork design. Nodes in different networks with the same color indicate an orthologous relationship.

each species from transcriptomic data as described in section "Materials and Methods." All four networks are scale-free networks (Supplementary Figure S1) (Romero-Campero et al., 2013, 2016) and have comparable sizes in terms of number of nodes, but there are remarkable differences in the way they are connected (**Table 1**). The *A. thaliana* network contains more than twice as many edges than the others, the average degree of its nodes (average number of connections) is one order of magnitude higher and its average shortest path length (average number of nodes between two random nodes) is lower. Even though the number of genes of each species represented in the networks is similar, in some species they are more connected, possibly due to differences in their endogenous regulation and the availability of experimental data. For that reason, we decided to do every comparative analysis between the different species in relative terms.

To be able to compare the co-expression networks of the different species, we first identified the orthologous nodes in each of them using the OrthoMCL method (Li et al., 2003). Up to 17,053 groups of genes were obtained. Genes in the same group were considered orthologs or paralogs if they belonged to different or the same species, respectively. The four species were represented unequally, as both *A. thaliana* and *S. lycopersicum* genes were present in ca. 70% of the groups, while *P. patens* genes were found in little more than 50% of them, and only ca. 30% of the groups contained genes from *C. reinhardtii*

TABLE 1 | General parameters in co-expression networks.

	C. reinhardtii			*P. patens*			*S. lycopersicum*			*A. thaliana*		
	Full	**Neigh**	**Ortho**	**Full**	**Neigh**	**Ortho**	**Full**	**Neigh**	**Ortho**	**Full**	**Neigh**	**Ortho**
Nodes	8652	48	658	8564	448	1503	7851	1314	2885	5663	2070	2949
Edges	145903	78	1173	295317	15078	19828	287409	153396	169171	593730	460951	512042
Average degree	33.73	3.25	3.57	68.97	67.31	26.38	73.22	233.48	117.28	209.69	445.36	347.26
Average shortest path length	7.37	1.91	8.71	13.11	1.39	12.01	13.78	1.67	5.63	4.28	2.15	3.09
Diameter	23	4	24	46	4	41	44	6	25	20	9	12

Parameters of networks and subnetworks used in this study. Full, full gene co-expression network; Neigh, first neighbors subnetwork; Ortho, orthologs subnetwork; C. reinhardtii, Chlamydomonas reinhardtii; P. patens, Physcomitrella patens; S. lycopersicum, Solanum lycopersicum; A. thaliana, Arabidopsis thaliana.

(**Figure 1B**). This was already expected, given the evolutionary distance among these species and the genomic complexity of each one.

To assess the possible contribution of DELLA proteins to co-expression networks architecture, we created subnetworks based on reported DELLA interactors known to act as transcriptional regulators. First, we compiled a list of all published DELLA interactors (**Supplementary Table S1**), obtained their orthologs for the four species, and localized them in their respective networks. Since most of the interactions have been described for *A. thaliana*, the corresponding orthologs in the other species are only "putative interactors of the DELLA proteins" (PIDs), and the first neighbors of AtDELLA interactors and PIDs are their putative expression targets. Second, we built two different subnetworks using this information. The first one, called "Neighbors" subnetwork (abbreviated as AtNeigh, SlNeigh, PpNeigh, and CrNeigh), is composed of the DELLA interactors (or the corresponding PIDs) and their first neighbors (**Figure 1C** and **Supplementary Table S2**). The second one, called "Orthologs" subnetwork (abbreviated as AtOrtho, SlOrtho, PpOrtho, and CrOrtho), contains the orthologs of all the first neighbors of PIDs in all the species (**Figure 1C** and **Supplementary Table S3**). For a given species, the "Neighbors" subnetwork provides a good approximation to its actual DELLA-dependent transcriptome, while the "Orthologs" subnetwork represents the full landscape of potential transcriptional targets for DELLAs, since it includes orthologs of genes that are DELLA transcriptional targets in other species (**Figure 2**).

DELLA-Associated Subnetworks Reflect Increased Relevance of DELLAs after Being Recruited by GA Signaling

It is important to take into account a circumstance that affects the construction of subnetworks: OrthoMCL does not always retrieve orthologs for some of the genes, because either they do not exist in the other species, or the method does not provide high-confidence results. This results in a particular bias toward smaller subnetwork sizes with increasing phylogenetic distance (**Table 1**). However, the impact of this bias can be disregarded when analyzing relative parameters. Hence, regardless of the absolute sizes, we observed that the average degree in the Neighbor subnetworks increased dramatically in SlNeigh and AtNeigh with respect to their full networks (more than threefold and twofold, respectively), while this parameter did not change in PpNeigh, and it actually decreased in CrNeigh (**Table 1**). Similarly, the Orthologs subnetworks displayed an equivalent behavior as the Neighbors subnetworks: their diameter and average shortest path length decreased considerably more in SlOrtho and AtOrtho with respect to the full networks; and the same happened with the increase of the average degree. In summary, both subnetworks showed a higher compaction and interconnection of nodes in relative terms in the case of *S. lycopersicum* and *A. thaliana* compared with *P. patens* and *C. reinhardtii*, indicating that the putative interactors and targets of the DELLAs become more connected in those species presenting GA-regulated DELLAs.

A confirmation of the impact of GA regulation on the relevance of DELLA function is found in the analysis of neighborhood conservation. **Figure 3A** shows the percentage of genes with a significantly overlapping neighborhood in each comparison (see Materials and Methods). When comparing *P. patens* with the other species, there are no substantial differences between the full network and the Orthologs subnetwork. On the contrary, SlOrtho and AtOrtho contain a considerably higher proportion of genes with conserved neighborhood than their corresponding full networks (15% vs. 10%). Between *S. lycopersicum* and *A. thaliana*, the regulation of the putative DELLA targets is more conserved than for the network in general, so this group of genes seems to have a cohesive element in the two species.

Furthermore, we examined gene–gene co-expression values, as a measure of the conservation of individual edges. For every pair of linked genes in one species, if the corresponding orthologs are also linked in a second species, it is considered that gene–gene co-expression is conserved. Therefore, the calculation of conserved links between two subnetworks is a measure of functional conservation of a regulatory module. Interestingly, we observed that gene links between PpOrtho and SlOrtho were less conserved than in the full networks, and almost unaltered between PpOrtho and AtOrtho (**Figure 3B**). However, the gene–gene co-expression was three times more conserved between SlOrtho and AtOrtho than between their full networks (11% vs. 3.5%). In other words, these data are compatible with the proposition that the presence of GA-regulated DELLAs (in *S. lycopersicum* and *A. thaliana*) provides stronger links between transcriptional programs, not detected in an organism with GA-independent DELLAs (*P. patens*).

Efficiency of Transcriptional Regulation Is a DELLA-Associated Parameter

The efficiency of a transcriptional regulatory mechanism can be evaluated through two additional parameters in gene co-expression networks: shortest path length distribution and motif frequency. In network theory, average shortest-path length is defined as the average number of steps along the shortest paths for all possible pairs of network nodes. It is a measure of the efficiency of information propagation on a network, with a shorter average path length being more efficient (Vragovic et al., 2005). When we compared the distribution of shortest path lengths in full and Orthologs subnetworks, we observed a clear tendency toward shorter path lengths in the Orthologs subnetworks of organisms possessing DELLAs (*S. lycopersicum*, *A. thaliana*, and *P. patens*) compared with the situation in an organism without DELLAs (*C. reinhardtii*) (**Figure 4**).

Network motifs are small recurring patterns involving a few nodes that appear more frequently in biological networks than in randomized ones. They consist of a certain level of regulation which connects small sets of nodes with a particular topology. Motifs characterize a network, as some of them are useful for the regulation of determined functions, and thus conserved along evolution (Kashtan and Alon, 2005). After measuring the frequency of the eight common motifs

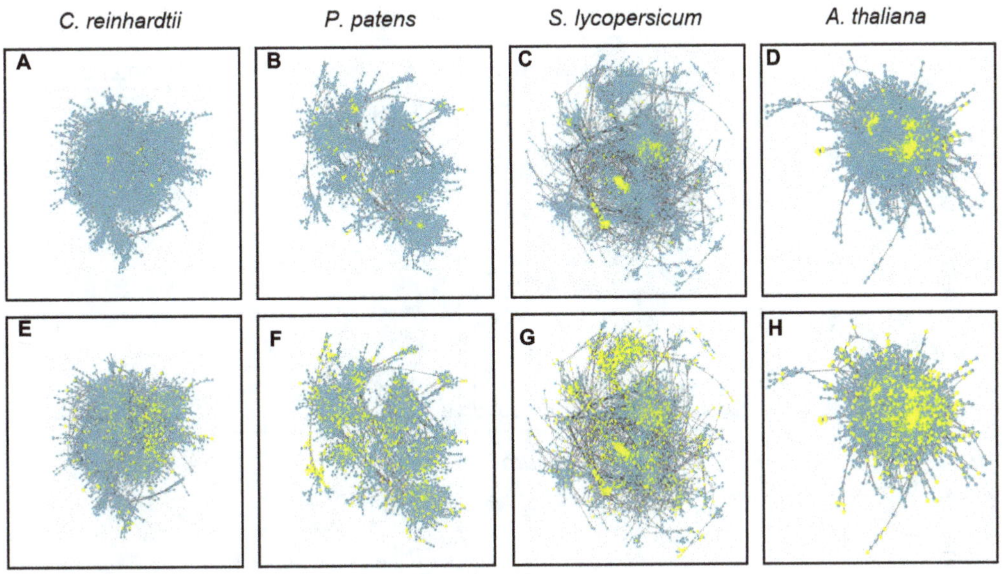

FIGURE 2 | Gene co-expression networks. Full *Chlamydomonas reinhardtii* **(A,E)**, *Physcomitrella patens* **(B,F)**, *Solanum lycopersicum* **(C,G)**, and *Arabidopsis thaliana* **(D,H)** gene co-expression networks. Neighbors subnetworks are comprised of yellow-marked nodes in A-D. Orthologs subnetworks are comprised of yellow-marked nodes in **(E–H)**.

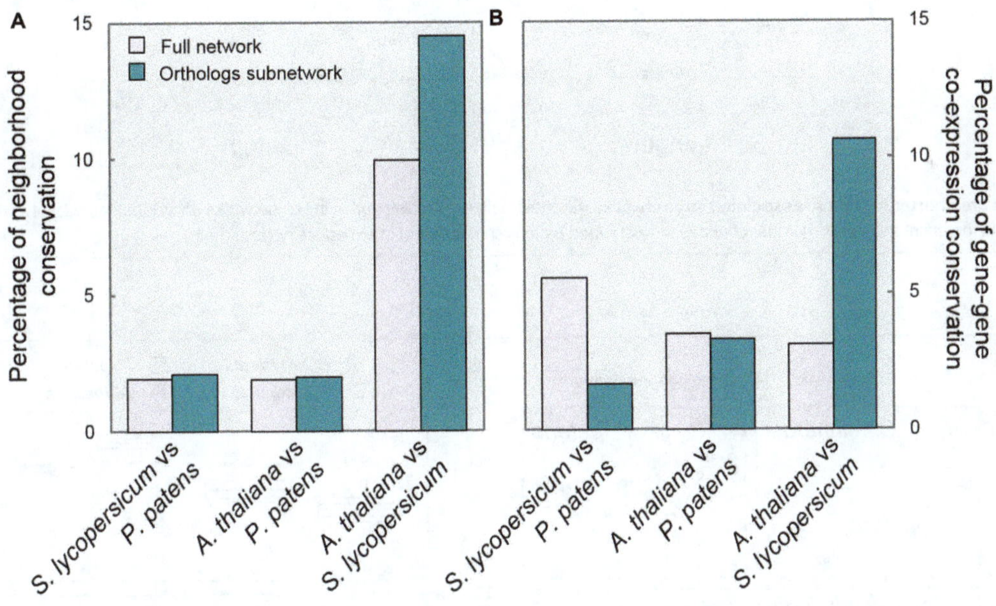

FIGURE 3 | Gene connections are more conserved in species with GA-regulated DELLAs. Pairwise comparisons of *P. patens*, *S. lycopersicum,* and *A. thaliana* Full networks and Ortho subnetworks regarding: **(A)** Percentage of genes with significantly overlapping neighborhoods; **(B)** Percentage of conserved gene–gene links.

composed of three and four nodes in the full networks, we found that there was no relative enrichment of any particular motif between species when comparing the full networks or the Orthologs subnetworks (**Figure 5A**). However, the AtOrtho, SlOrtho, and PpOrtho subnetworks displayed a clear enrichment in virtually every motif, compared with their respective full networks (**Figure 5B**). Given that the function of this sort of motifs is to allow coordinated expression of a group of genes with shared function (Alon, 2007), the increase in the proportion of small regulatory patterns among all the putative DELLA targets in species that do contain DELLAs indicates an increase in the complexity of gene regulation, in which DELLAs might mediate the coordination of transcriptional programs.

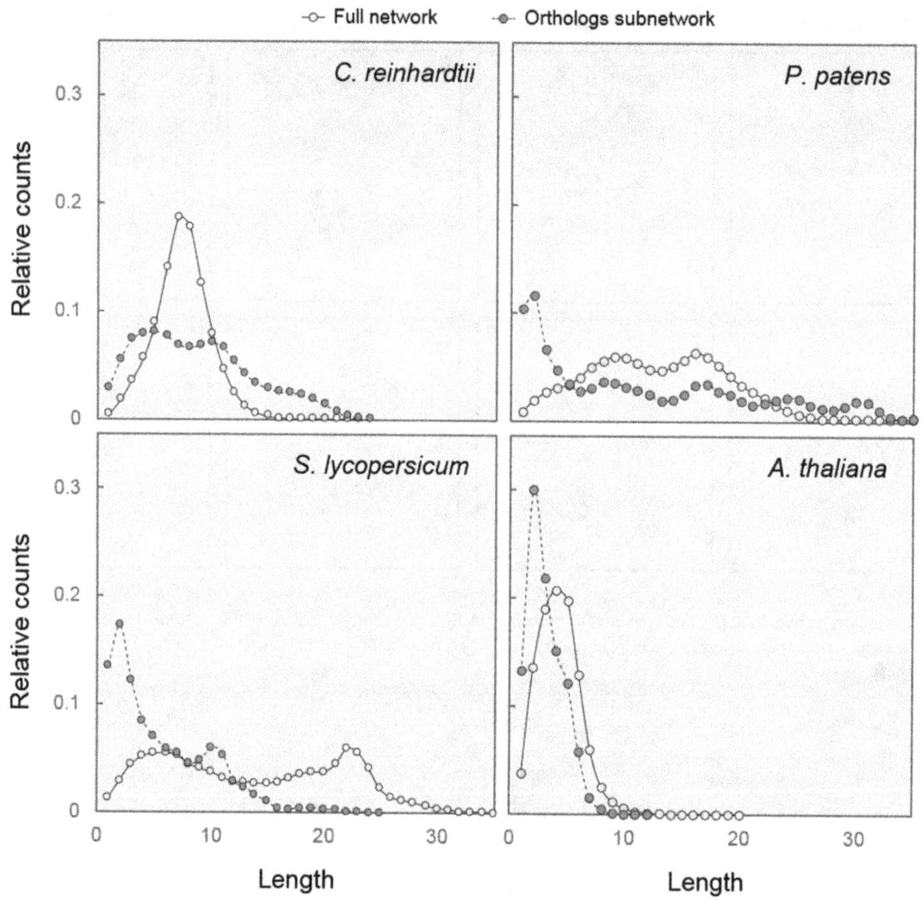

FIGURE 4 | Paths are shorter in DELLA-associated subnetworks. Shortest path length distribution in Full networks and Orthologs subnetworks from the four species. The graphs represent the relative number of nodes (*y*-axis) joined by a given number of intermediate nodes (*x*-axis).

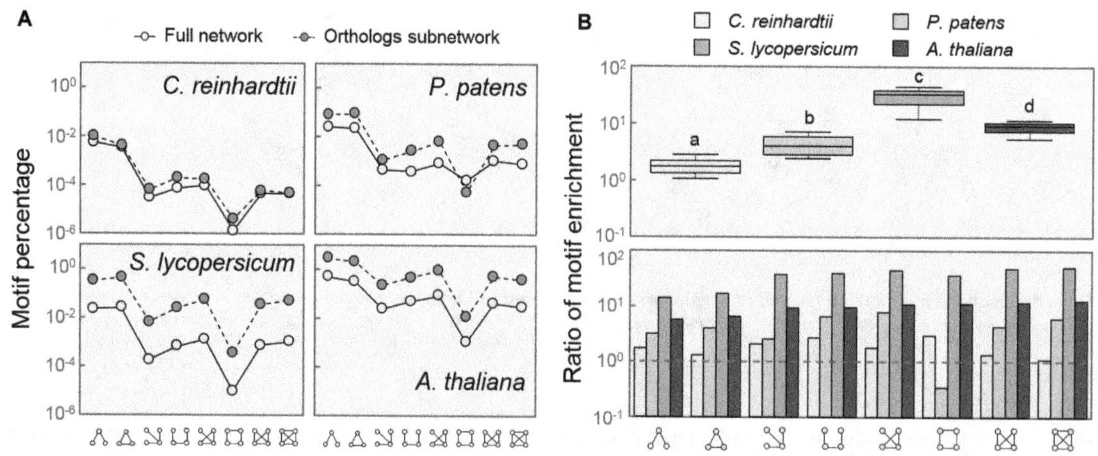

FIGURE 5 | Network motifs are enriched in DELLA related networks. (A) Percentage of motifs found in each network compared to possible combinations of three and four nodes. **(B)** Ratio of motif enrichment comparing Orthologs subnetworks to Full networks per species (upper panel), and per motif (lower panel). Dashed lines in **(B)** mark a ratio of 1. Motifs are as depicted in *X*-axis. Letters indicate significant differences between groups, $p < 0.01$ (One way ANOVA, Tukey HSD *Post Hoc* test). Box-plot whiskers are Tukey-defined (extended 1.5 times the IQR from the box edges).

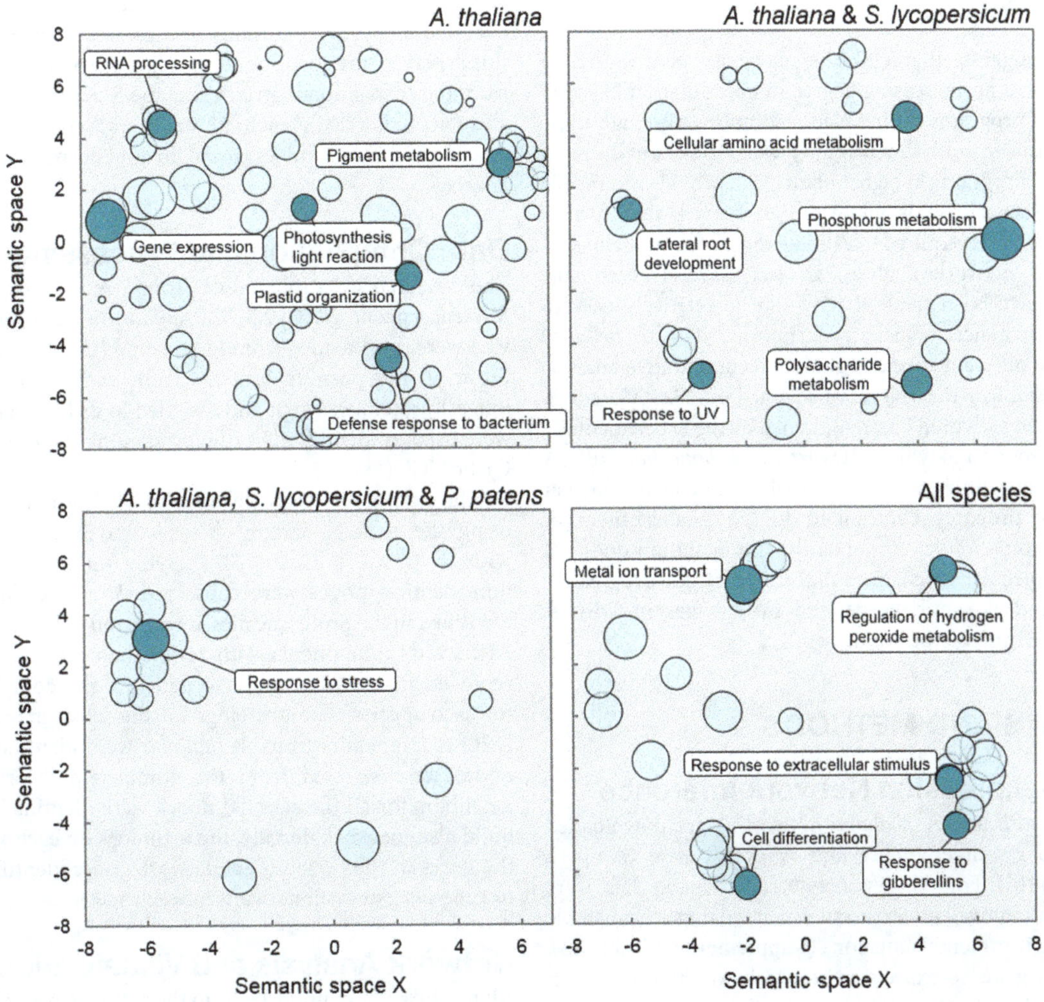

FIGURE 6 | Gene Ontology terms enriched in Neighbors subnetworks. Scatterplots show cluster representatives after redundancy reduction in a two dimensional space derived by applying multidimensional scaling to a matrix of the GO categories semantic similarities. Bubble size is proportional to *p*-value significance of GO enrichment.

The Regulation of the Stress Response: A Likely Role of Ancestral DELLA Proteins

The results shown above suggest that the origin of DELLAs in land plants would be associated to an increase in the co-expression between genes that are putative targets of DELLA-interacting TFs, both in terms of size of the gene set and degree of the co-expression value. Therefore, DELLAs would have helped in the coordination of certain transcriptional circuits, and their recruitment to mediate GA signaling later in development would have further expanded their coordination capacity. To reveal the most likely functions ultimately regulated by DELLAs in the common ancestor of land plants, we carried out Gene Ontology (GO) analyses on each of the Neighbor subnetworks, with the idea that the terms shared by those in *S. lycopersicum*, *A. thaliana*, and *P. patens* could represent likely functions regulated by the ancestral DELLA proteins.

Not surprisingly, given the larger size of AtNeigh (**Table 1**), GO analysis rendered a much larger number of terms significantly enriched in this subnetwork, compared to those from the other three organisms (**Supplementary Table S4**). Terms referring to chloroplast function, such as plastid organization, photosynthesis, or pigment biosynthesis (including chlorophyll) were specifically enriched among the putative DELLA targets in *A. thaliana* only (**Figure 6**). This result might reflect functions whose regulation by DELLA has been acquired more recently, or it could simply be a bias of the analysis, caused by the big difference in size of the analyzed sets in the different species. On the contrary, the finding that terms comprised under general 'response to stress' were significantly over-represented in the subnetworks of the three land plants, but not *C. reinhardtii*, suggests that this function might have been the primary target of the regulation by ancestral DELLAs through their interaction with specific TFs.

CONCLUSION

Our analysis suggests that DELLAs may have contributed to the acquisition of an increasing degree of coordination between transcriptional programs during plant evolution. Although these results are consistent with the current view of DELLAs as 'hubs' in transcriptional programs in higher plants, and provide a plausible evolutionary scenario, it is important to remark that further experimental work is required to validate most of the conclusions from *in silico* network analysis. In fact, several reasonable assumptions have been made that would be relatively easy to confirm. For instance, actual transcriptomic data of *dellaKO* mutants in the different species, coupled to comparative analysis would help establish the role of ancestral DELLAs. Moreover, our current analysis would be strengthened by the experimentally obtained information of which PIDs are in fact *bona-fide* DELLA interactors in the different species. Finally, the conclusion that DELLAs have probably contributed to the establishment of new co-regulatory circuits during land-plant evolution does not explain the molecular mechanism that supports this progressive acquisition, and it can be generated by changes in DELLA proteins, in their interactors, or in both.

MATERIALS AND METHODS

Gene Co-expression Network Inference

The *C. reinhardtii* and *A. thaliana* networks were downloaded from the web resources of previous work (Romero-Campero et al., 2013, 2016). For the new networks, RNA-seq data were selected from equivalent experiments involving comparable tissues and environmental situations (**Supplementary Table S5**). The *P. patens* gene co-expression network was inferred from the RNA-seq data freely available from the Gene Expression Omnibus identified with accession numbers GSE19824, GSE33279, GSE36274, and GSE25237. The *S. lycopersicum* network was constructed based on the RNA-seq data identified with the accession numbers GSE45774, GSE64665, GSE64981, GSE68018, and GSE77340 in the Gene Expression Omnibus. In both cases, RNA-seq data was processed using the Tuxedo protocol (Trapnell et al., 2012) to obtain gene expression levels measured as FPKM. Briefly, short reads were mapped to the corresponding reference genome using Tophat, transcripts were assembled using Cufflinks and expression levels were computed using Cuffdiff. The Bioconductor R package cummeRbund (Goff et al., 2013) was used for subsequent analysis of the results generated by the Tuxedo protocol. In order to reduce noise in our analysis only genes that were detected as differentially expressed in at least one of the studies integrated in this work were considered. Differentially expressed genes were determined comparing each condition with the corresponding control within each study using a fold-change threshold of two. For each species, a matrix containing the expression levels of the selected genes was extracted. The Pearson correlation coefficient between every pair of gene expression profiles was computed using the *cor* function from the stats R package to generate a correlation matrix. Two genes were assumed to be co-expressed when the Pearson correlation coefficient between their expression profiles over the analyzed conditions was greater than 0.95. Following this criterion, the corresponding adjacency matrix was generated from the correlation matrix. Using the R package igraph[1] (Csardi and Nepusz, 2006), each network was constructed from its adjacency matrix and exported in gml formal for subsequent analysis.

Data Compilation and Processing

The reference proteomes from *A. thaliana* TAIR10, *S. lycopersicum* iTAGv2.3, *C. reinhardtii* v5.5, and *P. patens* v3.3 were downloaded from Phytozome (Goodstein et al., 2012). From all the possible proteins from each locus tag only the longest protein was kept and assigned to its locus tag. These files were used to identify the orthologs among the four species with OrthoMCL (Li et al., 2003).

The networks were converted to SIF format and processed using the package igraph[1] (Csardi and Nepusz, 2006) made with R[2] (R Core Team, 2016). Only the edges between two non-identical nodes were conserved. If a given node was not identified in the proteome files, it was removed from the network. Afterward, components with fewer than seven elements were removed from the network to generate the complete network for each species. The orthologs for the set of manually curated DELLA interactors from *A. thaliana* were identified, and these nodes were selected from the complete networks. The first neighbors for all the selected nodes were identified and used to build a subnetwork. Finally, the orthologs on each species for all the genes in the previous subnetworks were identified and used to generate a new subnetwork for each species.

Network Analysis and Visualization

All networks were imported into the software package Cytoscape (Smoot et al., 2011) for their visualization using the Prefuse Force Directed layout.

The measures of network topology were calculated using both predefined and custom made functions. The gene–gene co-expression and neighborhood conservation were determined following the approach described by Netotea et al. (2014), using Fisher exact tests to check for statistical significance.

Gene Ontology analysis on Neigh subnetworks was made with AgriGO (Du et al., 2010), and represented with ReviGO (Supek et al., 2011).

AUTHOR CONTRIBUTIONS

AB-M, JH-G and MB conceived and designed the work. FR-C, JR, and FV constructed the co-expression networks. AB-M, CV-C, and JH-G performed network analyses. AB-M and MB wrote the first draft of the manuscript, to which all authors contributed.

[1] http://igraph.com

[2] http://www.Rproject.org

FUNDING

Work in the laboratories was funded by grants BFU2016-80621-P and BIO2014-52425-P of the Spanish Ministry of Economy, Industry and Competitiveness, and H2020-MSCA-RISE-2014-644435 of the European Union. AB-M and JH-G hold Fellowships of the Spanish Ministry of Education, Culture and Sport FPU14/01941 and FPU15/01756, respectively.

ACKNOWLEDGMENTS

We thank the members of the Hormone Signaling and Plasticity Lab at IBMCP (http://www.ibmcp.upv.es/BlazquezAlabadiLab/) for useful discussions and suggestions.

SUPPLEMENTARY MATERIAL

The Supplementary Material for this article can be found online at: http://journal.frontiersin.org/article/10.3389/fpls.2017.00626/full#supplementary-material

TABLE S1 | Compilation of DELLA interactors used in this study.

TABLE S2 | Genes included in the 'Neighbors' subnetworks.

TABLE S3 | Genes included in the 'Orthologs' subnetworks.

TABLE S4 | Gene Onthology categories enriched in the 'Neighbors' subnetworks.

TABLE S5 | RNA-seq datasets used for the construction of the new gene co-expression networks in *Physcomitrella patens* and *Solanum lycopersicum*.

REFERENCES

Achard, P., Cheng, H., De Grauwe, L., Decat, J., Schoutteten, H., Moritz, T., et al. (2006). Integration of plant responses to environmentally activated phytohormonal signals. *Science* 311, 91–94. doi: 10.1126/science.1118642

Achard, P., Gong, F., Cheminant, S., Alioua, M., Hedden, P., and Genschik, P. (2008a). The cold-inducible CBF1 factor-dependent signaling pathway modulates the accumulation of the growth-repressing DELLA proteins via its effect on gibberellin metabolism. *Plant Cell* 20, 2117–2129. doi: 10.1105/tpc.108.058941

Achard, P., Liao, L., Jiang, C., Desnos, T., Bartlett, J., Fu, X., et al. (2007). DELLAs contribute to plant photomorphogenesis. *Plant Physiol.* 143, 1163–1172. doi: 10.1104/pp.106.092254

Achard, P., Renou, J. P., Berthome, R., Harberd, N. P., and Genschik, P. (2008b). Plant DELLAs restrain growth and promote survival of adversity by reducing the levels of reactive oxygen species. *Curr. Biol.* 18, 656–660. doi: 10.1016/j.cub.2008.04.034

Alabadi, D., Blazquez, M. A., Carbonell, J., Ferrandiz, C., and Perez-Amador, M. A. (2009). Instructive roles for hormones in plant development. *Int. J. Dev. Biol.* 53, 1597–1608. doi: 10.1387/ijdb.072423da

Alabadí, D., Gil, J., Blázquez, M. A., and García-Martínez, J. L. (2004). Gibberellins repress photomorphogenesis in darkness. *Plant Physiol.* 134, 1050–1057. doi: 10.1104/pp.103.035451

Alon, U. (2007). Network motifs: theory and experimental approaches. *Nat. Rev. Genet.* 8, 450–461. doi: 10.1038/nrg2102

Aoki, K., Ogata, Y., and Shibata, D. (2007). Approaches for extracting practical information from gene co-expression networks in plant biology. *Plant Cell Physiol.* 48, 381–390. doi: 10.1093/pcp/pcm013

Arnaud, N., Girin, T., Sorefan, K., Fuentes, S., Wood, T. A., Lawrenson, T., et al. (2010). Gibberellins control fruit patterning in *Arabidopsis thaliana*. *Genes Dev.* 24, 2127–2132. doi: 10.1101/gad.593410

Bai, M. Y., Shang, J. X., Oh, E., Fan, M., Bai, Y., Zentella, R., et al. (2012). Brassinosteroid, gibberellin and phytochrome impinge on a common transcription module in *Arabidopsis*. *Nat. Cell Biol.* 14, 810–817. doi: 10.1038/ncb2546

Casal, J. J., Fankhauser, C., Coupland, G., and Blázquez, M. A. (2004). Signalling for developmental plasticity. *Trends Plant Sci.* 9, 309–314. doi: 10.1016/j.tplants.2004.04.007

Cheminant, S., Wild, M., Bouvier, F., Pelletier, S., Renou, J. P., Erhardt, M., et al. (2011). DELLAs regulate chlorophyll and carotenoid biosynthesis to prevent photooxidative damage during seedling deetiolation in *Arabidopsis*. *Plant Cell* 23, 1849–1860. doi: 10.1105/tpc.111.085233

Claeys, H., De Bodt, S., and Inze, D. (2014). Gibberellins and DELLAs: central nodes in growth regulatory networks. *Trends Plant Sci.* 19, 231–239. doi: 10.1016/j.tplants.2013.10.001

Colebrook, E. H., Thomas, S. G., Phillips, A. L., and Hedden, P. (2014). The role of gibberellin signalling in plant responses to abiotic stress. *J. Exp. Biol.* 217, 67–75. doi: 10.1242/jeb.089938

Crocco, C. D., Holm, M., Yanovsky, M. J., and Botto, J. F. (2010). AtBBX21 and COP1 genetically interact in the regulation of shade avoidance. *Plant J.* 64, 551–562. doi: 10.1111/j.1365-313X.2010.04360.x

Csardi, G., and Nepusz, T. (2006). The igraph software package for complex network research. *InterJournal Complex Syst.* 1695.

Daviere, J. M., and Achard, P. (2016). A pivotal role of DELLAs in regulating multiple hormone signals. *Mol. Plant* 9, 10–20. doi: 10.1016/j.molp.2015.09.011

Daviere, J. M., Wild, M., Regnault, T., Baumberger, N., Eisler, H., Genschik, P., et al. (2014). Class I TCP-DELLA interactions in inflorescence shoot apex determine plant height. *Curr. Biol.* 24, 1923–1928. doi: 10.1016/j.cub.2014.07.012

de Lucas, M., Davière, J. M., Rodríguez-Falcón, M., Pontin, M., Iglesias-Pedraz, J. M., Lorrain, S., et al. (2008). A molecular framework for light and gibberellin control of cell elongation. *Nature* 451, 480–484. doi: 10.1038/nature06520

Du, Z., Zhou, X., Ling, Y., Zhang, Z., and Su, Z. (2010). agriGO: a GO analysis toolkit for the agricultural community. *Nucleic Acids Res.* 38, W64–W70. doi: 10.1093/nar/gkq310

Feng, S., Martinez, C., Gusmaroli, G., Wang, Y., Zhou, J., Wang, F., et al. (2008). Coordinated regulation of *Arabidopsis thaliana* development by light and gibberellins. *Nature* 451, 475–479. doi: 10.1038/nature06448

Feurtado, J. A., Huang, D., Wicki-Stordeur, L., Hemstock, L. E., Potentier, M. S., Tsang, E. W., et al. (2011). The *Arabidopsis* C2H2 zinc finger INDETERMINATE DOMAIN1/ENHYDROUS promotes the transition to germination by regulating light and hormonal signaling during seed maturation. *Plant Cell* 23, 1772–1794. doi: 10.1105/tpc.111.085134

Fonouni-Farde, C., Tan, S., Baudin, M., Brault, M., Wen, J., Mysore, K. S., et al. (2016). DELLA-mediated gibberellin signalling regulates Nod factor signalling and rhizobial infection. *Nat. Commun.* 7:12636. doi: 10.1038/ncomms12636

Franco-Zorrilla, J. M., Lopez-Vidriero, I., Carrasco, J. L., Godoy, M., Vera, P., and Solano, R. (2014). DNA-binding specificities of plant transcription factors and their potential to define target genes. *Proc. Natl. Acad. Sci. U.S.A.* 111, 2367–2372. doi: 10.1073/pnas.1316278111

Fukazawa, J., Teramura, H., Murakoshi, S., Nasuno, K., Nishida, N., Ito, T., et al. (2014). DELLAs function as coactivators of GAI-ASSOCIATED FACTOR1 in regulation of gibberellin homeostasis and signaling in *Arabidopsis*. *Plant Cell* 26, 2920–2938. doi: 10.1105/tpc.114.125690

Gallego-Bartolomé, J., Arana, M. V., Vandenbussche, F., Zadnikova, P., Minguet, E. G., Guardiola, V., et al. (2011). Hierarchy of hormone action controlling apical hook development in Arabidopsis. *Plant J.* 67, 622–634. doi: 10.1111/j.1365-313X.2011.04621.x

Gallego-Bartolomé, J., Minguet, E. G., Grau-Enguix, F., Abbas, M., Locascio, A., Thomas, S. G., et al. (2012). Molecular mechanism for the interaction between gibberellin and brassinosteroid signaling pathways in *Arabidopsis*. *Proc. Natl. Acad. Sci. U.S.A.* 109, 13446–13451. doi: 10.1073/pnas.1119992109

Gallego-Bartolomé, J., Minguet, E. G., Marín, J. A., Prat, S., Blázquez, M. A., and Alabadí, D. (2010). Transcriptional diversification and functional conservation between DELLA proteins in Arabidopsis. *Mol. Biol. Evol.* 27, 1247–1256. doi: 10.1093/molbev/msq012

Goff, L., Trapnell, C., and Kelley, D. (2013). cummeRbund: Analysis, Exploration, Manipulation, and Visualization of Cufflinks High-Throughput Sequencing Data in: R Package Version 2.16.0. Burlington, MA: ScienceOpen, Inc.

Goodstein, D. M., Shu, S., Howson, R., Neupane, R., Hayes, R. D., Fazo, J., et al. (2012). Phytozome: a comparative platform for green plant genomics. *Nucleic Acids Res.* 40, D1178–D1186. doi: 10.1093/nar/gkr944

Heck, C., Kuhn, H., Heidt, S., Walter, S., Rieger, N., and Requena, N. (2016). Symbiotic fungi control plant root cortex development through the novel GRAS transcription factor MIG1. *Curr. Biol.* 26, 2770–2778. doi: 10.1016/j.cub.2016.07.059

Hedden, P., and Thomas, S. G. (2012). Gibberellin biosynthesis and its regulation. *Biochem. J.* 444, 11–25. doi: 10.1042/BJ20120245

Hirano, K., Nakajima, M., Asano, K., Nishiyama, T., Sakakibara, H., Kojima, M., et al. (2007). The GID1-mediated gibberellin perception mechanism is conserved in the lycophyte *Selaginella moellendorffii* but not in the bryophyte *Physcomitrella patens*. *Plant Cell* 19, 3058–3079. doi: 10.1105/tpc.107.051524

Hong, G. J., Xue, X. Y., Mao, Y. B., Wang, L. J., and Chen, X. Y. (2012). *Arabidopsis* MYC2 interacts with DELLA proteins in regulating sesquiterpene synthase gene expression. *Plant Cell* 24, 2635–2648. doi: 10.1105/tpc.112.098749

Hou, X., Lee, L. Y., Xia, K., Yan, Y., and Yu, H. (2010). DELLAs modulate jasmonate signaling via competitive binding to JAZs. *Dev. Cell* 19, 884–894. doi: 10.1016/j.devcel.2010.10.024

Hou, X., Zhou, J., Liu, C., Liu, L., Shen, L., and Yu, H. (2014). Nuclear factor Y-mediated H3K27me3 demethylation of the SOC1 locus orchestrates flowering responses of *Arabidopsis*. *Nat. Commun.* 5:4601. doi: 10.1038/ncomms5601

Huang, D., Wang, S., Zhang, B., Shang-Guan, K., Shi, Y., Zhang, D., et al. (2015). A gibberellin-mediated DELLA-NAC signaling cascade regulates cellulose synthesis in rice. *Plant Cell* 27, 1681–1696. doi: 10.1105/tpc.15.00015

Hyun, Y., Richter, R., Vincent, C., Martinez-Gallegos, R., Porri, A., and Coupland, G. (2016). Multi-layered regulation of SPL15 and cooperation with SOC1 integrate endogenous flowering pathways at the *Arabidopsis* shoot meristem. *Dev. Cell* 37, 254–266. doi: 10.1016/j.devcel.2016.04.001

Itoh, H., Ueguchi-Tanaka, M., Sato, Y., Ashikari, M., and Matsuoka, M. (2002). The gibberellin signaling pathway is regulated by the appearance and disappearance of SLENDER RICE1 in nuclei. *Plant Cell* 14, 57–70. doi: 10.1105/tpc.010319

Jaillais, Y., and Chory, J. (2010). Unraveling the paradoxes of plant hormone signaling integration. *Nat. Struct. Mol. Biol.* 17, 642–645. doi: 10.1038/nsmb0610-642

Jin, Y., Liu, H., Luo, D., Yu, N., Dong, W., Wang, C., et al. (2016). DELLA proteins are common components of symbiotic rhizobial and mycorrhizal signalling pathways. *Nat. Commun.* 7:12433. doi: 10.1038/ncomms12433

Josse, E. M., Gan, Y., Bou-Torrent, J., Stewart, K. L., Gilday, A. D., Jeffree, C. E., et al. (2011). A DELLA in disguise: SPATULA restrains the growth of the developing *Arabidopsis* seedling. *Plant Cell* 23, 1337–1351. doi: 10.1105/tpc.110.082594

Kashtan, N., and Alon, U. (2005). Spontaneous evolution of modularity and network motifs. *Proc. Natl. Acad. Sci. U.S.A.* 102, 13773–13778. doi: 10.1073/pnas.0503610102

Li, L., Stoeckert, C. J. Jr., and Roos, D. S. (2003). OrthoMCL: identification of ortholog groups for eukaryotic genomes. *Genome Res.* 13, 2178–2189. doi: 10.1101/gr.1224503

Li, M., An, F., Li, W., Ma, M., Feng, Y., Zhang, X., et al. (2016). DELLA proteins interact with FLC to repress flowering transition. *J. Integr. Plant Biol.* 58, 642–655. doi: 10.1111/jipb.12451

Li, Q. F., Wang, C., Jiang, L., Li, S., Sun, S. S., and He, J. X. (2012). An interaction between BZR1 and DELLAs mediates direct signaling crosstalk between brassinosteroids and gibberellins in *Arabidopsis*. *Sci. Signal.* 5, ra72. doi: 10.1126/scisignal.2002908

Lim, S., Park, J., Lee, N., Jeong, J., Toh, S., Watanabe, A., et al. (2013). ABA-INSENSITIVE3, ABA-INSENSITIVE5, and DELLAs interact to activate the expression of *SOMNUS* and other high-temperature-inducible genes in imbibed seeds in *Arabidopsis*. *Plant Cell* 25, 4863–4878. doi: 10.1105/tpc.113.118604

Locascio, A., Blazquez, M. A., and Alabadi, D. (2013a). Dynamic regulation of cortical microtubule organization through prefoldin-DELLA interaction. *Curr. Biol.* 23, 804–809. doi: 10.1016/j.cub.2013.03.053

Locascio, A., Blazquez, M. A., and Alabadi, D. (2013b). Genomic analysis of della protein activity. *Plant Cell Physiol.* 54, 1229–1237. doi: 10.1093/pcp/pct082

Ma, Z., Hu, X., Cai, W., Huang, W., Zhou, X., Luo, Q., et al. (2014). *Arabidopsis* miR171-targeted scarecrow-like proteins bind to GT *cis*-elements and mediate gibberellin-regulated chlorophyll biosynthesis under light conditions. *PLoS Genet.* 10:e1004519. doi: 10.1371/journal.pgen.1004519

Marin-de la Rosa, N., Pfeiffer, A., Hill, K., Locascio, A., Bhalerao, R. P., Miskolczi, P., et al. (2015). Genome wide binding site analysis reveals transcriptional coactivation of cytokinin-responsive genes by DELLA proteins. *PLoS Genet.* 11:e1005337. doi: 10.1371/journal.pgen.1005337

Marin-de la Rosa, N., Sotillo, B., Miskolczi, P., Gibbs, D. J., Vicente, J., Carbonero, P., et al. (2014). Large-scale identification of gibberellin-related transcription factors defines Group VII ETHYLENE RESPONSE FACTORS as functional DELLA partners. *Plant Physiol.* 166, 1022–1032. doi: 10.1104/pp.114.244723

Netotea, S., Sundell, D., Street, N. R., and Hvidsten, T. R. (2014). ComPlEx: conservation and divergence of co-expression networks in *A. thaliana, Populus and O. sativa*. *BMC Genomics* 15:106. doi: 10.1186/1471-2164-15-106

Oh, E., Zhu, J. Y., Bai, M. Y., Arenhart, R. A., Sun, Y., and Wang, Z. Y. (2014). Cell elongation is regulated through a central circuit of interacting transcription factors in the Arabidopsis hypocotyl. *Elife* 3:e03031. doi: 10.7554/eLife.03031

Park, J., Nguyen, K. T., Park, E., Jeon, J. S., and Choi, G. (2013). DELLA proteins and their interacting RING finger proteins repress gibberellin responses by binding to the promoters of a subset of gibberellin-responsive genes in *Arabidopsis*. *Plant Cell* 25, 927–943. doi: 10.1105/tpc.112.108951

Pimprikar, P., Carbonnel, S., Paries, M., Katzer, K., Klingl, V., Bohmer, M. J., et al. (2016). A CCaMK-CYCLOPS-DELLA complex activates transcription of *RAM1* to regulate arbuscule branching. *Curr. Biol.* 26, 987–998. doi: 10.1016/j.cub.2016.01.069

Qi, T., Huang, H., Wu, D., Yan, J., Qi, Y., Song, S., et al. (2014). *Arabidopsis* DELLA and JAZ proteins bind the WD-repeat/bHLH/MYB complex to modulate gibberellin and jasmonate signaling synergy. *Plant Cell* 26, 1118–1133. doi: 10.1105/tpc.113.121731

R Core Team (2016). *R: A Language and Environment for Statistical Computing*. Vienna: R Foundation for Statistical Computing.

Resentini, F., Felipo-Benavent, A., Colombo, L., Blazquez, M. A., Alabadi, D., and Masiero, S. (2015). TCP14 and TCP15 mediate the promotion of seed germination by gibberellins in *Arabidopsis thaliana*. *Mol. Plant* 8, 482–485. doi: 10.1016/j.molp.2014.11.018

Rombolá-Caldentey, B., Rueda-Romero, P., Iglesias-Fernandez, R., Carbonero, P., and Onate-Sanchez, L. (2014). *Arabidopsis* DELLA and two HD-ZIP transcription factors regulate GA signaling in the epidermis through the L1 box *cis*-element. *Plant Cell* 26, 2905–2919. doi: 10.1105/tpc.114.127647

Romero-Campero, F. J., Lucas-Reina, E., Said, F. E., Romero, J. M., and Valverde, F. (2013). A contribution to the study of plant development evolution based on gene co-expression networks. *Front. Plant Sci.* 4:291. doi: 10.3389/fpls.2013.00291

Romero-Campero, F. J., Perez-Hurtado, I., Lucas-Reina, E., Romero, J. M., and Valverde, F. (2016). ChlamyNET: a *Chlamydomonas* gene co-expression network reveals global properties of the transcriptome and the early setup of key co-expression patterns in the green lineage. *BMC Genomics* 17:227. doi: 10.1186/s12864-016-2564-y

Sarnowska, E. A., Rolicka, A. T., Bucior, E., Cwiek, P., Tohge, T., Fernie, A. R., et al. (2013). DELLA-interacting SWI3C core subunit of switch/sucrose nonfermenting chromatin remodeling complex modulates gibberellin responses and hormonal cross talk in Arabidopsis. *Plant Physiol.* 163, 305–317. doi: 10.1104/pp.113.223933

Schwechheimer, C. (2011). Gibberellin signaling in plants – the extended version. *Front. Plant Sci.* 2:107. doi: 10.3389/fpls.2011.00107

Shen, Q., Cui, J., Fu, X. Q., Yan, T. X., and Tang, K. X. (2015). Cloning and characterization of DELLA genes in *Artemisia annua*. *Genet. Mol. Res.* 14, 10037–10049. doi: 10.4238/2015.August.21.10

Smoot, M. E., Ono, K., Ruscheinski, J., Wang, P. L., and Ideker, T. (2011). Cytoscape 2.8: new features for data integration and network visualization. *Bioinformatics* 27, 431–432. doi: 10.1093/bioinformatics/btq675

Supek, F., Bosnjak, M., Skunca, N., and Smuc, T. (2011). REVIGO summarizes and visualizes long lists of gene ontology terms. *PLoS ONE* 6:e21800. doi: 10.1371/journal.pone.0021800

Trapnell, C., Roberts, A., Goff, L., Pertea, G., Kim, D., Kelley, D. R., et al. (2012). Differential gene and transcript expression analysis of RNA-seq experiments with TopHat and Cufflinks. *Nat. Protoc.* 7, 562–578. doi: 10.1038/nprot.2012.016

Usadel, B., Obayashi, T., Mutwil, M., Giorgi, F. M., Bassel, G. W., Tanimoto, M., et al. (2009). Co-expression tools for plant biology: opportunities for hypothesis generation and caveats. *Plant Cell Environ.* 32, 1633–1651. doi: 10.1111/j.1365-3040.2009.02040.x

Vera-Sirera, F., Gomez, M. D., and Perez-Amador, M. A. (2015). "DELLA proteins, a group of GRAS transcription regulators that mediate gibberellin signaling," in *Plant Transcription Factors; Evolutionary, Structural and Functional Aspects*, ed. D. H. González (San Diego, CA: Elsevier), 313–328.

Vragovic, I., Louis, E., and Diaz-Guilera, A. (2005). Efficiency of informational transfer in regular and complex networks. *Phys. Rev. E Stat. Nonlin. Soft Matter Phys.* 71:036122. doi: 10.1103/physreve.71.036122

Wang, Y., and Deng, D. (2014). Molecular basis and evolutionary pattern of GA-GID1-DELLA regulatory module. *Mol. Genet. Genomics* 289, 1–9. doi: 10.1007/s00438-013-0797-x

Wild, M., Daviere, J. M., Cheminant, S., Regnault, T., Baumberger, N., Heintz, D., et al. (2012). The *Arabidopsis* DELLA *RGA-LIKE3* is a direct target of MYC2 and modulates jasmonate signaling responses. *Plant Cell* 24, 3307–3319. doi: 10.1105/tpc.112.101428

Xie, Y., Tan, H., Ma, Z., and Huang, J. (2016). DELLA proteins promote anthocyanin biosynthesis via sequestering MYBL2 and JAZ suppressors of the MYB/bHLH/WD40 complex in *Arabidopsis thaliana*. *Mol. Plant* 9, 711–721. doi: 10.1016/j.molp.2016.01.014

Xu, F., Li, T., Xu, P. B., Li, L., Du, S. S., Lian, H. L., et al. (2016). DELLA proteins physically interact with CONSTANS to regulate flowering under long days in *Arabidopsis*. *FEBS Lett.* 590, 541–549. doi: 10.1002/1873-3468.12076

Yamaguchi, N., Winter, C. M., Wu, M. F., Kanno, Y., Yamaguchi, A., Seo, M., et al. (2014). Gibberellin acts positively then negatively to control onset of flower formation in *Arabidopsis*. *Science* 344, 638–641. doi: 10.1126/science.1250498

Yang, D. L., Yao, J., Mei, C. S., Tong, X. H., Zeng, L. J., Li, Q., et al. (2012). Plant hormone jasmonate prioritizes defense over growth by interfering with gibberellin signaling cascade. *Proc. Natl. Acad. Sci. U.S.A.* 109, E1192–E1200. doi: 10.1073/pnas.1201616109

Yasumura, Y., Crumpton-Taylor, M., Fuentes, S., and Harberd, N. P. (2007). Step-by-step acquisition of the gibberellin-DELLA growth-regulatory mechanism during land-plant evolution. *Curr. Biol.* 17, 1225–1230. doi: 10.1016/j.cub.2007.06.037

Ye, Y., Liu, B., Zhao, M., Wu, K., Cheng, W., Chen, X., et al. (2015). CEF1/OsMYB103L is involved in GA-mediated regulation of secondary wall biosynthesis in rice. *Plant Mol. Biol.* 89, 385–401. doi: 10.1007/s11103-015-0376-0

Yoshida, H., Hirano, K., Sato, T., Mitsuda, N., Nomoto, M., Maeo, K., et al. (2014). DELLA protein functions as a transcriptional activator through the DNA binding of the INDETERMINATE DOMAIN family proteins. *Proc. Natl. Acad. Sci. U.S.A.* 111, 7861–7866. doi: 10.1073/pnas.1321669111

Yu, N., Luo, D., Zhang, X., Liu, J., Wang, W., Jin, Y., et al. (2014). A DELLA protein complex controls the arbuscular mycorrhizal symbiosis in plants. *Cell Res.* 24, 130–133. doi: 10.1038/cr.2013.167

Yu, S., Galvao, V. C., Zhang, Y. C., Horrer, D., Zhang, T. Q., Hao, Y. H., et al. (2012). Gibberellin regulates the *Arabidopsis* floral transition through miR156-targeted SQUAMOSA PROMOTER BINDING-LIKE transcription factors. *Plant Cell* 24, 3320–3332. doi: 10.1105/tpc.112.101014

Zhang, Z. L., Ogawa, M., Fleet, C. M., Zentella, R., Hu, J., Heo, J. O., et al. (2011). Scarecrow-like 3 promotes gibberellin signaling by antagonizing master growth repressor DELLA in Arabidopsis. *Proc. Natl. Acad. Sci. U.S.A.* 108, 2160–2165. doi: 10.1073/pnas.1012232108

Zhou, X., Zhang, Z. L., Park, J., Tyler, L., Yusuke, J., Qiu, K., et al. (2016). The ERF11 transcription factor promotes internode elongation by activating gibberellin biosynthesis and signaling. *Plant Physiol.* 171, 2760–2770. doi: 10.1104/pp.16.00154

Conflict of Interest Statement: The authors declare that the research was conducted in the absence of any commercial or financial relationships that could be construed as a potential conflict of interest.

15

A Conserved Carbon Starvation Response Underlies Bud Dormancy in Woody and Herbaceous Species

Carlos Tarancón[1], Eduardo González-Grandío[1†], Juan C. Oliveros[2], Michael Nicolas[1] and Pilar Cubas[1]*

[1] Plant Molecular Genetics Department, Centro Nacional de Biotecnología (Consejo Superior de Investigaciones Científicas), Campus Universidad Autónoma de Madrid, Madrid, Spain, [2] Bioinformatics for Genomics and Proteomics Unit, Centro Nacional de Biotecnología (Consejo Superior de Investigaciones Científicas), Campus Universidad Autónoma de Madrid, Madrid, Spain

Edited by:
José M. Romero,
University of Seville, Spain

Reviewed by:
Nayelli Marsch-Martinez,
Centro de Investigación y de Estudios Avanzados del Instituto Politécnico Nacional, Mexico
Prakash Venglat,
University of Saskatchewan, Canada

***Correspondence:**
Pilar Cubas
pcubas@cnb.csic.es

†Present address:
Eduardo González-Grandío,
Plant Gene Expression Center, 800 Buchanan Street, Albany, CA, USA

Plant shoot systems give rise to characteristic above-ground plant architectures. Shoots are formed from axillary meristems and buds, whose growth and development is modulated by systemic and local signals. These cues convey information about nutrient and water availability, light quality, sink/source organ activity and other variables that determine the timeliness and competence to maintain development of new shoots. This information is translated into a local response, in meristems and buds, of growth or quiescence. Although some key genes involved in the onset of bud latency have been identified, the gene regulatory networks (GRNs) controlled by these genes are not well defined. Moreover, it has not been determined whether bud dormancy induced by environmental cues, such as a low red-to-far-red light ratio, shares genetic mechanisms with bud latency induced by other causes, such as apical dominance or a short-day photoperiod. Furthermore, the evolution and conservation of these GRNs throughout angiosperms is not well established. We have reanalyzed public transcriptomic datasets that compare quiescent and active axillary buds of Arabidopsis, with datasets of axillary buds of the woody species *Vitis vinifera* (grapevine) and apical buds of *Populus tremula x Populus alba* (poplar) during the bud growth-to-dormancy transition. Our aim was to identify potentially common GRNs induced during the process that leads to bud para-, eco- and endodormancy. In Arabidopsis buds that are entering eco- or paradormancy, we have identified four induced interrelated GRNs that correspond to a carbon (C) starvation syndrome, typical of tissues undergoing low C supply. This response is also detectable in poplar and grapevine buds before and during the transition to dormancy. In all eukaryotes, C-limiting conditions are coupled to growth arrest and latency like that observed in dormant axillary buds. Bud dormancy might thus be partly a consequence of the underlying C starvation syndrome triggered by environmental and endogenous cues that anticipate or signal conditions unfavorable for sustained shoot growth.

Keywords: bud dormancy, carbon starvation, gene regulatory networks, shoot architecture, plant evolution and development

INTRODUCTION

Shoot branching patterns define overall above-ground plant architecture. In angiosperms, shoots are formed from axillary meristems initiated at the base of leaves. These meristems grow and develop into axillary buds that contain, preformed, most of the elements of adult branches (shoot meristems, leaf primordia, reproductive meristems). Axillary buds can enter a quiescent state, rather than growing out immediately to give a branch. In this latent or dormant state their metabolic activity and cell division are very limited (Shimizu and Mori, 1998; Ruttink et al., 2007). Bud dormancy and bud activation are influenced by environmental signals such as nutrient and water availability, light quality, day-length and temperature, and by endogenous signals such as sink/source organ activity and hormone signaling. Once dormant, buds require changes in specific developmental and/or environmental cues to resume growth and generate an elongated branch. These cues are monitored in different organs, and inform the plant as to when develop new shoots. This information is transduced to the bud and translated into a gene response that leads to quiescence or growth activation (Rameau et al., 2015).

Bud dormancy is therefore an adaptive trait that allows plants to endure adverse situations until conditions are favorable for development of new shoots. It has great impact on plant reproductive success and productivity, and on survival in temperate woody species. Evolution of this trait might have allowed plants to colonize habitats with fluctuating conditions not always suitable for sustained, uninterrupted growth. Depending on the type of stimulus that promotes growth arrest, Lang et al. (1985, 1987) distinguished three types of bud dormancy. When dormancy is induced by environmental factors, it is termed *ecodormancy*; when promoted by other plant organs it is *paradormancy* or correlative inhibition, and when it is maintained by signals internal to the bud and can only be reversed under certain conditions it is defined as *endodormancy*. In woody plants, axillary buds undergo transitions between different dormant states throughout the year. Paradormant buds enter endodormancy in response to changes in daylength and temperature. Chilling promotes transition from endo- to ecodormancy, after which the buds are susceptible to grow in response to mild temperatures (Rohde and Bhalerao, 2007).

Transcriptomic studies have been carried out in several herbaceous and woody species to define expression changes in buds during the transitions into and out of different types of dormancy, in response to changes in daylength, light quality, and apical dominance, and in mutant genotypes in which bud growth is affected (e.g., Tatematsu et al., 2005; Ruttink et al., 2007; González-Grandío et al., 2013; Reddy et al., 2013; Ueno et al., 2013; Porto et al., 2015). The GRNs that act inside the bud to control the stages leading to dormancy nonetheless remain little known. It is also largely unknown whether different types of dormancy share common underlying genetic mechanisms. Even less is known about the degree of conservation and evolution of the genetic control of this process in different plant species. Comparative analyses to identify common themes among different types of dormancy, or across species, are scarce (González-Grandío and Cubas, 2014; Fennell et al., 2015; Howe et al., 2015; Hao et al., 2017). Such comparisons could help us determine whether eco-, para- and endodormancy are variations of a single ancestral genetic program or whether each type is controlled by unrelated GRNs. It also will help elucidate whether GRNs that cause bud growth arrest are conserved in different herbaceous and woody plant species.

The master regulators that locally control the dormancy onset are also largely unknown. The best characterized are the genes that encode the TCP transcription factors (TF) teosinte branched1 (Tb1, Doebley et al., 1997), BRANCHED1 (BRC1, Aguilar-Martínez et al., 2007; Finlayson, 2007) and their orthologs in mono- and dicotyledonous species, respectively. These widely conserved factors play a very important role in the regulation of para- and ecodormancy in herbaceous plants. These genes are expressed in axillary buds and promote bud dormancy in response to fluctuating environmental cues such as light quality and quantity, and endogenous signals such as apical dominance, sugar availability and hormone signaling (reviewed in Nicolas and Cubas, 2015). In *Arabidopsis thaliana*, *BRC1* controls transcription of several GRNs in buds; one, positively controlled by *BRC1*, leads to abscisic acid (ABA) accumulation and signaling, whereas another two that are downregulated by *BRC1* are enriched in ribosomal protein genes in one case, and in cell division and DNA replication genes in the other (González-Grandío et al., 2013, 2017). Additional GRNs controlled by *BRC1* remain to be characterized.

In this study our aim was to identify potentially common GRNs induced during the process that leads to bud para-, eco- and endodormancy. For that we compared publicly available transcriptomic data from active para- and ecodormant axillary buds of Arabidopsis, and found, induced in dormant buds, a shared transcriptomic response typical of tissue undergoing C starvation. We then detected this response also in *Populus tremula* × *Populus alba* (poplar) apical buds undergoing endodormancy and in *Vitis vinifera* (grapevine) axillary buds entering para-, endo- and ecodormancy. This C starvation transcriptional response, activated shortly after exposure to conditions leading to bud dormancy, anticipates and underlies the growth-to-dormancy transition in the three species. The C starvation syndrome entails a suite of interconnected transcriptional responses that include sugar signaling, sugar metabolism reprogramming, senescence, autophagy, catabolism, and ABA and ethylene signaling. It also involves downregulation of cytokinin (CK) signaling, inhibition of anabolism, and repression of protein/DNA synthesis and cell division, conditions typical of cells in dormant buds. This conserved starvation response, genetically connected to cell growth arrest, may be one of the underlying forces driving the growth-to-dormancy transition of axillary buds in response to suboptimal conditions in herbaceous and woody species.

RESULTS

Arabidopsis Bud Dormancy Is Associated With the Induction of Four GRN

Three independent transcriptomic analyses have compared active and dormant buds in Arabidopsis. One study compared the transcriptional profiling of (dormant) buds of intact plants and of (active) buds of decapitated plants at 24 h post-treatment (Tatematsu et al., 2005). Two additional experiments compared the transcripts of active vs. dormant buds of plants exposed to high red-to-far-red light ratio (R:FR, active buds) or low R:FR (dormant buds) (González-Grandío et al., 2013; Reddy et al., 2013). Here we define dormancy as a state in which bud growth is reversibly interrupted, regardless of the requirements to resume development. A search for genes upregulated in dormant buds relative to active buds in the three experiments identified 78 genes termed *bud dormancy* genes (Supplementary Figure S1 and Dataset S1; González-Grandío and Cubas, 2014). These genes correspond to the least common denominator of the three studies and were induced in para- and ecodormant buds by either correlative inhibition or low R:FR, respectively. They were also differentially expressed at 3 h (Reddy et al., 2013), 8 h (González-Grandío et al., 2013) and 24 h (Tatematsu et al., 2005) after treatment onset.

We evaluated the degree of coregulation of these genes using the most updated co-expression database of ATTED-II (15,275 microarray experiments; Obayashi et al., 2007). Hierarchical clustering analysis revealed four clusters of coregulated genes (14, 20, 13, and 31 genes; **Figure 1A** and Supplementary Dataset S1). We then searched for additional genes coregulated with each cluster using CoExSearch (ATTED-II, Obayashi et al., 2007) and obtained four lists of highly coregulated genes (Supplementary Dataset S1). Analysis of their fold change (FC) induction in the three active-vs.-dormant bud experiments confirmed that a significant proportion of the genes in each list were induced ($FC \geq 1.2$) in dormant buds in at least one experiment (**Figure 1B** and Supplementary Figure S2). We termed the gene lists that comprised the bud dormancy genes of the original clusters plus their coregulated genes (induced in dormant buds in at least one experiment, red dots in **Figure 1B**) *bud dormancy GRNI-IV*, with 297, 283, 271, and 295 genes respectively (Supplementary Dataset S1).

Bud Dormancy GRNs Are Related to Hormone Signaling, Stress, Catabolism and Starvation Response

To elucidate the biological processes in which these GRNs were involved, we searched for enrichment in gene ontology (GO) terms using the Panther Classification System (Mi et al., 2017; Supplementary Dataset S2), complemented with a MapMan bin analysis (Thimm et al., 2004; Supplementary Dataset S1). GRNI was significantly enriched in terms related to ethylene, auxin and gibberellin signaling and response; GRNII in terms related to ABA, catabolism and response to abiotic stress; GRNIII in terms related to lipid and amino acid catabolism, senescence, response to starvation and biotic stress; and GRNIV in terms related to protein ubiquitination and response to sucrose starvation.

We evaluated the degree of overlap between these GRNs by seeking common genes. GRNIII and GRNIV shared one-third of their genes; GRNII and GRNIII shared 30%, and GRNI and GRNIV had 26% genes in common (**Figure 1C**, Supplementary Figure S3 and Dataset S3). This suggested that these GRNs are not strictly independent, but correspond to related aspects of the same syndrome, probably coordinated or maintained by ethylene, auxin and ABA signaling (**Figure 1D**).

Bud Dormancy GRNs Are Enriched in Genes Typical of a C Starvation Response

We observed that three robust sugar starvation gene markers, *GIBBERELLIN-STIMULATED ARABIDOPSIS 6* (*GASA6*), *DORMANCY-ASSOCIATED PROTEIN-LIKE 1* (*DRM1/DYL1*) and *DARK INDUCIBLE 6* (*DIN6*) (Contento et al., 2004; Price et al., 2004; Gonzali et al., 2006; Zhong et al., 2015) were members of one or several GRNs (*GASA6*, GRNI; *DRM1*, GRNI, III and IV; *DIN6*, GRNIII and IV; Supplementary Dataset S3). As sugar has a prominent role in the control of shoot outgrowth (Mason et al., 2014; Barbier F. et al., 2015; reviewed in Barbier F.F. et al., 2015) and GRNIV was significantly enriched in terms related to sucrose starvation, we studied this response further. The C starvation syndrome, triggered under C-limiting conditions (e.g., an extended night), helps to obtain an alternative energy source and C skeletons. In Arabidopsis, it comprises a suite of interconnected events that result in changes in C balance and growth. They include reprogramming of sugar sensing, transport, signaling and metabolism, increased protein ubiquitination and degradation, amino acid and lipid catabolism, induction of ABA and ethylene signaling, and recycling of cell components via autophagy and senescence. In addition, CK signaling, ribosomal gene expression, DNA synthesis and cell division are inhibited (Contento et al., 2004; Lin and Wu, 2004; Thimm et al., 2004; Gonzali et al., 2006; Rolland et al., 2006; Rose et al., 2006). Remarkably, many of the GO terms and/or MapMan bins enriched in the four GRNs matched categories induced by C-limiting conditions (**Table 1** and Supplementary Datasets S1, S2).

To test the possibility that these GRNs correspond to a C starvation response, we compared the GRN genes with four lists of genes induced in C-limiting conditions: (i) 26 genes of a robust core of C-signaling response shared by 21 Arabidopsis accessions (Supplementary Dataset S4; Sulpice et al., 2009), (ii) 57 sugar-responsive genes, proposed upstream components of the transcriptional response to sucrose (Supplementary Dataset S4; Osuna et al., 2007), (iii) 429 dark-induced, sugar-repressed genes (Supplementary Dataset S4; Gonzali et al., 2006) and (iv) 507 genes responsive to *AKIN10*, a catalytic subunit of the SUCROSE-NON-FERMENTING-1-RELATED PROTEIN KINASE (SnRK1), which integrates stress and C signals to coordinate energy balance, metabolism and growth (Supplementary Dataset S4; Baena-González et al., 2007).

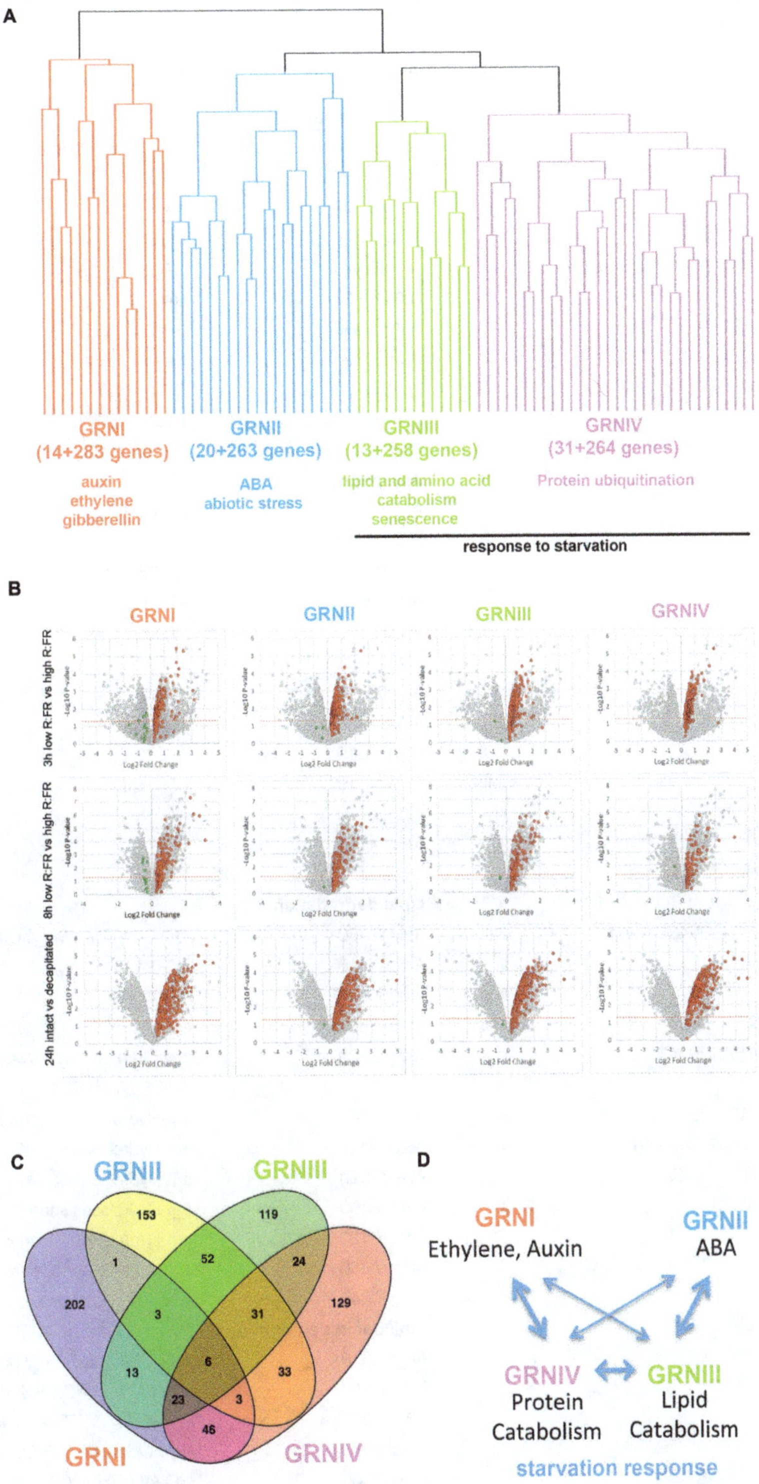

FIGURE 1 | Bud dormancy genes and GRN. (A) Hierarchical clustering representation of bud dormancy genes (González-Grandío and Cubas, 2014) based on their degree of coregulation in 15,275 microarray experiments (ATTED-II; Obayashi et al., 2007). The number of coregulated genes and GO terms enriched are indicated. **(B)** Volcano plots representing pval (–Log10 pval, vertical axis) and relative expression (Log2 fold change, horizontal axis) of all genes in each microarray. Normalized gene intensities in dormant buds vs. normalized gene intensities in active buds were compared in all experiments [3 h low R:FR (N-2 bud) vs. high R:FR (N-2 bud); 8 h low R:FR vs. High R:FR; intact plants vs. 24 h post-decapitation]. Bud dormancy genes and their coregulated genes are highlighted. In red and green, genes induced and repressed in dormant buds, respectively. Genes highlighted in red were attributed to Bud dormancy GRNI-IV (see Supplementary Dataset S1) and were used for subsequent analyses. **(C)** Venn diagram showing overlap between bud dormancy GRN. Number of common genes is indicated. **(D)** Model of the relationships between bud dormancy GRN. Line thickness indicates degree of overlap between GRN.

TABLE 1 | Bud dormancy genes from categories related to sugar sensing, transport, signaling and metabolism, protein ubiquitination and degradation, as well as amino acid and lipid catabolism, autophagy and senescence.

AGI	Name	GRN
Sugar		
Transport		
At3g48740	*SUCROSE EFLUX TRANSPORTER SWEET11*	I
At5g23660	*SUCROSE EFFLUX TRANSPORTER SWEET12*	I
At2g43330	*INOSITOL TRANSPORTER 1*	II
At1g22710	*SUCROSE TRANSPORTER 1 (SUT1/SUC2)*	I
At1g11260	*STP1*	I-III
At5g61520	*STP3*	I
At1g77210	*STP14*	I-III
At1g08920	*SUGAR TRANSPORTER ERD6-LIKE 3*	II
Sensing		
At5g64260	*EXL2*	II-III-IV
At5g09440	*EXL4*	ii
Signaling		
At5g21170	Snrkl submit β1 *(AKINBETA1)*	IV
At3g48530	Snrkl subunit γ1 *(KING1)*	II-III-IV
At1g68020	*TPS6*	II
At1g70290	*TPS8*	III-IV
At1g23870	*TPS9*	III-IV
At1g60140	*TPS10*	II-III-IV
At2g18700	*TPS11*	II-III-IV
At4g24040	*TREHALASE*	III
Metabolism		
At1g13700	*6-PHOSPHOGLUCONOLACTONASE 1*	IV
At1g54100	*ALDEHYDE DEHYDROGENASE H7B4*	II-III
At3g23920	*β-AMYLASE 1 (BAM 1)*	II
At5g55700	*BAM 4*	II-III
At2g42790	*CITRATE SYNTHASE 3*	II-III
At2g47180	*GALACTINOL SYNTHASE 1 (GOLS 1)*	II
At1g56600	*GOLS 2*	II
At1g08940	*PHOSPHOGLYCERATE MUTASE AT74H*	III
At4g15530	*PYRUVATE, PHOSPHATE DIKINASE 1*	III
At5g51970	*SORBITOL DEHYDROGENASE*	IV
At4g02280	*SUCROSE SYNTHASE 3*	II
Senescence		
At4g37790	*ABA INSENSITIVE GROWTH 1 (ABIG1/HAT22)*	II-III
At1g01720	*ATAF1*	II
At1g69490	*NAP*	I-II-III-IV
At5g39610	*ORE1*	I-II-III
At5g51070	*SAG15*	III
At3g10985	*SAG20*	III-IV
At4g02380	*SAG21*	III
At5g59220	*SAG113/HAI1*	II
At4g35770	*SEN1/DIN1*	I-III-IV
At1g62300	*WRKY6*	III
At2g42620	*ORE9/MAX2*	II-IV
Autophagy		
At3g51840	*ATG6*	IV
At4g21980	*ATG8A*	III
At4g04620	*ATG8B*	III
At1g62040	*ATG8C*	IV
At2g45170	*ATG8E*	I-III-IV
At4g16520	*ATG8F*	I-IV
At3g06420	*ATG8H*	II
At3g15580	*ATG81*	IV
At1g54210	*ATG12A*	IV
At1g54710	*ATG18H*	II
At3g62770	*ATG18A*	III
At5g54730	*ATG18F*	II-III
Lipid degradation		
At4g16760	*ACYL-COENZYME A OXIDASE 1 (ACX1)*	III
At5g65110	*ACX2*	III
At3g51840	*ACX4*	IV
At1g68620	*CARBOXYLESTERASE 6*	II-III
At1g48320	*1,4-D1HYDROXY-2-NAPHTHOYL-COA THIOESTERASE 1*	III
At5g18640	α/β-Hydrolases superfamily protein	I-IV
At2g39400	α/β-Hydrolases superfamily protein	I-III-IV
At1g73920	α/β-Hydrolases superfamily protein	II
At1g18460	α/β-Hydrolases superfamily protein	II
At5g16120	α/β-Hydrolases superfamily protein	III
At5g18630	α/β-Hydrolases superfamily protein	IV
At3g60340	α/β-Hydrolases superfamily protein	IV
Aminoacid degradation		
At1g55510	*2-OXOISOVALERATE DEHYDROGENASE β1*	IV
At4g33150	*α-AMINOADIPIC SEMIALDEHYDE SYNTHASE*	II-III
At5g53970	*AMINOTRANSFERASE TAT2*	II
At5g54080	*HOMOGENTISATE 1,2-DIOXYGENASE*	II-III-IV
At3g45300	*ISOVALERYL-COA DEHYDROGENASE*	II-III-IV
At1g64660	*METHIONINE GAMMA-LYASE*	II-III
At4g34030	*METHYLCROTONOYL-COA CARBOXYLASE β*	IV
At1g03090	*METHYLCROTONOYL-COA CARBOXYLASE α*	III-IV
At1g08630	*THREONINE ALDOLASE 1*	III

(Continued)

TABLE 1 | Continued

AGI	Name	GRN	AGI	Name	GRN
Protein degradation					
At5g57360	*ADAGIO 1*	II	At1g80440	*KMD1*	I-III-IV
At2g18915	*ADAGIO 2*	II	At1g15670	*KMD2*	III
At1g05840	*ASPARTYL PROTEASE*	IV	At2g44130	*KMD3*	I-III-IV
At1g21780	BTB/POZ domain-containing protein	II-IV	At3g59940	*KMD4/SKIP20*	I-IV
At5g18650	CHY-/CTCHY-/RING-type Zinc finger protein	IV	At4g24990	*ATGP4*	III
At5g22920	CHY-/CTCHY-/RING-type Zinc finger protein	IV	At1g23440	*PEPTIDASE C15*	III
At3g13550	*COP10*	IV	At4g02440	*PHYTOCHROME A-ASSOCIATED*	IV
At2g40880	Cysteine proteinase inhibitor 3	I	At1g60190	*PLANT U-BOX 19*	II
At5g05110	Cysteine proteinase inhibitor 7	II	At2g22690	RING finger protein	II
At4g39090	Cysteine proteinase RD19a	IV	At1g13195	RING/U-box superfamily protein	III
At5g04250	Cysteine proteinases superfamily protein	II	At1g14200	RING/U-box superfamily protein	II
At4g23450	E3 ubiquitin ligase	II	At1g24440	RING/U-box superfamily protein	II-IV
At5g42200	E3 ubiquitin ligase ATL23	IV	At1g26800	RING/U-box superfamily protein	II-IV
At1g74410	E3 ubiquitin ligase ATL24	II	At1g49850	RING/U-box superfamily protein	IV
At3g05200	E3 ubiquitin ligase ATL6	III	At1g55530	RING/U-box superfamily protein	II
At1g49210	E3 ubiquitin ligase ATL76	I	At1g75400	RING/U-box superfamily protein	II
At1g76410	E3 ubiquitin ligase ATL8	I-III-IV	At2g15580	RING/U-box superfamily protein	I
At3g09770	E3 ubiquitin ligase LOG2	II	At2g37150	RING/U-box superfamily protein	II
At4g11360	E3 ubiquitin ligase RHA1B	I-III-IV	At3g02340	RING/U-box superfamily protein	II-III-IV
At5g22000	E3 ubiquitin-protein ligase RHF2A	II-III	At3g05250	RING/U-box superfamily protein	IV
At4g03510	E3 ubiquitin-protein ligase RMA1	I-III-IV	At3g13430	RING/U-box superfamily protein	III
At4g28270	E3 ubiquitin-protein ligase RMA2	IV	At3g15070	RING/U-box superfamily protein	IV
At4g27470	E3 ubiquitin-protein ligase RMA3	IV	At3g47160	RING/U-box superfamily protein	I-IV
At5g02880	E3 ubiquitin-protein ligase UBC4	II	At3g61180	RING/U-box superfamily protein	IV
At2g04240	E3 ubiquitin-protein ligase XERICO	II-IV	At4g33940	RING/U-box superfamily protein	II-III
At5g25350	EIN3-binding F-box protein 2	I-III	At5g01520	RING/U-box superfamily protein	II
At1g23390	F-box protein	IV	At5g03180	RING/U-box superfamily protein	II
At1g26930	F-box protein	III	At5g10650	RING/U-box superfamily protein	II
At1g30200	F-box protein	IV	At5g19430	RING/U-box superfamily protein	II-IV
At1g51550	F-box protein	I-IV	At5g24870	RING/U-box superfamily protein	II
At1g55000	F-box protein	IV	At5g55970	RING/U-box superfamily protein	II-III-IV
At1g70590	F-box protein	II	At4g00335	*RING-H2 FINGER B1A*	IV
At3g12350	F-box protein	II	At5g01880	*RING-H2 FINGER PROTEIN ATL7*	II
At5g27920	F-box protein	III	At2g17450	*RING-H2 FINGER PROTEIN ATL44*	I-IV
At5g43190	F-box protein	III	At2g18670	*RING-H2 FINGER PROTEIN ATL56*	IV
At1g21760	F-box protein 7	II-IV	At5g66160	*RMR1*	IV
At2g42620	F-box protein MAX2	II, IV	At3g60300	RWD domain-containing protein	IV
At4g03030	F-box protein OR23	II-III-IV	At2g22980	*SERINE CARBOXYPEPTIDASE 13*	I
At1g80110	F-box protein PP2-B11	II	At1g01650	*SIGNAL PEPTIDE PEPTIDASE 4*	II-III
At5g57900	F-box protein SKIP1	II-III-IV	At1g45976	*S-RIBONUCLEASE BINDING PROTEIN 1*	II-III-IV
At1g21410	F-box protein SKP2A	II	At3g06380	*TUBBY-LIKE F-BOX PROTEIN 9*	IV
At1g77000	F-box protein SKP2B	II	At1g63800	*UBIQUITIN CONJUGATING ENZYME E2 5 (UBC5)*	I-IV
At4g10925	F-box protein SKIP8	IV	At5g41700	*UBC8*	I
At3g26000	F-box protein SKIP14	II	At4g36410	*UBC17*	I
At4g21510	F-box protein SKIP27	III	At1g64230	*UBC28*	IV
At5g45360	F-box protein SKIP31	IV	At3g17000	*UBC32*	II-III

The GRN to which each gene belongs to is indicated.

Genes from these sets appeared in the GRNs at a much higher frequency than expected in a random list (pval 4.5E-11 to 7.5E-215; **Table 2** and Supplementary Figure S4), indicating that the bud dormancy GRNs were very highly enriched in genes typical of a C starvation response.

GSEA Analyses Confirm a C Starvation Response in Dormant Buds

We assessed this potential C starvation response by performing a Gene Set Enrichment Analysis (GSEA) using all transcribed

TABLE 2 | C starvation genes are overrepresented in the bud dormancy GRNs.

Geneset	N	Freq. (Gen) x100	NExp. (GRN)	NObs. (GRN)	Freq. (GRN) x100	pval	GRN
Core C-signaling	26	0.08	0.230	8	2.7	4.5E-11	GRNI
Sugar-responsive	57	0.17	0.504	18	6.1	1.4E-23	
Dark-induced, sugar-repressed	429	1.28	3.792	95	32.0	7.9E-107	
AKIN10-responsive	507	1.51	4.481	56	18.9	2.8E-44	
Core C-signaling	26	0.08	0.219	6	2.1	6.6E-08	GRNII
Sugar-responsive	57	0.17	0.480	10	3.5	4.5E-11	
Dark-induced, sugar-repressed	429	1.28	3.613	72	25.4	2.4E-72	
AKIN10-responsive	507	1.51	4.270	63	22.3	1.5E-54	
Core C-signaling	26	0.08	0.210	10	3.7	4.6E-15	GRNIII
Sugar-responsive	57	0.17	0.460	18	6.6	2.7E-24	
Dark-induced, sugar-repressed	429	1.28	3.460	91	33.6	1.8E-104	
AKIN10-responsive	507	1.51	4.089	67	24.7	2.6E-61	
Core C-signaling	26	0.08	0.228	19	6.4	2.9E-34	GRNIV
Sugar-responsive	57	0.17	0.500	31	10.5	3.5E-49	
Dark-induced, sugar-repressed	429	1.28	3.766	152	51.5	7.5E-215	
AKIN10-responsive	507	1.51	4.451	120	40.7	1.7E-141	

Four sets of genes induced under conditions of C starvation (Core C-signaling, Sulpice et al., 2009; Sugar-responsive, Osuna et al., 2007; Dark-induced, sugar-repressed, Gonzali et al., 2006; AKIN10-responsive, Baena-González et al., 2007) were tested for overrepresentation in the bud dormancy GRNs performing a hypergeometric test. All gene sets were very significantly overrepresented in the GRNs. N(Gen), number of genes of the gene set; Freq.(Gen), frequency of gene sets in the Arabidopsis genome (33602 genes); NExp.(GRN), number of expected genes in the GRN; NObs.(GRN), number of observed genes in the GRN; Freq.(GRN), frequency in the GRN.

genes from each experiment, rather than focusing on the bud dormancy GRNs. GSEA is a statistical approach that allows identification of overrepresented gene sets among differentially up- or downregulated genes of a transcriptomic experiment (Subramanian et al., 2005). For each "active-vs.-dormant bud" experiment, we generated a ranked gene list using relative gene expression levels and False Discovery Rate (FDR) values. We then tested whether gene sets related to a potential C starvation syndrome (sugar-, darkness- and *AKIN10*-responsive genes, ABA, ethylene and CK markers, ribosomal genes, cell cycle and cell division genes; Supplementary Dataset S4) were found toward the top (upregulated) or the bottom (downregulated) of the ranked gene lists. Analyses[1] confirmed significant overrepresentation of C-signaling, sugar-repressed, *AKIN10*-induced and ABA and ethylene marker genes among those upregulated in the three experiments (**Figures 2A, 3**). In contrast, CK markers, ribosomal genes and S-phase genes were overrepresented among the downregulated genes (**Figures 2B, 3**). Other cell division markers such as M-phase genes, histones and kinesins were overrepresented only in the 8 and 24 h experiments, which suggests they are downregulated at later stages of the process (**Figures 2B, 3**).

All these results suggest that a C starvation syndrome is induced early in the growth-to-dormancy transition in para- and ecodormant axillary buds in Arabidopsis.

Regulation of the Bud Dormancy GRNs

To find potential master regulators of the C starvation-related bud dormancy GRNs, we searched for overrepresented motifs in the gene promoters (1 kilobase upstream of the transcription start site) of each GRN using Oligo-analysis and Pattern assembly (Rsat; Medina-Rivera et al., 2015). In GRNI and GRNIV, tcTTATCCAc was the most-overrepresented motif (Supplementary Figure S5 and Dataset S5); it contains the sucrose-repressible element TATCCA, bound in rice by the MYB factors OsMYBS1, 2, and 3, which mediate sugar-regulated gene expression (Lu et al., 1998, 2002). We looked for TFs within GRNI and IV that could bind this motif, based on DNA affinity purification sequencing (DAP-Seq) data (O'Malley et al., 2016) or chromatin immunoprecipitation sequencing (ChIP-Seq) data (Song et al., 2016), and that could act as master regulators of the GRNs. The Arabidopsis OsMYBS2 ortholog MYBS2 (At5g08520), which has a role in sugar and ABA signaling (Chen et al., 2016), pertains to GRNI and might bind this motif (Supplementary Figure S6A and Dataset S5). In GRNIV, MYBS2 and three other MYB-related proteins, MYBH/KUA1 (At5g47390), At1g19000 and At1g74840 could bind this motif (Supplementary Figure S6A and Dataset S5). *MYBH/KUA1* has a critical role in dark-induced leaf senescence (Huang et al., 2015). These sugar-regulated genes could be instrumental in coordinating gene expression in GRNI and GRNIV.

The most overrepresented motifs in GRNII and III, gaCACGTGtc, tgaCACGT and gACACGT, overlap with the G-box (CACGTG), which is bound by bZIP, bHLH and NAC proteins (Supplementary Figure S5 and Dataset S5). These motifs are overrepresented in the promoters of C starvation response genes (Cookson et al., 2016). In GRNII, the master regulators of ABA signaling GBF2, GBF3, ABF3 and ABF4, and the senescence-inducing NAC factors NAP, NAC6/ORE1 and ATAF1 as well as NAC047 and NAC3 could bind these motifs (Supplementary Figure S6A and Dataset S5; Guo and Gan, 2006; Balazadeh et al., 2010; Garapati et al., 2015; O'Malley et al., 2016; Song et al., 2016). In GRNIII, NAP, NAC6/ORE1, NAC047, NAC3, NAC102, RD26 (Supplementary Figure S6A

[1]http://bioinfogp.cnb.csic.es/files/projects/tarancon_et_al_2017_supp/

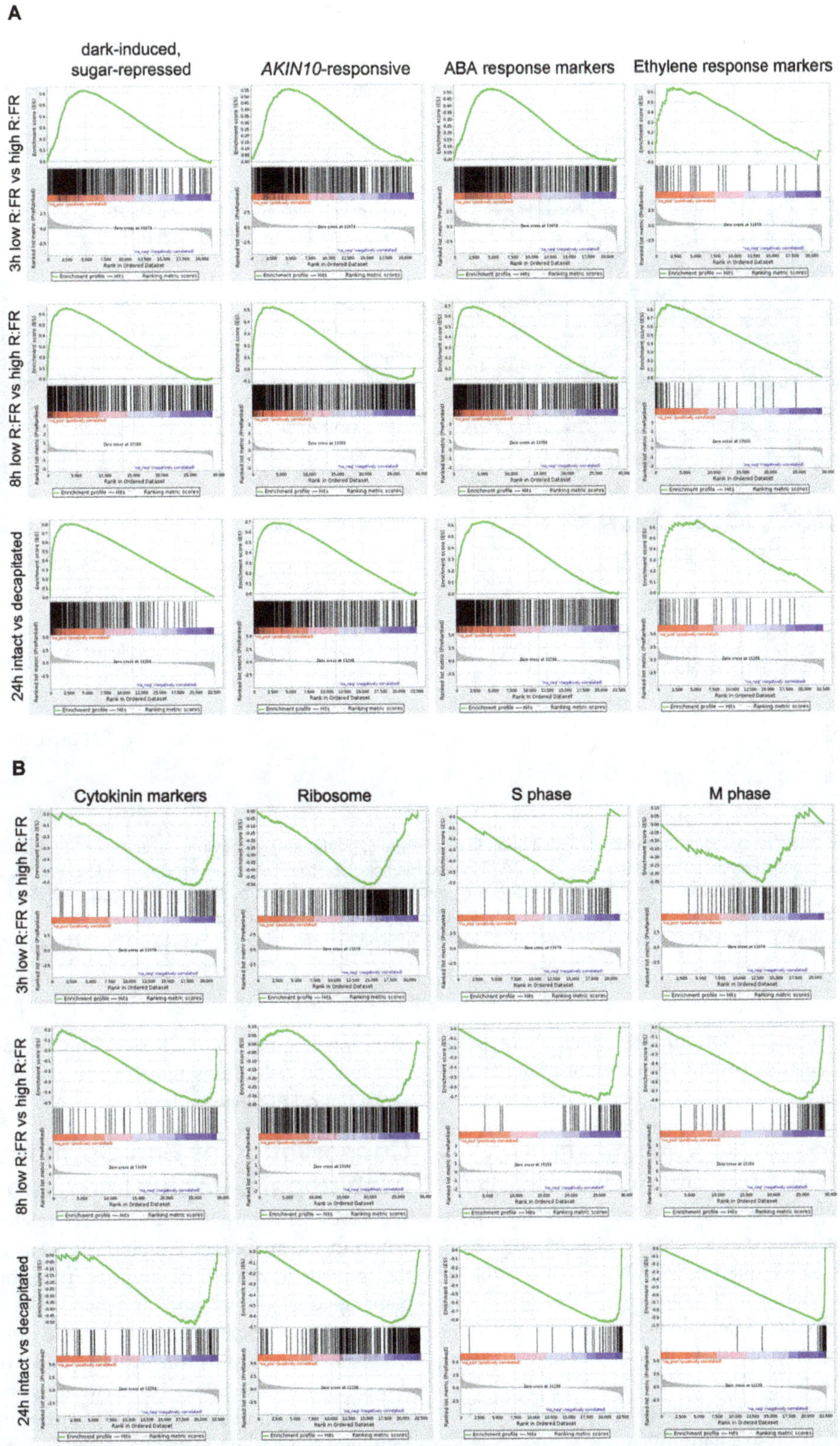

FIGURE 2 | Gene Set Enrichment Analysis (GSEA) analyses of dormant *vs.* active bud experiments. (A,B) Enrichment Scores (ES; green line) of selected gene sets that illustrate significant overrepresentation among up- **(A)** or down-regulated genes **(B)**. Barcode-like vertical black lines represent logRatios of genes of each gene set in the ranked ordered data sets. Left (positive logRatios), genes induced in dormant buds; right (negative logRatios), genes repressed in dormant buds.

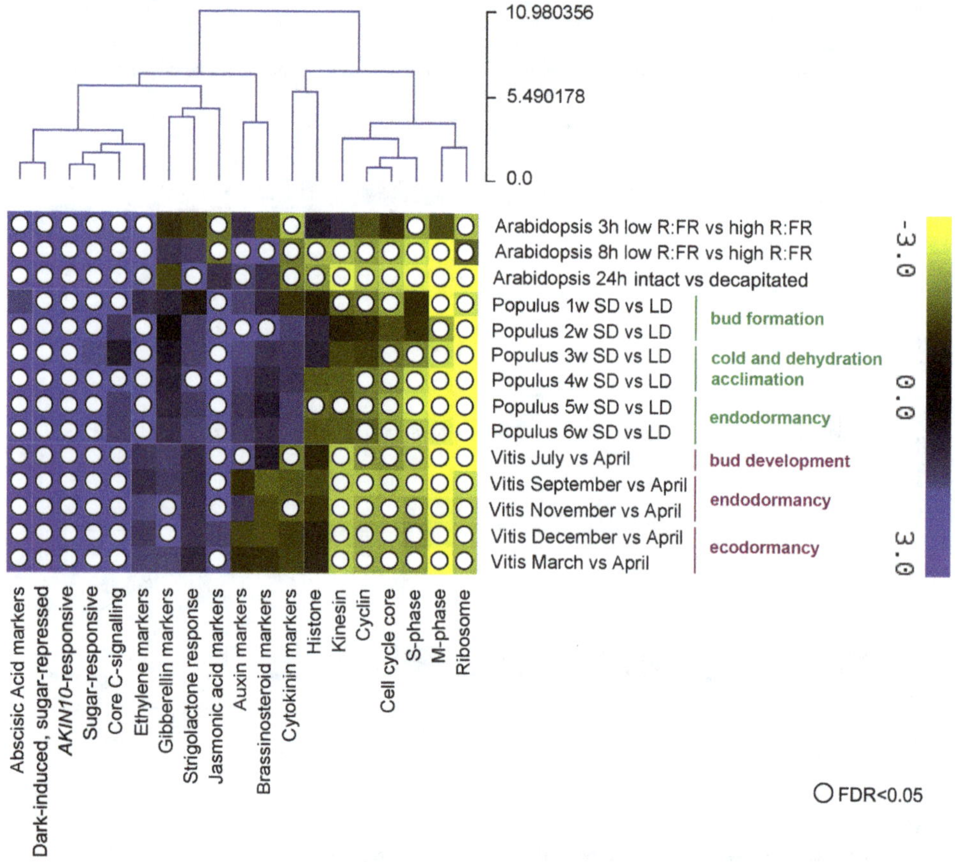

FIGURE 3 | Summary of GSEA analyses of dormant *vs.* active buds in Arabidopsis, poplar and grapevine. Clustering of GSEA results for all transcriptomic samples and gene sets based on their normalized enrichment score (NES) for each sample. Complete results are in http://bioinfogp.cnb.csic.es/files/projects/tarancon_et_al_2017_supp/. Positive NES values (blue) are for gene sets overrepresented in the "dormant bud" condition. Negative NES values (yellow) are for gene sets overrepresented in the "active bud" condition. White circles indicate gene sets with a significant statistical overrepresentation (FDR < 0.05). Core C-signaling and Brassinolide (BL) markers in poplar, and ethylene and strigolactone (SL) markers in grapevine contained <10 genes each, which may have prevented obtaining significant results.

and Dataset S5) and possibly NAC19, for which there is no available binding information, might regulate these motifs and promote gene expression.

We confirmed significant enrichment of the GRNs in the target genes of these TFs by using DAP-Seq and ChIP-Seq data (O'Malley et al., 2016; Song et al., 2016); their numbers in the GRNs were significantly higher than expected in a random gene list (pval < 0.01). For instance, the number of gene targets for NAC102, RD26 and ABF4 was 6, 3.2, and 3.4 times higher, respectively; for the remainder, this value was between 1.7 and 2.7 times higher than predicted (Supplementary Figure S6B).

All these results indicate that four interrelated GRNs associated to a C starvation response are induced in para- and ecodormant Arabidopsis buds. MYB-related, bZIP and NAC TFs could have a key role in the regulation of these GRNs. A large proportion of the genes in the GRNs are rapidly repressed by sugar and upregulated by *AKIN10*. They are tightly coregulated with or directly involved in sugar signaling and metabolism, autophagy, senescence, catabolism of lipids and proteins, and ABA and ethylene signaling. This response is also associated with downregulation of CK signaling, protein synthesis and cell division, all conditions that lead to the cell and tissue growth arrest typical of dormant buds.

Conservation of Bud Dormancy GRNs in Arabidopsis, Poplar and Grapevine

We investigated whether the GRNs related to a C starvation syndrome identified in Arabidopsis were also induced during the growth-to-dormancy transition in buds of the woody plant species, poplar and grapevine. We studied two public transcriptomic experiments in which apical buds of poplar (Ruttink et al., 2007) or axillary buds of grapevine (Díaz-Riquelme et al., 2012) underwent dormancy. To induce dormancy, shoot apices of poplar plants grown in long days (LD, 16 h light-8 h darkness) were exposed to 1–6 weeks of short days (w SD, 8 h light-16 h darkness) (Ruttink et al., 2007). During treatment, the shoot apices developed into buds (1–3 w SD), grew adapted to dehydration and cold (3–6 w SD), and became dormant (5–6 w SD). Samples were collected weekly. Díaz-Riquelme et al. (2012) collected monthly samples of axillary buds of grapevine plants grown in natural conditions in the northern hemisphere. Grapevine axillary buds are formed between April and May; in July and August they grow, undergo flowering and develop inflorescence meristems, enter endodormancy at the end of September, and exit dormancy by the end of November. They remain ecodormant throughout December, until environmental conditions become benign around March, when they sprout (Martìnez de Toda Fernaàndez, 1991; Díaz-Riquelme et al., 2012).

We analyzed the expression patterns of the poplar and grapevine orthologs of the GRNI-IV genes. Of 838 Arabidopsis genes in these GRNs, we identified 390 poplar and 421 grapevine orthologs (Supplementary Dataset S6). In both species, we studied gene expression relative to levels in the "active bud" sample (LD in poplar, April in grapevine). In general, a large proportion of the bud dormancy gene orthologs were significantly induced at most time points in poplar and grapevine buds (Supplementary Figure S7), which supports a conservation, during the growth-to-dormancy transition in these woody species, of the responses found in Arabidopsis. In poplar, the global induction appeared to increase over the weeks in SD, especially for GRNII and III genes. In contrast, grapevine gene induction was detectable throughout the year (Supplementary Figure S7).

A group of genes showed high expression levels from the earliest stages (1–3 w SD in poplar and July in grapevine), weeks/months before endodormancy onset, and throughout the experiment (**Figure 4** and Supplementary Dataset S6). In poplar, these were *DRM1, HIS1-3, GID1C, COR413IM1, ALANINE:GLYOXYLATE AMINOTRANSFERASE 3 (AGT3), SUCROSE SYNTHASE3 (SUS3), TEMPRANILLO1 (TEM1)* and *SEIPIN* (**Figure 4A**). In grapevine, they were *DRM1, HIS1-3, DIN6, PIF4, CBP1/MEE14, HSPRO2, EXL4, BCAT2* and *ATY13/MYB31, ERF2*, ABA receptor *PYL9/ABI1*, the protein phosphatases 2C *HAI1/SAG113* and *AIP1/HAI2*, involved in ABA signaling and sucrose sensitivity (Lim et al., 2012), *DOF5.4, HSFC1, PLANT U-BOX 19, GOLS1*, senescence factor *NAP, RD26, ALUMINUM SENSITIVE 3 (ALS3)*, and oxidative stress-related *At3g10020* (**Figure 4B**).

Other bud dormancy genes were induced exclusively between 1 and 3 w SD in poplar, and in July in grapevine. Early poplar genes were *GASA6*, the sugar-responsive gene *CBP1/MEE14* (Bi et al., 2005), *AKINBETA1* and *LSD ONE LIKE* 1 (*LOL1*), a positive regulator of cell death (Epple et al., 2003) (**Figure 4A** and Supplementary Dataset S6). July-induced grapevine genes were sugar transporters *STP14* and *STP1, KISS ME DEADLY 4/SKP20*, which encodes an F-box protein that negatively regulates the CK response, autophagy factor *ATG8I*, chaperone *DNAJ11, PLASMA-MEMBRANE ASSOCIATED CATION-BINDING PROTEIN 1 (PCaP1), ASD1*, involved in cell wall remodeling (Chávez Montes et al., 2008), and ABA-responsive *MYB3* (**Figure 4B** and Supplementary Dataset S6).

In summary, a large proportion of the genes orthologous to Arabidopsis bud dormancy genes are also induced, either early and transiently or early and constantly during the growth-to-dormancy transition in poplar and grapevine, which supports their functional conservation in these woody species.

The C Starvation Response Is Conserved in Poplar and Grapevine Buds

To obtain a general view of the transcriptomic responses in these experiments, we performed GSEA similar to that for Arabidopsis, using all genes with proposed Arabidopsis orthologs (8023 genes in poplar and 8390 in grapevine) (Ruttink et al., 2007; Díaz-Riquelme et al., 2012).

The sugar- and *AKIN10*-responsive gene sets were overrepresented among upregulated genes from 1 w SD in poplar, and July in grapevine, and were also induced throughout the treatment/year (**Figure 3**). This finding confirms that the C starvation response begins early, long before endodormancy onset, and underlies the entire process. The ribosomal gene set was constitutively overrepresented among downregulated genes in all three species, which confirmed that inhibition of protein synthesis is an early and sustained response in buds entering dormancy. General downregulation of cell cycle and cell division genes was also observed in grapevine, whereas in poplar, cell division gene sets were repressed more gradually and reached maximum repression at 5 w SD. In contrast to Arabidopsis, histones were not significantly downregulated in the woody species. Nevertheless, C starvation response gene sets (upregulated) and cell growth-related gene sets (downregulated) clustered together in the three species.

Unlike the gene sets discussed above, hormone responses did not appear to be strongly conserved among species, suggesting more relaxed evolution of these pathways. Whereas ABA-related genes were induced constitutively in grapevine, ABA and ethylene responses were induced from 2 w SD onward in poplar, in accordance with previous observations (Ruttink et al., 2007). CK signaling is repressed in Arabidopsis, but not notably in poplar or grapevine. An early, extended response to the senescence-associated hormone jasmonate (MJ) was repressed in two Arabidopsis experiments, and was induced in most poplar and grapevine samples (**Figure 3**).

This results indicate that an early and sustained sugar-starvation response associated with downregulation of ribosomal and cell cycle proteins is conserved in buds of Arabidopsis, poplar and grapevine, and might constitute a core response of buds entering dormancy in the angiosperms.

Cell Type-Specific Gene Expression of Bud Dormancy Genes in the Shoot Apex

To further analyze the function of the genes induced during the C starvation response in buds, we selected those most highly expressed in Arabidopsis, poplar and grapevine, to determine the cell types in which they are expressed. We used a high-resolution gene expression database of the Arabidopsis shoot apex, which contains the same tissues as axillary buds: meristem and leaf primordia. This database comprises gene expression profiles of different cell populations obtained by fluorescence-activated cell sorting (L1, L2 and L3 layers, central (CZ) and peripheral zone (PZ), leaf primordia, xylem and phloem) (Yadav et al., 2014). As we cannot rule out that the expression levels of these genes change in dormant axillary buds, we used this database for qualitative rather than quantitative analysis, to

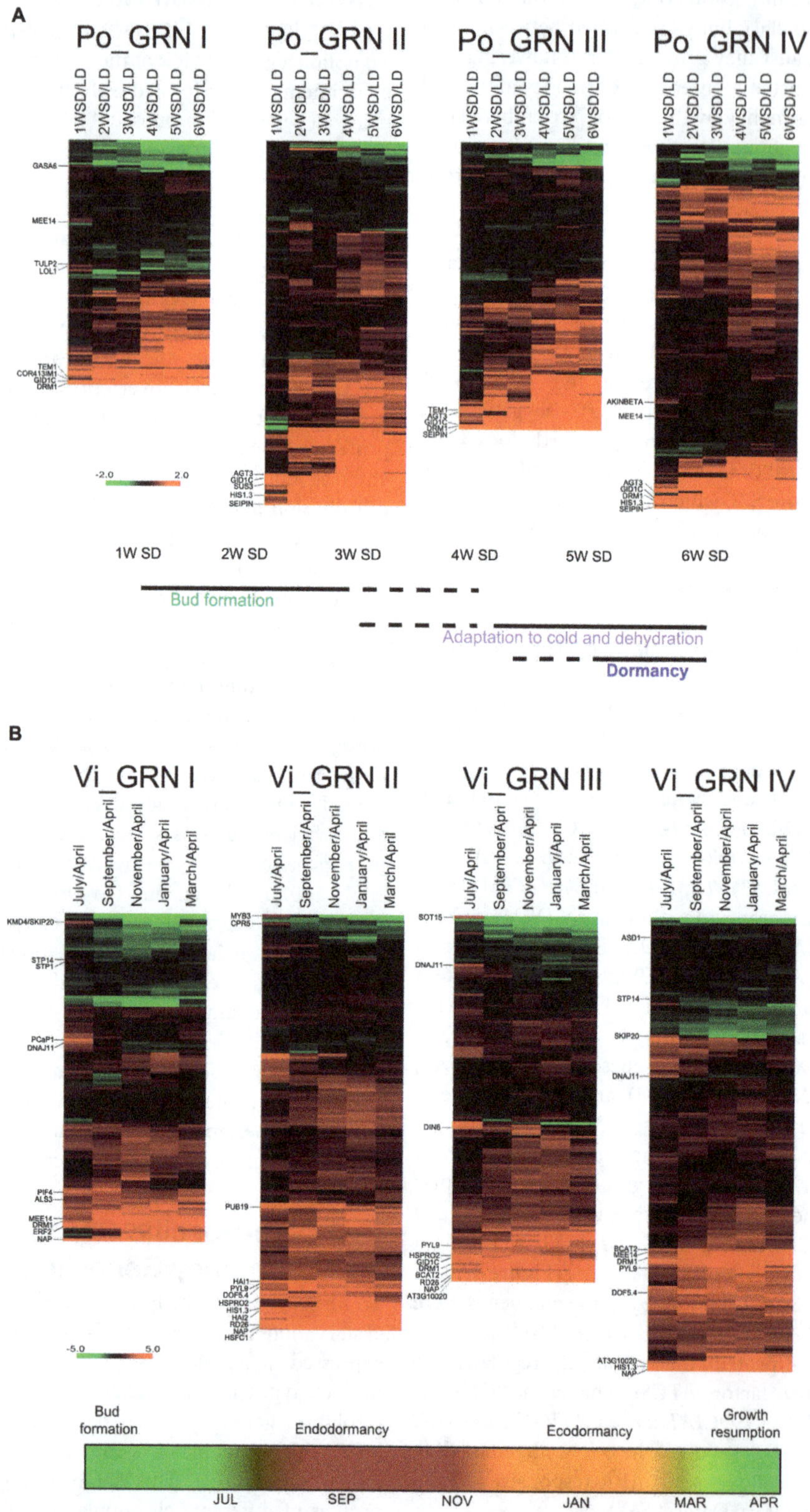

FIGURE 4 | Expression profiles of bud dormancy genes in poplar and grapevine. Heatmap of gene expression for poplar (**A**, Po_GRN) and grapevine (**B**, Vi_GRN) orthologs of the Arabidopsis GRN genes. Log2 ratios of normalized gene intensities in each time point vs. normalized gene intensities on the active bud sample are indicated. For poplar and grapevine the "active bud" sample are LD and April, respectively. In red and green, genes up- and downregulated in dormant buds respectively. Genes mentioned in the text are indicated. Schematic representations based on information from Ruttink et al. (2007) **(A)** and Díaz-Riquelme et al. (2012) **(B)**, below indicate the proposed developmental stage of buds in each time point.

identify the cell types in which these genes were expressed most abundantly.

Many of the most highly induced genes in buds were expressed preferentially in the vasculature (**Figures 5A–D**); sugar-related genes *SUC2, STP3, TPS11, GOLS1*, and TF *HB-40, HB-7, HB-12, PAT1* and *CDF2* were expressed exclusively in phloem (**Figure 5A**). Many *ABA*-related genes (*PYL9, HAI1, NAP, NAC055, NAC002, NFYA1, HAT22, RVE6, LEA4-5, DOF5.4, HIS1-3* and *TSPO*), *SCR*, the F box-encoding genes *MAX2* and *KMD2, CBP1/MEE14, AGT3*, and *CBSX5* were expressed almost exclusively in xylem (**Figure 5B**). Other *ABA*-related genes (*ABF4, HAB1, HAI2, GBF3, RD26, MYB31, RAP2.3*) as well as *KING1, TPS10, TRE, XERICO, EXL2, KMD3, VOZ1, SIS* and *PUB19* were expressed in both xylem and phloem (**Figures 5C,D**). In the *CLV3/WUS* expression domain, we found *TEM1*, protein kinase *CIPK14* that interacts with SnRK1 (Yan et al., 2014), *SUS3, UDP-GLUCOSYL TRANSFERASE 87A2* (*UGT87A2*) and *6 PHOSPHOGLUCONOLACTONASE 1 (PGL1)* (**Figure 5E**). The genes *PIF1, PIF4, EXL4, ADAGIO (ADO1), ETR2* and *COR413 IM1* accumulated preferentially in leaf primordia (**Figure 5F**). Other strongly expressed genes such as *DRM1, PCAP1, KMD1,* and *AFP3* were found in both vascular tissue and leaf primordia, and *GID1C* in xylem and the peripheral zone of the meristem. The autophagy genes *ATG8F, ATG8C, ATG18A, ATG18F, ATG18H*, the senescence gene *SAP3*, as well as *ERF2, GID1B, BYPASS* and *HISTONE DEACETYLASE 8* were widely expressed throughout the meristem.

In summary, whereas ABA signaling occurs mostly in the xylem, sugar signaling in the phloem, and ethylene in the meristem proper (Yadav et al., 2014), autophagy and arrest of cell growth take place throughout the meristem. This suggests that cell-to-cell communication and movement of signaling molecules, hormones and proteins must take place across different cell types in buds entering dormancy.

Bud Dormancy Early Markers

It is of great interest to identify robust, universal markers that allow diagnosis of axillary bud status. These markers should be induced early, and their expression be sustained in para-, eco-, and endodormant buds throughout angiosperms. Based on our analysis, several genes met these criteria. We tested further four of them: *DRM1, HIS1-3, GID1C*, and *NAP*. *DRM1* and *HIS1-3* were expressed at high levels in the Arabidopsis, poplar and grapevine experiments. *DRM1* is a well-known dormancy marker in herbaceous and woody plants species (Stafstrom et al., 1998; Park and Han, 2003; Tatematsu et al., 2005; Aguilar-Martínez et al., 2007; Kebrom et al., 2010; Wood et al., 2013). It is repressed by sugar, which supports its strong association to low sugar levels and dormancy. ABA-responsive *HIS1-3* is upregulated before ABA signaling in poplar and grapevine buds. *GID1C*, which encodes a gibberellin receptor, is expressed at high levels in the three Arabidopsis experiments and in poplar (**Figure 4A**). The senescence-promoting gene *NAP* is expressed at very high levels in Arabidopsis and throughout the year in grapevine (**Figure 4B**). Both *GID1C* and NAP belong to the four bud dormancy GRNs (Supplementary Dataset S1).

We tested whether expression of these genes also correlated with bud dormancy in axillary buds of potato (*Solanum tuberosum*, Solanaceae, Asteridae), a species only distantly related to Arabidopsis, poplar (both Rosidae) and grapevine (basal angiosperms). We identified the potato ortholog genes for the four candidates. In the case of *NAP*, we found two potato paralogs (*NAPa* and *NAPb*). We studied their mRNA levels in buds of plants treated for 10 h with white light (W) or with W supplemented with far red light (W+FR), a treatment that promotes axillary bud dormancy in potato. We also compared mRNA levels in buds of intact and decapitated plants. Whereas *DRM1* and *HIS1-3* were confirmed as reliable markers of bud dormancy in potato, *GID1C, NAPa* and *NAPb* did not respond as anticipated in decapitated plants; *GID1C* was not upregulated in low R:FR and *NAPa/NAPb* were highly induced after decapitation (**Figure 6**).

DISCUSSION

Carbon Availability, a Key Signal for Growth

In all eukaryotes, cell proliferation and growth demand high carbohydrate levels for energy generation and macromolecule synthesis. Correspondingly, low C availability promotes a reduction in growth rate in order to retain sufficient C to support essential maintenance functions (Rolland et al., 2006). In Arabidopsis, sugar availability has a great influence on growth and development, both in seedlings and adult plants. C-limiting conditions (e.g., sucrose depletion, night extensions, short-day photoperiods, starchless mutants) trigger a suite of transcriptional responses that lead to growth cessation, and which include repression of genes involved in anabolism, protein synthesis, cell division, cell cycle, and DNA synthesis and repair (Moore et al., 2003; Smith and Stitt, 2007; Wiese et al., 2007).

Regarding shoot branching, it has been shown that sugar availability to buds plays a major role in its control in pea and rose (Mason et al., 2014; Barbier F. et al., 2015). In agreement, we have found that induction of GRNs typical of tissues undergoing C starvation precedes and underlies the bud growth-to-dormancy transition in Arabidopsis, poplar and grapevine. This is concomitant with transcriptional repression of ribosomal and cell-cycle genes, responses typical of buds entering dormancy as well as of tissues undergoing C limitations (e.g., Thimm et al., 2004; Smith and Stitt, 2007; González-Grandío et al., 2013). Indeed it is possible that bud dormancy is a manifestation and a consequence of the observed C starvation syndrome.

Dormancy-Promoting Stimuli and the C Starvation Response

How is this C starvation response induced? In apical dominance it has been proposed that the growing shoot apex acting as a sugar sink might limit sugar availability to axillary buds so this can be the direct trigger of the response (Mason et al., 2014). Tre6P, a metabolite that acts as a proxy for C status, may also promote signaling in addition to, or instead of direct sugar sensing (Paul

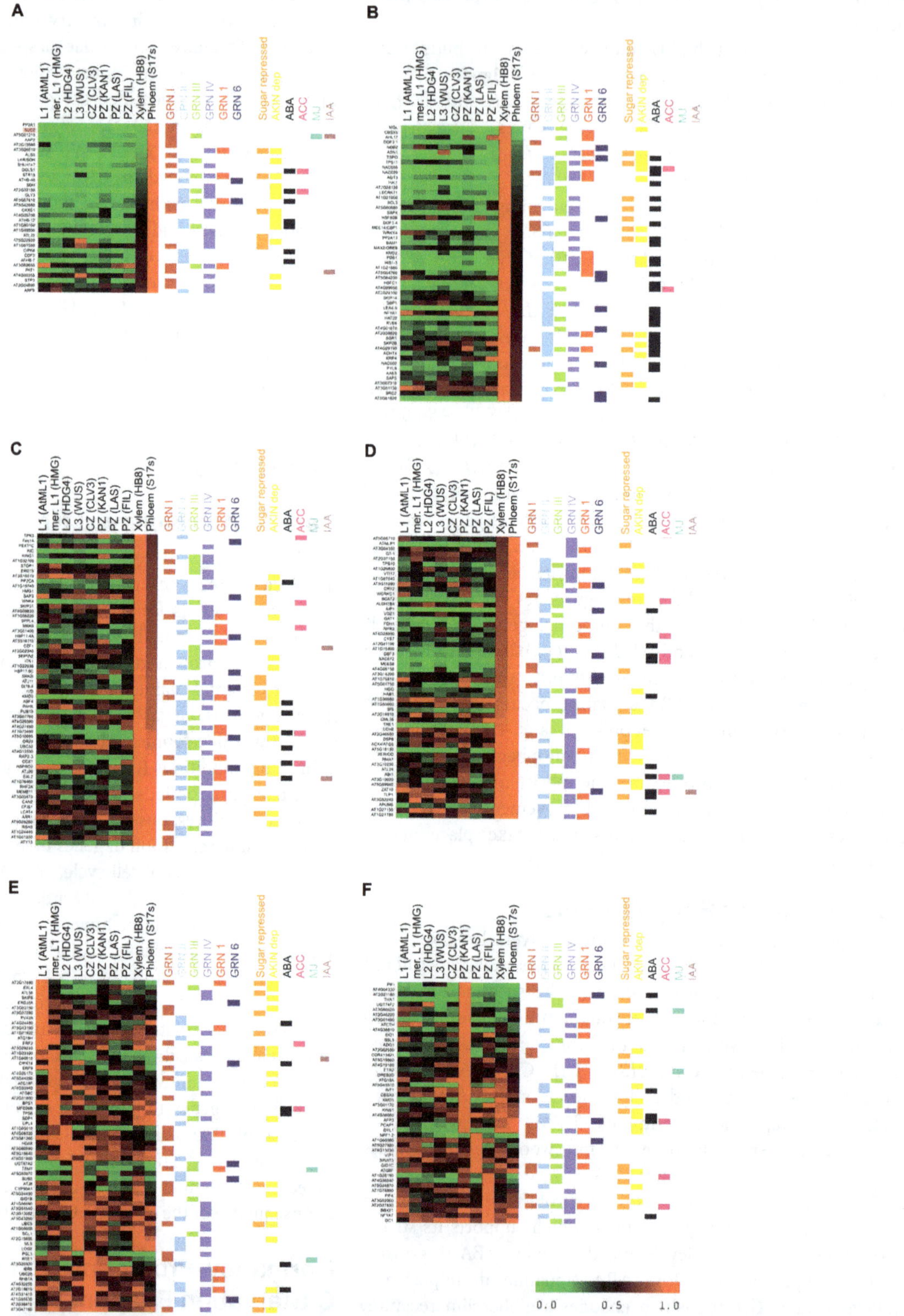

FIGURE 5 | Cell type-specific expression of bud dormancy genes in the Arabidopsis shoot apical meristem (SAM). Heatmap of bud dormancy gene expression in SAM cell types, normalized for each gene relative to the cell type with the highest expression levels (1.0, red). No expression (0.0), green. Different cell types are indicated on top. Horizontal color lines on the right indicate gene sets to which each gene belongs. **(A)** Genes expressed almost exclusively in phloem. **(B)** Genes expressed almost exclusively in xylem. **(C,D)** Genes expressed preferentially in phloem and xylem. **(E)** Genes expressed preferentially in layers 1, 2, 3 (L1, L2, L3), meristem (mer) and central zone (CZ). **(F)** Genes expressed preferentially in peripheral zone (PZ).

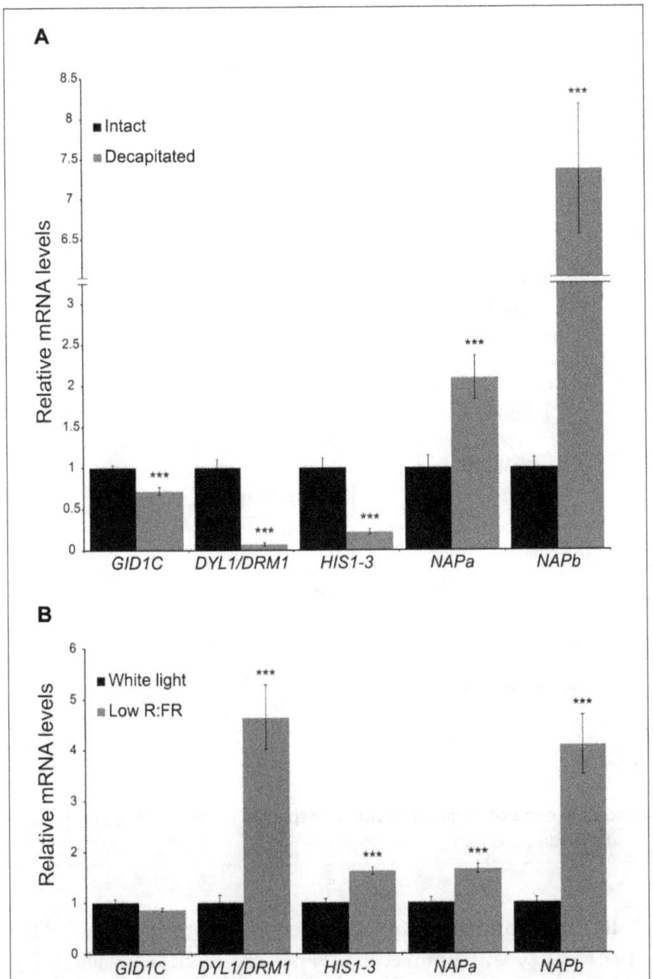

FIGURE 6 | Expression levels of candidate bud dormancy markers in potato. Transcript levels of *DRM1*, *HIS1-3*, *GID1C* and *NAP* potato ortholog genes in aerial axillary buds were analyzed by quantitative PCR. **(A)** Intact and decapitated plants, 10 h after treatment. **(B)** Plants exposed for 10 h to white light or to a low red:far-red light ratio. Values are mean ± SEM ($N_A = 5$, $N_B = 4$). Each biological replicate is a pool of 16 axillary buds. **pval < 0.01; ***pval < 0.001; two-tailed Student's *t*-test.

et al., 2008; Lunn et al., 2014). This would be in agreement with the observation that plants that express microbial trehalose-phosphate synthase (*TPS)* genes show increased shoot branching (Goddijn et al., 1997) and maize plants with a mutation in the trehalose-phosphate phosphatase gene *RAMOSA3* have altered inflorescence branching (Satoh-Nagasawa et al., 2006).

Nevertheless, it is likely that the syndrome is not only induced by an actual sugar shortfall, but also by cues that inform of current or future suboptimal conditions which may affect energy availability and/or interfere with respiration and C assimilation (Baena-González et al., 2007; Baena-González and Sheen, 2008). Seasonal environmental changes that perturb these processes (e.g., daylength shortening, light levels, temperature, water availability) may trigger acclimatory signaling pathways that anticipate C limitations (Smith and Stitt, 2007). Those and other stimuli could feed into regulatory networks that economize resources locally, to result in a moderation of growth rate in axillary meristems and buds.

In two of the Arabidopsis experiments examined, the dormancy-inducing stimuli was an exposure to low R:FR light ratio. Low R:FR light is interpreted by plants as a situation with limited light available for photosynthesis. It severely reduces the expression of photosynthesis-related genes (Cagnola et al., 2012) and induces cell-wall remodeling in stem and petioles, which may divert carbohydrates away from axillary buds (Sasidharan et al., 2010). Furthermore, low R:FR promotes ethylene and ABA signaling and CK degradation (Carabelli et al., 2007; Cagnola et al., 2012), hormonal responses tightly linked to the C starvation response (see below).

In the poplar and grapevine studies, the sugar-repressed networks are induced in buds soon after beginning of daylength shortening: in poplar at 1 w SD; in grapevine, in July, when daylength shortening has just begun (June 21), even though buds are still growing. Short-day photoperiod leads to localized flower and seed abortion associated with low levels of C in Arabidopsis (Lauxmann et al., 2016). Likewise, in poplar a measurable shortage of sugar availability is detectable after 1 w SD (Ruttink et al., 2007). Under the natural conditions in which grapevine plants are grown, daylength shortening and C limitations are progressive, but relatively small changes in C balance may trigger the response. Indeed in Arabidopsis even minor alterations in C status, well before C starvation, lead to notable changes in C-related signaling and response (Usadel et al., 2008). In addition, genetic pathways that sense photoperiod might help anticipate and adapt to impending C-limiting conditions in short days. These pathways, controlled by phytochromes, circadian clock, and genes controlling flowering time (Horvath, 2009), may regulate and establish crosstalk with the C starvation response. Indeed, sugars affect the expression of clock genes (Haydon et al., 2013) and conversely, the clock regulates carbohydrate metabolism (Smith and Stitt, 2007). Phytochromes, which monitor changes in R:FR and in day-length, also regulate SD-induced endodormancy in woody species (Johnson et al., 1994; Reed et al., 1994; Olsen et al., 1997; Neff and Chory, 1998; Monte et al., 2003; Ruonala et al., 2008; Franklin and Quail, 2010). Changes in low R:FR light ratio or photoperiod might therefore trigger partially overlapping responses, including potential anticipation of a C-limiting situation.

Although it has not been analyzed in this work, coordination between C and N metabolic pathways probably affect this process as well, as sugar responses depend significantly on the N status of the plant.

The C Starvation Syndrome in Axillary Buds: Sugar Signaling

The C starvation syndrome comprises a cascade of transcriptomic events that culminate in changes in growth and C balance (**Figure 7**). These events include induction of genes involved in transcriptional regulation, sugar sensing, transport and signaling, catabolism (i.e., amino acid and lipid degradation), protein ubiquitination and degradation, hormone

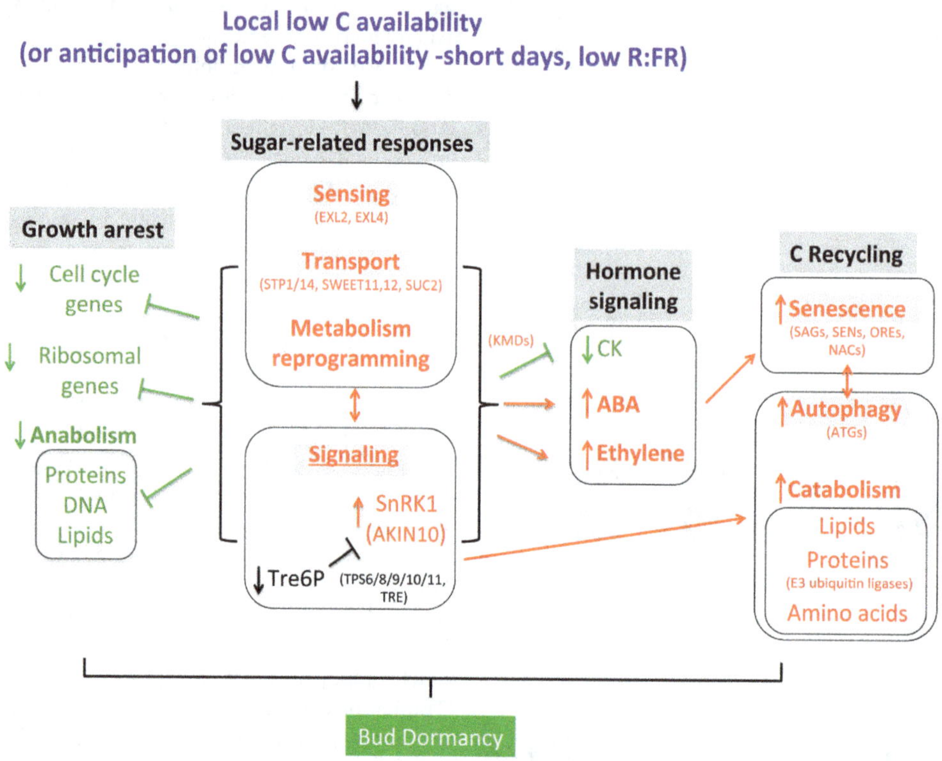

FIGURE 7 | Summary of responses observed in buds entering dormancy in the context of a potential C starvation response. Relationships are based on data obtained in Arabidopsis (see Discussion). Some relevant Bud dormancy genes are indicated.

signaling, autophagy and senescence. Genes required for growth, such as ribosomal, cell cycle and anabolism-related genes become downregulated (Thimm et al., 2004). Buds entering dormancy in Arabidopsis, poplar and grapevine show induction of genes of the former categories and repression of genes of the latter categories.

EXORDIUM-like (*EXL*)*2* and *EXL4* are bud dormancy genes potentially involved in sugar sensing. They are induced in extended night treatments in Arabidopsis seedlings, in accordance with a role under C-limiting conditions (Schröder et al., 2012). Their close paralogs, *EXORDIUM* (*EXO*) and *EXL1*, are proposed to integrate apoplastic C status with intracellular responses (*EXO*) (Lisso et al., 2013) and to control primary and long-term adaptation to C starvation (*EXL1*) (Schröder et al., 2011, 2012).

Several sugar transporters are also induced in dormant buds. These are *STP1*, one of the most rapidly and prominently downregulated genes in response to sugars (Price et al., 2004; Cordoba et al., 2015), *STP14*, which is strongly repressed by sugars (Büttner, 2010) and the sucrose efflux transporters *SWEET11* and *SWEET12*, which act with the sucrose/proton symporter *SUT1/SUC2* for phloem loading and long-distance transport (Chen et al., 2012).

Sugar signaling involves Tre6P (Paul et al., 2008; Lunn et al., 2014) and four class-II *TPS*, *TPS8*, *TPS9*, *TPS10*, and *TPS11*, are induced in buds entering dormancy. They belong to a core C-signaling response, are usually strongly upregulated in C starvation, and are *AKIN10*-responsive (Contento et al., 2004; Price et al., 2004; Thimm et al., 2004; Baena-González et al., 2007). Although their proteins may be catalytically inactive, they might modulate other TPSs or act as Tre6-P sensors (Lunn, 2007). In addition, the SnRK1 protein-kinase, a central regulator of growth in response to C availability (Baena-González et al., 2007), is likely to have a key role in the induction of bud dormancy (see below).

Ethylene, ABA, CK, Senescence and Autophagy during Bud Dormancy

Ethylene and ABA signaling are induced during the bud dormancy transition in Arabidopsis, which agrees with studies that associated these hormones with bud dormancy in many other herbaceous and woody species (Suttle, 1998; Ruonala et al., 2006; Rohde and Bhalerao, 2007; Ruttink et al., 2007; Horvath et al., 2008; Díaz-Riquelme et al., 2012; González-Grandío et al., 2013, 2017; Reddy et al., 2013; González-Grandío and Cubas, 2014; Yao and Finlayson, 2015). We propose that ethylene and ABA responses are closely connected to the C starvation syndrome. Indeed, many mutants with altered responses to sugars have impaired ethylene or ABA signaling (Zhou et al., 1998; Arenas-Huertero et al., 2000; Finkelstein and Lynch, 2000; Huijser et al., 2000; Laby et al., 2000; Rook et al., 2001; Cheng et al., 2002; Yuan and Wysocka-Diller, 2006), and there is compelling evidence of crosstalk between sugar sensing and ethylene and ABA response. C starvation leads to induction of ethylene and ABA-related genes, whereas sugar treatment has the opposite effect (Laby et al., 2000; Rook et al., 2001; Brocard et al., 2002; León and Sheen, 2003; Yanagisawa et al., 2003; Thimm et al., 2004; Buchanan-Wollaston et al., 2005; Rolland et al., 2006). Thus C starvation signaling could trigger ethylene and ABA responses in buds. Indeed, the reduction in sugar levels detected in poplar buds during the 1 w SD is suggested to induce ethylene signaling, followed by ABA signaling (Ruttink et al., 2007).

One role of ABA and ethylene in the C starvation syndrome is induction of senescence, a genetically programmed process that promotes degradation of cell components and macromolecules, remobilizes nutrients, and optimizes resources to supply energy and C skeletons. Ethylene and ABA activate senescence-related genes and senescence induces ABA signaling (Abeles et al., 1988; Zeevaart and Creelman, 1988; Zacarias and Reid, 1990; Reid and Wu, 1992; Weaver et al., 1998; Seo et al., 2000; Yang et al., 2003; Buchanan-Wollaston et al., 2005; Lim et al., 2007). It is noteworthy that the potential master regulators of GRNII and III, *ATAF1, ORE1/NAC6* and *NAP*, are ABA-induced factors that control senescence. *ATAF1* induces a C starvation transcriptome and ABA biosynthesis (Jensen et al., 2013; Garapati et al., 2015). *NAP* activates *SAG113/HAI1* and controls expression of *ABSCISIC ALDEHYDE OXIDASE3* (*AAO3*), encoding an enzyme that catalyzes the final steps of ABA synthesis (Guo and Gan, 2006; Yang et al., 2014). *ORE1* controls the expression of at least 78 SAGs and might also promote DNA degradation (Balazadeh et al., 2010; Matallana-Ramirez et al., 2013; Kim et al., 2014). *HAT22, GBF2, GBF3, ABF3* and *ABF4* are additional bud dormancy genes related both to senescence and ABA (Lin and Wu, 2004; Rivero et al., 2007; Song et al., 2016). Remarkably, *MAX2/ORE9*, which encodes an F-box involved directly in strigolactone perception and signaling and has a critical role in the control of shoot branching, also promotes senescence (Woo et al., 2001; Stirnberg et al., 2002). Finally, the MYB genes *At1g19000* and *At1g74840*, proposed to be master regulators of GRNI and IV, are responsive to dark-induced senescence (Lin and Wu, 2004).

In contrast, CK signaling is antagonistic to senescence, and a reduction in CK levels is a key signal for senescence initiation in Arabidopsis (Gan and Amasino, 1995; Kim et al., 2006). Consistent with this, in Arabidopsis dormant buds CK signaling is reduced, and in other species CK levels have also been reported to be reduced relative to active buds (Turnbull et al., 1997; Dun et al., 2012; Roman et al., 2016). Consistently, four genes encoding F-box proteins that promote the ubiquitination and degradation of ARR factors [*KISS ME DEADLY* (*KMD*)1-4] are bud dormancy genes.

Autophagy is another process induced by C starvation (Izumi et al., 2013) and whose markers (*ATG* genes) are upregulated in buds entering dormancy. This is a process by which cytoplasmic components and organelles are transported to the vacuole, where they are broken down and recycled. Under C-limiting conditions it contributes to plant energy availability (Aubert et al., 1996; Rose et al., 2006; Izumi et al., 2013). Autophagy is associated with induction of lipid degradation and upregulation of E2- and E3-ubiquitin ligase components, which promote proteasomal-dependent protein degradation (Thompson and Vierstra, 2005). We have found a remarkable number of bud dormancy genes related to autophagy, ubiquitination, protein degradation and lipid catabolism, many of them controlled by SnRK1 (see below).

SnRK1 Could Have a Pivotal Role during the Bud Growth-to-Dormancy Transition

SnRK1, a protein-kinase active in low energy conditions, promotes catabolism and represses anabolism, cell division and growth. Our transcriptomic data indicates that it may play an important role during the bud growth-to-dormancy transition. SnRK1 affects expression of robust dormancy markers such *HIS1.3* and *DRM1*, and the potential master regulator of GRNI and GRNIV, *MYBH/KUA1*. In buds entering dormancy, the SnRK1 β subunit *AKINBETA1*, whose mRNA levels correlate directly with night duration (Pokhilko et al., 2014), is induced. Most importantly, our GSEA analysis indicates that the transcriptional network downstream of the catalytic SnRK1 α subunit, *AKIN10*, is significantly induced from the earliest stages of growth-to-dormancy transition in Arabidopsis, poplar and grapevine buds, and is maintained in para-, eco- and endodormant buds. Many of the abovementioned genes involved in sugar sensing, signaling, autophagy and repression of CK signaling are *AKIN10*-dependent, including *EXL4, STP1/14, SWEET11/12, TPS8/9/19/11, AKINBETA, ATG8E/F/G/H, ATG18F/G*, and F-box genes *KMD1, 3, 4*. SnRK1 also causes downregulation of a large number of ribosomal genes, another conserved significant effect detected by our GSEA analysis. SnRK1 could also be responsible for at least part of the observed induction of the ubiquitination machinery and lipid degradation.

A Conserved Core C Starvation Response Underlies Bud Dormancy in Angiosperms

Bud dormancy is an adaptive response present in all angiosperms. It prevents shoot development when endogenous or environmental conditions are unfavorable for sustained growth. It has great impact on reproductive success, productivity and survival, and must have been influential in the colonization of habitats with fluctuating conditions.

We have found induction of a conserved C starvation syndrome that precedes and underlies the growth-to-dormancy transition in buds of three distantly-related species, one herbaceous (Arabidopsis) and two woody (poplar and grapevine). This transcriptional response, composed by several interconnected GRNs, comprises ortholog genes in Arabidopsis, poplar and grapevine, as gene sets generated in Arabidopsis were used to detect the response in the woody species. Furthermore, this syndrome has been observed is several unrelated experiments, regardless the stimulus that promoted dormancy, either environmental (low R:FR, short-day photoperiods) or endogenous (apical dominance). This remarkable conservation suggests that a syndrome aimed at adapting to C-limiting situations is deeply rooted in the control of shoot meristem and bud development across angiosperms. Bud dormancy might thus be an ancestral response directly resulting from this C starvation syndrome, coordinated by different pathways that sense and/or anticipate situations on low C availability and feed into this core response to prevent untimely growth and development.

MATERIALS AND METHODS

Identification of Coregulated Genes in Bud Dormancy GRNs

Bud dormancy genes (Supplementary Figure S1) were obtained from González-Grandío and Cubas (2014). Coregulation of the 78 bud dormancy genes was analyzed by hierarchical clustering (Hcluster, ATTED-II, Obayashi et al., 2007). Additional coregulated genes were obtained using CoEx-Search (ATTED-II, Obayashi et al., 2007). The 300 genes most coregulated with each cluster were selected. These genes were validated for induction in dormant buds in the original arrays (Tatematsu et al., 2005; González-Grandío et al., 2013; Reddy et al., 2013). Only genes upregulated (positive fold change FC $\geq$ 1.2) in at least one experiment in dormant buds were included in the lists of bud dormancy GRNs (Supplementary Figure S8).

Functional Annotation of Bud Dormancy GRNs

Automated function prediction for the GRNs was carried out using GO analyses. The PANTHER classification system (Mi et al., 2017) was used to identify overrepresented biological process ontologies using a statistical overrepresentation test followed by Bonferroni correction for multiple testing. TAIR10 version of *Arabidopsis thaliana* genome was used as reference. We selected ontologies with a pval < 0.05. In addition, Mapman bins (Thimm et al., 2004) were added to all the genes in Supplementary Dataset S1.

Gene Set Enrichment Analysis

Gene Set Enrichment Analysis (Subramanian et al., 2005) was used to identify gene sets whose genes are overrepresented in different conditions. The GSEA method evaluates whether these genes occur preferentially toward the top or bottom of a ranked list. Enrichment scores are calculated using "weighted" statistics. For each sample, we calculated the log2 ratios of normalized gene intensities vs. normalized gene intensities of the "active bud" sample: white light-treated buds for González-Grandío et al. (2013); High R:FR-treated n-2 buds for Reddy et al. (2013); Buds of decapitated plants for Tatematsu et al. (2005); Buds of LD-grown poplars for Ruttink et al. (2007), April grapevine buds for Díaz-Riquelme et al. (2012). Genes were ranked by their log2 ratios calculated as the difference between normalized log intensity in the "dormant bud" condition minus normalized log2 intensity in the "active bud" condition. Intensity expression values were obtained from the references above. Gene sets for hormone markers were obtained from Nemhauser et al. (2006) and Mashiguchi et al. (2009). Gene sets related to sugar and *AKIN10* responses were obtained from Sulpice et al. (2009) (Core C-signaling), Osuna et al. (2007) (Sugar-responsive), Gonzali et al. (2006) (Dark-induced, sugar-repressed), and Baena-González et al. (2007) (*AKIN10*-responsive). The other gene sets are from González-Grandío et al. (2013). The GSEA Normalized Enrichment Score for all gene sets in all comparisons were clustered with TM4 Multi Experiment Viewer (MeV, Saeed et al., 2003). Tree was generated by the hierarchical clustering method (HCL) using Euclidean distance and average linkage options. Complete results are in http://bioinfogp.cnb.csic.es/files/projects/tarancon_et_al_2017_supp/.

Promoter Motif Analysis

Sequences (1 kb) 5′ of the transcription start site of the bud dormancy GRN genes were retrieved with Sequence Bulk Download[2]. Overrepresented 6-8mer motifs were identified with Motif discovery (RSAT, Medina-Rivera et al., 2015). The oligo-analysis tool was used to find significantly overrepresented motifs, which were assembled into frequency matrices with pattern-assembly and default parameters. Matrices were converted into consensus motifs with convert-matrix and represented using WebLogo (Crooks et al., 2004).

Generation and Visualization of Poplar and Grapevine Expression Datasets

For each time point we calculated the log2 ratios of normalized gene intensities vs. normalized gene intensities on LD (active buds). Expression data was visualized and clustered with MeV. Tree was generated by HCL, using Euclidean distance and average linkage options.

Cell-Type Specific Shoot Apex Expression of Bud Dormancy GRN Genes

For each sample, we calculated the log2 ratios of normalized gene intensities vs. normalized gene intensities of the "active bud" sample: LD for poplar, April for grapevine. Expression data for selected bud dormancy genes obtained from Yadav et al. (2014) was visualized and clustered with MeV. Trees were generated by HCL using Euclidean distance and average linkage options.

Identification of *Solanum tuberosum* Orthologs

The putative orthologs of Arabidopsis genes were identified by a tblastn search with protein sequences as query in the Spud DB Potato Genomics Resource website[3]. cDNAs showing a high similarity *e*-value with the query were selected. Proteins were aligned with those of Arabidopsis and phylogenetic trees (BioNeighbor joining method, 500 replicates; Gascuel, 1997) were built to identify the most likely orthologs, which were selected for expression studies (Supplementary Figure S9).

[2]www.arabidopsis.org

[3]solanaceae.plantbiology.msu.edu/index.shtml

Quantitative-PCR Expression Analyses in *Solanum tuberosum*

Plant growth conditions, experimental design, light treatments, techniques and expression level normalization were as described in Nicolas et al. (2015). For each biological replicate, 8 axillary buds from node 2, and 8 from node 3 were dissected (node 1 = lowest plant node); 4–5 biological replicates were collected for each condition. Primers used are listed in Suplemmentary Table S1.

AUTHOR CONTRIBUTIONS

CT, EG-G, JO, and MN, performed experiments. PC, performed experiments, and wrote the manuscript.

ACKNOWLEDGMENTS

We thank Desmond Bradley and Elena Baena for constructive criticisms of the manuscript and Catherine Mark for editorial assistance. PC is supported by a MINECO grant (BIO2014-57011-R). CT is a La Caixa predoctoral fellow. EG-G was a predoctoral fellow of Fundación Ramón Areces and a CSIC JAE-Predoc fellow. MN is a Excelence Severo Ochoa (MINECO) postdoctoral researcher.

SUPPLEMENTARY MATERIAL

The Supplementary Material for this article can be found online at: http://journal.frontiersin.org/article/10.3389/fpls.2017.00788/full#supplementary-material

REFERENCES

Abeles, F. B., Dunn, L. J., Morgens, P., Callahan, A., Dinterman, R. E., and Schmidt, J. (1988). Induction of 33-kD and 60-kD Peroxidases during Ethylene-induced senescence of cucumber cotyledons. *Plant Physiol.* 87, 609–615. doi: 10.1104/pp.87.3.609

Aguilar-Martínez, J. A., Poza-Carrión, C., and Cubas, P. (2007). Arabidopsis BRANCHED1 acts as an integrator of branching signals within axillary buds. *Plant Cell* 19, 458–472. doi: 10.1105/tpc.106.048934

Arenas-Huertero, F., Arroyo, A., Zhou, L., Sheen, J., and León, P. (2000). Analysis of Arabidopsis glucose insensitive mutants, gin5 and gin6, reveals a central role of the plant hormone ABA in the regulation of plant vegetative development by sugar. *Genes Dev.* 14, 2085–2096.

Aubert, S., Gout, E., Bligny, R., Marty-Mazars, D., Barrieu, F., Alabouvette, J., et al. (1996). Ultrastructural and biochemical characterization of autophagy in higher plant cells subjected to carbon deprivation: control by the supply of mitochondria with respiratory substrates. *J. Cell Biol.* 133, 1251–1263. doi: 10.1083/jcb.133.6.1251

Baena-González, E., Rolland, F., Thevelein, J. M., and Sheen, J. (2007). A central integrator of transcription networks in plant stress and energy signalling. *Nature* 448, 938–942. doi: 10.1038/nature06069

Baena-González, E., and Sheen, J. (2008). Convergent energy and stress signaling. *Trends Plant Sci.* 13, 474–482. doi: 10.1016/j.tplants.2008.06.006

Balazadeh, S., Siddiqui, H., Allu, A. D., Matallana-Ramirez, L. P., Caldana, C., Mehrnia, M., et al. (2010). A gene regulatory network controlled by the NAC transcription factor ANAC092/AtNAC2/ORE1 during salt-promoted senescence. *Plant J.* 62, 250–264. doi: 10.1111/j.1365-313X.2010.04151.x

Barbier, F., Péron, T., Lecerf, M., Perez-Garcia, M.-D., Barrière, Q., Rolčík, J., et al. (2015). Sucrose is an early modulator of the key hormonal mechanisms controlling bud outgrowth in Rosa hybrida. *J. Exp. Bot.* 66, 2569–2582. doi: 10.1093/jxb/erv047

Barbier, F. F., Lunn, J. E., and Beveridge, C. A. (2015). Ready, steady, go! A sugar hit starts the race to shoot branching. *Curr. Opin. Plant Biol.* 25, 39–45. doi: 10.1016/j.pbi.2015.04.004

Bi, Y.-M., Zhang, Y., Signorelli, T., Zhao, R., Zhu, T., and Rothstein, S. (2005). Genetic analysis of Arabidopsis GATA transcription factor gene family reveals a nitrate-inducible member important for chlorophyll synthesis and glucose sensitivity. *Plant J.* 44, 680–692. doi: 10.1111/j.1365-313X.2005.02568.x

Brocard, I. M., Lynch, T. J., and Finkelstein, R. R. (2002). Regulation and role of the Arabidopsis abscisic acid-insensitive 5 gene in abscisic acid, sugar, and stress response. *Plant Physiol.* 129, 1533–1543. doi: 10.1104/pp.005793

Buchanan-Wollaston, V., Page, T., Harrison, E., Breeze, E., Lim, P. O., Nam, H. G., et al. (2005). Comparative transcriptome analysis reveals significant differences in gene expression and signalling pathways between developmental and dark/starvation-induced senescence in Arabidopsis. *Plant J.* 42, 567–585. doi: 10.1111/j.1365-313X.2005.02399.x

Büttner, M. (2010). The Arabidopsis sugar transporter (AtSTP) family: an update. *Plant Biol.* 12, 35–41. doi: 10.1111/j.1438-8677.2010.00383.x

Cagnola, J. I., Ploschuk, E., Benech-Arnold, T., Finlayson, S. A., and Casal, J. J. (2012). Stem transcriptome reveals mechanisms to reduce the energetic cost of shade-avoidance responses in tomato. *Plant Physiol.* 160, 1110–1119. doi: 10.1104/pp.112.201921

Carabelli, M., Possenti, M., Sessa, G., Ciolfi, A., Sassi, M., Morelli, G., et al. (2007). Canopy shade causes a rapid and transient arrest in leaf development through auxin-induced cytokinin oxidase activity. *Genes Dev.* 21, 1863–1868. doi: 10.1101/gad.432607

Chávez Montes, R. A., Ranocha, P., Martinez, Y., Minic, Z., Jouanin, L., Marquis, M., et al. (2008). Cell wall modifications in Arabidopsis plants with altered alpha-L-arabinofuranosidase activity. *Plant Physiol.* 147, 63–77. doi: 10.1104/pp.107.110023

Chen, L.-Q., Qu, X.-Q., Hou, B.-H., Sosso, D., Osorio, S., Fernie, A. R., et al. (2012). Sucrose efflux mediated by SWEET proteins as a key step for phloem transport. *Science* 335, 207–211. doi: 10.1126/science.1213351

Chen, Y.-S., Chao, Y.-C., Tseng, T.-W., Huang, C.-K., Lo, P.-C., and Lu, C.-A. (2016). Two MYB-related transcription factors play opposite roles in sugar signaling in Arabidopsis. *Plant Mol. Biol.* 93, 299–311. doi: 10.1007/s11103-016-0562-8

Cheng, W.-H., Endo, A., Zhou, L., Penney, J., Chen, H.-C., Arroyo, A., et al. (2002). A unique short-chain dehydrogenase/reductase in Arabidopsis glucose signaling and abscisic acid biosynthesis and functions. *Plant Cell* 14, 2723–2743. doi: 10.1105/tpc.006494

Contento, A. L., Kim, S.-J., and Bassham, D. C. (2004). Transcriptome profiling of the response of Arabidopsis suspension culture cells to Suc starvation. *Plant Physiol.* 135, 2330–2347. doi: 10.1104/pp.104.044362

Cookson, S. J., Yadav, U. P., Klie, S., Morcuende, R., Usadel, B., Lunn, J. E., et al. (2016). Temporal kinetics of the transcriptional response to carbon depletion and sucrose readdition in *Arabidopsis* seedlings. *Plant. Cell Environ.* 39, 768–786. doi: 10.1111/pce.12642

Cordoba, E., Aceves-Zamudio, D. L., Hernandez-Bernal, A. F., Ramos-Vega, M., and Leon, P. (2015). Sugar regulation of SUGAR TRANSPORTER PROTEIN 1 (STP1) expression in *Arabidopsis thaliana*. *J. Exp. Bot.* 66, 147–159. doi: 10.1093/jxb/eru394

Crooks, G. E., Hon, G., Chandonia, J.-M., and Brenner, S. E. (2004). WebLogo: a sequence logo generator. *Genome Res.* 14, 1188–1190. doi: 10.1101/gr.849004

Díaz-Riquelme, J., Grimplet, J., Martínez-Zapater, J. M., and Carmona, M. J. (2012). Transcriptome variation along bud development in grapevine (*Vitis vinifera* L.). *BMC Plant Biol.* 12:181. doi: 10.1186/1471-2229-12-181

Doebley, J., Stec, A., and Hubbard, L. (1997). The evolution of apical dominance in maize. *Nature* 386, 485–488. doi: 10.1038/386485a0

Dun, E. A., de Saint Germain, A., Rameau, C., and Beveridge, C. A. (2012). Antagonistic action of strigolactone and cytokinin in bud outgrowth control. *Plant Physiol.* 158, 487–498. doi: 10.1104/pp.111.186783

Epple, P., Mack, A. A., Morris, V. R. F., and Dangl, J. L. (2003). Antagonistic control of oxidative stress-induced cell death in Arabidopsis by two related, plant-specific zinc finger proteins. *Proc. Natl. Acad. Sci. U.S.A.* 100, 6831–6836. doi: 10.1073/pnas.1130421100

Fennell, A. Y., Schlauch, K. A., Gouthu, S., Deluc, L. G., Khadka, V., Sreekantan, L., et al. (2015). Short day transcriptomic programming during induction of dormancy in grapevine. *Front. Plant Sci.* 6:834. doi: 10.3389/fpls.2015.00834

Finkelstein, R. R., and Lynch, T. J. (2000). The Arabidopsis abscisic acid response gene ABI5 encodes a basic leucine zipper transcription factor. *Plant Cell* 12, 599–609. doi: 10.1105/tpc.12.4.599

Finlayson, S. A. (2007). Arabidopsis Teosinte Branched1-like 1 regulates axillary bud outgrowth and is homologous to monocot Teosinte Branched1. *Plant Cell Physiol.* 48, 667–677. doi: 10.1093/pcp/pcm044

Franklin, K. A., and Quail, P. H. (2010). Phytochrome functions in Arabidopsis development. *J. Exp. Bot.* 61, 11–24. doi: 10.1093/jxb/erp304

Gan, S., and Amasino, R. M. (1995). Inhibition of leaf senescence by autoregulated production of cytokinin. *Science* 270, 1986–1988. doi: 10.1126/science.270.5244.1986

Garapati, P., Feil, R., Lunn, J. E., Van Dijck, P., Balazadeh, S., and Mueller-Roeber, B. (2015). Transcription factor Arabidopsis activating factor1 integrates carbon starvation responses with trehalose metabolism. *Plant Physiol.* 169, 379–390. doi: 10.1104/pp.15.00917

Gascuel, O. (1997). BIONJ: an improved version of the NJ algorithm based on a simple model of sequence data. *Mol. Biol. Evol.* 14, 685–695. doi: 10.1093/oxfordjournals.molbev.a025808

Goddijn, O. J., Verwoerd, T. C., Voogd, E., Krutwagen, R. W., de Graaf, P. T., van Dun, K., et al. (1997). Inhibition of trehalase activity enhances trehalose accumulation in transgenic plants. *Plant Physiol.* 113, 181–190. doi: 10.1104/pp.113.1.181

González-Grandío, E., and Cubas, P. (2014). Identification of gene functions associated to active and dormant buds in Arabidopsis. *Plant Signal. Behav.* 9, 19–21. doi: 10.4161/psb.27994

González-Grandío, E., Pajoro, A., Franco-Zorrilla, J. M., Tarancón, C., Immink, R. G. H., and Cubas, P. (2017). Abscisic acid signaling is controlled by a BRANCHED1/HD-ZIP I cascade in Arabidopsis axillary buds. *Proc. Natl. Acad. Sci. U.S.A.* 114, E245–E254. doi: 10.1073/pnas.1613199114

González-Grandío, E., Poza-Carrión, C., Sorzano, C. O. S., and Cubas, P. (2013). BRANCHED1 promotes axillary bud dormancy in response to shade in Arabidopsis. *Plant Cell* 25, 834–850. doi: 10.1105/tpc.112.108480

Gonzali, S., Loreti, E., Solfanelli, C., Novi, G., Alpi, A., and Perata, P. (2006). Identification of sugar-modulated genes and evidence for in vivo sugar sensing in Arabidopsis. *J. Plant Res.* 119, 115–123. doi: 10.1007/s10265-005-0251-1

Guo, Y., and Gan, S. (2006). AtNAP, a NAC family transcription factor, has an important role in leaf senescence. *Plant J.* 46, 601–612. doi: 10.1111/j.1365-313X.2006.02723.x

Hao, X., Yang, Y., Yue, C., Wang, L., Horvath, D. P., and Wang, X. (2017). Comprehensive transcriptome analyses reveal differential gene expression profiles of *Camellia sinensis* axillary buds at Para-, Endo-, ecodormancy, and bud flush stages. *Front. Plant Sci.* 8:553. doi: 10.3389/fpls.2017.00553

Haydon, M. J., Mielczarek, O., Robertson, F. C., Hubbard, K. E., and Webb, A. A. R. (2013). Photosynthetic entrainment of the *Arabidopsis thaliana* circadian clock. *Nature* 502, 689–692. doi: 10.1038/nature12603

Horvath, D. (2009). Common mechanisms regulate flowering and dormancy. *Plant Sci.* 177, 523–531. doi: 10.1016/j.plantsci.2009.09.002

Horvath, D. P., Chao, W. S., Suttle, J. C., Thimmapuram, J., and Anderson, J. V. (2008). Transcriptome analysis identifies novel responses and potential regulatory genes involved in seasonal dormancy transitions of leafy spurge (*Euphorbia esula* L.). *BMC Genomics* 9:536. doi: 10.1186/1471-2164-9-536

Howe, G. T., Horvath, D. P., Dharmawardhana, P., Priest, H. D., Mockler, T. C., and Strauss, S. H. (2015). Extensive transcriptome changes during natural onset and release of vegetative bud dormancy in *Populus*. *Front. Plant Sci.* 6:989. doi: 10.3389/fpls.2015.00989

Huang, C.-K., Lo, P.-C., Huang, L.-F., Wu, S.-J., Yeh, C.-H., and Lu, C.-A. (2015). A single-repeat MYB transcription repressor, MYBH, participates in regulation of leaf senescence in Arabidopsis. *Plant Mol. Biol.* 88, 269–286. doi: 10.1007/s11103-015-0321-2

Huijser, C., Kortstee, A., Pego, J., Weisbeek, P., Wisman, E., and Smeekens, S. (2000). The Arabidopsis SUCROSE UNCOUPLED-6 gene is identical to ABSCISIC ACID INSENSITIVE-4: involvement of abscisic acid in sugar responses. *Plant J.* 23, 577–585. doi: 10.1046/j.1365-313x.2000.00822.x

Izumi, M., Hidema, J., Makino, A., and Ishida, H. (2013). Autophagy contributes to nighttime energy availability for growth in Arabidopsis. *Plant Physiol.* 161, 1682–1693. doi: 10.1104/pp.113.215632

Jensen, M. K., Lindemose, S., Masi, F., de Reimer, J. J., Nielsen, M., Perera, V., et al. (2013). ATAF1 transcription factor directly regulates abscisic acid biosynthetic gene NCED3 in *Arabidopsis thaliana*. *FEBS Open Bio* 3, 321–327. doi: 10.1016/j.fob.2013.07.006

Johnson, E., Bradley, M., Harberd, N. P., and Whitelam, G. C. (1994). Photoresponses of light-grown phyA mutants of Arabidopsis (phytochrome a is required for the perception of daylength extensions). *Plant Physiol.* 105, 141–149. doi: 10.1104/pp.105.1.141

Kebrom, T. H., Brutnell, T. P., and Finlayson, S. A. (2010). Suppression of sorghum axillary bud outgrowth by shade, phyB and defoliation signalling pathways. *Plant Cell Environ.* 33, 48–58. doi: 10.1111/j.1365-3040.2009.02050.x

Kim, H. J., Hong, S. H., Kim, Y. W., Lee, I. H., Jun, J. H., Phee, B.-K., et al. (2014). Gene regulatory cascade of senescence-associated NAC transcription factors activated by ETHYLENE-INSENSITIVE2-mediated leaf senescence signalling in Arabidopsis. *J. Exp. Bot.* 65, 4023–4036. doi: 10.1093/jxb/eru112

Kim, H. J., Ryu, H., Hong, S. H., Woo, H. R., Lim, P. O., Lee, I. C., et al. (2006). Cytokinin-mediated control of leaf longevity by AHK3 through phosphorylation of ARR2 in Arabidopsis. *Proc. Natl. Acad. Sci. U.S.A.* 103, 814–819. doi: 10.1073/pnas.0505150103

Laby, R. J., Kincaid, M. S., Kim, D., and Gibson, S. I. (2000). The Arabidopsis sugar-insensitive mutants sis4 and sis5 are defective in abscisic acid synthesis and response. *Plant J.* 23, 587–596. doi: 10.1046/j.1365-313x.2000.00833.x

Lang, G. A., Early, J. D., Arroyave, N. J., Darnell, R. L., Martin, G. C., and Stutte, G. W. (1985). Dormancy: toward a reduced, universal terminology. *HortScience* 20, 809–811.

Lang, G. A., Early, J. D., Martin, G. C., and Darnell, R. L. (1987). Endo-, para-, and ecodormancy: physiological terminology and classification for dormancy research. *HortScience* 22, 371–377.

Lauxmann, M. A., Annunziata, M. G., Brunoud, G., Wahl, V., Koczut, A., Burgos, A., et al. (2016). Reproductive failure in *Arabidopsis thaliana* under transient carbohydrate limitation: flowers and very young siliques are jettisoned and the meristem is maintained to allow successful resumption of reproductive growth. *Plant Cell Environ.* 39, 745–767. doi: 10.1111/pce.12634

León, P., and Sheen, J. (2003). Sugar and hormone connections. *Trends Plant Sci.* 8, 110–116. doi: 10.1016/S1360-1385(03)00011-6

Lim, C. W., Kim, J.-H., Baek, W., Kim, B. S., and Lee, S. C. (2012). Functional roles of the protein phosphatase 2C, AtAIP1, in abscisic acid signaling and sugar tolerance in Arabidopsis. *Plant Sci.* 187, 83–88. doi: 10.1016/j.plantsci.2012.01.013

Lim, P. O., Kim, H. J., and Gil Nam, H. (2007). Leaf Senescence. *Annu. Rev. Plant Biol.* 58, 115–136. doi: 10.1146/annurev.arplant.57.032905.105316

Lin, J.-F., and Wu, S.-H. (2004). Molecular events in senescing *Arabidopsis* leaves. *Plant J.* 39, 612–628. doi: 10.1111/j.1365-313X.2004.02160.x

Lisso, J., Schröder, F., and Müssig, C. (2013). EXO modifies sucrose and trehalose responses and connects the extracellular carbon status to growth. *Front. Plant Sci.* 4:219. doi: 10.3389/fpls.2013.00219

Lu, C.-A., Ho, T. D., Ho, S.-L., and Yu, S.-M. (2002). Three novel MYB proteins with one DNA binding repeat mediate sugar and hormone regulation of alpha-amylase gene expression. *Plant Cell* 14, 1963–1980. doi: 10.1105/tpc.001735

Lu, C. A., Lim, E. K., and Yu, S. M. (1998). Sugar response sequence in the promoter of a rice alpha-amylase gene serves as a transcriptional enhancer. *J. Biol. Chem.* 273, 10120–10131. doi: 10.1074/jbc.273.17.10120

Lunn, J. E. (2007). Gene families and evolution of trehalose metabolism in plants. *Funct. Plant Biol.* 34, 550. doi: 10.1071/fp06315

Lunn, J. E., Delorge, I., Figueroa, C. M., Van Dijck, P., and Stitt, M. (2014). Trehalose metabolism in plants. *Plant J.* 79, 544–567. doi: 10.1111/tpj.12509

Martìnez de Toda Fernaàndez, F. (1991). *Biología de la Vid: Fundamentos Biológicos de la Viticultura*. Madrid: Mundi-Prensa.

Mashiguchi, K., Sasaki, E., Shimada, Y., Nagae, M., Ueno, K., Nakano, T., et al. (2009). Feedback-regulation of strigolactone biosynthetic genes and strigolactone-regulated genes in Arabidopsis. *Biosci. Biotechnol. Biochem.* 73, 2460–2465. doi: 10.1271/bbb.90443

Mason, M. G., Ross, J. J., Babst, B. A., Wienclaw, B. N., and Beveridge, C. A. (2014). Sugar demand, not auxin, is the initial regulator of apical dominance. *Proc. Natl. Acad. Sci. U.S.A.* 111, 6092–6097. doi: 10.1073/pnas.1322045111

Matallana-Ramirez, L. P., Rauf, M., Farage-Barhom, S., Dortay, H., Xue, G. P., Dröge-Laser, W., et al. (2013). NAC transcription factor ORE1 and doi: senescence-induce BIFUNCTIONAL NUCLEASE1 (BFN1) constitute a regulatory cascade in Arabidopsis. *Mol. Plant* 6, 1438–1452. doi: 10.1093/mp/sst012

Medina-Rivera, A., Defrance, M., Sand, O., Herrmann, C., Castro-Mondragon, J. A., Delerce, J., et al. (2015). RSAT 2015: regulatory sequence analysis tools. *Nucleic Acids Res.* 43, W50–W56. doi: 10.1093/nar/gkv362

Mi, H., Huang, X., Muruganujan, A., Tang, H., Mills, C., Kang, D., et al. (2017). PANTHER version 11: expanded annotation data from gene ontology and reactome pathways, and data analysis tool enhancements. *Nucleic Acids Res.* 45, D183–D189. doi: 10.1093/nar/gkw1138

Monte, E., Alonso, J. M., Ecker, J. R., Zhang, Y., Li, X., Young, J., et al. (2003). Isolation and characterization of phyC mutants in Arabidopsis reveals complex crosstalk between phytochrome signaling pathways. *Plant Cell* 15, 1962–1980. doi: 10.1105/tpc.012971

Moore, B., Zhou, L., Rolland, F., Hall, Q., Cheng, W.-H., Liu, Y.-X., et al. (2003). Role of the Arabidopsis glucose sensor HXK1 in nutrient, light, and hormonal signaling. *Science* 300, 332–336. doi: 10.1126/science.1080585

Neff, M. M., and Chory, J. (1998). Genetic interactions between phytochrome A, phytochrome B, and cryptochrome 1 during Arabidopsis development. *Plant Physiol.* 118, 27–35. doi: 10.1104/pp.118.1.27

Nemhauser, J. L., Hong, F., and Chory, J. (2006). Different plant hormones regulate similar processes through largely nonoverlapping transcriptional responses. *Cell* 126, 467–475. doi: 10.1016/j.cell.2006.05.050

Nicolas, M., and Cubas, P. (2015). "TCP transcription factors: evolution, structure, and biochemical function," in *Plant Transcription Factors*, ed. D. H. Gonzalez (Amstredam: Elsevier), 250–267.

Nicolas, M., Rodríguez-Buey, M. L., Franco-Zorrilla, J. M., and Cubas, P. (2015). A recently evolved alternative splice site in the BRANCHED1a gene controls potato plant architecture. *Curr. Biol.* 25, 1799–1809. doi: 10.1016/j.cub.2015.05.053

Obayashi, T., Kinoshita, K., Nakai, K., Shibaoka, M., Hayashi, S., Saeki, M., et al. (2007). ATTED-II: a database of co-expressed genes and cis elements for identifying co-regulated gene groups in Arabidopsis. *Nucleic Acids Res.* 35, D863–D869. doi: 10.1093/nar/gkl783

Olsen, J. E., Junttila, O., Nilsen, J., Eriksson, M. E., Martinussen, I., Olsson, O., et al. (1997). Ectopic expression of oat phytochrome A in hybrid aspen changes critical daylength for growth and prevents cold acclimatization. *Plant J.* 12, 1339–1350. doi: 10.1046/j.1365-313x.1997.12061339.x

O'Malley, R. C., Huang, S. C., Song, L., Lewsey, M. G., Bartlett, A., Nery, J. R., et al. (2016). Cistrome and epicistrome features shape the regulatory DNA landscape. *Cell* 165, 1280–1292. doi: 10.1016/j.cell.2016.04.038

Osuna, D., Usadel, B., Morcuende, R., Gibon, Y., Bläsing, O. E., Höhne, M., et al. (2007). Temporal responses of transcripts, enzyme activities and metabolites after adding sucrose to carbon-deprived Arabidopsis seedlings. *Plant J.* 49, 463–491. doi: 10.1111/j.1365-313X.2006.02979.x

Park, S., and Han, K.-H. (2003). An auxin-repressed gene (RpARP) from black locust (*Robinia pseudoacacia*) is posttranscriptionally regulated and negatively associated with shoot elongation. *Tree Physiol.* 23, 815–823. doi: 10.1093/treephys/23.12.815

Paul, M. J., Primavesi, L. F., Jhurreea, D., and Zhang, Y. (2008). Trehalose metabolism and signaling. *Annu. Rev. Plant Biol.* 59, 417–441. doi: 10.1146/annurev.arplant.59.032607.092945

Pokhilko, A., Flis, A., Sulpice, R., Stitt, M., and Ebenhöh, O. (2014). Adjustment of carbon fluxes to light conditions regulates the daily turnover of starch in plants: a computational model. *Mol. Biosyst.* 10, 613. doi: 10.1039/c3mb70459a

Porto, D. D., Bruneau, M., Perini, P., Anzanello, R., Renou, J.-P., dos Santos, H. P., et al. (2015). Transcription profiling of the chilling requirement for bud break in apples: a putative role for FLC-like genes. *J. Exp. Bot.* 66, 2659–2672. doi: 10.1093/jxb/erv061

Price, J., Laxmi, A., St Martin, S. K., and Jang, J.-C. (2004). Global transcription profiling reveals multiple sugar signal transduction mechanisms in Arabidopsis. *Plant Cell* 16, 2128–2150. doi: 10.1105/tpc.104.022616

Rameau, C., Bertheloot, J., Leduc, N., Andrieu, B., Foucher, F., and Sakr, S. (2015). Multiple pathways regulate shoot branching. *Front. Plant Sci.* 5:741. doi: 10.3389/fpls.2014.00741

Reddy, S. K., Holalu, S. V., Casal, J. J., and Finlayson, S. A. (2013). Abscisic acid regulates axillary bud outgrowth responses to the ratio of red to far-red light. *Plant Physiol.* 163, 1047–1058. doi: 10.1104/pp.113.221895

Reed, J. W., Nagatani, A., Elich, T. D., Fagan, M., and Chory, J. (1994). Phytochrome A and Phytochrome B have overlapping but distinct functions in arabidopsis development. *Plant Physiol.* 104, 1139–1149. doi: 10.1104/pp.104.4.1139

Reid, M. S., and Wu, M.-J. (1992). Ethylene and flower senescence. *Plant Growth Regul.* 11, 37–43. doi: 10.1007/BF00024431

Rivero, R. M., Kojima, M., Gepstein, A., Sakakibara, H., Mittler, R., Gepstein, S., et al. (2007). Delayed leaf senescence induces extreme drought tolerance in a flowering plant. *Proc. Natl. Acad. Sci. U.S.A.* 104, 19631–19636. doi: 10.1073/pnas.0709453104

Rohde, A., and Bhalerao, R. P. (2007). Plant dormancy in the perennial context. *Trends Plant Sci.* 12, 217–223. doi: 10.1016/j.tplants.2007.03.012

Rolland, F., Baena-Gonzalez, E., and Sheen, J. (2006). Sugar sensing and signaling in plants:conserved and novel mechanisms. *Annu. Rev. Plant Biol.* 57, 675–709. doi: 10.1146/annurev.arplant.57.032905.105441

Roman, H., Girault, T., Barbier, F., Péron, T., Brouard, N., Pěnčík, A., et al. (2016). Cytokinins are initial targets of light in the control of bud outgrowth. *Plant Physiol.* 172, 489–509. doi: 10.1104/pp.16.00530

Rook, F., Corke, F., Card, R., Munz, G., Smith, C., and Bevan, M. W. (2001). Impaired sucrose-induction mutants reveal the modulation of sugar-induced starch biosynthetic gene expression by abscisic acid signalling. *Plant J.* 26, 421–433. doi: 10.1046/j.1365-313X.2001.2641043.x

Rose, T. L., Bonneau, L., Der, C., Marty-Mazars, D., and Marty, F. (2006). Starvation-induced expression of autophagy-related genes in Arabidopsis. *Biol. Cell* 98, 53–67. doi: 10.1042/BC20040516

Ruonala, R., Rinne, P. L. H., Baghour, M., Moritz, T., Tuominen, H., and Kangasjärvi, J. (2006). Transitions in the functioning of the shoot apical meristem in birch (*Betula pendula*) involve ethylene. *Plant J.* 46, 628–640. doi: 10.1111/j.1365-313X.2006.02722.x

Ruonala, R., Rinne, P. L. H., Kangasjärvi, J., and van der Schoot, C. (2008). CENL1 expression in the rib meristem affects stem elongation and the transition to dormancy in *Populus*. *Plant Cell* 20, 59–74. doi: 10.1105/tpc.107.056721

Ruttink, T., Arend, M., Morreel, K., Storme, V., Rombauts, S., Fromm, J., et al. (2007). A molecular timetable for apical bud formation and dormancy induction in poplar. *Plant Cell* 19, 2370–2390. doi: 10.1105/tpc.107.052811

Saeed, A. I., Sharov, V., White, J., Li, J., Liang, W., Bhagabati, N., et al. (2003). TM4: a free, open-source system for microarray data management and analysis. *Biotechniques* 34, 374–378.

Sasidharan, R., Chinnappa, C. C., Staal, M., Elzenga, J. T. M., Yokoyama, R., Nishitani, K., et al. (2010). Light quality-mediated petiole elongation in Arabidopsis during shade avoidance involves cell wall modification by Xyloglucan Endotransglucosylase/Hydrolases. *Plant Physiol.* 154, 978–990. doi: 10.1104/pp.110.162057

Satoh-Nagasawa, N., Nagasawa, N., Malcomber, S., Sakai, H., and Jackson, D. (2006). A trehalose metabolic enzyme controls inflorescence architecture in maize. *Nature* 441, 227–230. doi: 10.1038/nature04725

Schröder, F., Lisso, J., and Mussig, C. (2011). EXORDIUM-LIKE1 promotes growth during low carbon availability in Arabidopsis. *Plant Physiol.* 156, 1620–1630. doi: 10.1104/pp.111.177204

Schröder, F., Lisso, J., and Müssig, C. (2012). Expression pattern and putative function of *EXL1* and homologous genes in Arabidopsis. *Plant Signal. Behav.* 7, 22–27. doi: 10.4161/psb.7.1.18369

Seo, M., Koiwai, H., Akaba, S., Komano, T., Oritani, T., Kamiya, Y., et al. (2000). Abscisic aldehyde oxidase in leaves of *Arabidopsis thaliana*. *Plant J.* 23, 481–488. doi: 10.1046/j.1365-313x.2000.00812.x

Shimizu, S., and Mori, H. (1998). Analysis of cycles of dormancy and growth in pea axillary buds based on mRNA accumulation patterns of cell cycle-related genes. *Plant Cell Physiol.* 39, 255–262. doi: 10.1093/oxfordjournals.pcp.a029365

Smith, A. M., and Stitt, M. (2007). Coordination of carbon supply and plant growth. *Plant Cell Environ.* 30, 1126–1149. doi: 10.1111/j.1365-3040.2007.01708.x

Song, L., Huang, S. S. C., Wise, A., Castanon, R., Nery, J. R., Chen, H., et al. (2016). A transcription factor hierarchy defines an environmental stress response network. *Science* 354, aag1550. doi: 10.1126/science.aag1550

Stafstrom, J. P., Ripley, B. D., Devitt, M. L., and Drake, B. (1998). Dormancy-associated gene expression in pea axillary buds. *Planta* 205, 547–552. doi: 10.1007/s004250050354

Stirnberg, P., van de Sande, K., and Leyser, H. M. O. (2002). MAX1 and MAX2 control shoot lateral branching in Arabidopsis. *Development* 129, 1131–1141.

Subramanian, A., Tamayo, P., Mootha, V. K., Mukherjee, S., Ebert, B. L., Gillette, M. A., et al. (2005). Gene set enrichment analysis: a knowledge-based approach for interpreting genome-wide expression profiles. *Proc. Natl. Acad. Sci. U.S.A.* 102, 15545–15550. doi: 10.1073/pnas.0506580102

Sulpice, R., Pyl, E.-T., Ishihara, H., Trenkamp, S., Steinfath, M., Witucka-Wall, H., et al. (2009). Starch as a major integrator in the regulation of plant growth. *Proc. Natl. Acad. Sci. U.S.A.* 106, 10348–10353. doi: 10.1073/pnas.0903478106

Suttle, J. C. (1998). Involvement of ethylene in potato microtuber dormancy. *Plant Physiol.* 118, 843–848. doi: 10.1104/pp.118.3.843

Tatematsu, K., Ward, S., Leyser, O., Kamiya, Y., and Nambara, E. (2005). Identification of cis-elements that regulate gene expression during initiation of axillary bud outgrowth in Arabidopsis. *Plant Physiol.* 138, 757–766. doi: 10.1104/pp.104.057984

Thimm, O., Bläsing, O., Gibon, Y., Nagel, A., Meyer, S., Krüger, P., et al. (2004). MAPMAN: a user-driven tool to display genomics data sets onto diagrams of metabolic pathways and other biological processes. *Plant J.* 37, 914–939. doi: 10.1111/j.1365-313X.2004.02016.x

Thompson, A. R., and Vierstra, R. D. (2005). Autophagic recycling: lessons from yeast help define the process in plants. *Curr. Opin. Plant Biol.* 8, 165–173. doi: 10.1016/j.pbi.2005.01.013

Turnbull, C. G. N., Raymond, M. A. A., Dodd, I. C., and Morris, S. E. (1997). Rapid increases in cytokinin concentration in lateral buds of chickpea (*Cicer arietinum* L.) during release of apical dominance. *Planta* 202, 271–276. doi: 10.1007/s004250050128

Ueno, S., Klopp, C., Leplé, J. C., Derory, J., Noirot, C., Léger, V., et al. (2013). Transcriptional profiling of bud dormancy induction and release in oak by next-generation sequencing. *BMC Genomics* 14:236. doi: 10.1186/1471-2164-14-236

Usadel, B., Bläsing, O. E., Gibon, Y., Retzlaff, K., Höhne, M., Günther, M., et al. (2008). Global transcript levels respond to small changes of the carbon status during progressive exhaustion of carbohydrates in Arabidopsis rosettes. *Plant Physiol.* 146, 1834–1861. doi: 10.1104/pp.107.115592

Weaver, L. M., Gan, S., Quirino, B., and Amasino, R. M. (1998). A comparison of the expression patterns of several senescence-associated genes in response to stress and hormone treatment. *Plant Mol. Biol.* 37, 455–469. doi: 10.1023/A:1005934428906

Wiese, A., Christ, M. M., Virnich, O., Schurr, U., and Walter, A. (2007). Spatio-temporal leaf growth patterns of *Arabidopsis thaliana* and evidence for sugar control of the diel leaf growth cycle. *New Phytol.* 174, 752–761. doi: 10.1111/j.1469-8137.2007.02053.x

Woo, H. R., Chung, K. M., Park, J. H., Oh, S. A., Ahn, T., Hong, S. H., et al. (2001). ORE9, an F-box protein that regulates leaf senescence in Arabidopsis. *Plant Cell* 13, 1779–1790. doi: 10.1105/tpc.13.8.1779

Wood, M., Rae, G. M., Wu, R.-M., Walton, E. F., Xue, B., Hellens, R. P., et al. (2013). Actinidia DRM1–an intrinsically disordered protein whose mRNA expression is inversely correlated with spring budbreak in kiwifruit. *PLoS ONE* 8:e57354. doi: 10.1371/journal.pone.0057354

Yadav, R. K., Tavakkoli, M., Xie, M., Girke, T., and Reddy, G. V. (2014). A high-resolution gene expression map of the Arabidopsis shoot meristem stem cell niche. *Development* 141, 2735–2744. doi: 10.1242/dev.106104

Yan, J., Niu, F., Liu, W.-Z., Zhang, H., Wang, B., Yang, B., et al. (2014). Arabidopsis CIPK14 positively regulates glucose response. *Biochem. Biophys. Res. Commun.* 450, 1679–1683. doi: 10.1016/j.bbrc.2014.07.064

Yanagisawa, S., Yoo, S.-D., and Sheen, J. (2003). Differential regulation of EIN3 stability by glucose and ethylene signalling in plants. *Nature* 425, 521–525. doi: 10.1038/nature01984

Yang, J., Worley, E., and Udvardi, M. (2014). A NAP-AAO3 regulatory module promotes chlorophyll degradation via ABA biosynthesis in Arabidopsis leaves. *Plant Cell* 26, 4862–4874. doi: 10.1105/tpc.114.133769

Yang, J. C., Zhang, J. H., Wang, Z. Q., Zhu, Q. S., and Liu, L. J. (2003). Involvement of abscisic acid and cytokinins in the senescence and remobilization of carbon reserves in wheat subjected to water stress during grain filling. *Plant Cell Environ.* 26, 1621–1631. doi: 10.1046/j.1365-3040.2003.01081.x

Yao, C., and Finlayson, S. A. (2015). Abscisic acid is a general negative regulator of Arabidopsis axillary bud growth. *Plant Physiol.* 169, 611–626. doi: 10.1104/pp.15.00682

Yuan, K., and Wysocka-Diller, J. (2006). Phytohormone signalling pathways interact with sugars during seed germination and seedling development. *J. Exp. Bot.* 57, 3359–3367. doi: 10.1093/jxb/erl096

Zacarias, L., and Reid, M. S. (1990). Role of growth regulators in the senescence of *Arabidopsis thaliana* leaves. *Physiol. Plant.* 80, 549–554. doi: 10.1111/j.1399-3054.1990.tb05677.x

Zeevaart, J. A. D., and Creelman, R. A. (1988). Metabolism and physiology of abscisic acid. *Annu. Rev. Plant Physiol. Plant Mol. Biol.* 39, 439–473. doi: 10.1146/annurev.pp.39.060188.002255

Zhong, C., Xu, H., Ye, S., Wang, S., Li, L., Zhang, S., et al. (2015). Gibberellic acid-stimulated Arabidopsis6 Serves as an integrator of gibberellin, abscisic acid, and glucose signaling during seed germination in Arabidopsis. *Plant Physiol.* 169, 2288–2303. doi: 10.1104/pp.15.00858

Zhou, L., Jang, J. C., Jones, T. L., and Sheen, J. (1998). Glucose and ethylene signal transduction crosstalk revealed by an Arabidopsis glucose-insensitive mutant. *Proc. Natl. Acad. Sci. U.S.A.* 95, 10294–10299. doi: 10.1073/pnas.95.17.10294

Conflict of Interest Statement: The authors declare that the research was conducted in the absence of any commercial or financial relationships that could be construed as a potential conflict of interest.

16

Evolution and Expression Patterns of *TCP* Genes in Asparagales

Yesenia Madrigal[1], Juan F. Alzate[2] and Natalia Pabón-Mora[1]*

[1] *Facultad de Ciencias Exactas y Naturales, Instituto de Biología, Universidad de Antioquia, Medellín, Colombia,* [2] *Centro Nacional de Secuenciación Genómica, Sede de Investigación Universitaria, Facultad de Medicina, Universidad de Antioquia, Medellín, Colombia*

CYCLOIDEA-like genes are involved in the symmetry gene network, limiting cell proliferation in the dorsal regions of bilateral flowers in core eudicots. *CYC-like* and closely related *TCP* genes (acronym for *TEOSINTE BRANCHED1, CYCLOIDEA,* and *PROLIFERATION CELL FACTOR*) have been poorly studied in Asparagales, the largest order of monocots that includes both bilateral flowers in Orchidaceae (ca. 25.000 spp) and radially symmetrical flowers in Hypoxidaceae (ca. 200 spp). With the aim of assessing *TCP* gene evolution in the Asparagales, we isolated *TCP-like* genes from publicly available databases and our own transcriptomes of *Cattleya trianae* (Orchidaceae) and *Hypoxis decumbens* (Hypoxidaceae). Our matrix contains 452 sequences representing the three major clades of *TCP* genes. Besides the previously identified *CYC* specific core eudicot duplications, our ML phylogenetic analyses recovered an early *CIN-like* duplication predating all angiosperms, two *CIN-like* Asparagales-specific duplications and a duplication prior to the diversification of Orchidoideae and Epidendroideae. In addition, we provide evidence of at least three duplications of *PCF-like* genes in Asparagales. While *CIN-like* and *PCF-like* genes have multiplied in Asparagales, likely enhancing the genetic network for cell proliferation, *CYC-like* genes remain as single, shorter copies with low expression. Homogeneous expression of *CYC-like* genes in the labellum as well as the lateral petals suggests little contribution to the bilateral perianth in *C. trianae*. *CIN-like* and *PCF-like* gene expression suggests conserved roles in cell proliferation in leaves, sepals and petals, carpels, ovules and fruits in Asparagales by comparison with previously reported functions in core eudicots and monocots. This is the first large scale analysis of *TCP-like* genes in Asparagales that will serve as a platform for in-depth functional studies in emerging model monocots.

Keywords: *Cattleya trianae, CINCINNATA, CYCLOIDEA,* Hypoxidaceae, *Hypoxis decumbens,* floral symmetry, Orchidaceae, *PROLIFERATION CELL FACTOR*

Edited by:
José M. Romero,
University of Seville, Spain

Reviewed by:
Jill Christine Preston,
University of Vermont, USA
Tomotsugu Koyama,
Suntory Foundation for Life Sciences, Japan

****Correspondence:***
Natalia Pabón-Mora
lucia.pabon@udea.edu.co

INTRODUCTION

As currently circumscribed the order Asparagales is a species-rich group comprising ca. 50% of all monocots, corresponding to 10–15% of flowering plants (Chase et al., 2009, 2016; Chen et al., 2013; Givnish et al., 2016). The most recent phylogenetic analyses in the monocots place Orchidaceae as sister to all other Asparagales (Chen et al., 2013). The family is divided into five subfamilies: Apostasioideae, Vanilloideae, Cypripedioideae, Orchidoideae, and Epidendroideae (Chase et al., 2015; Endress, 2016). The floral groundplan in Asparagales varies primarily in the floral symmetry and the number of stamens (Simpson, 2006). The floral morphology of

Asparagales outside orchids consists of radially symmetrical, trimerous flowers with tepaloid perianth and free floral organs, although a few exceptions have been documented in *Aspidistra* (Asparagaceae), *Gethyllis* (Amaryllidaceae), *Neoastelia* (Asteliaceae) and *Pauridia* (Hypoxidaceae) (Rudall, 2002; Rudall and Bateman, 2002, 2004; Kocyan, 2007). Conversely, orchid flowers are variously bilateral and undergo extreme elaboration of some organs, including differentiation of perianth parts, stamen abortion, and fusion of floral parts from the same whorl or from different whorls (Rudall, 2002). In the bilateral resupinated orchid flowers the two dorsal petals are very similar to each other, whereas the ventral one (the lip or labellum) often undergoes extreme elaboration in shape, color, size and epidermal specializations (Rudall and Bateman, 2004; Pabón-Mora and González, 2008; Mondragón-Palomino and Theißen, 2009; Rudall et al., 2013; Endress, 2016). In the inner floral whorls bilateral symmetry is evident by the formation of a gynostemium that results from the congenital fusion between the single fertile stamen (sometimes two fertile stamens) and stigmas (Rudall and Bateman, 2002; Pabón-Mora and González, 2008; Endress, 2016). Such floral elaboration has been linked to extremely specialized pollination mechanisms and the exceedingly high diversification rates in Orchidaceae (Gong and Huang, 2009; Mondragón-Palomino and Theißen, 2009; Mondragón-Palomino, 2013).

The genetic network underlying bilateral floral symmetry has been assessed using *Antirrhinum majus* floral symmetry mutants (Luo et al., 1996). This network includes the differential dorsiventral expression of four transcription factors in the two-lipped flowers of this species. Three transcription factors, *CYCLOIDEA (CYC)*, *DICHOTOMA (DICH)*, and *RADIALIS (RAD)* regulate cell division on the dorsal portion of the flower primordium and during dorsal petal and stamen primordia initiation. Additionally, *RAD* outcompetes *DIVARICATA (DIV)* for binding proteins in the dorsal side of the flower, restricting *DIV* function to the ventral and lateral petals (Almeida et al., 1997; Galego and Almeida, 2002; Raimundo et al., 2013). Thus, *cyc/dich* mutants show radially symmetrical ventralized flowers (Luo et al., 1996, 1999). Both, *CYC* and *DICH* genes belong to the *TCP* gene family (acronym for *TEOSINTE BRANCHED 1 -TB1-* from *Z. mays*, *CYCLOIDEA -CYC-* from *A. majus y PROLIFERATION CELL FACTOR 1* and *2 -PCF1* and *PCF2-* from *Oryza sativa)* (Doebley et al., 1997; Kosugi and Ohashi, 1997; Luo et al., 1999). *RAD* and *DIV* belong to the *MYB* (*Myeloblastosis*) gene family (Luo et al., 1999; Galego and Almeida, 2002; Corley et al., 2005; Costa et al., 2005).

The *TCP* genes encode putative basic-Helix-Loop-Helix (bHLH) transcription factors (Cubas et al., 1999). The bHLH domain recognizes a consensus sequence GGNCCCAC/GTGGNCCC required for DNA binding and activation or repression of transcription (Kosugi and Ohashi, 2002; Martín-Trillo and Cubas, 2010). Gene evolution analyses have identified two large groups of *TCP* genes, namely Class I (which include *PCF* homologs) and Class II (containing the *CIN/CYC/TB1-like* genes) (Cubas et al., 1999; Damerval and Manuel, 2003; Reeves and Olmstead, 2003; Broholm, 2009; Mondragón-Palomino and Trontin, 2011). Additional large scale duplications (i.e., those occurring prior to the diversification of major inclusive hierarchical groupings) have been found within the *CYC* genes. Two rounds of duplication occurred specifically in core eudicots, resulting in *CYC1*, *CYC2*, and *CYC3* clades, and one duplication specific to monocots resulting in the *RETARDED PALEA 1 (REP1)* and *TEOSINTE BRANCHED 1 (TB1)* clades (Vieira et al., 1999; Damerval and Manuel, 2003; Howarth and Donoghue, 2006; Navaud et al., 2007; Yao et al., 2007; Mondragón-Palomino and Trontin, 2011). In addition, species specific duplications (i.e., those occurring in a single species) have also been reported, often linked to polyploidy (Ma et al., 2016). Non-core eudicot homologs are known as the *CYC-like* genes (Damerval et al., 2007; Preston and Hileman, 2012; Horn et al., 2015). Functional characterization has concentrated in *CYC2* orthologs in eudicots, including Asterales, Brassicales, Dipsacales, Fabales, Lamiales and Malpighiales, among others (Busch and Zachgo, 2007; Gao et al., 2008; Preston et al., 2009; Wang et al., 2010; Zhang et al., 2010, 2013; Howarth et al., 2011; Tähtiharju et al., 2012; Yang et al., 2012). These studies have found *CYC2* expression restricted to the same dorsal floral domain and a conserved role as cell proliferation repressors resulting in bilateral symmetry (reviewed in Hileman, 2014). Fewer studies have been made in basal eudicots, but dissymmetric Fumarioids (Papaveraceae) do have asymmetric expression of *CYC-like* genes, suggesting that *CYC-like* recruitment to form bilateral flowers has occurred independently several times in eudicots (Damerval et al., 2007, 2013).

Less is known about the role of pre-duplication *CYC-like* genes in monocots (Bartlett and Specht, 2011; Mondragón-Palomino and Trontin, 2011; Preston and Hileman, 2012). Expression analyses of *CYC-like* genes in *Costus* (Costaceae; Zingiberales) and *Commelina* (Commelinaceae; Commelinales) suggest that they play a role in bilateral symmetry (Bartlett and Specht, 2011; Preston and Hileman, 2012). Functional studies in *O. sativa* (Poaceae, Poales) confirm that these genes contribute to the asymmetric growth of the dorsal *versus* the ventral portions of the flower, as shown by the *rep1* mutants which exhibit a smaller palea due to cell division arrest (Yuan et al., 2009). The only two studies available in Orchidaceae are particularly intriguing as they show very different expression patterns of *CYC/TB1-like* orthologs. Whereas, the only copy of *CYC/TB1-like* in *Orchis italica (OitaTB1)* is expressed exclusively in leaves (De Paolo et al., 2015), two of three *CYC/TB1-like* copies in *Phalaenopsis equestris, PeCYC1* and *PeCYC2,* seem to be expressed in higher levels (2–10 times more) in the dorsal sepals and the labellum compared to the ventral sepal and the lateral petals (Lin et al., 2016). Furthermore, some authors have hypothesized that the expression gradient of *TCP* genes is largely controlled by upstream expression of the *AP3/DEF* petal-stamen identity genes, resulting in higher concentrations of *CYC/TB1-like* genes in the dorsal floral regions; however, more experimental data is needed to support this (Mondragón-Palomino and Theißen, 2009).

It is unclear whether closely related *TCP-like CINCINNATA* (*CIN*) and *PROLIFERATION CELL FACTOR* (*PCF*) genes play any role in floral symmetry. *CIN* was originally characterized in *A. majus* and more recently in *Arabidopsis thaliana* (Crawford et al., 2004; Nag et al., 2009; Sarvepalli and Nath, 2011; Danisman et al., 2013). In both species *CIN* controls cellular proliferation in petals and cellular arrest in leaves (Crawford et al., 2004; Nag et al., 2009). On the other hand, *O. sativa PCF1* and *PCF2* are involved in axillary meristem repression, likely *via* the activation of *PROLIFERATING CELL NUCLEAR ANTIGEN (PCNA)*, which encodes a protein involved in DNA replication and repair, maintenance of chromatin structure, chromosome segregation, and cell-cycle progression (Kosugi and Ohashi, 1997). Other studies in *A. thaliana* suggest that *PCF-like* genes are also involved in gametophyte development, transduction of hormonal signals, mitochondrial biogenesis, leaf and flower morphogenesis, seed germination, branching, and even circadian clock regulation (Koyama et al., 2007; Pruneda-Paz et al., 2009; Giraud et al., 2010; Kieffer et al., 2011; Resentini et al., 2015). Recent studies in the model orchid *P. equestris* have found that *PeCIN8* (*CIN-like*) and *PePCF10* (*PCF-like*) control cell proliferation and cell shape in petals, ovules and leaves (Lin et al., 2016).

In order to study the contribution of *CYC/TB1-like* and the closely related *CIN-like* and *PCF-like* genes to floral patterning in Asparagales, we first determined copy number and characteristic protein motifs and then assessed gene lineage evolution including a vast sampling of *TCP-like* genes across angiosperms and particularly of Asparagales monocots. Next, we evaluated the expression patterns of all *TCP-like* genes in dissected floral organs, young leaves, and fruits of *Hypoxis decumbens* (Hypoxidaceae) which has typical asparagalean radial trimerous flowers with free parts, and *Cattleya trianae* (Orchidaceae), that has bilateral flowers, and a single fertile stamen fused with the tree stigmas (i.e., gynostemium). Finally, we propose hypotheses on functional evolution based on previous literature reports and comparisons with our results that suggest different trends among gene clades when comparing Asparagales to model core eudicots.

MATERIALS AND METHODS

Gene Isolation and Phylogenetic Analyses

In order to isolate putative *TCP-like* homologs in Asparagales, searches were performed using previously reported *TCP* genes from eudicots, monocots and in particular Orchidaceae as queries (Mondragón-Palomino and Trontin, 2011; Preston and Hileman, 2012; De Paolo et al., 2015; Horn et al., 2015). Searches included homologs from all the three main clades of *TCP* genes: *CYC-like*, *CIN-like* and *PCF-like*. Searches were done using BLAST tools (Altschul et al., 1990) in the orchid specific available databases including Orchidbase 2.0 (http://orchidbase.itps.ncku.edu.tw/) (Tsai et al., 2013), Orchidstra (http://orchidstra2.abrc.sinica.edu.tw/orchidstra2/index.php) (Su et al., 2013), as well as the more inclusive OneKP database (http://www.bioinfodata.org/Blast4OneKP/). All core eudicot sequences, were isolated from Phytozome (https://phytozome.jgi.doe.gov/pz/portal.html) and genbank (https://www.ncbi.nlm.nih.gov/genbank/).

In addition to the available databases we generated two transcriptomes from *C. trianae* (Orchidaceae) and *H. decumbens* (Hypoxidaceae). The transcriptome for each species was generated from mixed material from 3 biological replicates and included vegetative and reproductive meristems, floral buds, young leaves and fruits in as many developmental stages as possible. Total RNA was purified and used for the preparation of one mRNA (polyA) HiSeq library for each species. RNA-seq experiments were conducted using truseq mRNA library construction kit (Illumina, San Diego, California, USA) and sequenced in a HiSeq 2000 instrument reading 100 base paired end reads.

The transcriptome was assembled *de novo* with Trinity v2 following default settings. Read cleaning was performed with prinseq-lite v0.20.4 with a quality threshold of Q35 and a minimum read length of 50 bases. Contig metrics are as follows: (1) *H. decumbens* total assembled bases: 73,787,751; total number of contigs (>101 bases): 157,153; average contig length: 469 bp; largest contig: 15,554 bp; contig N50: 1075 bp; contig GC%: 46,42. (2) *C. trianae* total assembled bases: 63,287,862 bp; total number of contigs (>101 bases): 109,708; average contig length: 576 bp; largest contig: 9321 bp; contig N50: 1401 bp; contig GC%: 42,73. Homologous gene search was performed using BLASTN with the query sequences downloaded from GenBank and other databases (see above). In order to estimate the relative abundance of the assembled contigs, cleaned reads were mapped against the *de novo* assembled dataset and counted with two different strategies. The first one involved the mapping algorithm of the software Newbler v2.9 where only the unique matching read pairs were accepted as positive counts. The second one involved the mapping algorithm BOWTIE2 and raw reads counts as well as RPKM were calculated to each assembled contig (**Table 1**).

All sequences isolated were compiled with Bioedit (http://www.mbio.ncsu.edu/bioedit/bioedit.html). Sequences shorter than 200 bp lacking similarity with a region of the putative bHLH motif were discarded. Nucleotide sequences were subsequently aligned using the online version of MAFFT (http://mafft.cbrc.jp/alignment/software/) (Katoh et al., 2002) with a gap open penalty of 3.0, offset value of 1.0 and all other default settings. The alignment was then refined by hand using Bioedit considering as a reference the 60–70 aa reported as conserved in the TCP protein domain (Cubas et al., 1999). To better understand the evolution of the *TCP* gene lineage, and to integrate previous eudicot and monocot specific phylogenetic analyses (Damerval and Manuel, 2003; Hileman and Baum, 2003; Reeves and Olmstead, 2003; Howarth and Donoghue, 2006; Damerval et al., 2007; Bartlett and Specht, 2011; Mondragón-Palomino and Trontin, 2011; Preston and Hileman, 2012; Horn et al., 2015), we performed Maximum likelihood (ML) phylogenetic analyses using the nucleotide sequences with RaxML-HPC2 BlackBox through the CIPRES Science Gateway (https://www.phylo.org/) (Miller et al., 2010). Bootstrapping (BS) was performed according to the default criteria in RAxML (200–600 replicates). The *PCF-like* gene from *Amborella trichopoda (AtrTCP4)* as well as all other *PCF-like* sequences were used as the outgroup. To find the molecular evolution model that best fit our data, we used the jModelTest package implemented in MEGA6 (Posada and Crandall, 1998). Trees were observed and edited using FigTree v1.4.0 (http://tree.bio.ed.ac.uk/software/figtree/) (Rambaut, 2014). Newly isolated sequences from our own generated transcriptomes from *C. trianae* (Orchidaceae) and *H. decumbens* (Hypoxidaceae) can

TABLE 1 | Selected features of *TCP* and *AP3-like* gene contigs taken from sequences and the transcriptomes of *Cattleya trianae* and *Hypoxis decumbens*.

Gene CDS	Contig	CDS size (in bp)	Protein size (in AA)	URP* Newbler	RRC** RPKM	RPKM	RPKM rank
CtrAP3a	c19830_g1_i1	684	227	4047	3877	91,02338579	1689
CtrAP3b	c16430_g1_i2	696	231	15	719	16,64198541	7551
CtrAP3c	c16430_g1_i1	669	222	184	1002	23,72478436	5775
CtrAP3d	c14728_g2_i1	628	208	1218	1228	38,72393988	3786
CtrAP3e	c13983_g1_i1	601	199	321	269	9,445413314	11262
CtrCIN1	c10815_g1_i1	945	314	565	540	7,473218106	13078
CtrCIN2	c20571_g1_i1	975	324	5413	5455	58,71380149	2597
CtrCIN3	c20515_g2_i1	1158	385	2065	2091	28,27740553	4987
CtrCIN4	c20515_g1_i1	1116	371	1436	880	13,36057701	8882
CtrCIN5	c19791_g1_i1	1116	371	530	491	6,141835864	14794
CtrCIN6	c18629_g1_i1	957	319	453	411	8,008505294	12506
CtrPCF1	c14669_g1_i1	930	309	260	258	6,560658931	14228
CtrPCF2	c10284_g1_i1	759	252	306	308	8,318287473	12198
CtrPCF3	c22906_g1_i1	882	293	1109	1118	23,04243274	5910
CtrPCF4	c10751_g2_i1	561	187	315	230	11,37525956	9984
CtrPCF5	c21320_g2_i1	900	299	493	401	6,065023082	14927
CtrPCF6	c23464_g2_i1	828	275	5821	928	15,39001893	7984
CtrPCF7	c13613_g1_i1	684	227	1023	1050	25,7820899	5397
CtrPCF8	c8381_g1_i1	642	213	136	136	3,198372774	21993
CtrPCF9	c16850_g1_i1	1032	343	1021	529	6,68019211	14059
CtrPCF10	c24243_g4_i1	1107	368	269	140	2,935430718	23098
CtrPCF11	c20858_g1_i1	963	320	181	171	3,452818247	21033
CtrTB1	c62044_g1_i1/ c74262_g1_i1	330	110	6/15	7/14	0,507584894922/ 0,584445576	58492/ 52988
HydDEF1a	c29923_g1_i2	181	59	175	8	1,074179469	58639
HydDEF1b	c29923_g1_i4	684	227	80	84	3,992750518	25641
HydDEF2	c23342_g1_i1	672	223	533	513	19,98971549	6376
HydCIN1	c26829_g2_i1	717	239	605	420	12,7762255	9853
HydCIN2	c25446_g2_i3	1272	424	496	589	17,24524272	7370
HydCIN3	c25446_g2_i4	651	217	2655	338	17,37878284	7322
HydCIN4	c25446_g2_i2	858	286	1355	1314	44,60569165	2753
HydPCF1	c25819_g1_i2	1038	345	12	266	7,587936907	15736
HydPCF2	c25819_g1_i1	939	312	11	236	8,050262478	14962
HydPCF3	c29001_g5_i3	300	100	75	64	5,51807531	20195
HydPCF4	c29001_g5_i1	387	129	23	20	1,853254689	41814
HydPCF5	c29001_g5_i4	408	136	32	43	3,099053674	30449
HydPCF6	c29001_g5_i2	411	137	42	57	4,208770608	24720
HydPCF7	c27872_g4_i1	771	256	732	896	26,84914255	4743
HydPCF8	c27872_g3_i2	762	253	334	528	16,30864126	7759
HydPCF9	c27872_g3_i1	756	251	108	181	6,950127044	16867
HydPCF10	c27872_g2_i1	675	225	138	157	9,763071436	12576
HydTB1	c52607_g1_i1	258	85	3	3	0,364509027	119496

**Unique read pairs - number of specific/unique read pairs supporting each contig.*
*** Raw read counts.*

be found under Genbank numbers KY296315–KY296347. All sequences included in the phylogenetic analyses can be found in the **Supplementary Table 1**.

Identification of New Protein Motifs

In order to detect previously reported, as well as to identify new, conserved motifs, 77 *TCP-like* genes were selected representing major model eudicot and monocot groups of this study (i.e., *A. majus, O. sativa, Aloe vera, P. equestris, C. trianae* and *H. decumbens*). Sequences were permanently translated and uploaded as amino acids to the online MEME server (http://meme-suite.org/tools/meme) and run with all the default options (Bailey et al., 2006). Specific analyses for each of the *TCP* clades (i.e., *CYC-like, CIN-like, PCF-like*) were also performed.

Expression Analyses by RT-PCR

To examine and compare the expression patterns of *TCP-like* genes we used floral buds, dissected floral organs, leaves, and fruits of *C. trianae* and *H. decumbens*. Preanthetic floral buds of *H. decumbens* were dissected into sepals, petals, stamens and carpels. In addition whole floral buds, inmature fruits (F1, right after tepals shed off), mature fruits (F2, before lignification), and young leaves were also collected. Preanthetic floral buds of *C. trianae* were dissected into sepals, lateral petals, labellum (or lip), gynostemium, and ovary. Young leaves were also collected. Total RNA was isolated from each organ collected using the SV Total RNA Isolation System kit (Promega, Madison, WI, USA), and resuspended in 20 μl of DEPC water. RNA was treated with DNAseI (Roche, Basel, Switzerland) and quantified with a NanoDrop 2000 (Thermo Scientific, Waltham, MA) (Wilfinger et al., 1997). Three Micrograms of RNA were used as a template for cDNA synthesis (SuperScriptIII RT, Invitrogen) using OligodT primers. The cDNA was diluted 1:4 for amplification reactions by RT-PCR. Primers were designed in specific regions like portions flanking the conserved domains for each copy found in *C. trianae* and *H. decumbens* (**Supplementary Table 2**). Each amplification reaction incorporated 9 μl of EconoTaq (Lucigen, Middleton, WI), 6 μl of nuclease free water, 1 μl of BSA (5 μg/ml), 1 μl of Q solution (5 μg/μl) 1 μl fwd primer (10 mM), 1 μl rev primer (10 mM), and 1 μl of template cDNA for a total of 20 μl. Thermal cycling profiles followed an initial denaturation step (94°C for 30 s), an annealing step (50–59°C for 30 s) and an extension step with polymerase (72°C for 10 min), all by 30–38 amplification cycles. *ACTIN2* was used as a load control. PCR products were run on a 1.0% agarose gel stained with ethidium bromide and digitally photographed using a Whatman Biometra® BioDoc Analyzer.

RESULTS

Exhaustive search from available databases retrieved 452 *TCP* Class I and Class II sequences from flowering plants. From these 138 belong to the Orchidaceae, including 18 homologs from *C. trianae*; and 110 sequences belong to non-Orchidaceae Asparagales, including 15 homologs from *H. decumbens* (**Supplementary Table 1**).

ML analyses were performed using the complete nucleotide sequences of all *TCP-like* genes isolated. The *A. trichopoda AtrTCP4* together with all other isolated *PCF-like* genes were used as the outgroup. The analysis recovered two clades previously reported in *TCP* genes, namely the *CYC/TB1-like* clade (with a Bootstrap Support, BS = 98) and the *CIN-like* clade (BS = 67) (**Figure 1**). We will discuss our results for each clade separately.

CYC/TB1-like Gene Evolution

We were able to isolate 168 sequences belonging to the *CYC/TB1-like* clade (**Supplementary Table 1**). Our sampling includes 15 sequences from four species of Poales, 10 sequences from seven species of Asparagales (incl. Orchidaceae), eight from five species of Commelinales (*Commelina, Alstroemeria, Tradescantia*), 14 from nine species of basal angiosperms and 121 from 46 species of eudicots. Only one homolog from *Curculigo* spp. (*CurTB1*) and one homolog from *H. decumbens* (*HydTB1*) were recovered from our blast searches. Moreover, an exhaustive search in Orchidaceae specific databases resulted in eight additional copies: three from *P. equestris* (*PETCP06750, PETCP06749, PETCP11715*), two from *P. aphrodite* (*PaTB1, PaTCP06749*) and one from *C. trianae* (*CtrTB1*) (Epidendroideae); one homolog from *O. italica* (*OitaTB1*) (Orchidoideae); and one homolog from *Vanilla shenzhenica* (*VaTCP06749*) (Vanilloideae). *CYC/TB1-like* homologs seem to have undergone size reduction in Asparagales, and only the searches made with monocot *TB1* genes yielded positive hits. Interestingly, in the two transcriptomes newly obtained in the present research, the *CYC/TB1-like* contigs were supported by fewer reads (when compared to *MADS-box APETALA3* floral organ identity genes and other *TCP-like* genes; **Table 1**) suggesting low expression of these transcripts.

The resulting ML topology recovered the three previously established core eudicot subclades (Howarth and Donoghue, 2006) with very low support (BS < 50), namely *CYC1/TCP18, CYC2/TCP1,* and *CYC3/TCP12* (**Figure 1**). Our analysis also recovers the previously identified duplication of *CYC-like* genes in basal eudicots (Citerne et al., 2013), and another in Poales, the latter resulting in the *REP1/TB1* clades (Mondragón-Palomino and Trontin, 2011). Most basal angiosperms and many monocots outside Poales have single copy *CYC-like* genes that predate the independent duplications in eudicots and Poales (**Figure 1**). Intraspecific duplications in monocots have occurred in *Zea mays,* as well as in the orchids *P. equestris* and *P. aphrodite* (**Figure 1**; BS = 100). Outside of the monocots, local specific duplications have also occurred in the basal angiosperms *Aristolochia ringens* and *Persea americana*, in the basal eudicots *Circaeaster* and *Nelumbo,* and in core eudicots such as *Gerbera, Antirrhinum, Citrus, Glycine, Populus,* and *Gossypium* (**Figure 1**).

Members of the *CYC/TB1-like* clade show very little variation in the ca. 60 amino acid TCP domain (*sensu* Cubas et al., 1999) consisting of a putative basic-Helix-Loop-Helix (bHLH) domain (**Figure 2**). We were able to identify the highly conserved residues previously reported for the putative bipartite Nuclear Localization Signal (NLS) at the N-terminus flanking the bHLH, which provide hydrophobicity in the α-helices and in the loop region between the two. The loop itself is highly conserved in all CYC/TB1-like sequences except for AmDICH and AmCYC that have an A > P change at position 42. The second helix contains the LxxLL motif in all CYC2 proteins; this motif is modified into a VxWLx motif in other CYC-like proteins.

Our MEME analysis identified motifs 1 and 2 corresponding to the TCP domain (**Supplementary Figure 1**). At the start of Helix I, between positions 24–29 we found specific amino acids exclusive to CYC protein homologs. Outside the TCP domain, motifs 7 and 10 (reported also by Bartlett and Specht, 2011) and motifs 36–40, 42, and 43 (reported also by De Paolo et al., 2015) were recovered in our analysis as conserved in all CYC proteins. In addition, the protein interaction R domain (motif 11) putatively involved in hydrophilic α-helix formation in TCP Class II genes (shared between CYC and CIN proteins) was also identified (**Supplementary Figure 1**; Cubas et al., 1999). However, this motif is absent from CtrTB1 and REP-1 (**Supplementary Figure 1**; Yuan et al., 2009). Previously

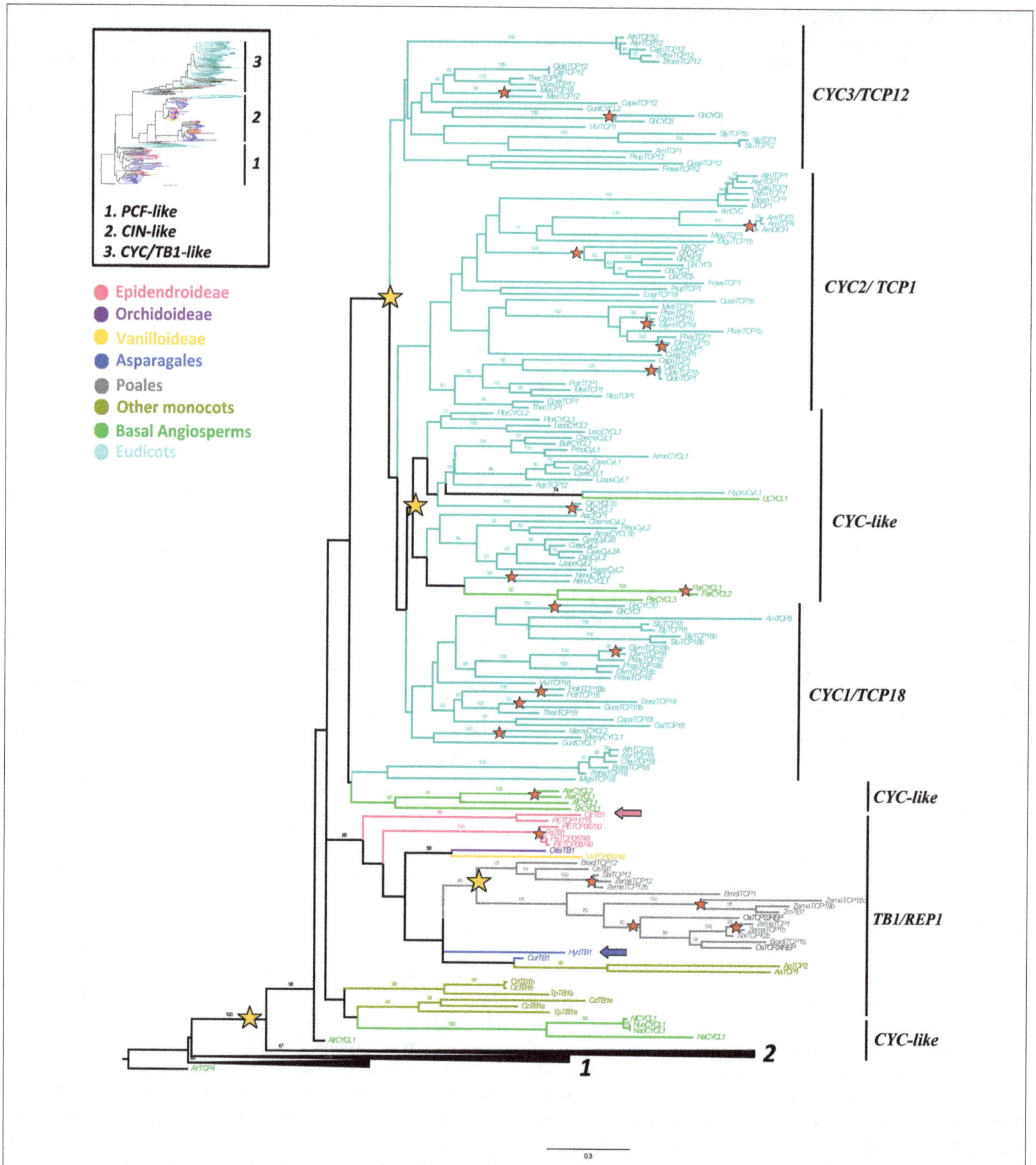

FIGURE 1 | ML analysis of *CYC/TB1-like* genes. Overview (upper left) summary Maximum Likelihood (ML) analysis of all *TCP-like* genes, where (1) corresponds to *PCF-like* homologs, (2) to the *CIN-like* clade and (3) to the *CYC/TB1-like* clade. Detailed (right) gene tree expanded in the *CYC/TB1-like* clade (3); yellow stars indicate large-scale duplication events in core, basal eudicots and monocots; resulting in the core eudicot clades *CYC3/TCP12*, *CYC2/TCP1* and *CYC1/TCP18*, the basal eudicot *CYC-like* clades and the *TB1* and *REP1* clades in Poales; red stars indicate species-specific duplication events; blue and pink arrows indicate *H. decumbens* and *C. trianae* homologs, respectively; branch and taxa colors correspond to those in the overview tree to the upper left. BS values ≥ 50 are shown.

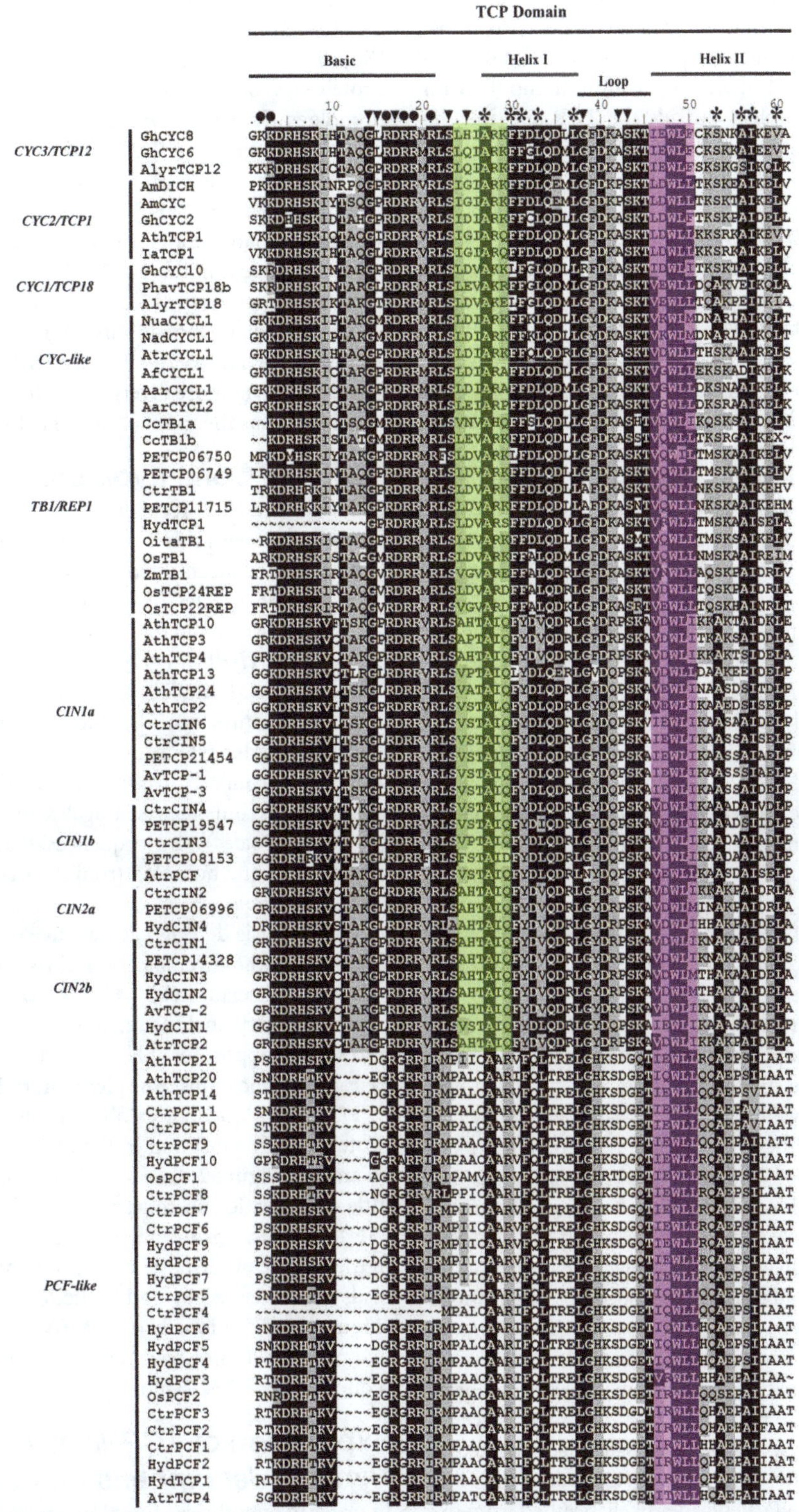

FIGURE 2 | TCP protein domain alignment with Asparagales representative sequences. *Oryza sativa* (Poales) was used for reference. Names to the left indicate the clade to which sequences belong according to **Figures 1, 4, 5** and **Supplementary Table 1**. The upper bars point to the putative structure bHLH (basic-Helix-Loop-Helix) at the TCP domain. Circles indicate residues forming part of the putative bipartite NLS; asterisks indicate conserved hydrophobic residues in the helices; black arrowheads point to residues (glycine or proline) that disrupt α-helix formation (modified following Cubas et al., 1999). The green box indicates changes in residues between CYC-like and CIN-like proteins. The pink box indicates the LxxLL motif with significant variations outside the CYC2 clade.

unidentified motifs include motif 4, exclusive to Epidendroideae and motif 5, exclusive to *Phalaenopsis* species. Whereas, most orchid CYC-like proteins (including *O. italica* OitaTB1 and *C. trianae* CtrTB1) do not share any common motifs with the canonical *A. majus* paralogs outside the TCP domain, the *Phalaenopsis* CYC-like homolog (PETCP11715) shares motifs 6, 9, and 12 with AmDICH or AmCYC.

CIN-like Gene Evolution

A total of 155 *CIN-like* homologs were recovered and unlike the *CYC-like* sampling, most *CIN-like* sequences belong to the Asparagales (**Supplementary Table 1**). Our sampling contains 57 sequences from 35 non-Orchidaceae Asparagales species, including four paralogs from *H. decumbens* labeled *HydCIN1-HydCIN4*. A total of 78 *CIN-like* homologs were isolated, including four homologs from two Apostasioideae species, nine homologs from three Vanilloideae species, five homologs from three Cypripedioideae species, 22 homologs from seven Orchidoideae species and 38 homologs from 11 Epidendroideae species. Furthermore, six paralogs were identified in *C. trianae* labeled *CtrCIN1-CtrCIN6*. Searches outside Asparagales were restricted to six homologs from *O. sativa*, two from *A. trichopoda* and 12 homologs from five eudicots species, including the canonical *A. majus CINCINNATA (AmCIN)*.

The *CIN-like* ML analysis shows a duplication (BS = 75) that predates the diversification of angiosperms resulting in the *CIN1* (BS = 79) and *CIN2* clades (BS = 99) (**Figure 3**). This is confirmed by the position of the two *A. trichopoda CIN* paralogs, *AtrPCF* and *AtrTCP2*, each in its own clade (**Figure 3**). Additional support for this early duplication is found in the topology yielded by a second complementary analysis that includes 11 Solanaceae *CIN* homologs, where *CIN1* and *CIN2* clades have monocot and core eudicot representatives and, at least *CIN2* is well supported (BS = 93) (**Supplementary Figure 2**). The *CIN1* clade has undergone at least two additional duplications resulting in the *CIN1a-c* clades. It is likely that the duplication resulting in *CIN1a* and *CIN1b/c* occurred exclusively in monocots, although the exact timing is unclear. The other duplication resulting in *CIN1b* and *CIN1c* appears to be Orchidaceae-specific, prior to the diversification of Orchidoideae and Epidendroideae (**Figure 3**). On the other hand, the *CIN2* clade underwent an independent duplication predating the diversification of Asparagales, resulting in *CIN2a* and *CIN2b* subclades. Intraspecific duplications were identified in *Hesperaloe, Disporopsis, Maianthemum, Rhodophiaia, Sansevieria* and *Yucca*, (**Figure 3**). Poales *CIN* homologs form a clade, with a low BS, in the first analysis, with the exception of *OsPCF5* clustered with *AmCIN* (**Figure 3**). However, our second analysis shows two Poales clades nested in each angiosperm paralogous *CIN1* and *CIN2* clades (with low BS), suggesting that the two Poales clades likely resulted from the angiosperm *CIN1/2* duplication (**Supplementary Figure 2**).

CIN-like sequences show high degree of conservation at the N-flank of the TCP domain (**Figure 2**). The only changes with respect to the key aminoacids in the bHLH domain in CYC proteins are at the second helix where the LxxLL motif shifts to V/IxxLL (**Figure 2**). Toward the 3′ end of the TCP domain proteins are highly variable, except for motifs 13, 14, 16–18 and 21, reported also by De Paolo et al. (2015) (**Supplementary Figure 1**). The R domain (motif 11) in CIN proteins is only present in the *CIN1a* clade and *O. sativa* homologs *OsTCP21, OSTCP8* and *OsTCP10*. Motif 14, which corresponds to the *miR319* binding site is present in most *CIN-like* sequences. The *miR319* binding motif is lacking in *HydCIN2, HydCIN3, OsTCP21, OsTCP27,* and *OsTCP10*. All *CIN1* subclade sequences share motifs 13, 17, 19, 20 21, and 25; the *CIN1a* subclade shares motifs 11, 28, 29, 34 and 35, while the *CIN1b* subclade shares motifs 15, 23, 26, and 30. Synapomorphies for the *CIN2* subclade include motifs 16, 18, 22, 24 and only motif 27 is exclusive to Orchidaceae. Finally, *CIN2b* homologs share motifs 32 and 33. The most divergent *C. trianae* sequence is *CtrCIN6*, which only has the motifs 1, 2, 3, 11, 13, 14, 17, and 21.

PCF-like Gene Evolution

Our analysis recovered 129 *PCF-like* homologs (**Supplementary Table 1**). Similarly to *CIN-like* genes most sampling is concentrated in Asparagales, thus 53 homologs belong to 35 species of non-Orchidaceae Asparagales and 52 sequences correspond to Orchidaceae. *H. decumbens* has 10 *PCF-like* copies (*HydPCF1–HydPCF10*). Sampling in Orchidaceae includes one homolog from one Vanilloideae species, 14 homologs from seven Orchidoideae species and 37 homologs from 11 Epidendroideae species. We recovered 11 *PCF-like* homologs from *C. trianae* (*CtrPCF1–CtrPCF11*). Sampling in monocots outside Asparagales include 10 homologs from *O. sativa* (Poales) and sequences that are not monocots are restricted to one homolog from *A. trichopoda* and 13 homologs from *A. thaliana*.

Our analysis detected at least five duplication events of *PCF-like* prior to the diversification of Asparagales, however support is low for all clades (BS = < 50) (**Figure 4**). In addition, our complementary analysis including 18 Solanaceae *PCF-like* genes also shows support for at least two rounds of core eudicot specific *PCF-like* duplications (**Supplementary Figure 2**).

PCF-like sequences exhibit the shortest basic motif in TCP proteins, with a deletion in the bipartite NLS between positions 10 and 13 (**Figure 2**; Cubas et al., 1999). The TCP domain shows little conservation in comparison to the TCP II class (CYC and CIN) proteins. For instance, there is a four amino acid deletion in the middle of the basic motif, and only 12 out of the 23 amino acids characterized in both helices and the loop are conserved (**Figure 2**). Additionally our MEME analysis show that PCF-like proteins do not have an R domain (motif 11), nor a target sequence for *miR319* (motif 14) (**Supplementary Figure 1**).

Expression of *TCP-like* Homologs from *Hypoxis decumbens* and *Cattleya trianae*

In order to hypothesize functional roles for the Asparagales *TCP-like* homologs, the expression patterns of all homologs isolated from transcriptomic analysis in *H. decumbens* and *C. trianae* were evaluated (**Figure 5**). Although in both species we were able to dissect floral organs in preanthesis and young leaves, we were not able to find fruits of *C. trianae*, thus only young and old fruits of *H. decumbens* were included in the expression study.

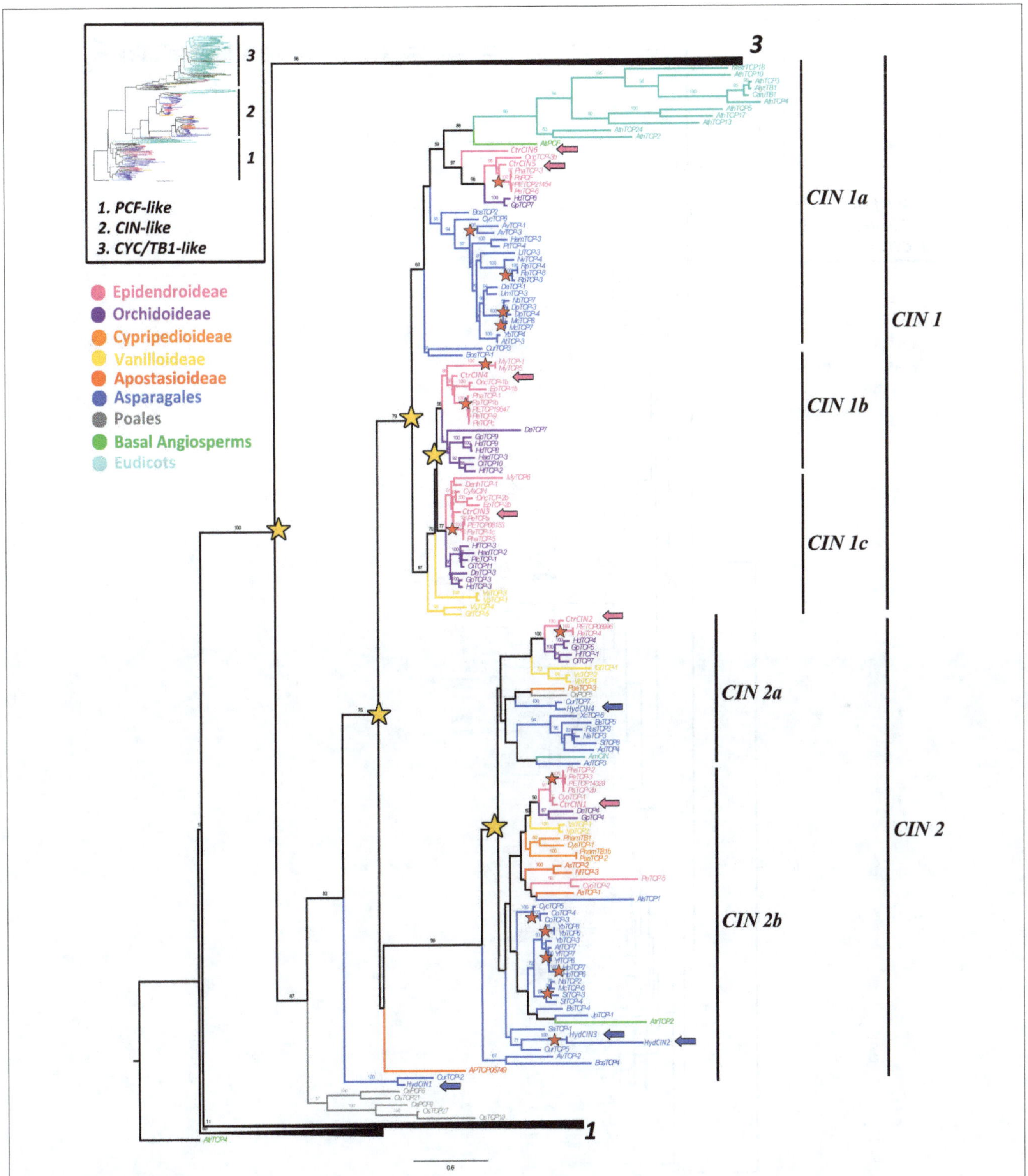

FIGURE 3 | ML analysis of *CIN-like* genes. Overview (upper left), summary tree as in **Figure 1**. To the right, ML phylogenetic analysis of *TCP* genes expanded to show the *CIN-like* clade (2). Yellow stars indicate large scale duplication events: one prior to the diversification of angiosperms yielding clades CIN1 and CIN2, two before the radiation of Asparagales and one prior to the origin of Epidendroideae and Orchidoideae; red stars indicate species-specific duplication events; blue and pink arrows indicate *H. decumbens* and *C. trianae* homologs, respectively. Branch and taxa colors correspond to those in the conventions to the left. BS values ≥ 50 are shown.

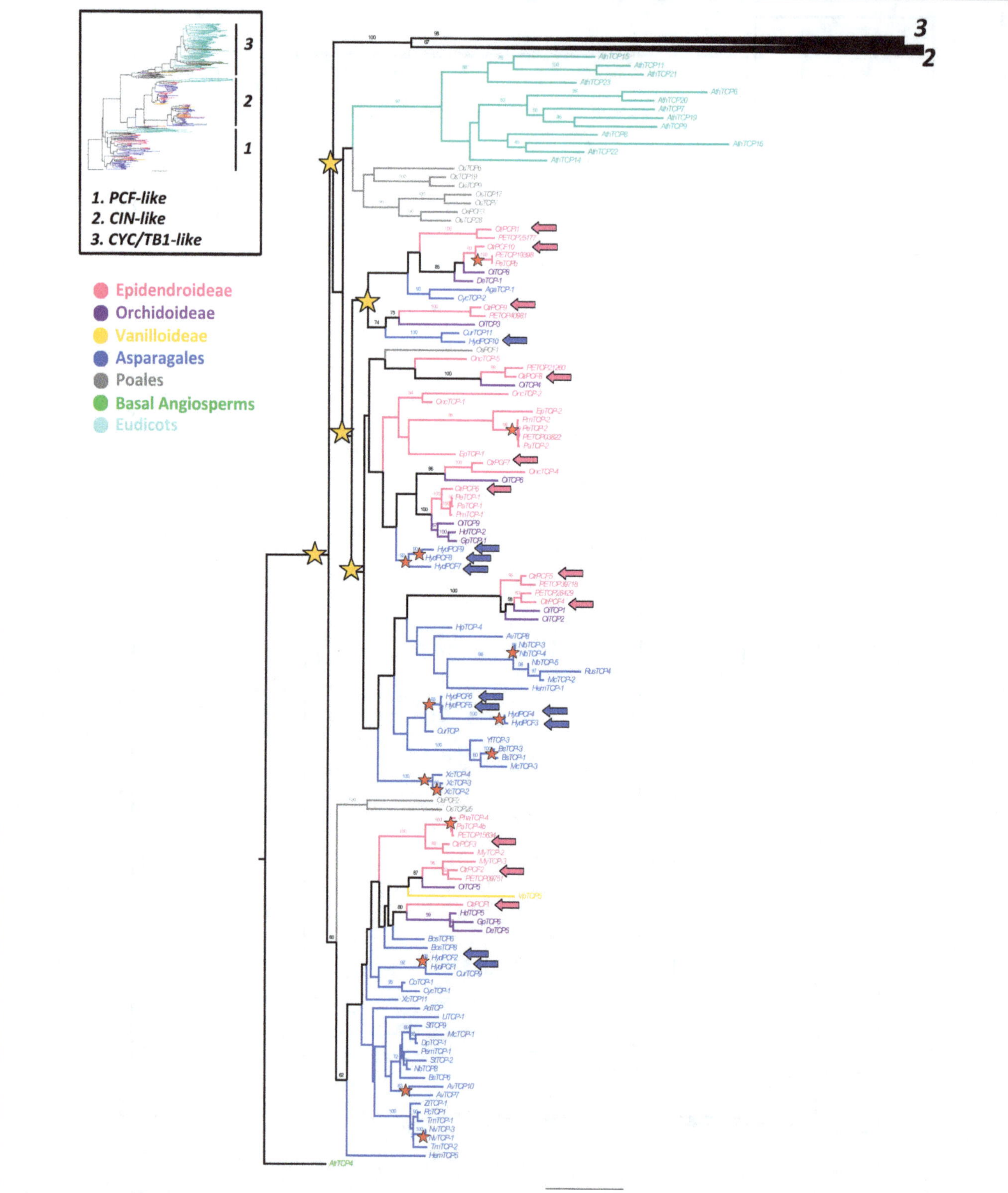

FIGURE 4 | ML analysis of *PCF-like* genes. Overview (upper left) summary tree as in **Figure 1**. To the right, ML phylogenetic analysis of *TCP* genes expanded to show the *PCF-like* clade (1); yellow stars indicate large scale duplication events at least three before the diversification of Asparagales; red stars indicate species-specific duplication events; blue and pink arrows indicate *H. decumbens* and *C. trianae* homologs, respectively. Branch and taxa colors correspond to those in the conventions to the left. BS values ≥ 50 are shown.

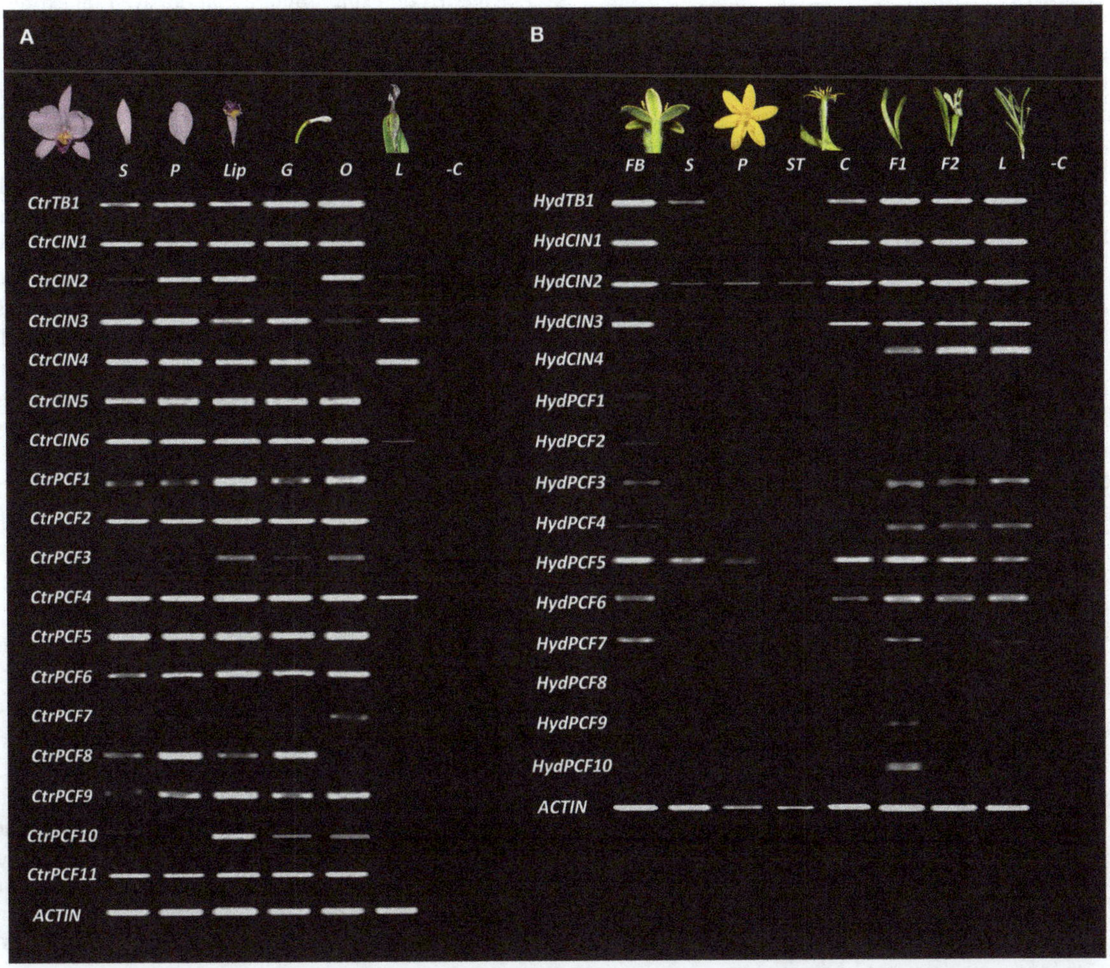

FIGURE 5 | *TCP-like* expression analysis in Asparagales. *ACTIN* was used as positive control. **(A)** Expression of all isolated *TCP-like* transcription factors in dissected floral organs and leaves of *C. trianae* (Orchidaceae). **(B)** Expression of all isolated *TCP-like* transcription factors in dissected floral organs, fruits and leaves of *H. decumbens* (Hypoxidaceae). BF, flower bud; C, carpels; F1, immature fruit; F2, mature fruit; G, gynostemium; L, Leaves; Lip, lip; O, Ovary; P, petals; S, sepals; ST, stamens. −C Indicates the amplification reaction of PCR without cDNA (negative control).

The *CYC/TB1-like* homologs have very different expression patterns in *C. trianae* and *H. decumbens*. *HydTB1* is expressed in the floral bud, sepals, carpels, young and mature fruits and leaves, whereas *CtrTB1* is expressed in all floral whorls and is not expressed in leaves (**Figure 5**). *CIN-like* homologs also exhibit different expression patterns in *C. trianae* and *H. decumbens*. *HydCIN1*, *HydCIN2* and *HydCIN3* are expressed in the floral bud, carpels, fruits and leaves. Only *HydCIN2* expression is extended to sepals, petals and stamens at very low levels. Interestingly both *HydCIN2* and *HydCIN3* lack the *miR319* binding site. *HydCIN4* expression is restricted to fruits and leaves. *CtrCIN1*, *CtrCIN2*, *CtrCIN5*, and *CtrCIN6* are expressed in sepals, petals, lip, gynostemium, and ovary, although, *CtrCIN2* expression in sepals and gynostemium occurs at low levels. *CtrCIN3* and *CtrCIN4* have similar expression patterns to *CtrCIN1*, however, they are poorly expressed in the ovary and are the only *CIN-like* homologs that extend their expression to leaves.

Similar to CYC/TB1-like and CIN-like genes, the expression of PCF-like homologs varies dramatically between *C. trianae* and *H. decumbens* (**Figure 5**). *CtrPCF1*, *CtrPCF2*, *CtrPCF3*, *CtrPCF4*, *CtrPCF5*, *CtrPCF6*, *CtrPCF9*, and *CtrPCF11* are expressed in all floral organs and the only copy with expression in leaves is *CtrPCF4*. *CtrPCF3* has low expression in sepals and petals. *CtrPCF7* is only expressed in petals and ovary. Finally, *CtrPCF10* has expression in all floral whorls except in petals. In *H. decumbens* only *HydPCF3*, *HydPCF4*, *HydPCF5*, and *HydPCF6* are expressed in the floral buds, carpels, fruits and leaves, although *HydPCF3* and *HydPCF4* have low expression in carpels. *HydPCF5* is also expressed in the perianth with higher expression in sepals than in petals. *HydPCF7* has restricted expression to the floral buds and the young fruits. *HydPCF9* and *HydPCF10* are only expressed in young fruits and the remaining copies (*HydPCF1*, *HydPCF2*, and *HydPCF8*) are only expressed at very low levels in the floral buds.

DISCUSSION

Most functional studies on *TCP* genes have focused on identifying their contribution to floral symmetry and plant architecture, as expected by the functions of the canonical *CYC* and *DICH* from *A. majus* and *AtTCP1* from *A. thaliana* respectively (Luo et al., 1996, 1999; Costa et al., 2005). Studies on the evolution of TCP transcription factors have concentrated on core eudicots and particularly on *CYC2* homologs (Damerval and Manuel, 2003; Howarth and Donoghue, 2006; Preston and Hileman, 2009; Martín-Trillo and Cubas, 2010; Mondragón-Palomino and Trontin, 2011; Sarvepalli and Nath, 2011; Danisman et al., 2012, 2013; Uberti-Manassero et al., 2012; Aguilar-Martínez and Sinha, 2013; Das Gupta et al., 2014; Lin et al., 2016). Our matrix includes sampling from the *Phalaenopsis* genome as well as all transcriptomes available for Asparagales. The phylogenetic analysis made with the full-length coding sequences, allowed us to identify a number of large scale as well as local *TCP* gene duplications and changes in protein sequences linked to these duplications. Moreover, this is the first large scale analysis of *CIN-like* and *PCF-like* genes. We are able to report a comparative expression pattern in two Asparagales species, representing the two floral groundplans in the order, and present hypotheses on the putative role of these genes in floral patterning in representative Asparagales.

Asparagales *CYC/TB1* Homologs Are Found Predominantly as Single Copies and Have Divergent Expression Patterns in *Hypoxis decumbens* and *Cattleya trianae*

Our study detected a single copy *CYC/TB1* gene in each of the species investigated in the Asparagales. The Asparagales *CYC/TB1* homologs fall outside of the Poales (TB1/REP clades) or Commelinales identified local duplications, described before and recovered here (Doebley et al., 1995, 1997; Yuan et al., 2009; Mondragón-Palomino and Trontin, 2011). The tree topology suggests that independent duplications have occurred in *CYC-like* genes in the monocots (Yuan et al., 2009; Mondragón-Palomino and Trontin, 2011; Hileman, 2014). Species-specific duplications in Asparagales were found only in *Phalaenopsis* (Orchidaceae). Moreover, expression data of *CYC/TB1* genes in Asparagales show remarkable differences between *H. decumbens* and *C. trianae,* even though sampled organs correspond to fairly well developed tissues. *CtrTB1* is expressed homogeneously in all floral whorls while *HydTB1* is expressed predominantly in carpels, fruits and leaves (**Figure 5**). Homogeneous expression of *CYC-like* genes in dorsal and ventral floral organs in *C. trianae* suggest that *CtrTB1* is likely not playing important roles in maintenance of bilateral symmetry in orchid flowers (see also Horn et al., 2015). However, only pre-anthethic floral buds were sampled and earlier stages are needed to test whether *CtrTB1* can be contributing to the establishment of bilateral symmetry in the flower primordia. Comparative expression studies of *CYC/TB1-like* genes in other Orchidaceae point to significant variations. For instance, while the *O. italica* homolog *OitaTB1* is only expressed in leaves (De Paolo et al., 2015), two *CYC/TB1* genes in *P. equestris* do exhibit differential dorsiventral expression in the floral bud suggesting species specific roles in bilateral symmetry establishment (Lin et al., 2016).

Expression of *CYC/TB1* genes in Orchidaceae contrasts with studies in Commelinales, Zingiberales, and Alstroemeriaceae that show differential dorsiventral expression of *CYC/TB1* genes and hence support convergent recruitment of *CYC* homologs in the acquisition of bilateral symmetry in different monocots (Bartlett and Specht, 2011; Preston and Hileman, 2012; Hoshino et al., 2014). Within core eudicots, only the *CYC2/TCP1* clade members have been linked to shifts toward bilateral floral symmetry. This has been extensively documented for *CYC* and *DICH* in *A. majus* (recent paralogs within the *CYC2/TCP1* clade) which are expressed in the dorsal regions of the floral meristems and negatively regulate cell proliferation (Luo et al., 1996, 1999). However, many other *CYC2* orthologs have been shown to control bilateral symmetry in Asterales, Brassicales, Dipsacales, Fabales, Lamiales, and Malpighiales (Busch and Zachgo, 2007; Gao et al., 2008; Preston et al., 2009; Wang et al., 2010; Zhang et al., 2010, 2013; Howarth et al., 2011; Tähtiharju et al., 2012; Yang et al., 2012; Ma et al., 2016).

Recruitment of CYC2 homologs in bilateral symmetry is likely facilitated by conserved protein-protein interactions mediated by the LxxLL motif (Heery et al., 1997; Damerval and Manuel, 2003; Reeves and Olmstead, 2003; Howarth and Donoghue, 2006; Li et al., 2009; Preston et al., 2009; Tähtiharju et al., 2012; Parapunova et al., 2014; Ma et al., 2016). If so, it is possible that different CYC-like proteins can form specific homo- and heterodimers, or even have unique partners, and thus protein motifs can provide clues to protein affinity and functional specificity (Kosugi and Ohashi, 2002). Our MEME analysis rescues motifs 6, 9 and 12, shared only between *P. equestris* PETCP11715 and the *A. majus AmCYC* and *AmDICH*, which are not present in other orchid CYC/TB1 proteins (i.e., *C. trianae* and *O. italica*) and hence putatively involved in establishing early bilateral floral symmetry in some Orchidaceae species (**Supplementary Figure 1**). When full length CtrTB1 is compared to the three *P. equestris* CYC/TB1 proteins very little similarity is detected (PETCP11715–0.38; PETCP06750–0.22; PETCP06750–0.24); for instance, CtrTB1 lacks motif 11 (R-domain). Nevertheless, the TCP domain is extremely conserved (0.91, 0.80, and 0.84 respectively). These results suggest that besides the key bHLH amino acids that target conserved genes, there are likely important motifs in the flanking regions allowing unique interactions and downstream partners in each species.

Conversely, the expression of *HydTB1* in *H. decumbens* is indicative of exclusive roles in carpels, fruits and leaves. This data suggests that *HydTB1* may have similar roles to other *CYC* genes that do not participate in establishing floral bilateral symmetry. Such is the case of the *A. thaliana, AtTCP1,* which promotes shoot growth and regulates leaf lamina size (Costa et al., 2005; Guo et al., 2010; Koyama et al., 2010b). To date less attention has been given to the putative role of *CYC* genes in the development of leaves, carpels and fruits, despite the fact that expression has been detected in these organs in other core eudicots. This is the case for *CYC* homologs in *Citrullus, Gerbera, Gossypium, Lotus, Solanum,* and some Papaveraceae that have been detected in seedlings, young leaves, and immature fruits (Damerval et al., 2007; Wang et al., 2010; Parapunova et al., 2014; Ma et al., 2016; Shi et al., 2016).

CIN-like Genes Have Undergone Numerous Duplications in Angiosperms, Monocots and Orchidaceae and Show Broad Expression Patterns in *Cattleya trianae* When Compared to *Hypoxis decumbens*

Here we show the first comprehensive phylogenetic analysis of *CIN-like* genes. Our results point to the occurrence of at least one duplication event predating angiosperm diversification, at least one duplication occurring prior to the origin of the Asparagales and one specific duplication prior to the diversification of Orchidoideae and Epidendroideae (**Figure 3**, **Supplementary Figure 2**; Floyd and Bowman, 2007; Martín-Trillo and Cubas, 2010). By comparison to *CYC* genes, *CIN* genes functional characterization is restricted to model species only. The canonical *CINCINNATA* in *A. majus* has dual roles in limiting the growth of leaf margins while promoting epidermal cell differentiation in petals (Nath et al., 2003; Crawford et al., 2004). The *cin* mutant in *A. majus*, as well as the *tcp4* mutant in *A. thaliana*, exhibit curly leaves as a result of excessive growth in leaf margins (Crawford et al., 2004; Koyama et al., 2010a). *CIN* regulates leaf shape through direct or indirect negative regulation of the boundary *CUP-SHAPED COTYLEDON (CUC)* genes, likely via the activation of *ASSYMETRIC LEAVES 1 (AS1)*, *miR164*, *INDOLE-3-ACETIC ACID3/SHORT HYPOCOTYL2 (IAA3/SHY2)*, and *SMALL AUXIN UP RNA (SAUR)* (Koyama et al., 2007, 2010a). *CIN* also controls leaf development through the regulation of cell proliferation by activating *miR396*, *CYCLIN-DEPENDENT KINASE INHIBITOR/KIP RELATED PROTEIN 1 (ICK1/KRP1)* and jasmonate biosynthesis (Schommer et al., 2014). Similarly, *AtTCP4* regulates the transition between cell proliferation and differentiation, controlling cytokinin and auxin receptors (Efroni et al., 2013; Das Gupta et al., 2014). *AtTCP4* also controls leaf senescence, maintains petal growth, and regulates early embryo development and seed viability, and finally it regulates *jasmonic acid (JA)* biosynthesis by the activation of *LIPOXIGENASE2 (LOX2)* (Schommer et al., 2008; Nag et al., 2009; Sarvepalli and Nath, 2011; Danisman et al., 2012). Leaf patterning is also controlled by *CIN-like* homologs in tomato and rice (Ori et al., 2007; Yang et al., 2013; Zhou et al., 2013; Ballester et al., 2015).

In addition to the roles of *CIN* genes in leaf patterning, other functions in carpel and fruit development have been identified. *AtTCP2* and *AtTCP3* are known to activate *NGATHA* genes that regulate carpel apical patterning. *NGA* orthologs from *A. thaliana*, rice, tomato and bean have conserved putative TCP binding site suggesting that this regulation is conserved in monocots and dicots (Ballester et al., 2015). Moreover, the *tcp3* mutant has shorter siliques with a crinkled surface (Koyama et al., 2007; Ballester et al., 2015). Additionally, expression analyses in different Solanaceae species suggest an important contribution of these genes to fruit development and maturation, as they are downstream targets of key ripening regulators including *RIPENING INHIBITOR* (*RIN*), *COLORLESS NON-RIPENING* (*CNR*) and *SlAP2a* (**Supplementary Figure 3**; Crawford et al., 2004; Parapunova et al., 2014). More recently *PeCIN8*, a *P. equestris CIN-like* homolog, was shown to be broadly expressed and to have roles in leaf cell proliferation, determining the fruit final size and controlling proper embryo and ovule development (Lin et al., 2016). In summary, *CIN-like* genes are pleiotropic regulators of cell division and differentiation in leaves, petals, carpels, ovules, fruits and seeds across angiosperms.

This study identified significant changes in the helix I residues and the loop between Orchidaceae sequences and other Asparagales CIN-like proteins (**Figure 2**). For instance, a number of motifs involved in protein interaction, including the R domain characteristic of class II TCP proteins were only identified in the CIN1a clade, but are absent in all other paralogous clades (**Supplementary Figure 1**, Cubas et al., 1999; Damerval and Manuel, 2003). Here we have also identified conserved motifs like the *miR319* binding site, previously reported in *CIN-like* homologs from *O. italica* (Nag et al., 2009; De Paolo et al., 2015), for all *CIN-like* sequences in Asparagales. This indicates, that Asparagales homologs are also regulated by *miR319*, similar to *AtTCP4*, and other *CIN-like Arabidopsis* paralogs, including *AtTCP2/3/10* and *24* (Palatnik et al., 2003; Koyama et al., 2007; Schommer et al., 2008, 2012; Koyama et al., 2010a; Danisman et al., 2012).

In comparison with reported functional data from model eudicots and monocots, the *CIN-like* differential expression observed in *C. trianae* and *H. decumbens* homologs points to three testable hypotheses. (1) While in *H. decumbens* all *CIN-like* genes are expressed in young leaves and may play roles in leaf development, only two of the six paralogs identified in *C. trianae* (*CtrCIN3/4*) are likely involved in cell division and differentiation during leaf development, similar to the *P. equestris PeCIN8* (Lin et al., 2016). (2) While *CIN-like* genes are likely playing key roles in both perianth and fertile organs development and growth in *C. trianae*, their contribution to perianth development and growth is less clear in *H. decumbens*, perhaps only with *HydCIN2* involved in cell proliferation in the perianth. (3) In both species *CIN-like* genes are strongly expressed in carpels and fruits, suggesting that their role in carpel patterning, ovule development as well as fruit maturation is likely conserved in Asparagales. Nevertheless, mRNA expression data for all *CIN-like* genes must be interpreted with caution given the putative conserved *miR139* postranscriptional regulation.

PCF-like Genes Are Extensively Duplicated and Have Overlapping Expression Patterns with *CIN-like* Genes in Asparagales

Our results on the evolution of *PCF-like* genes points to numerous duplication events within Asparagales in comparison to the *CYC-like* and the *CIN-like* clades. In addition, the topology recovered suggests that *PCF-like* gene duplications in monocots are independent from the one that occurred in core eudicots (**Figure 4**, **Supplementary Figure 2**). Functional data available for *PCF-like* genes suggest redundancy with *CIN-like* genes (Aguilar-Martínez and Sinha, 2013; Danisman et al., 2013). For instance, both *PCF-like* and *CIN-like* genes control leaf development through regulation of *LOX2*. However, while *AtTCP20 (PCF-like)* inhibits, *AtTCP4 (CIN-like)* induces the expression of *LOX2* (Danisman et al., 2012). *PCF-like* genes are also involved in the regulation of circadian clock genes, that is the case of *CCA1 Hiking Expedition CHE (AtTCP21)* (Pruneda-Paz et al., 2009; Giraud et al., 2010). In addition dimers formed between AtTCP15 and other TCP-like proteins (AtTCP2, AtTCP3, AtTCP11) are known to regulate circadian cycles, cell proliferation in floral organs and leaves, and to promote seed germination (Koyama et al., 2007; Kieffer et al., 2011; Resentini et al., 2015). *PCF-like* genes are also expressed in ovule, seed and fruit development in *Solanum lycopersicum*, *Solanum tuberosum*, *O. sativa* and in *P. equestris*, suggesting that they mediate cell proliferation in the carpel to fruit transition in both eudicots and monocots (**Supplementary Figure 3**; Kosugi and Ohashi, 1997; Yao et al., 2007; Parapunova et al., 2014; Lin et al., 2016).

Expression detected here of *PCF-like* homologs in *C. trianae* and *H. decumbens* show broad expression of most paralogs in *C. trianae* (except *CtrPCF7*) contrasting with a restricted expression of most paralogs in *H. decumbens* (except *HydPCF5*; **Figure 5**). Such expression patterns are in accordance with all putative functions identified in other monocots and in core eudicots including, but not restricted to, cell proliferation control in leaves, carpels, ovules, fruits and seeds (Lin et al., 2016). All data available point to pleiotropic roles of *TCP-like* genes with a high degree of redundancy among paralogs. Interestingly, when compared to *CYC-like and CIN-like* genes, the *PCF-like* gene *HydPCF5* in *H. decumbens* is more likely to be playing cell proliferation roles in the perianth.

In conclusion, the Asparagales, unlike core eudicot model plants, have a reduction in the number of *CYC-like* homologs and an increase of *CIN-like* and *PCF-like* copies. Characteristic key amino acids in the bHLH domain, flanking motifs and binding sites are for the most part conserved in Asparagales CIN-like and PCF like proteins suggesting similar conserved mechanisms of post-transcriptional regulation and interacting partners. The most notorious exception to this is the lack of a ***miR319*** binding site in *HydCIN2/3*. Nevertheless, the CYC-like proteins in Asparagales seem to be poorly expressed and have undergone important shifts in protein domains, suggesting changes in regulation as well as in protein-protein interactions. The orchid included in this study has similar numbers of TCP copies when compared to Asparagales with radially symmetrical flowers, and only the Epidendrioideae and Orchidoideae seem to have an additional *CIN* paralog lacking in other Orchidaceae and Asparagales. Expression data suggests different roles of *TCP-like* genes in *C. trianae* and *H. decumbens*, pointing to: (1) a possible decoupling of *TB1* homologs from bilateral symmetry in *C. trianae*; (2) conserved roles of *CIN-like* and *PCF-like* genes in the control of cell proliferation in carpels, ovules and fruits in both species; and (3) preferential leaf expression of *CIN-like* and *PCF-like* genes in *H. decumbens* and perianth expression in *C. trianae*. Here we have performed the first large scale analysis of *TCP* genes in Asparagales which will provide a platform for in depth comparative expression analyses as well as much needed functional studies of these genes in emerging model orchids.

AUTHOR CONTRIBUTIONS

YM and NP planned and designed the research. YM and NP conducted fieldwork. YM, JA, and NP performed experiments. YM, JA, and NP analyzed the data and wrote and approved the final version of the manuscript. All authors read and approved the final manuscript.

ACKNOWLEDGMENTS

We thank the recognition obtained by the Premio Fomento a la Investigación, Alcaldia de Medellín, 2016. We thank the OneKP repository staff (University of Alberta), the Orchidstra staff (Academia Sinica) and Orchidbase staff (National Cheng Kung University) for facilitating access to the online databases. We thank Favio González (Universidad Nacional de Colombia) and Mariana Mondragón-Palomino (Universität Regensburg) for their comments to the manuscript and to preliminary results presented in meetings, respectively. We thank Ricardo Callejas (Universidad de Antioquia) for allowing us to use office and laboratory space and Cecilia Zumajo-Cardona (The New York Botanical Garden) for laboratory assistance. We thank Victor Acosta, Francisco Villegas and the staff at Vivero Sol Rojo for maintaining cultivated plant material of *C. trianae*. Finally we thank two reviewers for their insightful comments on the manuscript.

SUPPLEMENTARY MATERIAL

The Supplementary Material for this article can be found online at: http://journal.frontiersin.org/article/10.3389/fpls.2017.00009/full#supplementary-material

Supplementary Figure 1 | Conserved motifs in and Orchidaceae and non-Orchidaceae Asparagales TCP-like proteins. Model core eudicots and monocots used as reference include *Arabidopsis thaliana*, *Antirrhinum majus*, and *Oryza sativa*. Motifs 1, 2, and 3 correspond to the conserved TCP domain. Motif 11 indicates the characteristic R domain in Class II TCP-like genes. Motif 14 corresponds to the *miR319* binding site in *CIN-like* genes.

Supplementary Figure 2 | ML analysis of *TCP-like* genes with extended Solanaceae sampling. ML phylogenetic analysis of *TCP-like* genes with reduced sampling including only model organisms like *Arabidopsis thaliana, Oryza sativa, Solanum lycopersicum, Solanum tuberosum,* two Orchidaceae species, *Cattleya trianae, Orchis italica,* one non-Orchidaceae Asparagales, *Hypoxis decumbens* and the early diverging angiosperm *Amborella trichopoda*. Yellow stars indicate large scale duplication events. TCP major clades are labeled to the right. Branch and taxa colors correspond to those in the conventions. BS values $\geq$ 50 are shown.

Supplementary Figure 3 | Expression patterns of TCP-like homologs in *Arabidopsis thaliana* (Brassicaceae), *Solanum lycopersicum* and *Solanum tuberosum* (Solanaceae). (A) *Arabidopsis thaliana* AthTCP24 (CIN-like). **(B)** *Arabidopsis thaliana* AthTCP20 (PCF-like). **(C)** Solanum lycopersicum SlyTCP24 (CIN-like). **(D)** *Solanum lycopersicum* SlyTCP11 (PCF-like). **(E)** *Solanum tuberosum* StuTCP2 (CIN-like). **(F)** *Solanum tuberosum* StuTCP11 (PCF-like). Expression patterns show here were selected for the genes that showed the broader, most comprehensive expression patterns after comparing expression for all paralogs, and are used to summarize putative expression patterns for each gene clade. Expression levels vary in intensity between different copies of CIN-like and PCF-like. Taken from the eFP browser, last accessed Sep 5/2016 (http://bar.utoronto.ca/efp_tomato/cgi-bin/efpWeb.cgi and http://bar.utoronto.ca/efp_potato/cgi-bin/efpWeb.cgi).

Supplementary Table 1 | List of sequences used in this study.

Supplementary Table 2 | Primers used for *TCP-like gene* expression analyses.

REFERENCES

Aguilar-Martínez, J. A., and Sinha, N. (2013). Analysis of the role of Arabidopsis class I *TCP* genes *AtTCP7, AtTCP8, AtTCP22*, and *AtTCP23* in leaf development. *Front. Plant Sci.* 4:406. doi: 10.3389/fpls.2013.00406

Almeida, J., Rocheta, M., and Galego, L. (1997). Genetic control of flower shape in *Antirrhinum majus*. *Development* 124, 1387–1392.

Altschul, S. F., Gish, W., Miller, W., Myers, E. W., and Lipman, D. J. (1990). Basic local alignment search tool. *J. Mol. Biol.* 215, 403–410. doi: 10.1016/S0022-2836(05)80360-2

Bailey, T. L., Williams, N., Misleh, C., and Li, W. W. (2006). MEME: discovering and analyzing DNA and protein sequence motifs. *Nucleic Acids Res.* 34, W369–W373. doi: 10.1093/nar/gkl198

Ballester, P., Navarrete-Gómez, M., Carbonero, P., Oñate-Sánchez, L., and Ferrándiz, C. (2015). Leaf expansion in Arabidopsis is controlled by a TCP-NGA regulatory module likely conserved in distantly related species. *Physiol. Plant* 155, 21–32. doi: 10.1111/ppl.12327

Bartlett, M. E., and Specht, C. D. (2011). Changes in expression pattern of the *TEOSINTE BRANCHED1*- like genes in the Zingiberales provide a mechanism for evolutionary shifts in symmetry across the order. *Am. J. Bot.* 98, 227–243. doi: 10.3732/ajb.1000246

Broholm, S. (2009). *The Role of MADS and TCP Transcription Factors in Gerbera Hybrida Flower Development*. Available online at: https://helda.helsinki.fi/handle/10138/22335

Busch, A., and Zachgo, S. (2007). Control of corolla monosymmetry in the *Brassicaceae Iberis* amara. *Proc. Natl. Acad. Sci. U.S.A.* 104, 16714–16719. doi: 10.1073/pnas.0705338104

Chase, M. W., Cameron, K. M., Freudenstein, J. V., Pridgeon, A. M., Salazar, G., van den Berg, C., et al. (2015). An updated classification of Orchidaceae. *Bot. J. Linn. Soc.* 177, 151–174. doi: 10.1111/boj.12234

Chase, M. W., Christenhusz, M. J. M., Fay, M. F., Byng, J. W., Judd, W. S., Soltis, D. E., et al. (2016). An update of the Angiosperm Phylogeny Group classification for the orders and families of flowering plants: APG IV. *Bot. J. Linn. Soc.* 181, 1–20. doi: 10.1111/boj.12385

Chase, M. W., Reveal, J. L., and Fay, M. F. (2009). A subfamilial classification for the expanded asparagalean families Amaryllidaceae, Asparagaceae and Xanthorrhoeaceae. *Bot. J. Linn. Soc.* 161, 132–136. doi: 10.1111/j.1095-8339.2009.00999.x

Chen, S., Kim, D. K., Chase, M. W., and Kim, J. H. (2013). Networks in a large-scale phylogenetic analysis: reconstructing evolutionary history of Asparagales (Lilianae) based on four plastid genes. *PLoS ONE* 8:e59472. doi: 10.1371/journal.pone.0059472

Citerne, H. L., Le Guilloux, M., Sannier, J., Nadot, S., and Damerval, C. (2013). Combining phylogenetic and syntenic analyses for understanding the evolution of TCP ECE genes in eudicots. *PLoS ONE* 8:e74803. doi: 10.1371/journal.pone.0074803

Corley, S. B., Carpenter, R., Copsey, L., and Coen, E. (2005). Floral asymmetry involves an interplay between TCP and MYB transcription factors in Antirrhinum. *Proc. Natl. Acad. Sci. U.S.A.* 102, 5068–5073. doi: 10.1073/pnas.0501340102

Costa, M. M. R., Fox, S., Hanna, A. I., Baxter, C., and Coen, E. (2005). Evolution of regulatory interactions controlling floral asymmetry. *Development* 132, 5093–5101. doi: 10.1242/dev.02085

Crawford, B. C. W., Nath, U., Carpenter, R., and Coen, E. S. (2004). CINCINNATA controls both cell differentiation and growth in petal lobes and leaves of Antirrhinum. *Plant Physiol.* 135, 244–253. doi: 10.1104/pp.103.036368

Cubas, P., Lauter, N., Doebley, J., and Coen, E. (1999). The TCP domain: a motif found in proteins regulating plant growth and development. *Plant J.* 18, 215–222. doi: 10.1046/j.1365-313X.1999.00444.x

Damerval, C., Citerne, H., Le Guilloux, M., Domenichini, S., Dutheil, J., De Craene, L. R., et al. (2013). Asymmetric morphogenetic cues along the transverse plane: shift from disymmetry to zygomorphy in the flower of fumarioideae. *Am. J. Bot.* 100, 391–402. doi: 10.3732/ajb.1200376

Damerval, C., Le Guilloux, M., Jager, M., and Charon, C. (2007). Diversity and evolution of *CYCLOIDEA-like* TCP genes in relation to flower development in Papaveraceae. *Plant Physiol.* 143, 759–772. doi: 10.1104/pp.106.090324

Damerval, C., and Manuel, M. (2003). Independent evolution of Cycloidea-like sequences in several angiosperm taxa. *Comptes Rendus Palevol* 2, 241–250. doi: 10.1016/S1631-0683(03)00031-9

Danisman, S., van der Wal, F., Dhondt, S., Waites, R., de Folter, S., Bimbo, A., et al. (2012). Arabidopsis class I and class II TCP transcription factors regulate jasmonic acid metabolism and leaf development antagonistically. *Plant Physiol.* 159, 1511–1523. doi: 10.1104/pp.112.200303

Danisman, S., Van Dijk, A. D. J., Bimbo, A., van der Wal, F., Hennig, L., De Folter, S., et al. (2013). Analysis of functional redundancies within the Arabidopsis TCP transcription factor family. *J. Exp. Bot.* 64, 5673–5685. doi: 10.1093/jxb/ert337

Das Gupta, M., Aggarwal, P., and Nath, U. (2014). CINCINNATA in *Antirrhinum majus* directly modulates genes involved in cytokinin and auxin signaling. *New Phytol.* 204, 901–912. doi: 10.1111/nph.12963

De Paolo, S., Gaudio, L., and Aceto, S. (2015). Analysis of the TCP genes expressed in the inflorescence of the orchid *Orchis italica*. *Sci. Rep.* 5:16265. doi: 10.1038/srep16265

Doebley, J., Stec, A., and Gustus, C. (1995). teosinte branched1 and the origin of maize: evidence for epistasis and the evolution of dominance. *Genetics* 141, 333–346.

Doebley, J., Stec, A., and Hubbard, L. (1997). The evolution of apical dominance in maize. *Nature* 386, 485–488. doi: 10.1038/386485a0

Efroni, I., Han, S. K., Kim, H. J., Wu, M. F., Steiner, E., Birnbaum, K. D., et al. (2013). Regulation of leaf maturation by chromatin-mediated modulation of cytokinin responses. *Dev. Cell* 24, 438–445. doi: 10.1016/j.devcel.2013.01.019

Endress, P. K. (2016). Development and evolution of extreme synorganization in angiosperm flowers and diversity: a comparison of Apocynaceae and Orchidaceae. *Ann. Bot.* 117, 749–767. doi: 10.1093/aob/mcv119

Floyd, S. K., and Bowman, J. L. (2007). The ancestral developmental tool kit of land plants. *Int. J. Plant Sci. Spec. Issue Discern. Homol. Gene Expr.* 168, 1–35. doi: 10.1086/509079

Galego, L., and Almeida, J. (2002). Role of *DIVARICATA* in the control of dorsoventral asymmetry in *Antirrhinum* flowers. *Genes Dev.* 16, 880–891. doi: 10.1101/gad.221002

Gao, Q., Tao, J. H., Yan, D., Wang, Y. Z., and Li, Z. Y. (2008). Expression differentiation of *CYC*-like floral symmetry genes correlated with their protein sequence divergence in *Chirita heterotricha* (Gesneriaceae). *Dev. Genes Evol.* 218, 341–351. doi: 10.1007/s00427-008-0227-y

Giraud, E., Ng, S., Carrie, C., Duncan, O., Low, J., Lee, C. P., et al. (2010). TCP transcription factors link the regulation of genes encoding mitochondrial proteins with the circadian clock in *Arabidopsis thaliana*. *Plant Cell* 22, 3921–3934. doi: 10.1105/tpc.110.074518

Givnish, T. J., Zuluaga, A., Marques, I., Lam, V. K. Y., Gomez, M. S., Iles, W. J. D., et al. (2016). Phylogenomics and historical biogeography of the monocot order Liliales: out of Australia and through Antarctica. *Cladistics* 32, 581–605. doi: 10.1111/cla.12153

Gong, Y.-B., and Huang, S.-Q. (2009). Floral symmetry: pollinator-mediated stabilizing selection on flower size in bilateral species. *Proc. R. Soc. B Biol. Sci.* 276, 4013–4020. doi: 10.1098/rspb.2009.1254

Guo, Z., Fujioka, S., Blancaflor, E. B., Miao, S., Gou, X., and Li, J. (2010). TCP1 modulates brassinosteroid biosynthesis by regulating the expression of the key biosynthetic gene DWARF4 in *Arabidopsis thaliana*. *Plant Cell* 22, 1161–1173. doi: 10.1105/tpc.109.069203

Heery, D. M., Kalkhoven, E., Hoare, S., and Parker, M. G. (1997). A signature motif in transcriptional co-activators mediates binding to nuclear receptors. *Nature* 387, 733–736. doi: 10.1038/42750

Hileman, L. C. (2014). Trends in flower symmetry evolution revealed through phylogenetic and developmental genetic advances. *Philos. Trans. R. Soc. Lond. B. Biol. Sci.* 369, 1–10. doi: 10.1098/rstb.2013.0348

Hileman, L. C., and Baum, D. A. (2003). Why do paralogs persist? Molecular evolution of *CYCLOIDEA* and related floral symmetry genes in Antirrhineae (Veronicaceae). *Mol. Biol. Evol.* 20, 591–600. doi: 10.1093/molbev/msg063

Horn, S., Pabón-Mora, N., Theuß, V. S., Busch, A., and Zachgo, S. (2015). Analysis of the CYC/TB1 class of TCP transcription factors in basal angiosperms and magnoliids. *Plant J.* 81, 559–571. doi: 10.1111/tpj.12750

Hoshino, Y., Igarashi, T., Ohshima, M., Shinoda, K., Murata, N., Kanno, A., et al. (2014). Characterization of *CYCLOIDEA-like* genes in controlling floral zygomorphy in the monocotyledon *Alstroemeria. Sci. Hortic. (Amsterdam).* 169, 6–13. doi: 10.1016/j.scienta.2014.01.046

Howarth, D. G., and Donoghue, M. J. (2006). Phylogenetic analysis of the "ECE" (CYC/TB1) clade reveals duplications predating the core eudicots. *Proc. Natl. Acad. Sci. U.S.A.* 103, 9101–9106. doi: 10.1073/pnas.0602827103

Howarth, D. G., Martins, T., Chimney, E., and Donoghue, M. J. (2011). Diversification of *CYCLOIDEA* expression in the evolution of bilateral flower symmetry in Caprifoliaceae and *Lonicera* (Dipsacales). *Ann. Bot.* 107, 1521–1532. doi: 10.1093/aob/mcr049

Katoh, K., Misawa, K., Kuma, K., and Miyata, T. (2002). MAFFT: a novel method for rapid multiple sequence alignment based on fast Fourier transform. *Nucleic Acids Res.* 30, 3059–3066. doi: 10.1093/nar/gkf436

Kieffer, M., Master, V., Waites, R., and Davies, B. (2011). TCP14 and TCP15 affect internode length and leaf shape in *Arabidopsis. Plant J.* 68, 147–158. doi: 10.1111/j.1365-313X.2011.04674.x

Kocyan, A. (2007). The discovery of polyandry in *Curculigo* (Hypoxidaceae): implications for androecium evolution of asparagoid monocotyledons. *Ann. Bot.* 100, 241–248. doi: 10.1093/aob/mcm091

Kosugi, S., and Ohashi, Y. (1997). PCF1 and PCF2 specifically bind to cis elements in the rice proliferating cell nuclear antigen gene. *Plant Cell* 9, 1607–1619. doi: 10.1105/tpc.9.9.1607

Kosugi, S., and Ohashi, Y. (2002). DNA binding and dimerization specificity and potential targets for the TCP protein family. *Plant J.* 30, 337–348. doi: 10.1046/j.1365-313X.2002.01294.x

Koyama, T., Furutani, M., Tasaka, M., and Ohme-Takagi, M. (2007). TCP transcription factors control the morphology of shoot lateral organs via negative regulation of the expression of boundary-specific genes in *Arabidopsis. Plant Cell* 19, 473–484. doi: 10.1105/tpc.106.044792

Koyama, T., Mitsuda, N., Seki, M., Shinozaki, K., and Ohme-Takagi, M. (2010a). TCP transcription factors regulate the activities of ASYMMETRIC LEAVES1 and miR164, as well as the auxin response, during differentiation of leaves in *Arabidopsis. Plant Cell* 22, 3574–3588. doi: 10.1105/tpc.110.075598

Koyama, T., Sato, F., and Ohme-Takagi, M. (2010b). A role of TCP1 in the longitudinal elongation of leaves in *Arabidopsis. Biosci. Biotechnol. Biochem.* 74, 2145–2147. doi: 10.1271/bbb.100442

Li, S., Lauri, A., Ziemann, M., Busch, A., Bhave, M., and Zachgo, S. (2009). Nuclear activity of ROXY1, a glutaredoxin interacting with TGA factors, is required for petal development in *Arabidopsis thaliana. Plant Cell* 21, 429–441. doi: 10.1105/tpc.108.064477

Lin, Y. F., Chen, Y. Y., Hsiao, Y. Y., Shen, C. Y., Hsu, J. L., Yeh, C. M., et al. (2016). Genome-wide identification and characterization of TCP genes involved in ovule development of *Phalaenopsis equestris. J. Exp. Bot.* 67, 5051–5066. doi: 10.1093/jxb/erw273

Luo, D., Carpenter, R., Copsey, L., Vincent, C., Clark, J., and Coen, E. (1999). Control of organ asymmetry in flowers of Antirrhinum. *Cell* 99, 367–376. doi: 10.1016/S0092-8674(00)81523-8

Luo, D., Carpenter, R., Vincent, C., Copsey, L., and Coen, E. (1996). Origin of floral asymmetry in *Antirrhinum. Nature* 383, 794–799. doi: 10.1038/383794a0

Ma, J., Liu, F., Wang, Q., Wang, K., Jones, D. C., and Zhang, B. (2016). Comprehensive analysis of TCP transcription factors and their expression during cotton (Gossypium arboreum) fiber early development. *Sci. Rep.* 6:21535. doi: 10.1038/srep21535

Martín-Trillo, M., and Cubas, P. (2010). TCP genes: a family snapshot ten years later. *Trends Plant Sci.* 15, 31–39. doi: 10.1016/j.tplants.2009.11.003

Miller, M. A., Pfeiffer, W., and Schwartz, T. (2010). "Creating the CIPRES science gateway for inference of large phylogenetic trees," in *Gateway Computing Environments Workshop, GCE* (San Diego, CA).

Mondragón-Palomino, M. (2013). Perspectives on MADS-box expression during orchid flower evolution and development. *Front. Plant Sci.* 4:377. doi: 10.3389/fpls.2013.00377

Mondragón-Palomino, M., and Theißen, G. (2009). Why are orchid flowers so diverse? Reduction of evolutionary constraints by paralogues of class B floral homeotic genes. *Ann. Bot.* 104, 583–594. doi: 10.1093/aob/mcn258

Mondragón-Palomino, M., and Trontin, C. (2011). High time for a roll call: gene duplication and phylogenetic relationships of TCP-like genes in monocots. *Ann. Bot.* 107, 1533–1544. doi: 10.1093/aob/mcr059

Nag, A., King, S., and Jack, T. (2009). miR319a targeting of TCP4 is critical for petal growth and development in *Arabidopsis. Proc. Natl. Acad. Sci. U.S.A.* 106, 22534–22539. doi: 10.1073/pnas.0908718106

Nath, U., Crawford, B. C. W., Carpenter, R., and Coen, E. (2003). Genetic control of surface curvature. *Am. Assoc. Adv. Sci.* 299, 1404–1407. doi: 10.1126/science.1079354

Navaud, O., Dabos, P., Carnus, E., Tremousaygue, D., and Hervé, C. (2007). TCP transcription factors predate the emergence of land plants. *J. Mol. Evol.* 65, 23–33. doi: 10.1007/s00239-006-0174-z

Ori, N., Cohen, A. R., Etzioni, A., Brand, A., Yanai, O., Shleizer, S., et al. (2007). Regulation of LANCEOLATE by miR319 is required for compound-leaf development in tomato. *Nat. Genet.* 39, 787–791. doi: 10.1038/ng2036

Pabón-Mora, N., and González, F. (2008). Floral ontogeny of *Telipogon* spp. (Orchidaceae) and insights on the perianth symmetry in the family. *Int. J. Plant Sci.* 169, 1159–1173. doi: 10.1086/591982

Palatnik, J. F., Allen, E., Wu, X., Schommer, C., Schwab, R., Carrington, J. C., et al. (2003). Control of leaf morphogenesis by microRNAs. *Nature* 425, 257–263. doi: 10.1038/nature01958

Parapunova, V., Busscher, M., Busscher-Lange, J., Lammers, M., Karlova, R., Bovy, A. G., et al. (2014). Identification, cloning and characterization of the tomato TCP transcription factor family. *BMC Plant Biol.* 14:157. doi: 10.1186/1471-2229-14-157

Posada, D., and Crandall, K. A. (1998). MODELTEST: testing the model of DNA substitution. *Bioinformatics* 14, 817–818. doi: 10.1093/bioinformatics/14.9.817

Preston, J. C., and Hileman, L. C. (2009). Developmental genetics of floral symmetry evolution. *Trends Plant Sci.* 14, 147–154. doi: 10.1016/j.tplants.2008.12.005

Preston, J. C., and Hileman, L. C. (2012). Parallel evolution of TCP and B-class genes in Commelinaceae flower bilateral symmetry. *Evodevo* 3:6. doi: 10.1186/2041-9139-3-6

Preston, J. C., Kost, M., A.„ and Hileman, L. C. (2009). Conservation and diversification of the symmetry developmental program among close relatives of snapdragon with divergent floral morphologies. *New Phytol.* 182, 751–762. doi: 10.1111/j.1469-8137.2009.02794.x

Pruneda-Paz, J. L., Breton, G., Para, A., and Kay, S. A. (2009). A functional genomics approach reveals CHE as a component of the Arabidopsis circadian clock. *Science* 323, 1481–1485. doi: 10.1126/science.1167206

Raimundo, J., Sobral, R., Bailey, P., Azevedo, H., Galego, L., Almeida, J., et al. (2013). A subcellular tug of war involving three MYB-like proteins underlies a molecular antagonism in *Antirrhinum* flower asymmetry. *Plant J.* 75, 527–538. doi: 10.1111/tpj.12225

Rambaut, A. (2014). *FigTree: Tree Figure Drawing Tool.*

Reeves, P., A., and Olmstead, R. G. (2003). Evolution of the TCP gene family in asteridae: cladistic and network approaches to understanding regulatory gene family diversification and its impact on morphological evolution. *Mol. Biol. Evol.* 20, 1997–2009. doi: 10.1093/molbev/msg211

Resentini, F., Felipo-Benavent, A., Colombo, L., Blázquez, M. A., Alabadí, D., and Masiero, S. (2015). TCP14 and TCP15 mediate the promotion of seed germination by gibberellins in *Arabidopsis thaliana. Mol. Plant* 8, 482–485. doi: 10.1016/j.molp.2014.11.018

Rudall, P. J. (2002). Unique floral structures and iterative evolutionary themes in asparagales: insights from a morphological cladistic analysis. *Bot. Rev.* 68, 488–509. doi: 10.1663/0006-8101(2002)068[0488:UFSAIE]2.0.CO;2

Rudall, P. J., and Bateman, R. M. (2002). Roles of synorganisation, zygomorphy and heterotopy in floral evolution: the gynostemium and labellum of orchids and other lilioid monocots. *Biol. Rev. Camb. Philos. Soc.* 77, 403–441. doi: 10.1017/S1464793102005936

Rudall, P. J., and Bateman, R. M. (2004). Evolution of zygomorphy in monocot flowers: iterative patterns and developmental constraints. *New Phytol.* 162, 25–44. doi: 10.1111/j.1469-8137.2004.01032.x

Rudall, P. J., Perl, C. D., and Bateman, R. M. (2013). Organ homologies in orchid flowers re-interpreted using the Musk Orchid as a model. *PeerJ* 1, 1–23. doi: 10.7717/peerj.26

Sarvepalli, K., and Nath, U. (2011). Hyper-activation of the TCP4 transcription factor in *Arabidopsis thaliana* accelerates multiple aspects of plant maturation. *Plant J.* 67, 595–607. doi: 10.1111/j.1365-313X.2011.04616.x

Schommer, C., Bresso, E. G., Spinelli, S. V., and Palatnik, J. F. (2012). MicroRNAs in plant development and stress responses. *Screen* 15, 29–47. doi: 10.1007/978-3-642-27384-1

Schommer, C., Debernardi, J. M., Bresso, E. G., Rodriguez, R. E., and Palatnik, J. F. (2014). Repression of cell proliferation by miR319-regulated TCP4. *Mol. Plant* 7, 1533–1544. doi: 10.1093/mp/ssu084

Schommer, C., Palatnik, J. F., Aggarwal, P., Chételat, A., Cubas, P., Farmer, E. E., et al. (2008). Control of jasmonate biosynthesis and senescence by miR319 targets. *PLoS Biol.* 6:e230. doi: 10.1371/journal.pbio.0060230

Shi, P., Guy, K. M., Wu, W., Fang, B., Yang, J., Zhang, M., et al. (2016). Genome-wide identification and expression analysis of the *ClTCP* transcription factors in *Citrullus lanatus*. *BMC Plant Biol.* 16:85. doi: 10.1186/s12870-016-0765-9

Simpson, M. G. (2006). *Plant Systematics*. San Diego, CA: Elsevier Academic Press.

Su, C. L., Chen, W. C., Lee, A. Y., Chen, C. Y., Chang, Y. C. A., Chao, Y. T., et al. (2013). A modified ABCDE model of flowering in orchids based on gene expression profiling studies of the moth orchid *Phalaenopsis aphrodite*. *PLoS ONE* 8:e80462. doi: 10.1371/journal.pone.0080462

Tähtiharju, S., Rijpkema, A. S., Vetterli, A., Albert, V. A., Teeri, T. H., and Elomaa, P. (2012). Evolution and diversification of the *CYC/TB1* gene family in asteraceae-a comparative study in Gerbera (mutisieae) and sunflower (heliantheae). *Mol. Biol. Evol.* 29, 1155–1166. doi: 10.1093/molbev/msr283

Tsai, W. C., Fu, C. H., Hsiao, Y. Y., Huang, Y. M., Chen, L. J., Wang, M., et al. (2013). OrchidBase 2.0: comprehensive collection of Orchidaceae floral transcriptomes. *Plant Cell Physiol.* 54, 1–8. doi: 10.1093/pcp/pcs187

Uberti-Manassero, N. G., Lucero, L. E., Viola, I. L., Vegetti, A. C., and Gonzalez, D. H. (2012). The class i protein AtTCP15 modulates plant development through a pathway that overlaps with the one affected by CIN-like TCP proteins. *J. Exp. Bot.* 63, 809–823. doi: 10.1093/jxb/err305

Vieira, C. P., Vieira, J., and Charlesworth, D. (1999). Evolution of the cycloidea gene family in Antirrhinum and Misopates. *Mol. Biol. Evol.* 16, 1474–1483. doi: 10.1093/oxfordjournals.molbev.a026059

Wang, J., Wang, Y., and Luo, D. (2010). LjCYC genes constitute floral dorsoventral asymmetry in *Lotus japonicus*. *J. Integr. Plant Biol.* 52, 959–970. doi: 10.1111/j.1744-7909.2010.00926.x

Wilfinger, W. W., Mackey, K., and Chomczynski, P. (1997). NanoDrop and design are registered trademarks of NanoDrop Technologies 260/280 and 260/230 Ratios NanoDrop® ND-1000 and ND-8000 8-Sample Spectrophotometers. *Biotechniques* 22, 474–481.

Yang, C., Li, D., Mao, D., Liu, X., Ji, C., Li, X., et al. (2013). Overexpression of microRNA319 impacts leaf morphogenesis and leads to enhanced cold tolerance in rice (*Oryza sativa*L.). *Plant Cell Environ.* 36, 2207–2218. doi: 10.1111/pce.12130

Yang, X., Pang, H. B., Liu, B. L., Qiu, Z. J., Gao, Q., Wei, L., et al. (2012). Evolution of double positive autoregulatory feedback loops in *CYCLOIDEA2* clade genes is associated with the origin of floral zygomorphy. *Plant Cell* 24, 1834–1847. doi: 10.1105/tpc.112.099457

Yao, X., Ma, H., Wang, J., and Zhang, D. (2007). Genome-wide comparative analysis and expression pattern of TCP gene families in *Arabidopsis thaliana* and *Oryza sativa*. *J. Integr. Plant Biol.* 49, 885–897. doi: 10.1111/j.1744-7909.2007.00509.x

Yuan, Z., Gao, S., Xue, D.-W., Luo, D., Li, L.-T., Ding, S.-Y., et al. (2009). RETARDED PALEA1 controls palea development and floral zygomorphy in rice. *Plant Physiol.* 149, 235–244. doi: 10.1104/pp.108.128231

Zhang, W., Kramer, E. M., and Davis, C. C. (2010). Floral symmetry genes and the origin and maintenance of zygomorphy in a plant-pollinator mutualism. *Proc. Natl. Acad. Sci. U.S.A.* 107, 6388–6393. doi: 10.1073/pnas.0910155107

Zhang, W., Steinmann, V. W., Nikolov, L., Kramer, E. M., and Davis, C. C. (2013). Divergent genetic mechanisms underlie reversals to radial floral symmetry from diverse zygomorphic flowered ancestors. *Front. Plant Sci.* 4:302. doi: 10.3389/fpls.2013.00302

Zhou, M., Li, D., Li, Z., Hu, Q., Yang, C., Zhu, L., et al. (2013). Constitutive expression of a miR319 gene alters plant development and enhances salt and drought tolerance in transgenic creeping bentgrass. *Plant Physiol.* 161, 1375–1391. doi: 10.1104/pp.112.208702

Conflict of Interest Statement: The authors declare that the research was conducted in the absence of any commercial or financial relationships that could be construed as a potential conflict of interest.

Permissions

All chapters in this book were first published in EGRNPD, by Frontiers Media SA; hereby published with permission under the Creative Commons Attribution License or equivalent. Every chapter published in this book has been scrutinized by our experts. Their significance has been extensively debated. The topics covered herein carry significant findings which will fuel the growth of the discipline. They may even be implemented as practical applications or may be referred to as a beginning point for another development.

The contributors of this book come from diverse backgrounds, making this book a truly international effort. This book will bring forth new frontiers with its revolutionizing research information and detailed analysis of the nascent developments around the world.

We would like to thank all the contributing authors for lending their expertise to make the book truly unique. They have played a crucial role in the development of this book. Without their invaluable contributions this book wouldn't have been possible. They have made vital efforts to compile up to date information on the varied aspects of this subject to make this book a valuable addition to the collection of many professionals and students.

This book was conceptualized with the vision of imparting up-to-date information and advanced data in this field. To ensure the same, a matchless editorial board was set up. Every individual on the board went through rigorous rounds of assessment to prove their worth. After which they invested a large part of their time researching and compiling the most relevant data for our readers.

The editorial board has been involved in producing this book since its inception. They have spent rigorous hours researching and exploring the diverse topics which have resulted in the successful publishing of this book. They have passed on their knowledge of decades through this book. To expedite this challenging task, the publisher supported the team at every step. A small team of assistant editors was also appointed to further simplify the editing procedure and attain best results for the readers.

Apart from the editorial board, the designing team has also invested a significant amount of their time in understanding the subject and creating the most relevant covers. They scrutinized every image to scout for the most suitable representation of the subject and create an appropriate cover for the book.

The publishing team has been an ardent support to the editorial, designing and production team. Their endless efforts to recruit the best for this project, has resulted in the accomplishment of this book. They are a veteran in the field of academics and their pool of knowledge is as vast as their experience in printing. Their expertise and guidance has proved useful at every step. Their uncompromising quality standards have made this book an exceptional effort. Their encouragement from time to time has been an inspiration for everyone.

The publisher and the editorial board hope that this book will prove to be a valuable piece of knowledge for researchers, students, practitioners and scholars across the globe.

List of Contributors

Shucai Wang
Key Laboratory of Molecular Epigenetics of MOE and Institute of Genetics and Cytology, Northeast Normal University, Changchun, China

Ying Chang
College of Life Science, Northeast Agricultural University, Harbin, China

Brian Ellis
Michael Smith Laboratories, The University of British Columbia, Vancouver, BC, Canada

Africa Gomariz-Fernández, Verónica Sánchez-Gerschon, Chloé Fourquin and Cristina Ferrándiz
Instituto de Biología Molecular y Celular de Plantas, Consejo Superior de Investigaciones Científicas-Universidad Politécnica de Valencia, Valencia, Spain

María A. Bello, Inés Álvarez, Guillermo Sanjuanbenito and Javier Fuertes-Aguilar
Plant Evolutionary Biology Group, Real Jardín Botánico (CSIC), Madrid, Spain

Pilar Cubas
Department of Plant Molecular Genetics, Centro Nacional de Biotecnología, CSIC-Universidad Autónoma de Madrid, Madrid, Spain

Yan Zhang, Guiye Zhao, Yushun Li, Ning Mo, Jie Zhang and Yan Liang
College of Horticulture, Northwest A&F University, Yangling, China
State Key Laboratory of Crop Stress Biology in Arid Region, Northwest A&F University, Yangling, China

Chao Li, Yan Wang, Liang Xu, Shanshan Nie, Dongyi Liang, Xiaochuan Sun, Benard K. Karanja, Xiaobo Luo and Liwang Liu
National Key Laboratory of Crop Genetics and Germplasm Enhancement, College of Horticulture, Nanjing Agricultural University, Nanjing, China

Yinglong Chen
School of Earth and Environment, The UWA Institute of Agriculture, The University of Western Australia, Perth, WA, Australia

Sylvia R. Silveira and Adriana P. Martinelli
Laboratório de Biotecnologia Vegetal, Centro de Energia Nuclear na Agricultura, Universidade de São Paulo, Piracicaba, Brazil

Marcelo C. Dornelas
Departamento de Biologia Vegetal, Instituto de Biologia, Universidade Estadual de Campinas, Campinas, Brazil

Pengkai Wang
Ministry of Education, Key Laboratory of Forest Genetics and Biotechnology, Nanjing Forestry University, Nanjing, China
Suzhou Polytechnic Institute of Agriculture, Suzhou, China

Tielong Cheng, Guangxin Liu, Meiping Li, Jisen Shi, Ye Lu and Jinhui Chen
Ministry of Education, Key Laboratory of Forest Genetics and Biotechnology, Nanjing Forestry University, Nanjing, China

Mengzhu Lu
Laboratory of Biotechnology, Chinese Academy of Forestry, Beijing, China

Thomas Laux
Institute of Biology III, University of Freiburg, Freiburg, Germany

Selahattin Danisman
Molecular Cell Physiology, Faculty of Biology, Bielefeld University, Bielefeld, Germany

Roberto Ferrari, Luca Tadini, Alex Costa and Simona Masiero
Dipartimento di Bioscienze, Università degli studi di Milano, Milano, Italy

Fabio Moratti
Max-Planck-Institut für Molekulare Pflanzenphysiologie, Potsdam-Golm, Germany

Marie-Kristin Lehniger and Christian Schmitz-Linneweber
Molecular Genetics, Institute of Biology, Humboldt University of Berlin, Berlin, Germany

Fabio Rossi
Dipartimento di Biotecnologie Mediche e Medicina Traslazionale, Università degli studi di Milano, Milano, Italy

Monica Colombo
Centro Ricerca e Innovazione, Fondazione Edmund Mach, San Michele all'Adige, Italy

Paolo Pesaresi
Dipartimento di Scienze Agrarie e Ambientali - Produzione, Territorio, Agroenergia, Università degli studi di Milano, Milano, Italy

Peng Wu, Wenli Wang, Ying Li and Xilin Hou
State Key Laboratory of Crop Genetics and Germplasm Enhancement, Key Laboratory of Biology and Germplasm Enhancement of Horticultural Crops in East China, Ministry of Agriculture, Nanjing Agricultural University, Nanjing, China

Weike Duan
State Key Laboratory of Crop Genetics and Germplasm Enhancement, Key Laboratory of Biology and Germplasm Enhancement of Horticultural Crops in East China, Ministry of Agriculture, Nanjing Agricultural University, Nanjing, China
School of Life Science and Food Engineering, Huaiyin Institute of Technology, Huaian

Jorge Reyes-Rivera, Emilio Petrone, Alejandra Vasco and Teresa Terrazas
Departamento de Botánica, Instituto de Biología, Universidad Nacional Autónoma de México, Mexico City, Mexico

Gustavo Rodríguez-Alonso and Svetlana Shishkova
Departamento de Biología Molecular de Plantas, Instituto de Biotecnología, Universidad Nacional Autónoma de México, Cuernavaca, Mexico

Francisco Vergara-Silva
Jardín Botánico, Instituto de Biología, Universidad Nacional Autónoma de México, Mexico City, Mexico

Yapeng Han, Xiangyong Li, Lin Cheng, Yanchun Liu, Hui Wang, Danxia Ke, Hongyu Yuan and Lei Wang
College of Life Sciences, Xinyang Normal University, Xinyang, China
Institute for Conservation and Utilization of Agro-bioresources in Dabie Mountains, Xinyang Normal University, Xinyang, China

Liangsheng Zhang
Center for Genomics and Biotechnology, Haixia Institute of Science and Technology, Fujian Agriculture and Forestry University, Fuzhou, China
Key Laboratory of Genetics, Breeding and Multiple Utilization of Corps (Fujian Agriculture and Forestry University), Ministry of Education, Fujian Provincial Key Laboratory of Haixia Applied Plant Systems Biology, Fujian Agriculture and Forestry

Francisco J. Romero-Campero
Plant Development Unit, Institute for Plant Biochemistry and Photosynthesis, Consejo Superior de Investigaciones Científicas, Universidad de Sevilla, Seville, Spain
Department of Computer Science and Artificial Intelligence, Universidad de Sevilla, Seville, Spain

Pedro de los Reyes, M. Teresa Ruiz, José M. Romero and Federico Valverde
Plant Development Unit, Institute for Plant Biochemistry and Photosynthesis, Consejo Superior de Investigaciones Científicas, Universidad de Sevilla, Seville, Spain

Asier Briones-Moreno, Jorge Hernández-García and Miguel A. Blázquez
Instituto de Biología Molecular y Celular de Plantas, Consejo Superior de Investigaciones Científicas-Universidad Politécnica de Valencia, Valencia, Spain

Carlos Vargas-Chávez
Institute for Integrative Systems Biology (I2SysBio), University of Valencia, Valencia, Spain

Francisco J. Romero-Campero
Department of Computer Science and Artificial Intelligence, Universidad de Sevilla, Sevilla, Spain
Instituto de Bioquímica Vegetal y Fotosíntesis, Consejo Superior de Investigaciones Científicas - Universidad de Sevilla, Sevilla, Spain

José M. Romero and Federico Valverde
Instituto de Bioquímica Vegetal y Fotosíntesis, Consejo Superior de Investigaciones Científicas - Universidad de Sevilla, Sevilla, Spain

Carlos Tarancón, Eduardo González-Grandío, Michael Nicolas and Pilar Cubas
Plant Molecular Genetics Department, Centro Nacional de Biotecnología (Consejo Superior de Investigaciones Científicas), Campus Universidad Autónoma de Madrid, Madrid, Spain

Juan C. Oliveros
Bioinformatics for Genomics and Proteomics Unit, Centro Nacional de Biotecnología (Consejo Superior de Investigaciones Científicas), Campus Universidad Autónoma de Madrid, Madrid, Spain

Yesenia Madrigal and Natal Pabón-Mora
Facultad de Ciencias Exactas y Naturales, Instituto de Biología, Universidad de Antioquia, Medellín, Colombia

Juan F. Alzate
Centro Nacional de Secuenciación Genómica, Sede de Investigación Universitaria, Facultad de Medicina, Universidad de Antioquia, Medellín, Colombia

Index

www.ingramcontent.com/pod-product-compliance
Lightning Source LLC
LaVergne TN
LVHW060511210926
848485LV00005B/1063